ERETIC	Electronic REference To access In vivo Concentrations
ESI	Electrospray Ionization
FA	Factor Analysis
FABS	Fluorine Atoms for Biochemical Screening
FAXS	Fluorine chemical shift Anisotropy and eXchange for Screening
FBDD	Fragment-Based Drug Design
FDA	Food and Drug Administration
FGFR	Fibroblast-Growth-Factor-Receptor
FID	Free-Induction Decay
FIREMAT	FIve-π REplicated Magic-Angle Turning
FLAIR	FLuid Attenuated Inversion Recovery
FLASH	Fast Low Angle SHot
fMRI	functional Magnetic Resonance Imaging
FOV	Field Of View
FSLG	Frequency-Switched Lee-Goldburg
FWHH	Full Width at Half Height
GAP	GTP Activating Protein
GARP	Globally optimized Alternating-phase Rectangular Pulses
GC	Gas Chromatography
GDM	Genetically Determined Metabotype
GEF	Guanine nucleotide Exchange Factor
GIPAW	Gauge Including Projector Augmented Wave
GIT	GastroIntestinal Tract
GLP	Good Laboratory Practice
GMP	Good Manufacturing Practice
GPCR	G Protein-Coupled Receptor
GSD1a	Glycogen Storage Disease Type 1a
GWAS	Genome-Wide Association Study
H2BC	Heteronuclear 2 Bond Correlation
HET-STOCSY	HETeronuclear Statistical TOtal Correlation SpectroscopY
HILIC	Hydrophobic Interaction LIquid Chromatography
HLA	Human Lymphocyte Antigen
HMBC	Heteronuclear Multiple Bond Coherence
HMBC	Heteronuclear Multiple Bond Correlation
HMDB	Human Metabolite DataBase
HMPAA	Hydrophobically Modified PolyAcrylic Acid
HMQC	Heteronulcear Multiple-Quantum Correlation
HOBS	HOmonuclear Band-Selective decoupled
HOESY	Heteronuclear Overhauser Effect SpectroscopY
HPLC	High-Performance Liquid Chromatography
HR-MAS	High-Resolution Magic Angle Spinning
HR-MS	High-Resolution Mass Spectrometry
HSQC	Heteronuclear Single Quantum Correlation
HTS	High-Throughput Screening

ICA	
ICH	International Conference for Harmonization of Technical Requirements for the Registration of Pharmaceuticals for Human Use
ICQ	Imaging Capability Questionnaire
IDP	Intrinsically Disordered Protein
IEM	Inborn Errors of Metabolism
IGT	Impaired Glucose Tolerance
IMCL	IntraMyoCellular Lipid
IR	Inversion Recovery
IRMS	Isotope Ratio Mass Spectrometry
J-MOD	J-MODulated
KOL	Key Opinion Leader
LC	Liquid Chromatography
LDA	Linear Discriminant Analysis
LDL	Low-Density Lipoprotein
LE	Ligand Efficiency
LLE	Ligand Lipophilicity Efficiency
LMICs	Low- and Middle-Income Countries
LMW	Low-Molecular Weight
LMWH	Low Molecular Weight Heparin
LOD	Limit Of Detection
LOGSY	Ligand Observed by Gradient SpectroscopY
LOQ	Limit Of Quantitation
LSR	Lanthanide Shift Reagent
LVEF	Left Ventricular Ejection Fraction
MACS	Magic-Angle Coil Spinning
MALDI	Matrix-Assisted Laser Desorption Ionization
MAS	Magic Angle Spinning
MAT	Magic-Angle Turning
mCRC	metastatic ColoRectal Cancer
MDD	Major Depressive Disorder
MHS	Mean liver Histopathology Score
MMOA	Molecular Mechanism Of Action
MPIO	MicroParticles of Iron Oxide
MRA	Magnetic Resonance Angiography
MRC	Medical Research Council
MRI	Magnetic Resonance Imaging
MRS	Magnetic Resonance Spectroscopy
MSUD	Maple Syrup Urine Disease
MS	Mass Spectrometry
MTC	Magnetization Transfer Contrast
MVDA	MultiVariate Data Analysis
NCE	New Chemical Entities
NCI	National Cancer Institute
NF	National Formulary
NIHR	National Institute for Health Research
NMF	Nonnegative Matrix Factorization
NOE	Nuclear Overhauser Effect
NP	Natural Product

continued on back endpapers

NMR in Pharmaceutical Sciences

eMagRes Books

eMagRes (formerly the *Encyclopedia of Magnetic Resonance*) publishes a wide range of online articles on all aspects of magnetic resonance in physics, chemistry, biology and medicine. The existence of this large number of articles, written by experts in various fields, is enabling the publication of a series of *eMagRes* Books – handbooks on specific areas of NMR and MRI. The chapters of each of these handbooks will comprise a carefully chosen selection of *eMagRes* articles.

Published *eMagRes* Books

NMR Crystallography
Edited by Robin K. Harris, Roderick E. Wasylishen,
    Melinda J. Duer
ISBN 978-0-470-69961-4

Multidimensional NMR Methods for the Solution State
Edited by Gareth A. Morris, James W. Emsley
ISBN 978-0-470-77075-7

Solid-State NMR Studies of Biopolymers
Edited by Ann E. McDermott, Tatyana Polenova
ISBN 978-0-470-72122-3

NMR of Quadrupolar Nuclei in Solid Materials
Edited by Roderick E. Wasylishen, Sharon E. Ashbrook,
    Stephen Wimperis
ISBN 978-0-470-97398-1

RF Coils for MRI
Edited by John T. Vaughan, John R. Griffiths
ISBN 978-0-470-77076-4

MRI of Tissues with Short T_2s or T_2^*s
Edited by Graeme M. Bydder, Gary D. Fullerton, Ian R. Young
ISBN 978-0-470-68835-9

NMR Spectroscopy: A Versatile Tool for Environmental Research
Edited by Myrna J. Simpson, André J. Simpson
ISBN 978-1-118-61647-5

NMR in Pharmaceutical Sciences
Edited by Jeremy R. Everett, Robin K. Harris, John C. Lindon,
    Ian D. Wilson
ISBN 978-1-118-66025-6

Forthcoming *eMagRes* Books

Handbook of in vivo MRS
Edited by John R. Griffiths, Paul A. Bottomley
ISBN 978-1-118-99766-6

eMagRes

Edited by Roderick E. Wasylishen, Edwin D. Becker, Marina Carravetta, George A. Gray, John R. Griffiths, Ann E. McDermott, Tatyana Polenova, André J. Simpson, Myrna J. Simpson, Ian R. Young.

eMagRes (formerly the *Encyclopedia of Magnetic Resonance*) is based on the original publication of the *Encyclopedia of Nuclear Magnetic Resonance*, first published in 1996 with an updated volume added in 2000. The *Encyclopedia of Magnetic Resonance* was launched in 2007 online with all the existing published material, and was later relaunched as *eMagRes* in 2013. New updates are added on a regular basis through the year to keep content up to date with current developments. NMR and MRI have developed into two very different research and application areas with their own dynamic development. *eMagRes* addresses this development by presenting the content as two separate topics. Readers can browse through the extensive lists of either the NMR topics or the MRI topics.

For more information see: http://www.wileyonlinelibrary.com/ref/eMagRes

NMR in Pharmaceutical Sciences

Editors

Jeremy R. Everett
University of Greenwich, Kent, UK

Robin K. Harris
University of Durham, Durham, UK

John C. Lindon
Imperial College London, UK

Ian D. Wilson
Imperial College London, UK

This edition first published 2015
© 2015 John Wiley & Sons Ltd

Registered office
John Wiley & Sons Ltd, The Atrium, Southern Gate, Chichester, West Sussex,
PO19 8SQ, United Kingdom

For details of our global editorial offices, for customer services and for information about
how to apply for permission to reuse the copyright material in this book please see our
website at www.wiley.com.

The right of the authors to be identified as the authors of this work has been asserted in
accordance with the Copyright, Designs and Patents Act 1988.

All rights reserved. No part of this publication may be reproduced, stored in a
retrieval system, or transmitted, in any form or by any means, electronic, mechanical,
photocopying, recording or otherwise, except as permitted by the UK Copyright, Designs
and Patents Act 1988, without the prior permission of the publisher.

Wiley also publishes its books in a variety of electronic formats. Some content that
appears in print may not be available in electronic books.

Designations used by companies to distinguish their products are often claimed as
trademarks. All brand names and product names used in this book are trade names,
service marks, trademarks or registered trademarks of their respective owners. The
publisher is not associated with any product or vendor mentioned in this book. This
publication is designed to provide accurate and authoritative information in regard to the
subject matter covered. It is sold on the understanding that the publisher is not engaged
in rendering professional services. If professional advice or other expert assistance is
required, the services of a competent professional should be sought.

Library of Congress Cataloging-in-Publication Data

NMR in pharmaceutical sciences / editors Jeremy R. Everett, Robin K. Harris,
John C. Lindon, Ian D. Wilson.
    pages cm
  Nuclear magnetic resonance in pharmaceutical sciences
  Includes bibliographical references and index.
  ISBN 978-1-118-66025-6 (cloth)
  1. Drug development. 2. Nuclear magnetic resonance. 3. Pharmaceutical
technology. I. Everett, Jeremy R., editor. II. Harris, Robin K. (Robin Kingsley), editor.
III. Lindon, John C., editor. IV. Wilson, Ian D., editor. V. Title: Nuclear magnetic
resonance in pharmaceutical sciences.
  RM301.25.N63 2015
  615.1'9--dc23
                                                                    2015018210

A catalogue record for this book is available from the British Library.

ISBN-13: 978-1-118-66025-6

Cover image courtesy of the Laboratory of Neuro Imaging and Martinos Center
for Biomedical Imaging, Consortium of the Human Connectome Project:
www.humanconnectome.org

Set in 9.5/11.5 pt Times by SPi Global, Chennai, India
Printed and bound in Singapore by Markono Print Media Pte Ltd

eMagRes

Editorial Board

Editor-in-Chief

Roderick E. Wasylishen
University of Alberta
Edmonton, Alberta
Canada

Section Editors

SOLID-STATE NMR & PHYSICS

Marina Carravetta
University of Southampton
Southampton
UK

SOLUTION-STATE NMR & CHEMISTRY

George A. Gray
Varian Inc.
Palo Alto, CA
USA

BIOCHEMICAL NMR

Ann E. McDermott
Columbia University
New York, NY
USA

Tatyana Polenova
University of Delaware
Newark, DE
USA

ENVIRONMENTAL & ECOLOGICAL NMR

André J. Simpson
University of Toronto
Ontario
Canada

Myrna J. Simpson
University of Toronto
Ontario
Canada

MRI & MRS

John R. Griffiths
Cancer Research UK
Cambridge Research Institute
Cambridge
UK

Ian R. Young
Imperial College
London
UK

HISTORICAL PERSPECTIVES

Edwin D. Becker
National Institutes of Health
Bethesda, MD
USA

International Advisory Board

Robin K. Harris
(Chairman)
University of Durham
Durham
UK

Isao Ando
Tokyo Institute
    of Technology
Tokyo
Japan

David M. Grant
(Past Chairman - deceased)
University of Utah
Salt Lake City, UT
USA

Adriaan Bax
National Institutes of Health
Bethesda, MD
USA

Chris Boesch
University of Bern
Bern
Switzerland

Paul A. Bottomley
Johns Hopkins University
Baltimore, MD
USA

William G. Bradley
UCSD Medical Center
San Diego, CA
USA

Graeme M. Bydder
UCSD Medical Center
San Diego, CA
USA

Paul T. Callaghan
(deceased)
Victoria University
    of Wellington
Wellington
New Zealand

Melinda J. Duer
University of Cambridge
Cambridge
UK

James W. Emsley
University of Southampton
Southampton
UK

Richard R. Ernst
Eidgenössische Technische
    Hochschule (ETH)
Zürich
Switzerland

Ray Freeman
University of Cambridge
Cambridge
UK

Lucio Frydman
Weizmann Institute
    of Science
Rehovot
Israel

Bernard C. Gerstein
Ames, IA
USA

Maurice Goldman
Villebon sur Yvette
France

Harald Günther
Universität Siegen
Siegen
Germany

Herbert Y. Kressel
Harvard Medical School
Boston, MA
USA

Gareth A. Morris
University of Manchester
Manchester
UK

C. Leon Partain
Vanderbilt University Medical
    Center
Nashville, TN
USA

Alexander Pines
University of California
    at Berkeley
Berkeley, CA
USA

George K. Radda
University of Oxford
Oxford
UK

Hans Wolfgang Spiess
Max-Planck Institute
    of Polymer Research
Mainz
Germany

Bernd Wrackmeyer
Universität Bayreuth
Bayreuth
Germany

Charles P. Slichter
University of Illinois
    at Urbana-Champaign
Urbana, IL
USA

Kurt Wüthrich
The Scripps Research
    Institute
La Jolla, CA
USA

and

ETH Zürich
Zürich
Switzerland

John S. Waugh (deceased)
Massachusetts Institute
    of Technology (MIT)
Cambridge, MA
USA

Contents

Contributors

Thomas M. Bocan *US Army Medical Research Institute of Infectious Diseases, 1425 Porter Street, Frederick, MD 21702, USA*
The Geneva Foundation, 917 Pacific Ave, Suite 600, Tacoma, WA 98402, USA
Chapter 18: Applications of Preclinical MRI/MRS in the Evaluation of Drug Efficacy and Safety

Sarah K. Branch *Medicines and Healthcare Products Regulatory Agency (MHRA), 151 Buckingham Palace Road, Victoria, London SW1W 9SZ, UK*
Chapter 29: Pharmaceutical Industry: Regulatory Control and Impact on NMR Spectroscopy
Chapter 30: NMR Spectroscopy in the European and US Pharmacopeias

Stephen J. Byard *Covance Laboratories, Alnwick, UK*
Chapter 23: Pharmaceutical Technology Studied by MRI

Leo L. Cheng *Massachusetts General Hospital, Harvard Medical School, Pathology, MGH 149, 13th Street Charlestown, Boston, MA 02129, USA*
Chapter 9: High-resolution MAS NMR of Tissues and Cells

Muireann Coen *Computational and Systems Medicine, Department of Surgery and Cancer, Imperial College London, Sir Alexander Fleming Building, South Kensington, London SW7 2AZ, UK*
Chapter 20: Preclinical Drug Efficacy and Safety Using NMR Spectroscopy

Olivia Corcoran *Medicines Research Group, School of Health, Sport and Bioscience, University of East London, London E15 4LZ, UK*
Chapter 13: Hit Discovery from Natural Products in Pharmaceutical R&D

Helen Corns *Medicines and Healthcare Products Regulatory Agency (MHRA), 151 Buckingham Palace Road, Victoria, London SW1W 9SZ, UK*
Chapter 30: NMR Spectroscopy in the European and US Pharmacopeias

Paulo Dani *Former employee from Merck Sharp & Dohme (MSD), Oss, The Netherlands*
Chapter 31: NMR in Pharmaceutical Manufacturing

Anthony C. Dona *MRC-NIHR National Phenome Centre, Department of Surgery & Cancer, Imperial College London, South Kensington Campus, London SW7 2AZ, UK*
Chapter 3: Experimental NMR Methods for Pharmaceutical Research and Development

Jeremy R. Everett	*Medway Metabonomics Research Group, University of Greenwich, Chatham Maritime, Kent ME4 4TB, UK* Chapter 1: Drug Discovery and Development: The Role of NMR Chapter 25: NMR-based Pharmacometabonomics: A New Approach to Personalized Medicine
Paul D. Hockings	*Antaros Medical, BioVenture Hub, 43183 Mölndal, Sweden* Chapter 26: Clinical MRI Studies of Drug Efficacy and Safety
John C. Hollerton	*GlaxoSmithKline R&D, Gunnels Wood Road, Stevenage, Herts SG1 2NY, UK* Chapter 6: High-throughput NMR in Pharmaceutical R&D
Ulrike Holzgrabe	*Institute of Pharmacy and Food Chemistry, University of Würzburg, 97074 Würzburg, Germany* Chapter 5: Quantitative NMR Spectroscopy in Pharmaceutical R&D
Mark Jeeves	*School of Cancer Sciences, University of Birmingham, Birmingham B15 2TT, UK* Chapter 22: Structure-based Drug Design Using NMR
Beatriz Jiménez	*Clinical Phenotyping Centre, Division of Computational and Systems Medicine, Department of Surgery and Cancer, Imperial College London, London SW7 2AZ, UK* Chapter 24: NMR-based Metabolic Phenotyping for Disease Diagnosis and Stratification
Lauren Keith	*National Institute of Allergy and Infectious Disease, 8200 Research Plaza, Frederick, MD 21702, USA* Chapter 18: Applications of Preclinical MRI/MRS in the Evaluation of Drug Efficacy and Safety
Edwin Kellenbach	*Aspen Pharmacare, PO Box 20, 5340BH Oss, Oss, The Netherlands* Chapter 31: NMR in Pharmaceutical Manufacturing
Sarantos Kostidis	*Center for Proteomics and Metabolomics, Leiden University Medical Center, P.O. Box 9600, 2300RC Leiden, The Netherlands* Chapter 10: NMR Studies of Inborn Errors of Metabolism
John C. Lindon	*Computational and Systems Medicine, Department of Surgery and Cancer, Imperial College London, Sir Alexander Fleming Building, South Kensington, London SW7 2AZ, UK* Chapter 4: ^{19}F NMR Spectroscopy: Applications in Pharmaceutical Studies Chapter 15: Mixture Analysis in Pharmaceutical R&D Using Hyphenated NMR Techniques
Myriam Malet-Martino	*Groupe de RMN Biomédicale, Laboratoire SPCMIB (UMR CNRS 5068), Université Paul Sabatier, 118 route de Narbonne, 31062 Toulouse cedex 9, France* Chapter 28: Analysis of Counterfeit Medicines and Adulterated Dietary Supplements by NMR

Robert Martino
Groupe de RMN Biomédicale, Laboratoire SPCMIB (UMR CNRS 5068), Université Paul Sabatier, 118 route de Narbonne, 31062 Toulouse cedex 9, France
Chapter 28: Analysis of Counterfeit Medicines and Adulterated Dietary Supplements by NMR

James S. McKenzie
Computational and Systems Medicine, Department of Surgery and Cancer, Faculty of Medicine, Imperial College London, London SW7 2AZ, UK
Chapter 7: Multivariate Data Analysis Methods for NMR-based Metabolic Phenotyping in Pharmaceutical and Clinical Research

Emmanuel Mikros
Faculty of Pharmacy, University of Athens Panepistimiopolis, 15771 Zografou, Greece
Chapter 10: NMR Studies of Inborn Errors of Metabolism

Jeremy K. Nicholson
Computational and Systems Medicine, Department of Surgery and Cancer, Faculty of Medicine, Imperial College London, London SW7 2AZ, UK
Chapter 7: Multivariate Data Analysis Methods for NMR-based Metabolic Phenotyping in Pharmaceutical and Clinical Research

Michael Overduin
School of Cancer Sciences, University of Birmingham, Birmingham B15 2TT, UK
Chapter 22: Structure-based Drug Design Using NMR

John A. Parkinson
WestCHEM, Department of Pure and Applied Chemistry, NMR Spectroscopy Facility, University of Strathclyde, Glasgow G1 1XL, UK
Chapter 2: Modern NMR Pulse Sequences in Pharmaceutical R&D

Torren M. Peakman
torren@nmrpeak.co.uk
Chapter 14: NMR-based Structure Determination of Drug Leads and Candidates

Antonio Pineda-Lucena
Structural Biochemistry Laboratory, Advanced Therapies Program, Centro de Investigación Príncipe Felipe, Eduardo Primo Yúfera 3, 46012-Valencia, Spain
Chapter 12: Fragment-based Drug Design Using NMR Methods

Leonor Puchades-Carrasco
Structural Biochemistry Laboratory, Advanced Therapies Program, Centro de Investigación Príncipe Felipe, Eduardo Primo Yúfera 3, 46012-Valencia, Spain
Chapter 12: Fragment-based Drug Design Using NMR Methods

Lee Quill
School of Cancer Sciences, University of Birmingham, Birmingham B15 2TT, UK
Chapter 22: Structure-based Drug Design Using NMR

David G. Reid
Department of Chemistry, University of Cambridge, Lensfield Road, Cambridge CB2 1EW, UK
Chapter 23: Pharmaceutical Technology Studied by MRI
Chapter 26: Clinical MRI Studies of Drug Efficacy and Safety

Ricardo Riguera   *Department of Organic Chemistry and Center for Research in Biological Chemistry and Molecular Materials (CIQUS), University of Santiago de Compostela, E-15782 Santiago de Compostela, Spain*
Chapter 17: NMR Methods for the Assignment of Absolute Stereochemistry of Bioactive Compounds

Andrea Ruggiero   *Medicines and Healthcare Products Regulatory Agency (MHRA), 151 Buckingham Palace Road, Victoria, London SW1W 9SZ, UK*
Chapter 29: Pharmaceutical Industry: Regulatory Control and Impact on NMR Spectroscopy

Nadeem Saeed   *BioClinica, 72 Hammersmith Road, London W14 8UD, UK*
Chapter 26: Clinical MRI Studies of Drug Efficacy and Safety

Krishna Saxena   *Institute for Organic Chemistry and Chemical Biology, Center for Biomolecular Magnetic Resonance, Johann Wolfgang Goethe-University Frankfurt, Max-von-Laue-Str. 7, Frankfurt am Main, D-60438, Germany*
German Cancer Consortium (DKTK), Im Neuenheimer Feld 280, 69120 Heidelberg, Germany
Chapter 8: The Role of NMR in Target Identification and Validation for Pharmaceutical R&D

Harald Schwalbe   *Institute for Organic Chemistry and Chemical Biology, Center for Biomolecular Magnetic Resonance, Johann Wolfgang Goethe-University Frankfurt, Max-von-Laue-Str. 7, Frankfurt am Main, D-60438, Germany*
German Cancer Consortium (DKTK), Im Neuenheimer Feld 280, 69120 Heidelberg, Germany
Chapter 8: The Role of NMR in Target Identification and Validation for Pharmaceutical R&D

Jose M. Seco   *Department of Organic Chemistry and Center for Research in Biological Chemistry and Molecular Materials (CIQUS), University of Santiago de Compostela, E-15782 Santiago de Compostela, Spain*
Chapter 17: NMR Methods for the Assignment of Absolute Stereochemistry of Bioactive Compounds

Raman Sharma   *Pfizer-PDM (Pharmacokinetics Dynamics and Metabolism), MS 8220-4374, Eastern Point Road, Groton, CT 06340, USA*
Chapter 19: Practical Applications of NMR Spectroscopy in Preclinical Drug Metabolism Studies

Gary J. Sharman   *Eli Lilly & Company, Erl Wood Manor, Sunninghill Road, Windlesham, Surrey GU20 6PH, UK*
Chapter 16: Conformation and Stereochemical Analysis of Drug Molecules

Philip J. Sidebottom   *GlaxoSmithKline R&D, Chemical Sciences – Analytical Chemistry, Gunnels Wood Road, Stevenage, Herts SG1 2NY, UK*
Chapter 11: NMR-based Structure Confirmation of Hits and Leads in Pharmaceutical R&D

David M. Thomasson

National Institute of Allergy and Infectious Disease, 8200 Research Plaza, Frederick, MD 21702, USA

Tunnel Government Services, 6701 Democracy Boulevard, Suite 515, Bethesda, MD 20817, USA

Chapter 18: Applications of Preclinical MRI/MRS in the Evaluation of Drug Efficacy and Safety

Kirill A. Veselkov

Computational and Systems Medicine, Department of Surgery and Cancer, Faculty of Medicine, Imperial College London, London SW7 2AZ, UK

Chapter 7: Multivariate Data Analysis Methods for NMR-based Metabolic Phenotyping in Pharmaceutical and Clinical Research

Frederick G. Vogt

Morgan, Lewis & Bockius, LLP, 1701 Market St., Philadelphia, PA 19103-2921, USA

Chapter 21: Characterization of Pharmaceutical Compounds by Solid-state NMR

Chapter 27: The Role of NMR in the Protection of Intellectual Property in Pharmaceutical R&D

Gregory S. Walker

Pfizer-PDM (Pharmacokinetics Dynamics and Metabolism), MS 8220-4374, Eastern Point Road, Groton, CT 06340, USA

Chapter 19: Practical Applications of NMR Spectroscopy in Preclinical Drug Metabolism Studies

Ian D. Wilson

Computational and Systems Medicine, Department of Surgery and Cancer, Imperial College London, Sir Alexander Fleming Building, South Kensington, London SW7 2AZ, UK

Chapter 4: ^{19}F NMR Spectroscopy: Applications in Pharmaceutical Studies

Chapter 15: Mixture Analysis in Pharmaceutical R&D Using Hyphenated NMR Techniques

Chapter 20: Preclinical Drug Efficacy and Safety Using NMR Spectroscopy

Series Preface

The *Encyclopedia of Nuclear Magnetic Resonance* was published, in eight volumes, in 1996, in part to celebrate the 50th anniversary of the first publications in NMR in January 1946. Volume 1 contained an historical overview and 200 articles by prominent NMR practitioners, while the remaining 7 volumes were constituted by 500 articles on a wide variety of topics in NMR (including MRI). A ninth volume was brought out in 2000 and two 'spin-off' volumes incorporating the articles on MRI and MRS (together with some new ones) were published in 2002. In 2006, the decision was taken to publish all the articles electronically (i.e., on the World Wide Web) and this was carried out in 2007. Since then, new articles have been placed on the web every 3 months and some of the original articles have been updated. This process is continuing and to recognize the fact that the *Encyclopedia of Magnetic Resonance* is a true on-line resource, the web site has been redesigned and new functionalities added, with a relaunch in January 2013 in a new Volume and Issue format, under the new name *eMagRes*. In December, 2012, a print edition of the Encyclopedia was published in 10 volumes (6200 pages). This much needed update of the 1996 edition of the Encyclopedia, encompasses the entire field of NMR with the exception of medical imaging (MRI).

The existence of this large number of articles, written by experts in various fields, is enabling a new concept to be implemented, namely the publication of a series of printed handbooks on specific areas of NMR and MRI. The chapters of each of these handbooks will be constituted by a carefully chosen selection of Encyclopedia articles relevant to the area in question. In consultation with the Editorial Board, the handbooks are coherently planned in advance by specially selected editors, and new articles written (together with updating of some already existing articles) to give appropriate complete coverage of the total area. The handbooks are intended to be of value and interest to research students, postdoctoral fellows, and other researchers learning about the topic in question and undertaking relevant experiments, whether in academia or industry. Consult the *eMagRes* web site (www.wileyonlinelibrary.com/ref/eMagRes) for the latest news on magnetic resonance Handbooks.

Roderick E. Wasylishen

January 2015

Preface

During the first few years of NMR, following its discovery in 1945, it was the preserve of physicists. However, the discovery of the chemical shift soon moved it into the realm of chemistry and it has been for many years a vital tool for research that has rapidly extended to all branches of chemistry, biology, and medicine and, in particular, to research efforts to develop new pharmaceutical drugs in both university and industrial environments. When MRI was invented, this was also incorporated. It is now not feasible to market a new drug system without the involvement of NMR at all stages of its development. While some earlier articles in the *Encyclopedia of Magnetic Resonance* have addressed topics relevant to pharmaceutical applications of NMR, it was only when planning started for this handbook that efforts were made to provide comprehensive coverage of the area. We hope that this has now been achieved.

This handbook is firmly based on the key processes in drug discovery, development, and manufacture but underpinned by an understanding of fundamental NMR principles and the unique contribution that NMR (including MRI) can provide. After an introductory chapter, which constitutes an overview, this handbook is organized into five sections. The first section is on the basics of NMR theory and relevant experimental methods. The rest follow a sequence based on the chronology of drug discovery and development, firstly 'Idea to Lead', then 'Lead to Drug Candidate', followed by 'Clinical Development', and finally 'Drug Manufacture'. The 31 chapters herein, written by experts, cover a vast range of topics from analytical chemistry, including aspects involved in regulatory matters and in the prevention of fraud, to clinical imaging studies.

While the handbook will be essential reading for many scientists based in pharmaceutical and related industries, it should also be of considerable value to a large number of academics who have research connections with industry; for them, it will supply vital understanding of the industrial concerns (which usually prove to be the drivers of funding by industry for university-based research). As with all the handbooks in this series, the individual articles are also to be found, with minimal differences but subtly changed format, in the on-line quarterly reference *eMagRes*, which can be accessed at (early issues have the title 'Encyclopedia of Magnetic Resonance'). The on-line versions generally contain brief autobiographies of the article authors, a list of related articles, and acknowledgements. In addition, article abstracts and key words can be found on-line.

We are very grateful to all the authors involved for their agreements to write the on-line articles that preceded this handbook, for their expert texts and for their cooperation in reaching this stage. We also thank the people at Wiley (especially Stacey Woods and Elke Morice-Atkinson) and at Laserwords for all their hard work (and patience!) in bringing this handbook to the point of publication.

We offer this handbook to the wider scientific community in the hope that it will not only provide valuable information and a greater understanding of the role of NMR in pharmaceutical R&D and the pharmaceutical industry but that it will also stimulate further advances in relevant NMR procedures.

Jeremy R. Everett
University of Greenwich, London, UK

John C. Lindon
Imperial College London, UK

Robin K. Harris
University of Durham, UK

Ian D. Wilson
Imperial College London, UK

January 2015

PART A
Introduction

Chapter 1

Drug Discovery and Development: The Role of NMR

Jeremy R. Everett

Medway Metabonomics Research Group, University of Greenwich, Chatham Maritime, Kent ME4 4TB, UK

1.1 INTRODUCTION TO DRUG DISCOVERY AND DEVELOPMENT AND THE ROLE OF NMR

Drug discovery and development [also known as research and development (R&D)] is at a crossroads. Over the past 10 years, many major pharmaceutical companies have reduced the size of their R&D organizations as they struggled to achieve the increased outputs expected of them in this postgenomic era. Recent

NMR in Pharmaceutical Sciences. Edited by Jeremy R. Everett, John C. Lindon, Ian D. Wilson, and Robin K. Harris.
© 2015 John Wiley & Sons, Ltd. ISBN: 978-1-118-66025-6
Also published in eMagRes (online edition)
DOI: 10.1002/9780470034590.emrstm1389

analyses have suggested that the common strategy of companies merging and acquiring other companies has not led to the improvements in output that were expected.[1] Indeed, until recently, the number of new drugs being launched per year was falling in spite of increasing expenditure on R&D.[2] The key reasons for the low productivity across the industry have been the very significant project attrition at all stages of the R&D process due to lack of efficacy and lack of safety in new drug candidates. This attrition has been particularly acute in the critical Phase II clinical trials, where drugs are tested in patients for the first time. We are now in a new era of drug discovery and development, with far greater involvement from academic laboratories and better cooperation between companies and academia, as well as between companies themselves, especially in precompetitive areas such as the development of new safety testing methodologies.

Before moving on to consider the role of NMR in this challenging environment, it is worthwhile to consider the different stages of the drug R&D process. These stages can be represented as follows:

1. Disease selection
2. Target selection and validation (the target is the name for the biological macromolecule in the human body that the drug is designed to interact with to produce clinical benefit: this is usually a protein)
3. Hit discovery against that target
4. Hit-to-lead optimization

5. Lead-to-drug optimization
6. Phase I drug safety studies in human volunteers
7. Phase II drug efficacy and safety studies in small cohorts of patients
8. Phase III efficacy and safety studies in larger patient populations
9. Drug launch
10. Postlaunch support

The most critical stage (and the most difficult) is stage 2: target selection and validation. For most small molecule drug discovery projects, the choice of which protein target to prosecute (via its inhibition, agonism, antagonism, partial agonism, etc.) is absolutely vital. Many projects fail right from this stage (although the project team will not know this for some years!) because the protein target chosen will not produce sufficient clinical benefit in the patient, or will only produce clinical benefit together with unacceptable side effects. The reason for these problems is that many drug discovery targets are unprecedented; that is, at the time of launch of the project, there are no Phase II data showing that modulation of that target gives clinical benefit in the absence of side effects. Drug discovery project teams therefore have to make key go/no-go decisions on target selection using less-reliable genomics or animal data.

If drug discovery and development are in difficulties, NMR, by contrast, is not. In the 70 years or so since its discovery, NMR spectroscopy has progressed quickly to become the most powerful method of molecular structure elucidation in both the solution state and the amorphous or heterogeneous solid state, and an indispensable tool in chemistry and biochemistry. In addition, the more recent developments of whole body MRI and magnetic resonance spectroscopy (MRS) have enabled the influence of NMR to spread from chemistry and biochemistry to biology and medicine, so that NMR is now pivotal to drug R&D, and the array of NMR applications stretches across each of the 10 stages of the R&D process listed earlier, up to and beyond drug launch.

The reason that NMR spectroscopy has become so dominant in solution-state molecular structure elucidation is that the information it provides is rich, coherent, and highly stable and reproducible over time. Information from proton, carbon-13, and fluorine-19 NMR chemical shifts, for instance, can be interpreted in terms of the chemical environment of the atoms giving rise to those NMR signals. In addition, the phenomenon of spin–spin coupling enables the establishment of through-bond connectivities between different atoms in a molecule in a way that is unobtainable with any other technique. Finally, the relaxation properties of the nuclear spins under study give information on the motions, dynamics, and through-space connectivities in molecules that is critical for a thorough understanding of their properties and behavior in solution.

The drawbacks to the use of NMR principally arise from its lack of sensitivity relative to some other techniques such as mass spectrometry (MS). It is actually appropriate to tackle this principal issue with NMR spectroscopy straight away, as tremendous progress has been made in this area. First of all, superconducting magnets of increasing field strength have been developed over the past 20 years. In the 1980s, NMR spectrometers operating at 9.4 T (resonance frequency of 400 MHz for protons) were common. Magnets operating at 18.8 T (800 MHz for protons) emerged in the 1990s, and the first 1 GHz system was installed in France in 2009. As the signal-to-noise ratio of a spectrum increases approximately with the 3/2 power of the field strength used, these increases in field strength have resulted in significant gains in sensitivity (a doubling in sensitivity from 600 to 900 MHz for protons) but, unfortunately, at the expense of a considerable rise in spectrometer costs. Therefore, other developments have been pursued. One of these is the development of cryocooled probes, where the receiver coil and preamplifier are cooled down to temperatures as low as about 20 K using liquid helium. This has resulted in sensitivity gains of about three- to four-fold, relative to running the same sample on a conventional probe at room temperature. Although technically demanding, switching to cryoprobes is much more cost effective in terms of sensitivity increase than moving to higher magnetic field strengths. A further development has been in the area of microcoil probes. Reducing the size of the receiver coils in NMR probes is particularly important for samples that have limited mass or volume, as, for constant coil length-to-volume ratio, sensitivity is inversely proportional to coil diameter. A standard 5 mm NMR probe may require a sample volume of about 600 µl, whereas a microcoil probe may allow measurements on volumes as low as about 1 µl. This approach has also been exploited for the coils in flow-detection experiments such as liquid chromatography–nuclear magnetic resonance (LC–NMR), capillary electrophoresis–nuclear magnetic resonance (CE–NMR), or even lab-on-a-chip.[3] Significant improvements in spectrometer performance have also ensued from the development of pulse field gradient technologies that allow for the selection or nonselection of resonances with much greater fidelity than previous phase-cycling approaches. In the solid-state nuclear magnetic resonance (SSNMR) arena, the development of rapid-spinning technologies combined with new homonuclear decoupling methods has enabled the acquisition of high-quality ^{1}H SS-NMR spectra with reduced ^{1}H–^{1}H dipolar coupling and increased chemical shift resolution. An array of solid-state homonuclear and heteronuclear 2D NMR experiments, based on correlations via J-coupling and dipolar coupling, are now available, meaning that the study of molecular structure in the solid state can employ many of the same 2D NMR approaches used in the solution state. Finally, and importantly, new data collection methodologies have emerged, including nonlinear data sampling and ultrafast pulsing, that can reduce experiment times by orders of magnitudes.[4] All of these advances have meant that low sensitivity is no longer a major issue for modern NMR spectroscopy. See Chapters 2 and 3, as these both have further important information.

In the remainder of this article, we will cover the principal uses of NMR in drug R&D with a greater focus on new and emerging areas.

1.2 NMR SPECTROSCOPY IN DRUG DISCOVERY AND DEVELOPMENT

1.2.1 Molecular Structure Elucidation by NMR Spectroscopy

1.2.1.1 Structure Confirmation by NMR Spectroscopy

The key area of application for NMR spectroscopy in drug R&D is, and has always been, its role in small molecule structure elucidation. There are several aspects to this area, including the structure confirmation of small molecules made by medicinal chemistry teams in drug discovery and by synthetic chemistry teams in drug development. This work is relatively routine, assuming that the molecules are of low molecular weight and reasonably pure, especially now that modern NMR processing software allows the automated zero-filling, apodization, FT, baseline correction, integration, and multiplet analysis of simple proton NMR spectra. Indeed, much of this work is done on automated or open-access systems for nonexpert users of the spectrometer. Some NMR software such as MNova[5] will also allow the user to predict the NMR

spectrum of the molecule under study and give an indication of the match between the predicted and experimental spectra, thus giving increased confidence in the structure confirmation, assuming that the prediction calculations are precise enough (see Chapters 6, 11, and 14).

1.2.1.2 Structure Elucidation by NMR Spectroscopy: Natural Products

The de novo structure elucidation of natural products, when limited information is available on the expected structure, is much more challenging than structure confirmation of expected molecules. Typically, the structure confirmation of a molecule made by an established synthetic chemistry methodology can be performed with a simple 1D proton NMR spectrum (provides an identity check and purity information) and a mass spectrum of some kind (identity check), while ultraviolet (UV) spectroscopy (provides chromophore of the molecule), infrared (IR) spectroscopy (provides functional groups in the molecule), and elemental analysis (provides elemental formula and confirmation of good purity) information is optional but sometimes acquired. By contrast, the structure elucidation of a genuine unknown will require UV, IR, MS, and an array of NMR data. Typically, for a small molecule (of <500 Da molecular weight), the NMR data acquired would include the information shown in Table 1.1.

As an example of the use of NMR spectroscopy in natural product structure elucidation, Figure 1.1 shows the scanning electron micrograph of a streptomyces organism isolated from a soil sample under the River Mole in Surrey, UK. A whole series of novel anthelmintic (antiparasitic worm) milbemycin natural products were isolated from this organism,[6] including VM44857 shown in Figure 1.1(b). The expansion of the 2D ^{1}H COSY-45 NMR spectrum in Figure 1.1(c) shows some remarkable long-range proton-to-proton connectivities that were used to help piece fragments of the molecular structure together. Note in particular the four-bond and five-bond connectivities to CH_3-26 from H2, H3, and H5.

If a reasonable amount of the natural product is available, carbon-13 NMR spectroscopy could be used to directly interrogate the chemical environments of the carbon atoms in the molecule, but these experiments are significantly less sensitive and more time consuming, owing to the low natural abundance and low sensitivity of carbon-13 NMR compared with proton NMR.

Table 1.1. The different NMR experiments generally required for the elucidation of the molecular structure of an unknown small molecule and the information provided by those experiments

NMR experiment	Information provided
1D ^{1}H NMR	• Number of protons • Chemical type of protons • Proton-to-proton coupling connectivities in simple cases • Information on molecular conformations and stereochemistry through values of coupling constants
2D ^{1}H COSY (correlated spectroscopy)	Proton-to-proton coupling connectivities typically over two to four bonds, sometimes more
2D ^{13}C, ^{1}H HSQC	Carbon-13 to proton coupling connectivities over one bond
2D ^{13}C, ^{1}H HMBC	Carbon-13 to proton coupling connectivities over two to three bonds, allowing connections to be made over the so-called spectroscopically silent centers such as oxygen atoms and carbonyl groups

HSQC, heteronuclear single quantum coherence spectroscopy; HMBC, heteronuclear multiple bond correlation spectroscopy.

A major issue with drug discovery using natural product approaches, apart from the work in elucidating the structure of the molecules, has been the issue of dereplication, i.e., how do you know when you are working on a genuinely novel natural product and not one that has already been discovered previously? Some databases are available but this area is problematic. Approaches that combine classical natural product discovery with metabolite profiling methods are being pursued. Essentially, these approaches use multivariate statistical methods borrowed from metabonomics to identify the metabolites responsible for differences in biological activity between different, partially purified natural product extracts, avoiding the need for the isolation of pure, individual natural products initially (see Chapter 13).[7,8]

1.2.1.3 Structure Elucidation by NMR Spectroscopy: Impurities and Degradation Products

The structure elucidation challenges here are very similar to those for natural products and consequently very similar approaches are taken. The issue is that drugs can degrade in surprising ways and extremely unexpected structures can arise as degradation products. A good example of this is the study of the degradation of the penicillins oxacillin and nafcillin in aqueous solution.[9]

A golden yellow precipitate was observed during the study of the degradation of both oxacillin and nafcillin (Figure 1.2). A combination of UV, IR, MS,

and, critically, 2D NMR methodologies was used to establish that the degradation products were unusual thietan-2-one structures formed by dehydration and rearrangement in aqueous solution. The critical information to piece the structure together was provided by long-range hydrogen-to-carbon-13 connectivities that established the connections between carbons 2 to 8. An added complication to the structure elucidation was that these thietan-2-ones exist as a mixture of two isomers, *cis-* and *trans-*, in slow exchange about the C6C7 double bond. These structures were considered so unusual when proposed from the NMR data that X-ray crystallography of the nafcillin degradation product and two independent chemical syntheses of the oxacillin degradation product were used to confirm their identities. This work took several months: the identification of the thietan-2-ones was achieved in 48 h by a combination of spectroscopic methods but principally by NMR![9]

Before leaving this section, we should note that one of the great powers of NMR spectroscopy is its ability to discriminate between different isomers and tautomers of a compound. This could be seen in the abovementioned example where two isomers of the thietan-2-one degradation product about the C6C7 double bond were in dynamic exchange with one another. However, many other forms of isomerism are readily determined by NMR spectroscopy, including ring and chain positional isomers, diastereomers formed by two or more chiral centers in a molecule, sugar anomers, and *cis* and *trans* amide isomers.

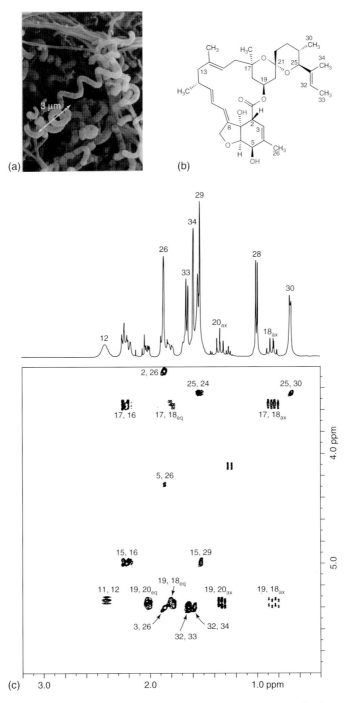

Figure 1.1. (a) A scanning electron micrograph of the *Streptomyces hygroscopicus aureolacrimosus* organism (code E225), isolated from an underwater soil sample in the River Mole, Surrey, that produced a number of milbemycins including VM44857. (b) The molecular structure of VM44857. (c) A portion of the 400 MHz 2D ^{1}H COSY-45 NMR spectrum of VM44857, displayed as a contour plot, showing some key long-range connectivities that helped elucidate the molecule structure. (Reproduced with permission from Ref. 6. © Nature Publishing Group, 1990)

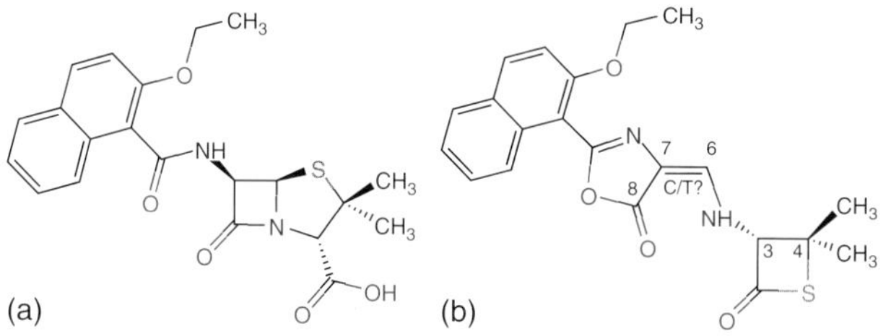

Figure 1.2. The molecular structure of nafcillin free acid (a) and nafcillin thietan-2-one degradation product (b). Note that the double bond between C6 and C7 in nafcillin thietan-2-one is in equilibrium between cis and trans forms

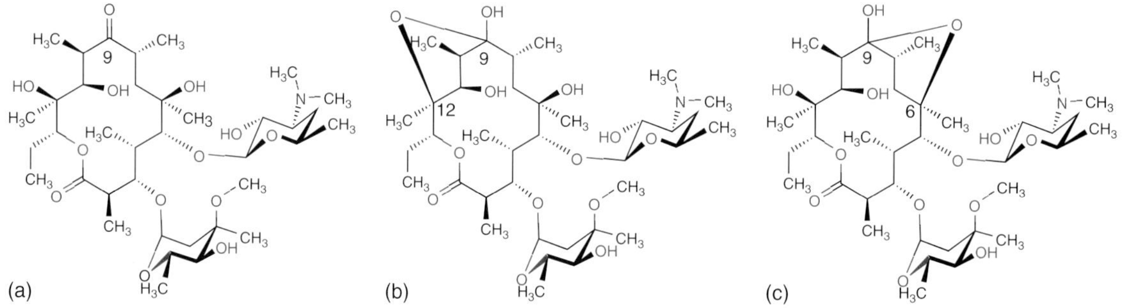

Figure 1.3. Three tautomers of erythromycin A: the C9 ketone tautomer (a), the 9,12-cyclic hemiacetal (b), and the 6,9-cyclic hemiacetal (c). Note that some bond lengths are distorted for clarity

In addition, NMR spectroscopy is uniquely powerful in establishing the tautomeric form in which drugs exist in solution, and this can be critical in the defense or assertion of drug patent claims (see Chapter 27). As an example of tautomerism in drug molecules, the antibiotic erythromycin A exists as a mixture of the dominant C9 ketone form (Figure 1.3) and two hemi-acetals; one 6,9-cyclic hemiacetal (9-deoxo-6-deoxy-9-hydroxy-6,9-epoxyerythromycin A) and one 9,12-cyclic hemiacetal (9-deoxo-12-deoxy-9-hydroxy-9,12-epoxyerythromycin A) formed by the interaction of the hydroxyl groups on C6 and C12, respectively, with the C9 ketone and subsequent cyclic hemiacetal formation.[10]

Finally, it is even possible to distinguish enantiomers using NMR spectroscopy. Ordinarily, NMR experiments are conducted in an achiral environment, and the signals from the two enantiomeric forms of a molecule, e.g., 2R, 3R-butanediol and 2S, 3S butanediol, will be identical (Figure 1.4).

However, in a chiral environment, the NMR signals of the two enantiomers will be different. This can be achieved by the addition of a chiral shift reagent to the solution under study. Note that there is a third isomer of

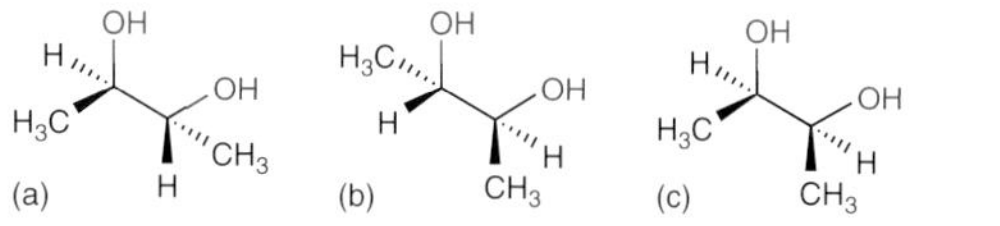

Figure 1.4. The molecular structures of 2R, 3R-butanediol (a), 2S, 3S-butanediol (b), and meso-butanediol (c).

2,3-butanediol: the meso isomer. This is a diastereomer of the 2R, 3R and 2S, 3S enantiomers and is distinguishable from both of them in an achiral environment. Note that for the meso isomer, the stereochemistry is 2R, 3S or 2S, 3R: these two molecules are identical! See Chapter 17 for additional information on this important area.

1.2.2 Additional Important Roles for NMR Spectroscopy in Drug Discovery and Development

There are many other significant roles for NMR spectroscopy in addition to molecular structure elucidation that are of key importance for drug discovery and development.

1.2.2.1 Conformational Analysis of Drug Molecules

In the previous section, there is a discussion of the phenomenal ability of NMR spectroscopy to distinguish between the various different isomeric forms that molecules can have. These isomers typically require bond breaking and making in order to convert one isomer into another. However, there are other forms of a molecule that can be interconverted by the simple act of rotation around single bonds and these are known as *conformational isomers* or *conformers*. One well-known example of this is the conformational equilibrium between the axial and equatorial conformations of 1-methylcyclohexane (Figure 1.5).

Rotation of the cyclohexane ring single bonds will convert the equatorial methyl conformation into the axial methyl conformation, which is less favored and of higher free energy owing to steric interactions with the axial hydrogen atoms at C3 and C5. At room temperature, the two conformations are in rapid exchange with one another, and NMR spectra of solutions of 1-methylcyclohexane at this temperature display a weighted average of the spectra. However, if a solution of 1-methylcyclohexane is cooled to below $-120\,^\circ$C, then the rate of exchange of the two conformations is slowed down to such an extent that separate NMR signals for the two conformations can start to be observed and their relative abundance directly quantified.[11]

Understanding the solution conformations of drug molecules is an important aspect of drug design. If a drug molecule has a great degree of conformational freedom, then there will be an entropic penalty to binding to their protein target.

On the other hand, if the drug molecule prefers to adopt a solution conformation that is different from that in the protein binding site, then there may be an enthalpic penalty to the binding. Both situations are nonoptimal, and medicinal chemists seek to design molecules that are preorganized into the preferred binding conformation, as this can lead to significant increases in potency. There is now a resurgence of interest in new NMR-based methods for the determination of the experimental solution-state conformations of leads or hits in drug discovery projects as this (i) avoids the need to conduct protein structure determination in the crystalline state (n.b., most structure-based drug design uses X-ray crystallographic approaches as they are faster and more versatile than the corresponding NMR-based approaches), (ii) avoids the need to rely on small-molecule X-ray crystal structures that may represent conformations completely different from those in solution, and (iii) avoids the need to rely on computational chemistry methods that have significant limitations in terms of the force fields available and the ability to treat solvation realistically.[12] The new methods are based on combining traditional NMR information such as NOEs and scalar coupling constants with residual dipolar coupling information arising from measurements in aligned media (see Chapter 16).

1.2.2.2 Quality Control of Small Molecule Drugs

On moving from a drug discovery to a drug development environment, the focus changes to include a greater concern with the quality of the drug molecule. In particular, it is important to know its purity and the levels and identities of the impurities from its synthesis or fermentation, and the products that arise from post-synthesis degradation of the drug (and its impurities!). Many excellent analytical methods are used in this endeavor, including high-performance liquid chromatography (HPLC), ultra performance liquid chromatography (UPLC), liquid chromatography–mass spectrometry (LC–MS), and UPLC–MS. However, chromatographic methods on their own are only of any use if the identities of all of the peaks in the chromatograms are known. Furthermore, even LC–MS and UPLC–MS are of limited use in novel impurity or degradation product identification, as these technologies are generally unable to discriminate between isomers. In these

Figure 1.5. The two stable conformations of methylcyclohexane: equatorial methyl conformation (low free energy, left) and axial methyl conformation (higher free energy, right)

circumstances, it has been common practice to isolate the impurities and degradation products of interest and to determine their structures using NMR spectroscopy (see Section 1.2.1.3). Increasingly these days, such studies are using hyphenated technologies such as LC–NMR–MS, enabling integrated MS and NMR data to be obtained on each peak of interest in a complex mixture (see Chapter 15).

The need to quantify both drug purity and the levels of impurities and degradation products is served very well by NMR spectroscopy, which is an inherently quantitative technique. Indeed, a whole branch of NMR dedicated to this area is now termed *qNMR* (quantitative nuclear magnetic resonance). See Chapter 5 for details of various approaches to quantification, all of which rely on the fact that (under suitable experimental conditions) the signal area in NMR spectra is directly proportional both to the quantity of the molecule giving rise to that signal in the sample and to the number of nuclei being observed in that signal, e.g., a CH_2 group in a given molecule will give an NMR signal twice as big as a CH group, provided that the experiment is set up correctly. Borrowing methods from metabonomics, qNMR is being applied to increasingly complex mixtures.[13]

1.2.2.3 *Quality Control of Biopharmaceuticals*

Exactly the same attributes that make NMR spectroscopy such a good technique for the quality control of small molecules also make it an excellent tool for the quality control of biopharmaceuticals, typically protein-based drugs. This area has been reviewed recently,[4] so this article will not go into great detail. The key applications include various methods to verify the molecular structure of the protein, especially features such as the glycosylation pattern for glycoproteins. In addition, the open-window, unbiased observational power of NMR spectroscopy lends itself extremely well to determine product quality and the absence of any adulteration with small molecules or other biomacromolecules.

1.2.2.4 *SSNMR Analyses of Leads and Drugs*

This is an important area of application of NMR and use of the technology spans from late discovery right through the development phase of drug R&D.

The principal use of SSNMR is to characterize the solid-state form of the drug. Many developments have occurred in the past decade in terms of new pulse sequences, faster sample rotation, and higher field spectrometers that have facilitated the success of these applications.[14] A typical use of SSNMR would be to investigate the occurrence of different polymorphs and solvates of a crystalline drug by ^{13}C SSNMR spectroscopy. In principle, each different polymorph or solvate will have a characteristic SSNMR spectrum, with different ^{13}C chemical shifts, owing to the unique packing arrangements of the molecules in the different crystal forms. The systematic exploration of these different forms by high-throughput crystallization methods is an important part of the early drug development process for two reasons. Firstly, it is critical that the solid-state form of the drug with optimal stability, solubility, formulation properties, and purity is taken forward for full development. Secondly, if a third-party competitor discovers a previously overlooked polymorph or solvate with significantly improved properties over the solid-state form of an existing agent, it is possible that they may derive intellectual property on that beneficial form of the molecule, even if they do not hold patent rights on the molecule itself, as first discovered.

It is important to mention here that SSNMR provides important information on the solid-state form of the drug not just in the drug substance but also in the drug product, when it is formulated with all of the excipients, and, of course, SSNMR can also give information on the solid-state form of all of the excipients. In addition, it can provide critical information on the stability of the drug solid-state form and can address the question of whether there is any transition between solid-state forms occurring on storage. Thus, SSNMR provides an important bridge between exploratory polymorph studies on the drug substance in the very early stages of development, right through to the demonstration that the solid-state form of the drug selected for final development is stable, both as the drug substance and as the drug product. SSNMR can, of course, also be used to monitor for degradation directly in the solid state without the need to use solution-state analytical methods.

The power of SSNMR in this area goes well beyond the examples cited earlier, however, and the technology is used in a host of different and important applications (see Chapter 21).

1.3 NEW APPLICATIONS OF NMR SPECTROSCOPY IN PHARMACEUTICAL R&D

1.3.1 Hit Discovery and Hit Follow-up Using NMR Spectroscopy

NMR Spectroscopy has always struggled as a tool for protein structure elucidation, although it has many attractive features such as the ability to provide information on dynamics as well as structure, because X-ray crystallography has been faster, provided higher resolution structures, avoided the requirement for isotope labeling of the proteins and required smaller samples. For example, the current protein data bank (http://www.wwpdb.org) has over 100 000 depositions, of which only about 10% derive from NMR studies, with most of the remainder from X-ray crystallography. However, NMR spectroscopy has become an important tool for hit finding against protein targets, which is itself a very important step close to the beginning of the drug discovery process.

Hit finding in the pharmaceutical industry has been based for a long time upon high-throughput screening (HTS) of large collections of compounds in compound libraries: the so-called screening files. Considerable investment went into HTS in the 1990s and 2000s, and the experience of large companies such as Pfizer was that HTS would deliver hits in >90% of screens and that it would initiate new hit-to-lead drug discovery projects in 50–70% of cases.[15] However, in spite of this success, HTS has acquired a bad press in some quarters, as it is unfairly blamed for the increasing molecular weight and lipophilicity of leads and drugs emerging from pharmaceutical R&D. This is inappropriate as the responsibility for this increase lies with the chemists designing those compounds and with the custodians of the screening files: the old maxim 'garbage in, garbage out' applies very well in HTS. Another issue with HTS is the expense of the screening equipment and material management operations required to support screening on a scale of hundreds of thousands or even millions of compounds in the screening files for each target screened. Here again, progress has been made in two areas to reduce the scale of these screening operations. Firstly, the development of the concept of molecular redundancy[15] has enabled the elimination of compounds that are too similar from the screening file, as these provide no new information and waste resources. Secondly, highly efficient subset screening methods such as plate-based diversity screening (PBDS) enable the coverage of the chemical space of the entire screening file while only screening a fraction of the screening plates: this innovation was developed simultaneously by groups in Pfizer and Novartis.[16] Nevertheless, there is a significant need for additional effective and efficient hit discovery methods, especially as academics become more involved in drug discovery and do not necessarily have access to the enormous resources and infrastructure that the pharmaceutical companies once did. In this context, huge efforts have gone into the development of a range of new hit-finding technologies that use NMR, X-ray crystallography, or other methods to detect binding of molecules to target proteins without recourse to the radioactive or fluorescent labeling of reagents required in conventional HTS.

The hit-finding methods using NMR can be divided into two categories: those detecting changes in the NMR signals from the protein when bound by a ligand and, conversely, those methods that detect changes in the NMR signals of a ligand when it experiences binding with a target protein molecule.

The ligand-detected hit-finding methods use ingenious NMR experiments such as saturation transfer difference (STD) and variants of this method like water–ligand observed via gradient spectroscopy (waterLOGSY) to determine which parts of a ligand are involved in binding with the protein binding site.[17] There are several attractive features to these methods including (i) the ability to screen ligands for binding interactions to a target protein in mixtures, improving the screening efficiency, (ii) the ability to screen at relatively low protein concentration, thus limiting protein reagent supply issues, (iii) the ability to screen the protein without the need for isotope labeling, which can be time consuming and expensive, (iv) the ability to derive at least some structural information on the mode of binding, and (v) the ability to derive binding constants directly from these NMR experiments.

The alternative approach is to titrate mixtures of ligands into a solution of an isotopically labeled protein and observe changes in the chemical shifts of protein resonances in contact with ligand molecules in the binding site. Typically, these experiments will require general or residue-selective ^{15}N, ^2H, or ^{13}C labeling of the protein in order to enable the observation of discrete resonances from, e.g., each amide N–H group in a uniformly ^{15}N-labeled protein. The observation of a binding event when screening mixtures will obviously require some deconvolution to determine which member of the ligand mixture was responsible for binding.

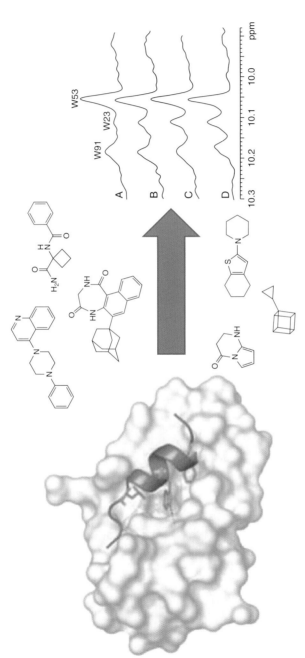

Figure 1.6. The use of antagonist-induced dissociation assay (AIDA) in order to find ligands that disrupt the binding of p53 (blue helix) to protein mdm2 (yellow surface). The proton NMR spectra labeled A to D show the signals for p53 in the presence of increasing ligand concentrations from A (no antagonist) and B to D (increasing concentrations of an antagonist present in the ligand mixture). As the protein–protein complex dissociates, the intensity of the signal due to tryptophan 23 (W23) in p53 increases significantly, as it is freed from its bound, buried position in the complex. (Reproduced with permission from Ref. 20. © John Wiley & Sons, Inc. 2010)

Chemical space is very large. Putting this another way, the number of drug-like, small molecules that can be designed is almost infinite and well beyond any current synthetic methodology. An issue with conventional ligand screening is that an enormous number of different molecules are required to cover a relatively small amount of chemical space. This has given rise to increased interest in fragment screening and then fragment-based drug design (FBDD), in which the molecules screened generally conform to the Rule of 3 rather than the well-known Rule of 5.[18] These fragment molecules are generally below molecular weight 300 Da, with a calculated partition coefficient clogP <3 and less than 3 hydrogen bond donors. The advantages of using fragments are that (i) fragment chemical space is much much smaller than drug space because of the smaller molecular size and it can thus be covered more completely with a reasonably sized library, (ii) fragments tend to be more promiscuous in their binding than larger molecules because they are unencumbered with lots of side chains and appendages that get in the way of a good binding interaction, and (iii) there is a perception that the ligand efficiency (binding energy per heavy atom) of fragments is higher than that of conventional, larger ligands, although in our experience that is not always the case. Utilizing fast pulsing methods, it has been reported that more than 200 2D heteronuclear correlation NMR spectra can be recorded in 24 h for fragment screening, using a 500 MHz NMR spectrometer with a cryoprobe, 40 µM protein concentration and a relatively small, globular protein of 15 kDa.[18]

These general methods can also be applied to the study of protein–carbohydrate interactions[19] and to the study of complexes with membrane proteins[20] (Figure 1.6).

This is an exciting and important area, and the first drug emerging from a fragment-based drug discovery approach, Vemurafenib (Figure 1.7), was approved in 2011 (see Chapter 12).

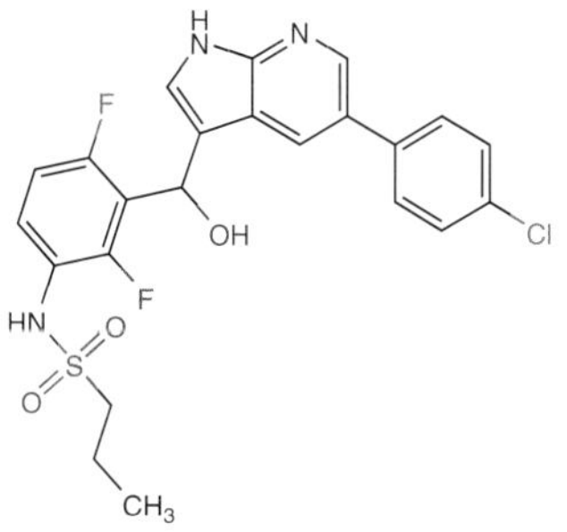

1.3.2 Metabolic Profiling, Metabonomics, and Pharmacometabonomics

Metabolic profiling is another new area that is increasing rapidly in importance for drug discovery and development. NMR spectroscopy has been an important tool for the study of metabolite profiles in biological fluids since the early 1980s.[21] The applications of the technology are manifold and include the direct

Figure 1.7. The molecular structure of Vemurafenib (V600E mutated BRAF inhibition), a B-Raf inhibitor for late-stage melanoma patients with B-Raf V600E mutation

study of drug metabolism for relatively high-dosage drugs,[22] the study of drug safety, and the study of disease states,[23] all of prime importance for drug discovery and development.

The increasing importance of metabolic profiling of biofluids was recognized in the late 1990s by the introduction of the term metabonomics to describe this activity[24] and to provide a conceptual framework for combined genomics, proteomics, and metabolic profiling studies that were underway at that time in Pfizer Global R&D. Metabonomics was defined as 'the study of the metabolic response of organisms to disease, environmental change, or genetic modification'. Some confusion has crept into the field with the later definition of the term metabolomics as a 'comprehensive analysis in which all the metabolites of a biological system are identified and quantified', which is quite a tall order, if not impossible.[25]

The increase in importance of metabonomics studies has arisen for several reasons, some of which are technological. Firstly, the development of efficient multivariate statistical methods that enable (i) the rapid analysis of large numbers of complex NMR spectra of biofluids (mainly) from different groups of subjects, (ii) the unbiased determination of differences between those groups based on differences in their spectral properties, and, most importantly, (iii) the unbiased derivation of the spectroscopic variables that drive those differences (see Chapter 7). Secondly, the availability of ever more reliable spectrometers that can perform demanding NMR experiments with excellent water suppression in full automation, often in conjunction with cooled sample changers that can keep samples stable for long periods before analysis, and automated sample preparation, enabling high throughput of samples for large studies. Thirdly, the development of powerful databases of metabolic data such as the Human Metabolite Database (HMDB, http://www.hmdb.ca), the Biological Magnetic Resonance Data Bank (BMRB, http://www.bmrb.wisc.edu/metabolomics/), and the Birmingham Metabolite Library (BML, http://www.bml-nmr.org) that facilitate the identification of metabolites that occur in the NMR spectra of biological fluids using information from 1D and 2D proton and carbon-13 NMR spectra. In addition, dedicated repositories for the deposition of metabonomics data are emerging such as MetaboLights (http://www.ebi.ac.uk/metabolights/), which will facilitate access to raw data throughout the metabolic

profiling community and help set standards for data deposition.

There are also scientific reasons for the increased importance of NMR-based metabolic profiling. Firstly, NMR spectroscopy is an unbiased detector that can provide fully quantitative data on all of the components in a complex mixture that are above the detection threshold (given certain caveats). Secondly, metabolite profiling gives information directly on the actual physiological status of the organism under study, in contrast to, say, genomics, which can only provide information on what might happen to an organism in the future: this is a really critical distinction. Thirdly, metabolic profiling of biological fluids such as urine provides

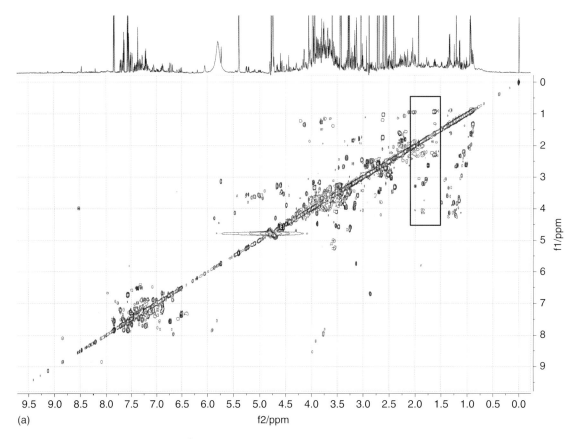

(a)

Figure 1.8. (a) The full 600 MHz 2D ^{1}H COSY NMR spectrum of the urine of a C57BL/6 mouse; (b) an expansion of the area of the COSY spectrum shown in the blue box in (a); and (c) the same expansion of the corresponding 600 MHz 2D ^{1}H J-resolved (JRES) spectrum of the same mouse urine, underneath the corresponding 1D ^{1}H NMR spectrum. The six peaks of the multiplet due to *N*-butyrylglycine across f1 are marked with blue arrows. This multiplet is a triplet of quartets but only six of the twelve lines are visible as the two coupling constants are very similar. All spectra are shown as contour plots and are referenced to TSP at 0 [Everett, 2014, unpublished, from work in collaboration with Dorsa Varshavi (University of Greenwich) and with Liz Shephard and Flora Scott, (University College London)]

information not only on the status of the organism and the expression of its genome but also on the activity of the various microbiomes in that organism. Again, this is a critical distinction and a difference between human genomics and human metabonomics studies.

The huge amount of high-quality data on metabolites that can be obtained from NMR spectroscopy studies of biological fluids at only modest NMR field strengths is illustrated in Figure 1.8, which shows representative 2D ^{1}H COSY and 2D ^{1}H J-resolved NMR spectra from a C57BL/6 mouse. In Figure 1.8(a), the full 600 MHz 2D ^{1}H COSY NMR spectrum is plotted underneath the corresponding 1D ^{1}H NMR spectrum.

A staggering array of proton-to-proton connectivity information is present in the 2D ^{1}H COSY NMR spectrum (Figure 1.8a), which becomes clearer on expansion (Figure 1.8b). Note that the expansion still contains a huge amount of data and information, although it represents less than about 3% of the entire spectrum! For instance, only four of the six lines of

the triplet of quartets multiplet signal for CH_2-3 of *N*-butyrylglycine (Figure 1.9) are visible in the 1D ^{1}H NMR spectrum at about 1.62 ppm (Figure 1.8b, top). All six peaks are however visible across the second dimension f1 of the 2D ^{1}H J-resolved (JRES) spectrum (Figure 1.8c), and the expanded COSY spectrum clearly shows well-resolved cross-peaks to both the CH_3-4 protons at about 0.925 ppm and the CH_2-2 protons at about 2.28 ppm (blue boxes in Figure 1.8b). More remarkably, the JRES spectrum shows two multiplets with at least 12 peaks each at about 1.845 and about 2.005 ppm, belonging to the CH_2-3 methylene protons of L-2-hydroxyglutaric acid. These peaks are almost invisible in the 1D ^{1}H NMR spectrum owing to the complexity of signals in this region, but are clearly defined across the second dimension, f1, of the JRES spectrum, and their mutual coupling connectivities to the CH-2 methyne proton at about 4.03 can clearly be seen in the COSY spectrum (green boxes).

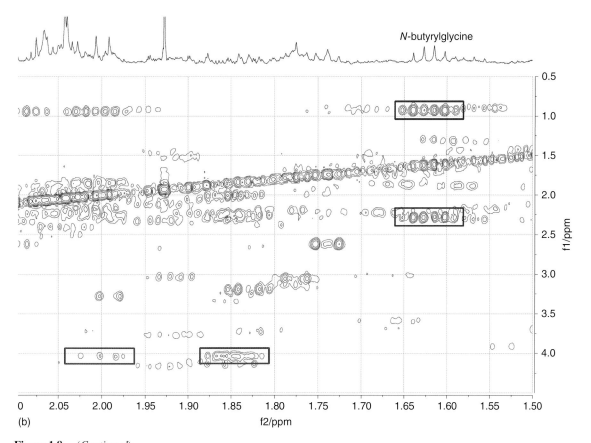

Figure 1.8.   (*Continued*)

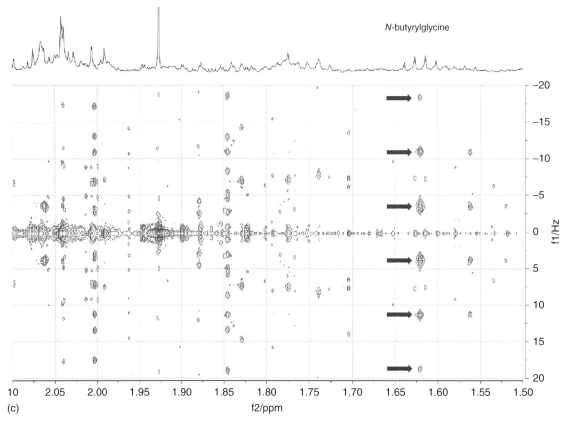

Figure 1.8. (*Continued*)

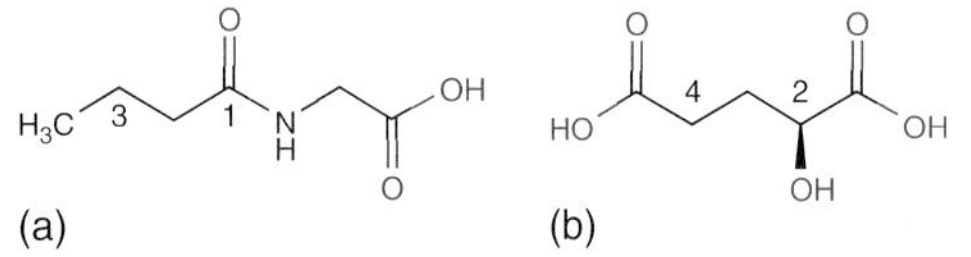

Figure 1.9. The molecular structures of *N*-butyrylglycine (a) and L-2-hydroxyglutaric acid (b)

A major new area of importance in medicine is that of personalized or precision medicine. The concept here is to try to stratify patients so that they receive treatments that will be both efficacious and safe for them. Unfortunately, oftentimes, drugs are either ineffective in certain patient groups or unsafe and this causes significant morbidity and mortality in the Western world. Efforts have been ongoing for decades to use genomic data to stratify patients and to predict the effects that drugs will have on the patients, or that the patients will have on the drug: this is known as *pharmacogenomics*.

Two landmark studies published in 2006 and 2009 demonstrated that *predose* metabolic profiles determined by NMR spectroscopy could be used to predict the outcome of drug treatment in animals in terms of drug metabolism and drug safety,[26] and also that *predose* metabolic profiles could also be used to predict drug metabolism in humans.[27] Moreover, in the latter study, the key biomarker indicating whether the particular drug acetaminophen would be sulfated or glucuronidated in humans was not of human origin, but came from the microbiome! This new area of application of metabolic profiling or metabonomics is called *pharmacometabonomics*[26] by analogy to pharmacogenomics. Many examples of this phenomenon have now been published.

See Chapters 10, 19, 20, 24, and 25, respectively, all of which have extensive additional material.

1.4 *IN VIVO* MRS AND MRI

MRI is now a very well-established technology for imaging in both clinical and preclinical settings, and it can also provide functional information, especially in the brain, through diffusion and flow measurements.[28,29] In one recent development of this area, MRI diffusion experiments are being used to define the human connectome (Figure 1.10), which is the map of neural connections in the brain. This will surely be of use in the development of future drugs for the treatment of central nervous system diseases especially those connected with neurodegeneration.

The term *MRS* is generally used to describe *in vivo* NMR spectroscopy experiments on animals and humans. MRS and MRI are two among a number of technologies being used to provide structural and functional information on various parts of human and animal bodies, particularly the brain, to aid drug discovery and development. MRS has the advantage of being able to measure the levels of a dozen or more high-abundance metabolites in localized tissue regions, such as a particular portion of the brain, in a completely noninvasive and safe manner (no ionizing

radiation involved), but suffers from low sensitivity and long acquisition times compared with other technologies. The most sensitive brain imaging technology is positron emission tomography (PET), which detects twin gamma rays emitted from annihilation events associated with positron-emitting isotopes such as carbon-11, fluorine-18, oxygen-15, and nitrogen-13. The PET imaging center needs to be colocated with, or very close to, the particle accelerator producing these isotopes and specialized chemistry facilities are needed to produce isotopically labeled forms of the drug of interest for PET studies. PET studies can give detailed information on the pharmacokinetics (PK) and distribution of labeled drug molecules with picomolar sensitivity. Unfortunately, while very sensitive (up to 10^8 times more sensitive than MRS), PET gives no information on the chemical nature of the molecules giving rise to the signals and the images may represent mixtures of signals from the drug plus any metabolites and degradation products. Single-photon emission computed tomography (SPECT) is an alternative imaging technology that utilizes the emission of single gamma rays from radioactive nuclei such as xenon-133, technetium-99,

www.humanconnectomeproject.org
© Laboratory of Neuro Imaging

Figure 1.10. White matter fiber architecture of the human brain from diffusion spectral imaging. The brain fibers are color-coded by direction: red = left to right, green = anterior to posterior, and blue = through the brain stem. (Image courtesy of the Laboratory of Neuro Imaging and Martinos Center for Biomedical Imaging, Consortium of the Human Connectome Project: www.humanconnectome.org)

or iodine-123 to give images that can be used to determine the PK and distribution of labeled drugs and ligands. While there is no requirement to be close to a synchrotron facility, the images from SPECT are of lower sensitivity and lower resolution than PET.

Although relatively insensitive, MRS is a field that has been developing since the 1980s and is now well established in preclinical research settings. MRS can provide information complementary to the anatomical information obtained from MRI in animal models of disease, for instance, on changes associated with disease progression or the effects of drug treatment. For example, in mouse models of Alzheimer's disease (AD), changes in the brain levels of *N*-acetylaspartate (NAA), and of *myo*-inositol, can be observed with good sensitivity and precision that enable the effects of treatment to be observed, and this also holds out the promise of translation to patients in the future.[30]

Challenges remain for the full utilization of MRS in clinical settings, however,[31] and some experts believe that this is more likely to occur when performed simultaneously with both PET and MRI. While it is possible to perform joint MRI and MRS studies, currently there are no commercial instruments available enabling the performance of PET and MRI/MRS in a single system.[32] Further technological developments in MRS that increase sensitivity, perhaps including DNP methods, will doubtlessly drive greater utility of MRS in both preclinical and clinical settings. It has to be taken as a given that any technology that can provide rapid, noninvasive, cost-effective, and safe chemical imaging of the human body will find increased usage in the future in drug discovery and development.

A similar situation applies for functional magnetic resonance imaging (fMRI), where in spite of the technology providing significant impact on the understanding of brain functioning and brain system phenotypes in many neurological and psychiatric diseases, there has been minimal benefit so far to individual patients.[28] Again, it is clear that this situation will change in the future with advances in technologies and their application to better diagnose disease states, recommend treatments, and predict outcomes. See Chapters 18 and 26 for further insights and expert review.

1.5 FUTURE DEVELOPMENTS

The technological progress of NMR spectroscopy and MRI over the past 30 years has been astonishing and whole areas of application exist now that were not even conceptualized 20 years ago. The future is notoriously difficult to predict but several areas seem ripe for significant change. Firstly, the identification of the rich array of metabolites present in metabonomics studies is still problematic and requires a significant amount of user knowledge and expertise. It is expected that this area will become increasingly amenable to computer-based expert systems over the next decade. Secondly, the emerging area of pharmacometabonomics is likely to become of much greater future importance for the prediction of drug effects and the delivery of personalized medicine, especially because of its ability to predict drug effects based on environmental (including microbiome) information, as well as genomic-derived information. Thirdly, a new area of predictive metabonomics, of which pharmacometabonomics is just a subset, will emerge, enabling prediction of patient outcomes over time, and as a result of varied interventions, not just drug effects: this has already begun. Finally, improvements in computer power will enable better prediction of NMR spectra based on quantum mechanical methods, giving better confidence in NMR-based structure elucidation by comparisons of actual spectra with precisely calculated spectra of the elucidated molecular structure and viable alternatives. The future is always exciting!

REFERENCES

1. J. L. LaMattina, *Nat. Rev. Drug Discovery*, 2011, **10**, 559.

2. S. M. Paul, D. S. Mytelka, C. T. Dunwiddie, C. C. Persinger, B. H. Munos, S. R. Lindborg, and A. L. Schacht, *Nat. Rev. Drug Discovery*, 2010, **9**, 203.

3. R. M. Fratila and A. H. Velders, *Annu. Rev. Anal. Chem.*, 2011, **4**, 227.

4. D. S. Wishart, *TrAC, Trends Anal. Chem.*, 2013, **48**, 96.

5. Mestrelab Research, The home page for MNova NMR software, Date accessed 5th November 2014, http://mestrelab.com/software/mnova/nmr/

6. G. Baker, R. Dorgan, J. R. Everett, J. Hood, and M. Poulton, *J. Antibiot.*, 1990, **43**, 1069.

7. D. A. Dias, S. Urban, and U. Roessner, *Metabolites*, 2012, **2**, 303.

8. M. Halabalaki, K. Vougogiannopoulou, E. Mikros, and A. L. Skaltsounis, *Curr. Opin. Biotechnol.*, 2014, **25**, 1.

9. K. Ashline, R. Attrill, E. Chess, J. Clayton, E. Cutmore, J. R. Everett, J. Nayler, D. Pereira, W. Smith, J. Tyler, M. Vieira, and M. Sabat, *J. Chem. Soc., Perkin Trans. 2*, 1990, 1559.

10. J. R. Everett, E. Hunt, and J. Tyler, *J. Chem. Soc., Perkin Trans. 2*, 1991, 1481.

11. H. Booth and J. R. Everett, *J. Chem. Soc., Chem. Commun.*, 1976, 278.

12. H. Finch, *Drug Discovery Today*, 2014, **19**, 320.

13. C. Simmler, J. G. Napolitano, J. B. McAlpine, S.-N. Chen, and G. F. Pauli, *Curr. Opin. Biotechnol.*, 2014, **25**, 51.

14. K. Paradowska and I. Wawer, *J. Pharm. Biomed. Anal.*, 2014, **93**, 27.

15. G. A. Bakken, A. S. Bell, M. Boehm, J. R. Everett, R. Gonzales, D. Hepworth, J. L. Klug-McLeod, J. Lanfear, J. Loesel, J. Mathias, and T. P. Wood, *J. Chem. Inf. Model.*, 2012, **52**, 2937.

16. A. S. Bell, J. Bradley, J. R. Everett, M. Knight, J. Loesel, J. Mathias, D. McLoughlin, J. Mills, R. E. Sharp, C. Williams, and T. P. Wood, *Mol. Diversity*, 2013, **17**, 319.

17. O. Cala, F. Guilliere, and I. Krimm, *Anal. Bioanal. Chem.*, 2014, **406**, 943.

18. M. J. Harner, A. O. Frank, and S. W. Fesik, *J. Biomol. NMR*, 2013, **56**, 65.

19. L. Unione, S. Galante, D. Diaz, F. Javier Canada, and J. Jimenez-Barbero, *MedChemComm*, 2014, **5**, 1280.

20. N. Yanamala, A. Dutta, B. Beck, B.van Fleet, K. Hay, A. Yazbak, R. Ishima, A. Doemling, and J. Klein-Seetharaman, *Chem. Biol. Drug Des.*, 2010, **75**, 237.

21. J. Lindon, J. Nicholson, and J. Everett, *Annu. Rep. NMR Spectrosc.*, 1999, **38**, 1.

22. J. R. Everett, K. Jennings, G. Woodnutt, and M. Buckingham, *J. Chem. Soc., Chem. Commun.*, 1984, 894.

23. A.-H. M. Emwas, R. M. Salek, J. L. Griffin, and J. Merzaban, *Metabolomics*, 2013, **9**, 1048.

24. J. Lindon, J. Nicholson, E. Holmes, and J. R. Everett, *Concepts Magn. Reson.*, 2000, **12**, 289.

25. O. Fiehn, *Plant Mol. Biol.*, 2002, **48**, 155.

26. T. Clayton, J. Lindon, O. Cloarec, H. Antti, C. Charuel, G. Hanton, J. Provost, J. Le Net, D. Baker, R. Walley, J. R. Everett, and J. Nicholson, *Nature*, 2006, **440**, 1073.

27. T. A. Clayton, D. Baker, J. C. Lindon, J. R. Everett, and J. K. Nicholson, *Proc. Natl. Acad. Sci. U.S.A.*, 2009, **106**, 14728.

28. E. T. Bullmore, *Neuroimage*, 2012, **62**, 1267.

29. D. Borsook, R. J. Hargreaves, and L. Becerra, *Expert Opin. Drug Discovery*, 2011, **6**, 597.

30. N. Beckmann, R. Kneuer, H.-U. Gremlich, H. Karmouty-Quintana, F.-X. Ble, and M. Mueller, *NMR Biomed.*, 2007, **20**, 154.

31. B. D. Ross, *Expert Opin. Drug Discovery*, 2013, **8**, 849.

32. W. Wolf, *Pharm. Res.*, 2011, **28**, 490.

PART B
NMR Theory & Experimental Methods

Chapter 2

Modern NMR Pulse Sequences in Pharmaceutical R&D

John A. Parkinson

WestCHEM, Department of Pure and Applied Chemistry, NMR Spectroscopy Facility, University of Strathclyde, Glasgow G1 1XL, UK

2.1 INTRODUCTION

Commercial NMR equipment evolution owes much to the demands imposed by users requiring ever more sophisticated experimental capabilities. One outcome is a bewildering array of pulse sequences, which are often subtle variations or improvements on a particular theme. They may depend on frequency-, band-, or spatially-selective excitation, entail diverse homo- and/or heteronuclear frequency correlation schemes, or use editing methods to simplify and enhance the resulting data. Across the breadth of all NMR-related

disciplines, such diversity can be overwhelming and confusing. In this article, I have selected pulse sequences that stand out as those most regularly used by NMR practitioners or which merit special emphasis owing to their future promise in routine spectroscopy. The methods yield rich information, are mostly straightforward to implement, and show robustness when applied in a modern spectrometer context. These are essentials in unsupervised, robotic settings. The methods yield data for understanding small molecules, their structures, behaviors, and interactions, whether as pure materials, contaminants, or components within complex mixtures. Here, I consider the sequences, use some examples of data they produce, and describe how they can be adjusted to function optimally.

2.2 INITIAL SAMPLE ADJUSTMENTS

Sample setup including modern approaches to optimizing magnetic field homogeneity (shimming) and calibration of the ^{1}H 90° pulse is an important initial consideration in a solution-phase context.

2.2.1 Shimming

The process of field optimization has evolved from one based on lock signal measurement using lengthy

NMR in Pharmaceutical Sciences. Edited by Jeremy R. Everett, John C. Lindon, Ian D. Wilson, and Robin K. Harris
© 2015 John Wiley & Sons, Ltd. ISBN: 978-1-118-66025-6
Also published in eMagRes (online edition)
DOI: 10.1002/9780470034590.emrstm1402

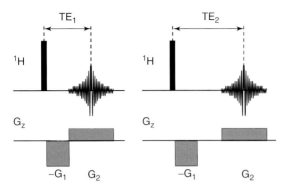

Figure 2.1. Gradient-echo imaging sequence where the total gradient strengths for G_1 and G_2 are equal and TE_1 and TE_2 are gradient-echo periods used to acquire different data sets. ^1H represents the RF channel where narrow black rectangles relate to 90° hard pulses. G_z relates to pulsed field gradients

iterative simplex procedures to the use of field mapping linked with resonance lineshape analysis and optimization,[1] which adds value to the rapid mapping procedure and provides an approach that virtually guarantees optimized field homogeneity under full automation. The classical gradient-echo imaging sequence (Figure 2.1) exploits the *z*-gradient coil of a modern solution-phase NMR probe for one-dimensional field mapping and in combination with room temperature *x* and *y* nonspin shims for three-dimensional field mapping.

Two experiments generate data using the echo periods TE_1 and TE_2, the difference between the resulting data yielding a phase map. When compared with reference data, the process allows shim values to be directly set, providing a more optimized field. The process is iterative but takes as little as 10–20 s for 1D field mapping and a little longer (10–15 min) for complete 3D field mapping. Rapid 1D field mapping with lineshape optimization is ideally suited to unsupervised sample analysis, relying as it does on one dominating resonance in the spectrum. For this reason, the lock solvent deuterium signal is often used in this process.

2.2.2 ^{1}H 90° Pulse Calibration

Accurate calibration of the ^{1}H 90° pulse has been traditionally measured by a time-consuming single pulse-acquire sequence with an array of pulse lengths.[2] This process is incompatible if required as

a feature of every data acquisition under automation. In an unsupervised setting, pulse missetting results in non-optimized data. The modern alternative relies on measuring nutation of the magnetization vector in a rapid, single shot.[3] The pulse sequence, akin to homonuclear decoupling, uses a train of pulses between each of which is acquired a single data point. The resulting sinusoidal data are Fourier transformed, generating two resonances (for a predominating response) separated by twice the nutation frequency, Δv. Data are acquired during the dwell period with a duty cycle, *d*, from which the 90° pulse is calculated according to $pw_{90} = d/(2\Delta v)$. The method can be fully automated and owing to its speed, can in theory be incorporated into all data acquisitions. In this way, all pulse sequences can be fully optimized for sensitivity and performance under full automation regardless of sample type or refined probe tuning, thus substantially improving data quality and reliability.

2.3   ONE-DIMENSIONAL METHODS

The single pulse-acquire sequence is the most basic NMR experiment. Delivered as a short burst (8–12 µs) of radiation at high power (typically in the order of 10–20 W), the pulse creates a non-crafted excitation profile across the expected data acquisition bandwidth. The pulse length is adjusted to yield magnetization vector rotation of any desirable angle. A small rotation angle of typically 30° is often desirable to promote restoration of equilibrium magnetization.

2.3.1   Editing Schemes

Crafted NMR data falls under the guise of frequency specific- or band-selective procedures.

2.3.1.1   *Frequency- and Band-selective Data Acquisition*

Similar to the single pulse-acquire scheme, frequency- or band-selective sequences make use of low-power, long-duration tailored pulses that narrow the excitation bandwidth to the region of interest, whether this be a single resonance (frequency selective) or a frequency region (band selective). Single resonance selection alone has limited use but is suitable for assessing pulse power calibration and frequency selectivity.

A range of pulse shapes has been developed which serve very wide ranging purposes.

2.3.1.2 One-dimensional Total Correlation Spectroscopy

The power of tailored excitation emerges when combined with sequences for revealing correlations between related spins. For instance, 1D TOCSY is useful in resolving data when signal overlap is extensive in some regions of data and resolved in others. Isolated signals are selected and manipulated to reveal spin-system partners at high digital resolution. The sequence is represented here in two forms (Figure 2.2). The benefit of incorporating procedures to eliminate, in a single transient, interfering effects caused by zero-quantum coherence is also illustrated.

An excitation construct is used to cleanly select the resonance from which magnetization is propagated to the remainder of the spin-system nuclei. Isotropic mixing by the DIPSI-2 spin-lock sequence is preferred for more efficient mixing of the magnetization and the use of lower pulse powers across the bandwidth of interest compared with non-isotropic mixing sequences.[4] The basic pulse sequence (Figure 2.2a) yields results typical of those shown (Figure 2.3b).

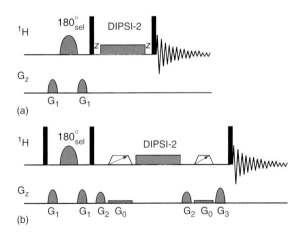

Figure 2.2. 1D TOCSY pulse sequences. (a) Classical 1D TOCSY using a DIPSI-2 spin-lock and *z*-filters. (b) 1D TOCSY sequence adapted to eliminate contributions from zero-quantum coherence pathways. The trapezoidal pulses with diagonal arrows represent frequency swept smoothed CHIRP adiabatic 180° pulses applied over spatially encoding gradients, G_0, with values 1–3% of maximum *z*-gradient strength

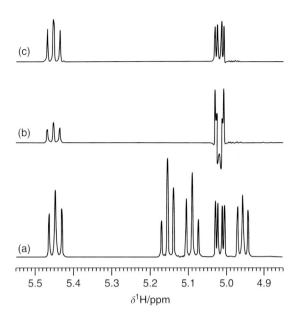

Figure 2.3. Example data for 1D TOCSY. (a) Region of data as reference acquired using a single pulse-acquire scheme. (b) Equivalent region to (a) but acquired with selective inversion of the signal at $\delta^1 H = 5.8$ ppm (not shown on the figure) using the pulse sequence shown in Figure 2.2(a). (c) Equivalent region to (a) but acquired by selective inversion of the same resonance at $\delta^1 H = 5.8$ ppm using the pulse sequence shown in Figure 2.2(b)

Interference from zero-quantum coherence ultimately contributes antiphase magnetization to the observed signal. Unwanted dispersive signals add to the pure absorption lineshape, playing havoc with the data. The popular alternative (Figure 2.2b) engages adiabatic pulses over weak field gradients to dephase unwanted zero-quantum coherence before and after the spin-lock period.[5] Z-filter delays of typical duration 2–3 ms are set to different values to avoid refocusing magnetization from unwanted coherence pathways. The cleaner result (Figure 2.3c) arises solely from pure absorption lineshape.

2.3.1.3 One-dimensional Nuclear Overhauser Effect Spectroscopy and One-dimensional Rotating Frame Overhauser Enhancement Spectroscopy

Tailored excitation has also benefited the development of the NOE experiment, which has evolved from the truncated-driven, difference method, requiring

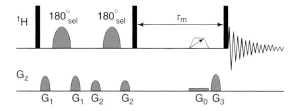

Figure 2.4. Typical scheme for 1D NOESY where τ_m is the mixing time

multiple transients to average out subtraction artifacts, to the modern single- or double-pulsed field gradient spin-echo (DPFGSE) approach that incorporates the same zero-quantum suppression construct described for 1D TOCSY (Figure 2.4).

In the preparation period, the DPFGSE feature cleanly selects only resonances of interest. A mixing pulse inverts the resulting spin magnetization from which evolves the NOE during the mixing period. The adiabatic 180° pulse over a weak field gradient eliminates the lineshape-distorting effects of zero-quantum coherence, which is then followed by a purging gradient and a 90° read pulse. The experiment works with relatively few transients compared with its historical counterpart. Variations on this theme exist which, for longer mixing times, require the distribution of hard 180° inversion pulses surrounded by opposing gradients within the mixing period to purge responses from unwanted coherences. 1D NOESY yields transient NOEs and is appropriate for also detecting chemical exchange processes as is the case with 2D NOESY.

1D ROESY requires adaptation of the 1D NOESY sequence by incorporating a spin-lock construct to replace the mixing sequence $90°-\tau_m-90°$. In all other respects, the sequence is the same but care must be taken to reduce the influence of unwanted TOCSY responses. As an alternative, 1D ROESY may better be derived from the robust 2D EasyROESY as further described in Section 2.4.2.2.

Techniques such as 1D NOESY, 1D TOCSY, and 1D ROESY when combined form homonuclear concatenated sequences that propagate magnetization transfer from spin to spin via dipolar- or spin/spin-coupling and/or chemical exchange mechanisms. Such spliced sequences can use similar constructs, e.g., 1D TOCSY–TOCSY, or different constructs, e.g., 1D TOCSY–ROESY. The combined methods help when teasing out assignments but are less robust in an automated setting.[6]

2.3.1.4 Homonuclear Broadband Decoupling

Spin–spin-coupling provides critical information about related nuclei but removing the resonance-spreading effects of J-coupling can be desirable to simplify NMR data and improve resolution. This is readily achieved in broadband (BB) heteronuclear decoupling for spin interactions between different types of nuclei such as ^{1}H and ^{13}C. However, broadband homonuclear decoupling is an entirely different avenue, having proved to be a significant historical challenge for ^{1}H NMR spectroscopy. Theoretically, the aim is to produce ^{1}H–$\{^1$H$\}_{BB}$ NMR data. Generating the skyline projection from fully processed 2D J-resolved NMR spectra is one way of visualizing this but the results are unsatisfactory owing to poor signal resolution and variable signal amplitude unrelated to true integral values. Other related methods have been published in the more recent literature in a bid to counter the integral issue but these prove unwieldy for routine practical use.[7] The prize for creating ^{1}H–$\{^1$H$\}_{BB}$ NMR data is increased signal resolution for no increase in static magnetic field strength. If achieved with reliable signal integration at full sensitivity, a new NMR milestone will have been reached. This goal is now within sight for routine use in an automated laboratory context. Two categories of pulse sequence now exist, which go some way toward achieving this, namely full ^{1}H sensitivity, homonuclear band-selective decoupled (HOBS) methods[8] and reduced ^{1}H sensitivity, homonuclear broadband decoupled (Pureshift) methods.[9,10] The former relies on band-selection for the special case where signals in the selected region do not J-couple to one another but rather to resonances beyond the selected region. This HOBS approach is ideally suited to biopolymers including peptides, proteins, and nucleic acids whose NMR spectra are regionalized owing to the polymeric nature of the molecules concerned. One-dimensional HOBS spectra can be produced by the sequence shown (Figure 2.5) from which a wide variety of other related sequences may be constructed.

The region-selective preparation construct uses a band-selective 180° pulse. The acquisition stage collects data noncontinuously as data packets. Between data collection, gradient-bracketed hard 180° and band-selective 180° pulses are inserted. These components repeatedly refocus ^{1}H spin–spin-couplings for those resonances in the band-selected region of interest for the special coupling condition described for biopolymers. Segments of FID are acquired for limited periods of up to ~30 ms (approximating to

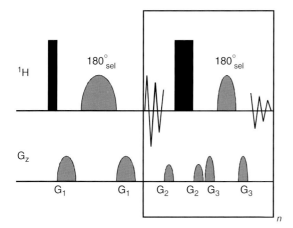

Figure 2.5. Pulse sequence for band-selective $^1H-\{^1H\}_{REG}$ NMR spectra (REG = region selection). The boxed element consists of an interrupted acquisition stage by which data are acquired between RF/gradient pulse elements. Complete data acquisition with homonuclear decoupling is achieved over n cycles of the acquisition stage

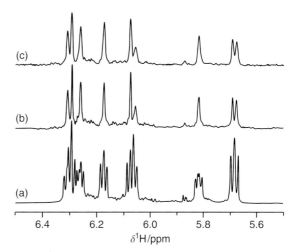

Figure 2.6. H1′/H5 resonance region of the ^1H NMR spectrum of a nucleic acid sample. (a) Standard 1D ^1H NMR spectrum; (b) ^1H NMR spectrum acquired using HOBS sequence shown in Figure 2.5. (c) ^1H NMR spectrum acquired using instantaneous Pureshift. (a) and (b) were acquired with 16 transients whereas (**c**) was acquired with 8192 transients

$1/3J_{HH}$). Chemical shifts evolve much faster than J-couplings meaning that chemical shift information is laid down in the FID segments while the effects of J-coupling are regularly removed through refocusing. A drawback to this technique is that during these refocusing periods, T_2-relaxation is active leading to some discontinuities in the FID yielding spectral artifacts. Looping the acquisition stage yields a FID with acquisition duration adequate for high-resolution spectra with acceptable lineshape (Figure 2.6b).

Full proton sensitivity is desirable and achieved for HOBS but the special coupling conditions make it less attractive for general purpose application, despite it being readily implemented under full automation for biopolymers. Related procedures for generating $^1H-\{^1H\}_{BB}$ NMR data are available using several approaches albeit with reduced sensitivity.[9–11] For routine use, the continuous (instantaneous) Pureshift data acquisition scheme is attractive.[10] The principal difference between this and HOBS lies in the application of a narrow band, frequency-selective shaped inversion pulse applied during a weak pulsed field gradient. The latter spatially encodes spin magnetization along the length of the NMR sample. If the field gradient is adjusted so that the entire ^1H NMR spectrum is excited by the selective inversion pulse along the sample length during the preparation period, a very acceptable high-resolution, broadband decoupled ^1H

NMR spectrum can be generated. Data quality compared with standard ^1H NMR spectra is affected in two ways. Firstly, sensitivity is compromised owing to the spatial encoding: only a fraction (~1%) of spins contribute to any particular signal. Secondly, the technique does not tolerate strong J-coupling. Alternative proposals for overcoming this have been published in a bid to address the insensitive nature of the technique.[11] These methods show merit in simplifying ^1H NMR spectra and their improved resolution 2D analogs are also promising.[12–14]

2.3.1.5 Diffusion- and T_2-editing

Diffusion and relaxation characteristics can also be used to simplify NMR data. Motional properties of molecules are exploited using DOSY, especially for studying mixtures of molecules whose size, shape, and molecular weights vary with respect to one another. Classical diffusion experiments use pulsed field gradients to label nuclei with their physical location in a sample. The diffusion period, Δ, is followed by a gradient refocusing construct that removes the spatial encoding. Diffusion of individual molecules during Δ causes the recovered magnetization to yield a lower than maximum response. A function of sample temperature, viscosity and the so-called friction factor, diffusion relies on molecular size, shape, and weight. In the context of complex mixtures of small and large molecules, a diffusion filter applied at the start of a pulse sequence draws out signals exclusively from large molecules such as proteins at the expense of those from small molecules. Similarly, T_2-filtering disposes of responses from large molecules to reveal more clearly those resonances from small molecules. Differences in T_2 relaxation time constants for different types of molecules are exploited using the classical Carr–Purcell–Meiboom–Gill (CPMG) sequence, which repeatedly refocuses chemical shift evolution by looping through a spin-echo sequence for a defined period of time during which T_2-relaxation acts (Figure 2.7).

Here, τ is short (1–2 ms) compared with $1/J_{HH}$. J-modulation of signals occurs during the spin-echo period and may be minimized but is not altogether eliminated using this sequence. The unwanted effect from RF heating caused by the short duty cycle arising from multiple spin-echoes also degrades data quality. To compensate, the introduction of a J-refocusing 90° pulse between two spin-echo periods (Figure 2.7b) removes both J-modulation artifacts and allows the delay period τ to be extended by at least an order of magnitude thus reducing the RF load. With $n > 1$, this PROJECT (periodic refocusing of J-evolution by coherence transfer)[15] sequence gives NMR spectra with clean multiplet structures while delivering broad background resonance suppression (Figure 2.8).

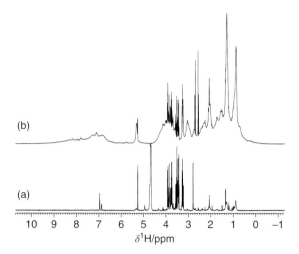

Figure 2.8. T_2-editing. (a) ^{1}H NMR spectrum of human blood serum edited using the pulse sequence shown in Figure 2.7(b). (b) Reference 1D ^{1}H NMR spectrum

2.3.2 Solvent Suppression

Samples studied in protonated solvents or very dilute samples studied in deuterated solvents give intense solvent signals that must be attenuated in the ^{1}H NMR spectrum to improve digitization without compromising solute responses. Many solvent suppression schemes are available with choice governed by application. No catch-all method exists but instead, each has its own advantages, disadvantages, and particular merits.

2.3.2.1 *Presaturation*

For routine use, presaturation, which is a solvent spin saturation technique, is the most straightforward, general purpose method typically used for handling aqueous samples. Solvent signal suppression operates by low-powered, continuous RF irradiation at the solvent resonance frequency during the long relaxation period. Following the saturation period, the sequence can be constructed from a number of different features (Figure 2.9).

Variants are designed to improve the quality of solvent suppression while leaving solute signals untouched. For aqueous solutions, the solvent frequency at the midpoint of the spectrum is made coincident with the transmitter offset and adjusted to minimize residual solvent signal. For other solvents whose ^{1}H

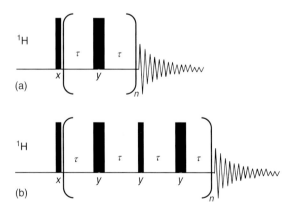

Figure 2.7. Spin echo. (a) Classical CPMG pulse sequence – the bracketed sequence is repeated n times to promote loss of signal through T_2 relaxation. (b) PROJECT pulse sequence where $n > 1$. Wide vertical bars represent 180° refocusing pulses

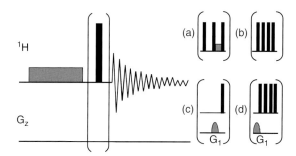

Figure 2.9. Presaturation pulse sequence variants. The standard procedure is shown as the main figure. Content in the bracketed region may be replaced by (a) 1D NOESYPRE-SAT, (b) composite 90° degree read pulse, (c) saturation period followed by spoiling gradient and 90° read pulse, and (d) spoiling gradient followed by composite 90° read pulse

resonance frequencies may not coincide with the center of the ^{1}H spectral window, alternative schemes are used or the frequency width of the spectrum extended and the transmitter offset frequency adjusted appropriately to coincide with the solvent resonance.

2.3.2.2 *Water Suppression Through Gradient-tailored Excitation (WATERGATE) and Double-pulsed Field Gradient Spin-echo (DPFGSE)*

Solvent suppression approaches that minimize saturation transfer effects suppress the solvent signal in a more tailored fashion. Basic WATERGATE (water suppression through gradient-tailored excitation, G_1-S-G_1, where $S \equiv$ 'soft-180°-hard-180°' and G represents pulsed field gradient)[16] and excitation sculpting[17] (DPFGSE, G_1-S-G_1-G_2-S-G_2) are variations on similar themes. Provided the transmitter offset coincides with the solvent resonance, the sequences dephase and eliminate the solvent resonance while refocusing the solute resonances that do not coincide with the region near the transmitter offset. With the exception of this central frequency window, data are acquired with uniform excitation across the desired region. DPFGSE yields a better phase response compared with the original Watergate but at the expense of longer spin manipulation periods. Reducing soft pulse and gradient periods helps to minimize saturation transfer effects. Variations on these methods use binomial pulse trains to further shorten the duration of the method. These show excitation nulls at regular frequency intervals including at the center of the

^{1}H NMR spectrum.[18] They are arguably less robust techniques that require some user intervention to ensure optimal performance. Watergate and DPFGSE suppress the solvent signal during an echo period *after* the read pulse, which contrasts with presaturation, when solvent suppression occurs as the first event in the pulse sequence. When combined with 2D methods that make use of extended mixing periods (as in TOCSY, ROESY, and NOESY), the excitation sculpting approach to suppressing the H_2O response is considered to yield superior benefits.

2.3.2.3 *Water Suppression Enhanced Through T_1 Effects (WET)*

Water suppression enhanced through T_1 effects (WET) makes use of frequency-shifted soft pulses and can be used together with simultaneous ^{13}C decoupling during the soft pulse and data acquisition periods to reduce the intensity of ^{13}C satellite signals from organic solvents, which can interfere with solute responses.[19] The method is ideal in the context of protonated organic solvents. WET incorporates a train of soft pulses followed by spoiling gradients. ^{13}C decoupling is applied using a second independent RF channel during each of the soft pulses. The procedure collapses the ^{13}C satellite response. When followed by a spoiling gradient, the combined solvent-selective pulse with simultaneous heteronuclear decoupling attenuates solvent signals from both ^{12}C- and ^{13}C-attached ^{1}H nuclei. The soft pulse can be tailored to suppress multiple resonances from the same solvent in one hit. A train of such pulses with gradients reinforces the suppression effect and eliminates excess solvent signal ahead of the read pulse (Figure 2.10). The net effect can be impressive in revealing solute resonances from behind ^{13}C satellite signals (Figure 2.11).

The sequence is particularly useful for reaction process monitoring with liquid chromatography (LC)-coupled NMR systems in which protonated organic solvents serve as the mobile phase.

2.3.3 Ligand Screening

NMR approaches to ligand screening against potential biological targets form a key stage in modern drug discovery and design and either focus on the biological target or the ligand. Structure–activity relationships (SAR) by NMR[20] focuses on the biological receptor and uses ^{15}N-labeled protein titrated

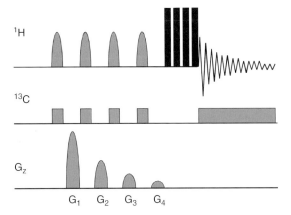

Figure 2.10. WET. Gradient ratio $G_1 : G_2 : G_3 : G_4 = 80 : 40 : 20 : 10$. ^{13}C decoupling takes place during each pulse and the acquisition period if organic solvents are used. The read pulse is presented as a composite 90° pulse

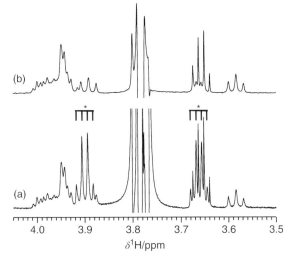

Figure 2.11. WET applied to a mixture of solutes dissolved in 60% H_2O and 40% CH_3CH_2OH (a) in the absence and (b) presence of ^{13}C decoupling according to the scheme of Figure 2.10. Note the ^{13}C doublet of quartet satellites * for the methylene proton signal of the ethanol solvent in (a). ^{13}C satellites of the CH_2 signal mask resonances from solute molecules, which are only revealed when WET is applied with ^{13}C decoupling during application of the selective pulse elements

with small molecule drug candidates. Conclusions are drawn from differences in ^1H and ^{15}N chemical shifts in 2D [^1H, ^{15}N] heteronuclear single quantum coherence (HSQC) NMR data acquired and compared for ligand-free and ligand-bound protein. In promising cases, the data allow the ligand binding site to be mapped on the receptor surface. The procedure is unsuitable for identifying candidate drugs or fragments by mixture screening from compound libraries and requires large quantities of ^{15}N-labeled protein target. Instead, diffusion editing, NOE pumping, Water-LOGSY (water–ligand observed by gradient spectroscopy), and saturation transfer difference (STD) methods are used. We focus here on Water-LOGSY and STD as the two more robust approaches to small molecule fragment screening.[21]

2.3.3.1 Saturation Transfer Difference (STD)

STD provides a fragment screening method for weakly binding ligands with K_D ~10 mM.[22] The process uses protein at concentrations well below the level of NMR detection, typically requiring 100 pmol of material. ^1H resonances from mixtures of small molecules added to this therefore dominate the NMR spectrum. A train of soft pulses applied for 1–2 s selectively saturates the $\delta^1H = 0.0$ to -1.0 ppm resonance region occupied only by unseen protein responses. Saturation propagates quickly by spin diffusion to all protein ^1H spins and to those of any bound ligand, regardless of how transient the binding event is. Spin saturation is then

carried with any binding ligand on its release into free solution. Control data, in which soft pulses are applied far off resonance, are used along with the on-resonance saturation data to generate difference spectra that only show responses for ligands, which bind to the protein target. The approach can be thought of as a special case of presaturation.

2.3.3.2 Water–Ligand Observed by Gradient Spectroscopy (Water-LOGSY)

Water-LOGSY (Figure 2.12)[23,24] also transfers magnetization to ligands by spin diffusion but the mechanism occurs between water–ligand and the protein–ligand complex. It relies on chemical exchange and NOE processes associated with bulk water, water associated with the ligand binding site and labile protons within the protein. The water resonance is cleanly selected using the first section of the pulse sequence. The mixing time, τ_m, typically of the order of 1–2 s, promotes magnetization transfer from water to ligands that are in exchange with the protein binding site. A 180° pulse midway into the

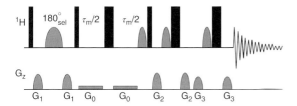

Figure 2.12.   Water-LOGSY

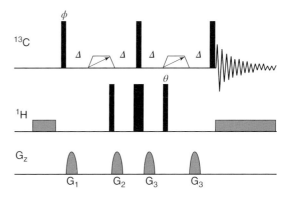

Figure 2.13.   DEPTQ. ϕ and θ indicate pulses of variable length, $\Delta = 1/(2J_{XH})$ and the trapezoids show adiabatic 180° refocusing pulses. $G_1 = G_2 = G_3$ yields all carbon signals, $G_1 = 0$, $G_2 = G_3$ yields only CH_n (no quaternary) and $G_1 = G_2 \neq G_3$ yields only quaternary carbon signals

mixing period suppresses unwanted artifacts. The water flip-back pulse applied at the end of the mixing period maintains the majority of water magnetization along the z-axis during acquisition and relaxation delays so that pulse sequence repetition rates can be accelerated. A DPFGSE scheme is applied following the read pulse to cleanly suppress residual solvent magnetization. The resulting one-dimensional spectra report on protein-bound species as positive signals and noninteracting species as negative signals.

2.3.4   Pulse Sequences for ^{13}C NMR Spectroscopy

Although the most sensitive and obvious approach to analyzing organic molecules in solution is by ^{1}H NMR spectroscopy, the framework of such molecules is by definition carbon based. The 2D methods described in later sections are powerful procedures by which bond connectivity through ^{1}H and ^{13}C spin networks can be established. Despite this, there remains an attraction toward direct observation of ^{13}C–{^{1}H} NMR data, partly as a response to the need to observe NMR signals from quaternary carbons at high resolution and partly through cryoprobe development optimized for maximum ^{13}C sensitivity.

2.3.4.1   *Distortionless Enhancement by Polarization Transfer with Retention of Quaternaries (DEPTQ)*

Distortionless Enhancement by Polarization Transfer with retention of quaternaries (DEPTQ)[25] is a more recent addition to the J-modulated (J-MOD) or attached proton test (APT),[26] DEPT, and INEPT sequences commonly used to edit ^{13}C spectra on the basis of spin multiplicity. DEPTQ improves data quality and robustness over other methods by inclusion of heteronuclear adiabatic 180° inversion pulses[27] (Figure 2.13).

This makes DEPTQ ideal for direct ^{13}C data acquisition in a high-throughput context. Saturation of protons during the relaxation period builds the ^{1}H–^{13}C NOE for enhancement of ^{13}C signal intensity prior to the initial ^{13}C ϕ pulse, which can be set to the Ernst angle for optimum sensitivity/pulse repetition. Adiabatic 180° ^{13}C pulses produce better inversion profiles for resonances far from the offset frequency (such as quaternary carbonyl carbon resonances) compared with standard hard 180° inversion pulses. These ensure that maximum signal sensitivity is achieved both close to and far from the transmitter offset. Gradients can be chosen to select different categories of signals (see caption, Figure 2.13) and the ^{1}H θ pulse angle is set to define the editing profile within the data: $\theta = 45°$ yields all positive signals, $\theta = 90°$ yields only resonances from CH centers, and $\theta = 135°$ results in patterns typical of DEPT ^{13}C NMR data with the added bonus of signals from quaternary carbon centers being present.

2.3.4.2   *^{13}C–{^{1}H} Uniform Driven Equilibrium Fourier Transform (UDEFT)*

A modern take on the DEFT[28] is also helpful for enhancing 1D ^{13}C–{^{1}H} NMR data. Originally designed to compensate for ^{13}C nuclei with particularly long T_1 relaxation times, DEFT allows significant shortening of the relaxation delay. It relies on a spin echo, during the first half of which the ^{13}C FID is acquired. The second half of the echo is used to

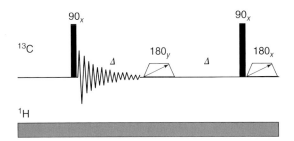

Figure 2.14. UDEFT. Data acquisition occurs following the first 90° pulse. Adiabatic pulses are used to precisely refocus and invert ^{13}C spin magnetization allowing short recycle times. ^1H decoupling may be applied throughout as shown in this example

refocus still-evolving magnetization, which is then returned to the z-axis with a 90° pulse of opposite phase to that used in preparing the initial magnetization. In its original form, DEFT was prone to all manner of imperfections arising from experimental setup including pulse missetting and field inhomogeneity effects causing the sequence to underperform and fall from favor. The modern version is uniform Driven Equilibrium Fourier Transform (UDEFT, Figure 2.14)[29] in which the features responsible for previous underperformance are corrected. In particular, the method is tolerant to pulse missetting. The adiabatic 180° pulse inverts the sense of ^{13}C spin-vector rotation and after a period 2Δ, ^{13}C magnetization is precisely refocused. Application of the remaining $90°_x - 180°_x$ (adiabatic) sequence perfectly restores ^{13}C z-magnetization. The repetition time can be reduced to $3-5$ s while generating quantitatively accurate data for nuclei with even the longest T_1 relaxation time constants. In the context of high-throughput laboratories, the technique is robust and can be used to replace the traditional power-gated $^{13}C-\{^1H\}$ data acquisition scheme.

2.4 PULSE SEQUENCES FOR 2D NMR SPECTROSCOPY

Sophisticated and powerful as they can be, pulse sequences that generate one-dimensional NMR spectra are only a fraction of the solution-phase NMR methods as a whole. Multidimensional methods provide access to a richer wealth of data. Modern methods allow such data to be accumulated at accelerated rates with tremendous reliability and accuracy. These are

considered under the separate headings of homonuclear and heteronuclear techniques.

2.4.1 2D Homonuclear Through-bond Correlations

Classical two-dimensional homonuclear correlations are traditionally applied to nondilute spins at natural abundance or to samples uniformly enriched in specific nuclei, a common feature of biomolecular NMR studies. They follow the general scheme *preparation–evolution–mixing–acquisition*. Preparation promotes spin relaxation, residual signal purging, restoration of z-magnetization, and solvent signal handling. Evolution yields indirect chemical shift labeling of spins as a function of incremental delays. Mixing allows cross talk between interacting spins and acquisition is the data collection period, which is increasingly taking on different forms in high-resolution liquids NMR[10].

2.4.1.1 Correlation Spectroscopy (COSY) and Total Correlation Spectroscopy (TOCSY)

Many variations of COSY exist from the simplest two-pulse, nonphase-sensitive experiment to sophisticated gradient-selected, MQ filtered, phase-sensitive variants. Absolute value COSY data can be acquired with a single transient per t_1 increment (Figure 2.15a) making it ideal for quick acquisitions on abundant nuclei when there is sufficient signal.

The MQ filtered analog (Figure 2.15b) selects the coherence pathway on the basis of gradient ratios. Phase-sensitive variants require the introduction of 180° pulses in association with gradient pulses to compensate for unwanted phase errors that arise through chemical shift evolution during the gradients. A zero-quantum suppression scheme may also be incorporated using the approach described previously and is followed by a purging gradient (Figure 2.15c). For comparison, 2D TOCSY yields strong responses with pure absorption lineshape, often revealing the identity of entire spin-system partners. Replacement of the selective RF pulse and related gradient pulse, G_1, in Figure 2.2(b) with a t_1 incremental delay results in the generalized 2D TOCSY with zero-quantum suppression. The results are analyzed in partnership with COSY and edited-HSQC/heteronulcear multiple-quantum correlation (HMQC) data to reduce

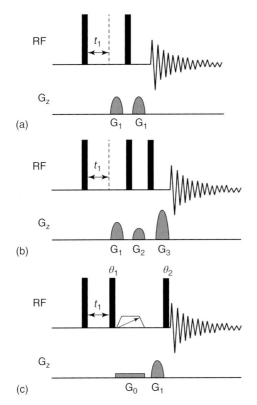

Figure 2.15. COSY. (a) Gradient-accelerated absolute value COSY. (b) MQ filtered absolute value COSY: G_1 is set for either 16% (DQ filter) or 4% (triple quantum filter) when G_2 and G_3 are set to 12 and 40%, respectively. (c) z-COSY with zero-quantum suppression. The small flip angle pulses (θ_1 and θ_2) give rise to z-COSY cross-peak and diagonal structures

the information content when establishing neighboring spin partner identities.

2.4.2 2D Homonuclear Through-space and Chemical Exchange Correlations

2.4.2.1 *Nuclear Overhauser Effect Spectroscopy (NOESY)*

Starting with the 1D NOESY sequence (Figure 2.4), removing the selective pulse components along with their associated gradients and replacing these by a t_1 incremental delay similarly generates a pulse sequence suitable for 2D NOESY data acquisition. The addition of preceding pulse constructs during the preparation

period or spin-echo components following the final read pulse allows ready adaptation of the sequence for solvent suppression. T_1 relaxation characteristics govern the choice of mixing time, τ_m, which, for small, nonaggregating molecules, typically lies in the range $0.5–1.0$ s and for larger molecules ($M_r > 2$ kDa) in the range $0.05–0.20$ s. 'Medium'-sized molecules ($1000 < M_r < 2000$), for which $\omega\tau_c \approx 1.12$ (where ω is the Larmor precession frequency and τ_c the molecular correlation time) require 2D ROESY for NOE observation in the rotating frame of reference.

2.4.2.2 *Rotating Frame Overhauser Enhancement Spectroscopy (ROESY)*

In contrast to NOESY as a straightforward and reliable method, ROESY has not been problem-free historically. The main issues have been data contamination caused by COSY and TOCSY cross-peaks and cross-peak intensity distortions arising from resonance offset effects. Some pulse schemes were developed to circumvent these issues to increase the reliability of ROESY but no single method proved to be a cure-all for the issues encountered until a method dubbed Easy-ROESY was recently proposed and demonstrated.[30]

EasyROESY uses adiabatic ramped pulses to transfer z-magnetization to and from the spin-lock axis at the beginning and end of each spin-lock period (Figure 2.16). The spin-lock offset frequency is shifted from a low value during the first half of the mixing period to a high value during the second half. Spin-lock frequency offsets are equally disposed about the transmitter offset and lie well outside the observation window, typical values being $\pm 5000 < \pm\nu_{offset} < \pm 6000$ Hz relative to a transmitter offset of 0 Hz. Purging gradients follow each spin-lock period. Using this approach, the offset dependency of the ROE signal intensity is eliminated and TOCSY artifacts are simultaneously reduced to a minimum. Resulting data are reliable for the purposes of integration and internuclear distance measurement in molecules of medium size.

2.4.3 Relaxation Parameter and Diffusion Coefficient Measurement

Although not formally two-dimensional, T_1, T_2, and diffusion coefficient (D) parameter measurements follow some two-dimensional data acquisition

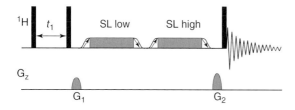

Figure 2.16. EasyROESY. Adiabatic ramped pulses are represented by the curves at either end of the spin-lock periods identified as SL low and SL high

principles. Time- or gradient-related parameters are systematically varied giving data attenuated as a function of the respective variable and stored as a two-dimensional array.

2.4.3.1 T_1 and T_2 Measurement

T_1, the spin–lattice relaxation time constant, traditionally measured by means of an inversion recovery sequence $180°_x$–vd–$90°_x$–acq (vd is a variable delay) is determined by fitting relevant data to an expression of the form $I_t = I_0 + Pe^{-t/T_1}$ (where I_t is signal intensity at vd = t, I_0 is signal intensity at vd = 0 and T_1 is the longitudinal spin–lattice relaxation time constant). T_2 data emerge from the application of those sequences shown (Figure 2.7) and are fitted to expressions of the form $I_t = I_0(1 - e^{-t/T_2})$, where T_2 is the transverse, spin–spin relaxation time constant. Although care needs to be taken in the choice of relaxation and interpulse delays, the methods are robust and easily implemented under automation.

2.4.3.2 Diffusion-ordered Spectroscopy (DOSY)

By contrast, quantitative evaluation of diffusion coefficients, D, under automation can be fraught with difficulty, requiring careful choices and a conservative evaluation of results. The principal competing issue is convection, which owes its presence, in an NMR context, to the design of sample temperature regulation systems. Addition of a constant convection velocity to diffusive motion severely distorts the resulting NMR data giving unreliable diffusion parameters. Some physical adjustments can help to eliminate convection flow including reduced diameter NMR tubes, viscous solvents, lower temperatures, and volume-restricted samples. Convection compensation can also be incorporated within the relevant pulse program (Figure 2.17), which is consequently extended compared with related sequences.

This stimulated echo approach spatially encodes the magnetization and stores it along the z-axis. Spatial encoding with gradients uses the bipolar pulse pair (BPP) sequence, which minimizes deuterium lock signal disturbance and lineshape distortions in the final NMR data. For the convection-compensated sequence shown, the self-diffusion period, Δ, is divided into two sections. Any phase error introduced by convection flow in the first half of the sequence is canceled in the second. The delay, T_e, of typical duration 5 ms, promotes eddy current dissipation prior to data acquisition. Pulse sequences such as this require extensive phase cycling to eliminate unwanted signals, meaning considerable signal averaging, which is not always desirable especially for strong samples. The alternative oneshot sequence only requires a single scan per spectrum[31] (Figure 2.18).

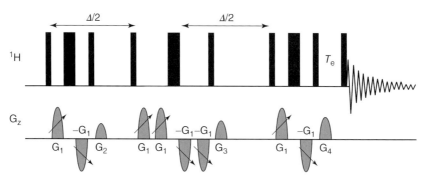

Figure 2.17. Convection-compensated stimulated-echo diffusion pulse sequence. Ramped gradients are shown with diagonal arrows. Δ is the diffusion period and T_e a delay to allow for eddy current dissipation

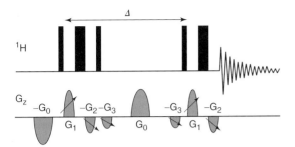

Figure 2.18. Oneshot DOSY

Oneshot uses BPPs where the ramped gradients are unbalanced by a factor α such that $(|G_1 - \alpha|) = (|G_2 + \alpha|)$. The gradient imbalance is countered by ramped gradient $|G_3| = 2\alpha$ applied during the diffusion period. The gradient imbalance dephases unwanted magnetization left over after application of the 180° pulses and also maintains lock stability. Purging gradients, G_0, are applied as balanced pairs during the relaxation delay and within the diffusion period Δ, maintaining the exclusive presence of z-magnetization during the diffusion period. Using these techniques, data acquired in pseudo-2D format as a function of gradient strength may be processed by regression fitting against the Stejskal–Tanner equation, modified where necessary depending on the pulse sequence used. 2D plots can be generated that carry chemical shift and diffusion coefficient information along orthogonal axes. Processing is technically nontrivial although this is mainly hidden behind user-friendly programs in most available software. The danger of this is that without care, false diffusion coefficient values can be interpreted as real. These can occur particularly for overlapped resonances or arise through the influence of chemical exchange. Methods such as PROJECTED (periodic refocusing of J-evolution by coherence transfer extended to diffusion-ordered spectroscopy)[32] can assist with the latter while the issue of resonance overlap may be addressed by combining DOSY and Pureshift methods to reduce the incidence of resonance overlap.[33]

2.4.4   2D Heteronuclear Correlations

Among the more powerful solution-phase NMR methods for small molecule structure elucidation, the data supply to automated structure verification (ASV) and the introduction of increased data

resolution are those that yield information from correlations between different types of nuclei. This can be through either $^nJ_{HX}$-couplings or dipolar couplings in the form of the heteronuclear Overhauser effect. Modern techniques benefit from many historical refinements to give artifact-free, signal- and resolution-maximizing procedures. Virtually exclusive use is made of proton-detection, which amplifies signals by factors of up to several orders of magnitude compared with the results obtained through direct observation of the heteronuclear spin. Further incremental advances in sensitivity are incorporated by means of sensitivity improvement schemes and recently through the introduction of Pureshift methods designed to decouple homonuclear proton couplings in the directly detected dimension.[34] The greatest advance in the quality of data arising from all of these methods owes its success to pulsed field gradients. These are used to best effect when selecting only the information required through gradient coherence selection. For organic molecules, this means the elimination of magnetization from protons attached either to spin-silent ^{12}C nuclei or quadrupolar spin-1 ^{14}N nuclei. This directly affects NMR spectrometer receiver gain: large responses that would otherwise be present from unwanted proton signals are eliminated prior to detection rather than being eliminated by phase cycling following detection, thus allowing optimized digitization of signals. Pulsed field gradients also help to eliminate other artifacts including t_1 ridges and thus improve the overall quality of data.

2.4.4.1   One-bond Heteronuclear Correlations

HMQC and HSQC are most commonly used for detecting correlations between ^1H and other spin-$^1/_2$ nuclei, particularly ^{13}C and ^{15}N, via $^1J_{HX}$. An advantage of HMQC is its robustness toward missetting of pulse lengths as well as delivering a limited number of RF pulses, thus shortening the sequence overall relative to HSQC. ^1H homonuclear coupling evolves during the heteronuclear chemical shift evolution period, producing split cross-peaks in the indirectly detected dimension, a drawback if resolution is an issue. HSQC does not suffer from this effect but is a longer pulse sequence that is not insensitive to pulse missetting. One optimized HSQC sequence is the gradient-selected version with sensitivity improvement and options for multiplicity editing (Figure 2.19), throughout which ^{13}C adiabatic 180° pulses are used in place of 180° hard pulses.

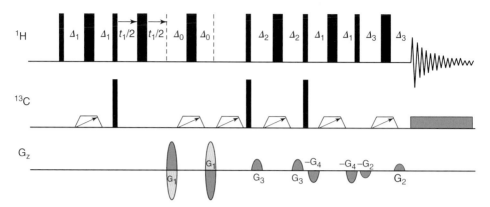

Figure 2.19.   Fully optimized gradient-selected, multiplicity-edited, phase-sensitive 2D [^{1}H, ^{13}C] HSQC

The adiabatic CHIRP (a pulse whose frequency changes during the duration of the pulse) 180° pulses invert ^{13}C magnetization and exactly refocus $^1J_{HC}$-couplings of different sizes. Phase distortion artifacts that would arise from incomplete refocusing of $^1J_{HC}$-couplings are minimized and double reverse INEPT at the end of the sequence generates sensitivity improvement.[35] The sequence can be tuned for the presence/absence and appearance of signals from CH, CH$_2$, and CH$_3$ groups by adjusting the delays Δ_x. For $\Delta_0 = 1/(2 \cdot {}^1J_{HC})$, CH and CH$_3$ signals appear inverted with respect to CH$_2$ signals. For $\Delta_2 = 1/(8 \cdot {}^1J_{HC})$, signals are present from all carbon types with the exception of quaternary carbons, whereas $\Delta_2 = 1/(4 \cdot {}^1J_{HC})$ gives only signals from CH groups. Gradient coherence selection is achieved by setting the gradient ratio $G_1 : G_2 = \gamma_H : \gamma_X$, which for [^{1}H, ^{13}C] HSQC means a ratio of 4 : 1. Typically, these gradients are at 80% and 20% of maximum strength and sufficient to suppress all unwanted magnetization arising from ^{12}C-attached ^{1}H nuclei. The conditions allow fast data acquisition with optimum digitization. Although complex in form, modern NMR systems are capable of handling the different elements of such a sequence with ease. Rapid calibration of pulse lengths under automation as described earlier make this method robust for operation without user intervention leading to excellent routine data generation. The classical elements of composite pulse decoupling during the data acquisition period and ^{1}H decoupling via the 180° inversion pulse midway into the ^{13}C chemical shift evolution period ensure that data are presented as simply as possible. Pureshift approaches

have recently yielded the Pureshift-HSQC[34] by which HSQC data are acquired with homonuclear broadband ^{1}H decoupling during the ^{1}H data acquisition period, along with ^{13}C broadband decoupling, in a segmented fashion that mirrors related methods. The data benefit by increased resolution in the ^{1}H dimension together with increased signal intensity. Where such techniques are implementable on modern NMR spectrometers, there is scope for simultaneous sensitivity and resolution improvement in routine data acquisition that should supersede many of the commonly adopted methods used for detecting heteronuclear $^1J_{HX}$ correlations.

2.4.4.2   Multiple-bond Heteronuclear Correlations

The partner to HMQC and HSQC is a method that correlates H and X through bonding networks according to $^nJ_{HX}$ where $n \geq 2$. This HMBC[36] (Figure 2.20) achieves direct suppression of signals arising from $^1J_{HX}$ and operates similarly to HMQC by generating responses through heteronuclear double and zero-quantum coherence.

Correlation through multiple bonds relies on the evolution of long-range $^nJ_{HX}$-coupling during delay Δ_{LR}. Optional low-pass J-filters eliminate signals from $^1J_{HX}$ prior to the long-range $^nJ_{HX}$-coupling evolution period. The delays $\Delta_1 - \Delta_3$ are matched to the range of possible $^1J_{HX}$-coupling constants. Gradient ratios during the low-pass filter period sum to zero meaning that evolution of the long-range couplings remains unaffected. In the example, sensitivity is enhanced compared with

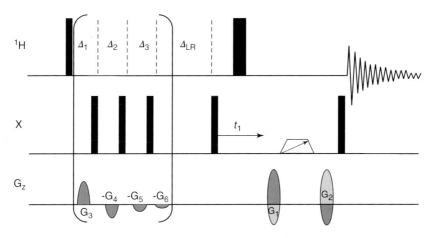

Figure 2.20.   Gradient-selected HMBC with an optional threefold low-pass $^1J_{HX}$ filter (bracketed region)

related methods and pure absorption lineshapes are generated in the indirectly detected dimension along with gradient coherence selection. A mixed lineshape is maintained in the directly detected dimension meaning that HMBC data must be presented in magnitude mode. In general for HMBC, no advantage is gained by decoupling the heteronucleus during signal detection. Long-range $^nJ_{HX}$-couplings revealed through HMBC are small (in the range 2–7 Hz) compared with the equivalent one-bond coupling constants. As Δ_{LR} corresponds to $1/(2 \cdot {}^nJ_{HX})$ for $n \geq 2$, this creates delays of typically 70–250 ms during the pulse sequence, which can be problematic if transverse relaxation plays a significant part. Widely differing values of Δ_{LR} are required to capture correlations associated with the possible range of small-sized $^nJ_{HX}$ values meaning more than one data set with different values of Δ_{LR} must be acquired. An alternative approach in the form of ACCORDIAN spectroscopy enables long-range couplings of different sizes to be revealed in a single data set.[37] HMBC and related methods do not immediately distinguish between $^2J_{HX}$, $^3J_{HX}$, or $^4J_{HX}$, these often being of similar size to one another. Care as well as judgement is therefore required for correct data interpretation. As a compromise, heteronuclear 2 bond correlation (H2BC) was devised to exclusively select correlations from $^2J_{HX}$-couplings.[38] This works by first exploiting $^1J_{HC}$ followed by $^3J_{HH}$ and is equivalent to running the concatenated sequence HMQC-COSY. The method is only suitable for revealing correlations where proton-attached heteronuclei are involved.

2.4.4.3   *Heteronuclear Single Quantum Coherence Total Correlation Spectroscopy (HSQC-TOCSY)*

2D HSQC-TOCSY extends this idea. Here the acquisition stage of the HSQC is replaced by a spin-lock construct. ^1H magnetization previously labeled with heteronuclear chemical shift information during the HSQC segment is spin-locked during the TOCSY mixing period. The effect is to propagate X-filtered ^1H magnetization to the associated spin-system partners under the condition of strong coupling brought about by the spin-lock. The resulting data show spin-system-related proton resonance networks at each heteronuclear chemical shift associated with each related proton. The technique disperses overlapped ^1H spin-systems along the orthogonal heteronuclear chemical shift dimension in a manner that cannot be matched using 2D [^1H, ^1H] TOCSY alone. When combined with data from 2D [^1H, ^1H] TOCSY and COSY, multiplicity-edited 2D [^1H, X] HSQC and 2D [^1H, X] HMBC, the 2D [^1H, X] HSQC-TOCSY can provide key confirmatory signal assignment evidence and is an invaluable tool in the NMR specialist's armory.

2.4.4.4   *Heteronuclear Overhauser Effect Spectroscopy (HOESY)*

Heteronuclear Overhauser effect spectroscopy (HOESY) is the final pulse sequence to be considered here. Like the homonuclear Overhauser

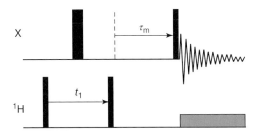

Figure 2.21. Basic HOESY configured for direct detection of the heteronuclear response

effect, the heteronuclear Overhauser effect occurs through a dipolar coupling mechanism between different types of nuclei, one of which is usually proton. The experiment (Figure 2.21) may be carried out in either heteronuclear- or proton-detected modes.

Choice in this is governed by sensitivity, resolution, T_1 relaxation times, and solvent considerations. ^{1}H-detection is preferable in the case of lower γ nuclei including ^{13}C and ^{15}N whereas some merit exists in X-nucleus detection, particularly if measuring heteronuclear Overhauser effects between ^{1}H and ^{19}F nuclei.[39] The added advantage of this for aqueous samples is that solvent suppression can be avoided. Particular care must be taken over the choice of mixing time in this context. ^{19}F T_1 relaxation times can be similar in size to ^{1}H T_1s but may be considerably shorter or considerably longer. The pulse sequence can be adapted in a number of ways including the incorporation of a spoiling gradient within the mixing period and adaptation as a gradient-selected procedure. The latter is particularly useful as the heteronuclear Overhauser effect response is small and nongradient-selected data can result in large artifacts arising from t_1 ridges.

2.5 CONCLUSIONS

In this article, I have highlighted pulse sequences that are current and routinely used within the setting of pharmaceutical or chemical NMR laboratories but I have also provided a nod toward those developing techniques that show significant promise for the future. While solution-phase sequences have formed the focus here, the importance of solid-state methods is not diminished given the consideration required of the solid-state form in many of the materials that

are studied in a pharmaceutical development context. The methods noted are meant to provide a toolkit of sequences that might be suitable for robust, routine, unsupervised, automated use but which are capable of providing top quality, reproducible results. The development of new pulse sequences, incremental improvements in tried and tested methods, and the revival of old procedures given a new twist due to modern hardware developments is a constant theme in the evolving field of NMR spectroscopy. My hope is that this article provides sufficient information for the reader to gain some knowledge of how the described techniques can benefit their work as well as providing some impetus to go further in exploring the much wider landscape of NMR methods.

RELATED ARTICLES IN EMAGRES

Carbon-13 Relaxation Measurements: Organic Chemistry Applications

COSY Spectra: Quantitative Analysis

COSY Two-Dimensional Experiments

Decoupling Methods

Diffusion Measurements by Magnetic Field Gradient Methods

Diffusion-Ordered Spectroscopy

Field Gradients and Their Application

Heteronuclear Shift Correlation Spectroscopy

INEPT

NOESY

Nuclear Overhauser Effect

Polarization Transfer Experiments via Scalar Coupling in Liquids

Relayed Coherence Transfer Experiments

ROESY

Selective NOESY

Selective Pulses

Shaped Pulses

TOCSY in ROESY and ROESY in TOCSY

Two-Dimensional *J*-Resolved Spectroscopy

Water Signal Suppression in NMR of Biomolecules

Ultrafast Multidimensional NMR: Principles and Practice of Single-Scan Methods

Metabonomics: NMR Techniques

TOCSY

Heteronuclear Multiple Bond Correlation (HMBC) Spectra

Multidimensional NMR: an Introduction

HPLC-NMR Spectroscopy

Shimming for High-Resolution NMR Spectroscopy

Zero-quantum Suppression Methods

Pure Shift NMR Spectroscopy

REFERENCES

1. M. Weiger, T. Speck, and M. Fey, *J. Magn. Reson.*, 2006, **182**, 38.

2. P. A. Keifer, *Concepts Magn. Reson.*, 1999, **11**, 165.

3. P. S. C. Wu and G. Otting, *J. Magn. Reson.*, 2005, **176**, 115.

4. O. W. Sørensen, M. Rance, and R. R. Ernst, *J. Magn. Reson.*, 1984, **56**, 527.

5. J. Keeler, *Understanding NMR Spectroscopy*, John Wiley and Sons: Chichester, 2005, 415.

6. D. Uhrín, in *Methods for Structure Elucidation by High-Resolution NMR: Applications to Organic Molecules of Moderate Molecular Weight*, eds G. Batta, K. Kover and C. Szantay Jr, Elsevier Science: Amsterdam, 1997, 51.

7. A. J. Pell, R. A. Edden, and J. Keeler, *Magn. Reson.Chem.*, 2007, **45**, 296.

8. L. Castañar, P. Nolis, A. Virgili, and T. Parella, *Chem. Eur. J.*, 2013, **19**, 17283.

9. J. A. Aguilar, S. Faulkner, M. Nilsson, and G. A. Morris, *Angew. Chem. Int. Ed.*, 2010, **49**, 3901.

10. N. H. Meyer and K. Zangger, *Angew. Chem. Int. Ed.*, 2013, **52**, 7143.

11. M. Foroozandeh, R. W. Adams, N. J. Meharry, D. Jeanerat, M. Nilsson, and G. A. Morris, *Angew. Chem. Int. Ed.*, 2014, **53**, 6990.

12. J. A. Aguilar, A. A. Colbourne, J. Cassani, M. Nilsson, and G. A. Morris, *Angew. Chem. Int. Ed.*, 2012, **51**, 6460.

13. V. M. R. Kakita and J. Bharatam, *Magn. Reson. Chem.*, 2014, **52**, 389.

14. J. Ying, J. Roche, and A. Bax, *J. Magn. Reson.*, 2014, **241**, 97.

15. J. A. Aguilar, M. Nilsson, G. Bodenhausen, and G. A. Morris, *Chem. Commun.*, 2012, **48**, 811.

16. M. Piotto, V. Saudek, and V. Sklenár, *J. Biomol. NMR*, 1992, **2**, 661.

17. T.-L. Hwang and A. J. Shaka, *J. Magn. Reson., Ser. A*, 1995, **112**, 275.

18. M. Liu, X. Mao, C. Ye, H. Huang, J. K. Nicholson, and J. C. Lindon, *J. Magn. Reson.*, 1998, **132**, 125.

19. S. H. Smallcombe, S. L. Patt, and P. A. Keifer, *J. Magn. Reson., Ser. A*, 1995, **117**, 295.

20. S. B. Shuker, P. J. Hajduk, R. P. Meadows, and S. W. Fesik, *Science*, 1996, **274**, 1531.

21. B. Meyer and T. Peters, *Angew. Chem. Int. Ed.*, 2003, **42**, 864.

22. D. W. Begley, S. O. Moen, P. G. Pierce, and E. R. Zartler, *Curr. Protoc. Chem. Biol.*, 2013, **5**, 251.

23. C. Dalvit, P. Pevarello, M. Tato, M. Veronesi, A. Vulpetti, and M. Sundström, *J. Biomol. NMR*, 2000, **18**, 65.

24. C. Dalvit, G. Fogliatto, A. Stewart, M. Veronesi, and B. Stockman, *J. Biomol. NMR*, 2001, **21**, 349.

25. R. Burger and P. Bigler, *J. Magn. Reson.*, 1998, **135**, 529.

26. S. L. Patt and J. N. Shoolery, *J. Magn. Reson.*, 1982, **46**, 535.

27. P. Bigler, R. Kümmerle, and W. Bermel, *Magn. Reson. Chem.*, 2007, **45**, 469.

28. E. D. Becker and T. C. Farrar, *J. Am. Chem. Soc.*, 1969, **91**, 7784.

29. M. Piotto, M. Bourdonneau, K. Elbayed, J.-M. Wieruszeski, and G. Lippens, *Magn. Reson. Chem.*, 2006, **44**, 943.

30. C. M. Thiele, K. Petzold, and J. Schleucher, *Chem. Eur. J.*, 2009, **15**, 585.

31. M. D. Pelta, G. A. Morris, M. J. Stchedroff, and S. J. Hammond, *Magn. Reson. Chem.*, 2002, **40**, S147.

32. J. A. Aguilar, R. W. Adams, M. Nilsson, and G. A. Morris, *J. Magn. Reson.*, 2014, **238**, 16.

33. M. Nilsson and G. A. Morris, *Chem. Commun.*, 2007, 933.

34. L. Paudel, R. W. Adams, P. Király, J. A. Aguilar, M. Foroozandeh, M. J. Cliff, M. Nilsson, P. Sándor, J. P. Waltho, and G. A. Morris, *Angew. Chem. Int. Ed.*, 2013, **52**, 11616.

35. P. K. Mandal and A. Majumdar, *Concepts Magn. Reson.*, 2004, **20A**, 1.

36. D. O. Cicero, G. Barbato, and R. Bazzo, *J. Magn. Reson.*, 2001, **148**, 209.

37. G. E. Martin and C. E. Hadden, *Magn. Reson. Chem.*, 2000, **38**, 251.

38. N. T. Nyberg, J. Ø. Duus, and O. W. Sørensen, *J. Am. Chem. Soc.*, 2005, **127**, 6154.

39. J. Battiste and R. A. Newmark, *Prog. Nucl. Magn. Reson. Spectrosc.*, 2006, **48**, 1.

FURTHER READING

T. D. W. Claridge (ed.), *High-resolution NMR Techniques in Organic Chemistry*, Elsevier Science Ltd.: Oxford, 1999.

Chapter 3

Experimental NMR Methods for Pharmaceutical Research and Development

Anthony C. Dona*

MRC-NIHR National Phenome Centre, Department of Surgery & Cancer, Imperial College London, South Kensington Campus, London SW7 2AZ, UK

3.1 INTRODUCTION

NMR spectroscopy, due to its ability to allow identification of small molecules in solution, has long been a key analytical technique in the field of pharmaceutical research. NMR methodologies have specific characteristics that make them particularly suited to use in pharmaceutical research and development. The hetero- and homonuclear experiments using one or more dimensions make up a powerful toolkit to probe the molecular structure, conformations, and dynamics of active pharmaceutical ingredients (APIs), excipients, and impurities. Furthermore, samples containing multiple constituents can be analyzed within a single run as NMR spectroscopy provides information on all the molecules containing the nucleus that is being detected.

In addition to small molecule solution structure NMR studies, many other important areas of pharmaceutical research have employed NMR, complementing the main drive to produce quality-controlled and physiologically safe pharmaceutical products. First of all, many successful studies have explored the binding of ligand molecules to macromolecules, namely biologically relevant proteins, through saturation transfer experiments. Following the saturation of the protons of the protein with an RF field, the magnetization is transferred to bound ligands through spin diffusion. By these methods, ligand binding can be measured.

NMR has also been successfully employed to understand the numerous polymorphs of crystalline structures. Arrangements can either be crystalline, with repeat structures over three spatial dimensions, or amorphous lacking the long-range periodicity, resulting in significantly different physical and chemical properties. Although amorphous structures often have favorable pharmaceutical formulation properties such as rapid dissolution rates, they are generally more

NMR in Pharmaceutical Sciences. Edited by Jeremy R. Everett, John C. Lindon, Ian D. Wilson, and Robin K. Harris
© 2015 John Wiley & Sons, Ltd. ISBN: 978-1-118-66025-6
Also published in eMagRes (online edition)
DOI: 10.1002/9780470034590.emrstm1386

*Current address - Northern Medical School, Kolling Institute of Medical Research, The University of Sydney, Royal North Shore Hospital, St. Leonards

unstable and so can be difficult to utilize in a pharmaceutical setting. Solid-state NMR spectroscopy has been used to measure polymorphic property differences including kinetic properties, thermodynamic properties, and packing properties (see Chapter 21).

In vivo methods of NMR spectroscopy [commonly referred to as magnetic resonance spectroscopy (MRS)] generally take advantage of the abundance, sensitivity, and presence of hydrogen (^{1}H) nuclei for practical applications in biomedicine, although ^{31}P NMR is often also used and ^{19}F NMR has been used for detection of fluorinated drugs and their metabolites. Unlike MRI, which exploits the presence of water and fatty tissue in the human body to noninvasively produce two-dimensional slice images, MRS signals of interest are those of metabolites. Metabolites are found in concentrations far less than those of water and lipids, so detection conditions need to meet stringent requirements. The magnetic field strength for human metabolic measurement is often in the order of 7 T and sometimes even 9.4 T. The magnetic field strength not only increases the sensitivity for this specific work but also decreases the amount of resonance overlap. Moreover, to avoid the water and lipid resonances overwhelming the entire spectrum, suppression techniques are critical to enable the reliable measurement of metabolites from a volume of interest, which can be as small at 1 cm^3.

Owing to its versatility from in vitro and in vivo studies, NMR spectroscopy is proving useful in the biomedical industry for not only complex mixture analysis such as from biofluids but also translationally for clinicians in diagnosis and pathology.

This article intends to provide the reader with a broad overview of the various experimental applications of NMR in the pharmaceutical industry and should be read in conjunction with Chapters 1 and 2.

3.2 NMR HARDWARE, ANALYTICAL CONDITIONS

3.2.1 NMR Hardware

Recent technical improvements in NMR hardware have resulted in vastly improved data quality and sensitivity. These improvements come from various aspects, including improvements in the magnet size and shielding, better field homogeneity control, more precise temperature control, digital electronics for acquisition control, new design of probes, improved performance preamplifiers, and automation hardware and software. Possibly, the most significant hardware advance in recent years has been the introduction of digital electronics.

The room-temperature shim system, mounted in the lower end of the magnet, is a set of current-carrying coils that create complex shaped magnetic fields that are used to correct field inhomogeneities in three dimensions and so optimize the field homogeneity within the probe by offsetting field gradients induced by the sample, probe, and the environment around the instrument. Automated routines to adjust the electrical current levels in each coil have become much more accurate and rapid over recent years (see Section 3.2.2).

Perhaps the most significant innovation for the advancement of pharmaceutical research, leading indirectly to the development of pulse field gradients (PFGs), is the development of actively shielded probe heads (see Section 3.2.2.3). Actively shielded gradient coils allow for the stray magnetic field generated to be shielded by a secondary coil that is located around the gradient coil. The field produced by the second coil almost exactly cancels out any stray fields outside the gradient, which would otherwise generate eddy currents in the magnet structure surrounding the probe and would severely degrade the recovery performance after the application of pulsed gradients. PFGs can be used to select the required spin coherence pathways, reduce the need for lengthy phase cycling, and thus reduce the required experimental time. They can also be used very effectively in combination with soft RF pulses for very selective excitation and for suppression of unwanted frequencies, for example, during either solvent suppression (see Section 3.2.2.3) or specific saturation transfer experiments (see Section 3.2.3.4). They are also necessary for the measurement of molecular diffusion properties.

3.2.2 High-throughput Hardware

To automate the process of screening many types of fragment molecules or even compounds undergoing high-throughput screening (HTS), many types of instrumentation are available to ensure as little human intervention is required for optimal procedural throughput and accuracy. Autosamplers allow for the unaided insertion of samples into the magnet. Modern-day probes along with modern procedures are able to optimize automatically all the required parameters including tuning and matching, temperature equilibration, locking, shimming, and pulse calibration to run a high-quality NMR experiment.

In many cases, workflows are created to automatically flag molecules that are screened and that either contain high levels of impurity or produce a spectrum that is not expected for the molecular structure. These workflows are built to optimize not only the accuracy but also the throughput of the system and are often designed to require short data acquisition times. The workflows are designed to optimize processes including sample receipt, sample preparation, analytical analysis, data quality management, and data interpretation.

3.2.2.1 Sample Receipt

On sample receipt, containers should be inspected visually for appropriate labeling and possible contamination. Raw materials should be stored in a suitable space and opened, sampled, and resealed in a manner that prevents contamination. Owing to the large number and type of samples often received, containers can be barcoded and placed in an external database detailing the sample receipt date and storage location. Before 'long-term' storage, samples are also often aliquoted or measured into smaller sized containers suitable for routine analysis within the laboratory. These samples are then to be barcoded and stored in the optimal solvent until time of analysis.

3.2.2.2 Sample Preparation of Proteins

For routine ^{1}H analysis of peptides or proteins, molecules need not be labeled although for heteronuclear NMR experimentation proteins must be labeled with ^{13}C and ^{15}N, and sometimes [2]H. In the pursuit of higher quality structures, particularly in larger proteins (>10 kDa), spectroscopists need to study such isotopically labeled molecules. These are prepared by growing cells and expressing the protein in media that has been enriched with [13]C- or ^{15}N-enriched components. This generally requires defined minimal media amounts (as labeled components are considerably more expensive than their naturally occurring isotopes) and the proteins must be cloned into high-expression microbial hosts. A number of excellent microbial expression systems have been developed and have been shown to provide consistently excellent yields of isotopically labeled protein. A number of protease-depleted *E. coli* strains have been developed to express proteins without the problem of proteolytic degradation and to permit residue-specific isotopic labeling.[1,2] Even after a protein has been purified (>90% pure), labeled, and placed into a tube, many challenges still exist. Protein precipitation obviously decreases the measured signal of an NMR experiment and also induces heterogeneity into a sample tube, reducing the quality of a collected spectrum. Often a lot of time is spent 'conditioning' the sample into solution at concentrations relevant to NMR analysis. In this case, a small scale is preferred to ensure that a minimal amount of protein is used, and this is generally achieved using microdialysis cells. These cells look much like buttons, and so the procedure was coined 'the button test'.[3] Conditions within the solution are changed by testing a range of pH, temperatures, salt concentrations, cosolvents, low-viscosity additives or detergents, and solubility-enhancement tags (SETs).[4–6]

3.2.2.3 NMR Analysis and Signal Identification

Biologically relevant proteins are soluble in water and so often pharmaceutically relevant experiments are conducted in solvents largely consisting of D_2O. Although deuterium signals do not interfere with ^{1}H, ^{13}C, or ^{15}N NMR experiments, labile hydrogens on the solute, such as amide protons, are in chemical exchange with the D_2O solvent deuterium atoms. Two major complications arise from the proton–deuterium exchange; firstly the exchange leads to the loss of the amide signals and secondly the detected water signal increases owing to the D_2O that has experienced exchange (obviously there is always some residual water proton NMR signal). Without proton signals from the amides, it is impossible to measure the NOE on their proton peak intensities to determine the protein structure or binding site interactions.

The water (H_2O or HDO) is generally present in a concentration such that its peak is much larger than those of individual protons from a protein, inevitably leading to problems of dynamic range in the NMR detector. The proton signal from water also occurs in a central position in the ^{1}H NMR spectrum (at a chemical shift near 4.8 ppm), thus interfering with many protein-derived peaks. There are a couple of experimental approaches to suppress the intense water peak. The first technique, known as *presaturation*, uses a long low-power continuous RF irradiation at the exact resonance frequency of the water signal. This pulse saturates the water signal, leading to a significant reduction in its intensity. Unfortunately, the saturation sequence also tends to reduce the peak intensities within a narrow range (approximately 0.3 ppm) of the water peak and in cases can lead to a variety of spectral artifacts. A further approach, known as *water suppression through gradient-tailored excitation* (WATERGATE),[7] exploits the fact that proteins, because of their size, have a much slower diffusion coefficient than water. Using a PFG, gradient pulses are applied to the water frequency without affecting other protons. In the case where protein signals are the signals of interest, this particular sequence works well. WATERGATE also has the advantage that it can be applied in a relatively small amount of time and can be easily appended to most 2D or 3D experiments; however, it can work less well with smaller molecule (ligand) signals of interest when the disparity in solvent–molecule size is less.

3.2.3   Analytical Conditions

3.2.3.1   *Quantitative NMR for Pharmaceutical Compound Libraries*

Screening in pharmaceutical research requires a compound library for targeted screening. It requires compounds that are identified, pure, and at a desired concentration. Over time, even pure compounds [kept in dimethyl sulfoxide (DMSO) stock solutions] in a repository can become unreliable owing to sample degradation, water absorption by the solvent, or precipitation. Proton NMR spectroscopy is considered the best way to measure molar concentrations; however, developing macros and scripts to automate the integration process and to locate a single peak that represents a single proton is not a trivial task. Performing quantitative nuclear magnetic resonance (qNMR) consists of aliquoting stock solutions from the compound library and diluting them with deuterated DMSO-d$_6$ usually in a ratio of 1 : 10. A molecule that is soluble and stable in DMSO is used as an external calibrant, which can be used to calculate the absolute concentration of other standards. For this purpose, often flavanone or acetylpyrazine is weighed out in large quantities (>10 mg) to minimize weighing errors and made up to a

standard solution. More details of the full use of qNMR spectroscopy are given in Chapter 5.

3.2.3.2   *High-throughput Screening*

The establishment of high-resolution NMR spectroscopy has made it a useful tool in characterizing the structure of biologically relevant proteins and nucleic acids (see Chapters 8 and 12). Furthermore, it is able to help understand the binding complexes and reversible binding of small molecules with macromolecules. Traditionally, the approach of randomly searching for compounds with activity in cell-based or biochemical functional assays is the first step in the drug discovery process. Typically, in the pharmaceutical industry, collections of compounds (often an entire library) known to have physiochemical properties similar to known drugs are screened in a high-throughput manner against a drug target. Hits are defined from the large number of compounds. The term *hit* can vary in meaning in different project requirements, but a molecule that has high potency of activity is often retested for confirmation. HTS assumes no prior knowledge of the nature of the chemotype likely to have activity at the target protein. Companies are, however, using computer-assisted analysis to reduce the numbers of compounds screened but still provide coverage across a wide chemical space.

Focused screens are also used by targeting a specific class of compounds that have similar structures of those which have been previously identified as targets. This HTS technique has the advantage of reducing the compound cohort size, but also inherently reducing the likelihood of finding an entirely novel structure.

Often after HTS, there are many compounds from which the group of researchers need to cluster into series and to reduce in number based on prior knowledge of HTS studies. The compounds known to be frequent hits are removed along with chemicals of structural similarity to one another ensuring that a broad range of chemical classes are represented in the list taken forward. Furthermore, during the refinement process, dose-response curves are used to narrow the selection based on their structure–activity relationships (SAR). The process identifies groups of compounds based on whether they have a section of structural similarity and clusters them into cohorts.

The defined clusters are then tested in a strategic manner to elucidate properties such as physicochemical and pharmacokinetic properties as well as rapidly generating SAR, absorption, distribution, metabolism, and excretion data. At this stage, NMR is often used to generate rapid knowledge on the structure of the target, the SAR, and possibly new binding sites on the target protein.

NMR spectroscopy is a useful technique as it can differentiate very easily between various small molecules in a simple mixture. This means that potential candidates for binding can be determined from a mixture of structurally different compounds without having to go through the synthesis and purification of single molecules. On the other hand, heteronuclear correlation NMR spectroscopy allows the nature of binding sites on proteins to be determined structurally, and more interestingly, the affinity of various sites can be affected by changes in the molecular conformations of proteins, and this can be assessed using NMR spectroscopy.

These two NMR techniques have been combined and exploited by many laboratories. Dubbed SAR-by-NMR, the procedure offers the advantage of being able to assign signals of substrates with sufficient binding affinity to a ^{15}N-enriched protein in solution from a heteronuclear single quantum correlation (HSQC) experiment. Within the same assays, SAR-by-NMR is able to check multiple small molecules in a combined standard solution and their binding affinity, making the process of finding potential binding candidates much cheaper and more rapid.

3.2.3.3 Metabolic Profiling

An increasing field of study in the pharmaceutical sector is the use of NMR spectroscopy for profiling the number and concentrations of endogenous metabolites in biofluids such as urine, plasma, and cell extracts. The application areas range from an improved understanding of preclinical toxicology to identifying surrogate biomarkers of beneficial or adverse effect in humans after drug administration (see Chapter 24).

For the production and comparison of large amounts of spectral data, procedural conditions must be obtained in the laboratory environment, which allow for intra- and interlaboratory comparisons. The three most important aspects to control during NMR analysis to ensure consistency across measurements on sample cohorts are temperature, pH, and ionic strength of a solution. The typical workflow is to stabilize the pH as much as possible using a simple buffer. The buffer routinely is made up of 90:10 $D_2O:H_2O$ (modern presaturation sequences are very capable of handling any percentage of protonated water, although a certain amount of D_2O is required for field locking) containing a small concentration (of the order of 0.1 mM) of sodium azide as an antimicrobial agent and an amount (generally ~0.1 mM) of an NMR calibrant. Three common molecules for calibration that currently exist are 4,4-dimethyl 4 silapentane-1-sulfonic acid (DSS), 4,4-dimethyl-4-silapentane-1-ammonium trifluoroacetate (DSA), and 3-(trimethylsilyl)-2,2′,3,3′-tetradeuteropropionic acid (TSP or TMSP); all of these are set to a chemical shift of 0.00 ppm. Each calibration molecule has advantages and disadvantages in its use, including its affinity to bind to other molecules (generally macromolecules), the solubility of the calibrant molecules in the solvent of interest, and other residual signals from the calibrant molecule obscuring spectral regions of interest. Each of these factors should be considered before large screening trials in order to optimize the outcome.

3.2.3.4 Saturation Transfer Difference (STD)-NMR Methods

Chemical exchange saturation transfer (CEST) is one of the more commonly exploited techniques used in pharmaceutical research. Saturation transfer difference (STD) by NMR involves subtracting a spectrum in which a protein was selectively saturated (much like water saturation; see Section 3.2.2.3) with signal intensities I_{SAT} (Figure 3.1), from one recorded without protein saturation, with signal intensities I_0 (i.e., the difference spectrum $I_{STD} = I_0 - I_{SAT}$ is measured). The resulting spectrum only reveals the signals of ligand(s), which received saturation transfer from the protein through the NOE. Ligand-based NMR screening and the NMR determination of the bound conformation of a ligand are important tools in the rational drug-discovery process. Not only can STD experiments confirm the cases of binding but also application of these experiments can be used to (i) map the interactions between the ligand–protein binding site and identify important ligand moieties[8,9] and (ii) measure the disassociation constant (K_D) between the protein and ligand of interest.

The STD-NMR experiment does have some limitations, and a few things should be considered before implementing the technique. STD experiments have a defined range of binding affinities that can be measured, making analysis particularly difficult in cases where ligands have particularly high affinities, or particularly weak affinities. When ligands have high

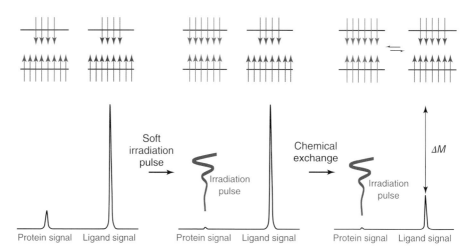

Figure 3.1.   Example of the exchange in irradiated proton signals during a saturation transfer difference experiment (I_{SAT}). In this particular case, the ligand molecule is bound to the binding site (region of saturation). Other compounds that are present however not bound to the receptor will not receive any saturation transfer and so the *on-resonance* and *off-resonance* spectra will be of equal intensity

affinities, saturated atoms cannot transfer magnetization effectively resulting in no observable effect, and on the other hand, when binding affinities are very weak, the probability of occupancy the ligand site becomes very low and so the observable effect is very weak (often invisible).

3.3   NMR-RELATED WORKFLOWS

3.3.1   Sample Preparation Workflow

Modern-day NMR spectrometers have many kinds of different robotic autosamplers for automatic spectrum measurement, which in turn has driven the development of sample preparation robots to prepare samples in the format required for the autosampler of choice. Classical autosamplers require regular 20 cm long conventional tubes to be inserted into a spinner manually at the correct height and either barcoded or the position manually recorded and placed into an autosampler. These systems all require a preparative robot that adds buffer or even volatile solvent to a compound (either liquid or solid) and dispenses the sample solution without air entrapment or bubbles into the bottom of a 5 mm diameter NMR tube and finally recaps the tube. Many solutions to this problem have been designed either using air pressure or solvent pressure to dispense given volumes of solution.

In all cases of robotic preparation, a detailed analysis of the variance in dispensing of sample volume, solvent volume, push solvent volume, and other variations introduced into the NMR tube must be understood to get a total possible error due to the preparative process.

Recently, new systems have been developed where NMR tubes (Match™ tubes) in a 96-well plate format rack can be loaded into an automated sampler (for example, the Bruker Biospin SampleJet) and automatically sampled from a rack into a shuttle (a spinner that has tubes automatically placed in and out of it). The tubes allow for preparation through a needle-sized hole in the lid, which can be sealed with a small ball after preparation. Using the Match tubes is advantageous as the rack is a standard 96-well rack format that takes up less preparative/storage space, caps never need to be removed or replaced, and spinners do not need to be manually placed on each sample. Sample racks are then placed into the autosampler where they are kept at a given temperature (often refrigerated for biological samples) until analysis.

3.3.2   Analytical Workflow

Many benefits of NMR spectroscopy exist, often the biggest advantage is due to the nondestructive nature of the technique; numerous experiments can therefore be performed on a single sample. When analyzing a cohort of samples, an array of experiments can be performed to tailor the elucidation of structural properties or binding properties for each particular study. Generally, single-dimensional experiments on hydrogen and carbon nuclei are acquired along with homonuclear (COSY, NOESY, and ROESY) and heteronuclear [HSQC and heteronuclear multiple-bond correlation (HMBC)] experiments if necessary to determine the exact structure/conformation of molecules in a given environment. If exact binding properties (such as binding coefficients or binding sites) are required on a cohort of samples, further experiments (often more lengthy in acquisition time) can be acquired on each sample to determine detailed binding properties.

Analytical data on small molecules often goes through two stages of quality assurance; the first stage is to ensure that the analytical data produced by the spectrometer is to the standard expected and the second stage is one that ensures that the sample has an expected structure, not undergone any physical change or any form of chemical reaction or degradation before experimental analysis.

3.3.2.1 *Instrument Calibration*

Within the probe head, there exists a sensor that facilitates the monitoring and regulation of the sample temperature inside the magnet. The sensor usually takes the form of a thermocouple that is placed such that its tip is placed very close to the sample within the stream of gas flow that is used to control the temperature of the sample. Owing to the finite heat capacity of the flowing gas, the recorded temperature may not reflect the true temperature of the sample. Calibration of the sample temperature is therefore only possible by use of a specific NMR parameter that has a known dependence on temperature. Two sample types (methanol for the range 175–310 K and 1,2-ethanediol for 300–400 K) have become accepted as standard for ^{1}H NMR calibration of temperature. Instruments should be calibrated approximately once a month or whenever there is a change in the operational temperature by simply placing a neat deuterated methanol sample in the instrument. In modern-day probes, an undeuterated solution produces a signal intensity that is too strong leading to radiation damping that broadens resonances. A proton spectrum should then be recorded on the residual methanol (CHD_2OH) and the difference in

the peak chemical shifts measured. Inherently, as the temperature increases, the intermolecular hydrogen bonding reduces, which in turn causes the hydroxyl resonance to move to a lower frequency. Using the appropriate calibration curve, the real temperature can be calculated.[10]

Following temperature calibration, a standard 2 mM sucrose sample (containing 0.5 mM TSP, 2 mM NaN$_3$ in 90% H_2O : 10% D_2O) is loaded into the magnet and is used to check the performance of the water suppression functionality if required for aqueous samples. At this stage, it is also worth running a gradient profile of the probe to check its field profile and integrity. Firstly, the center frequency (O1) is optimized using a 1D NMR experiment with presaturation, a long relaxation delay, and one scan. Subsequently, the water suppression performance is evaluated by acquiring a full-phase cycle (8 scans) experiment and relatively long delays (~10 s). On a modern 600 MHz spectrometer, the signal-to-noise ratio measured on the anomeric proton peak should be higher than around 300 : 1. Typically, the resolution should better than 15% (a measure of the height of the minimum of the anomeric proton peak as a percentage of the entire peak signal) and the water hump not bigger than 40/80 Hz (as measured at 100% and 50% of the TSP signal intensity, respectively). These parameters should be checked between each large study analysis to ensure that there are no major changes in the instrument performance.

Quantification based on reference to a synthetic signal in NMR data has been available for many years. Namely, the ERETIC (electronic reference to access in vivo concentrations) method provides a reference signal, synthesized by an electronic device, which can be used for the determination of absolute concentrations.[11] Lately, methodologies have been developing to allow even metabolites in complex biofluid matrices to be quantified by this method.

The final consideration is then to set up a sample of the type that will be run along with the parameter sequences that are intended to be used during the analysis of a large sample cohort. The major considerations to be made here are the length of the analysis and the quality of the data produced. It is often a delicate balancing act between having enough scanning time and relaxation delay to ensure quality data with ample signal-to-noise ratio, while still producing a sample analysis time that is reasonable considered the amount of samples to be analyzed.

3.3.3   Data Quality Management Workflow

There are many components to a quality assurance program, and it is important to have a detailed management structure in place for the laboratory. This effectively allows for each role in the laboratory to be well defined and dedicated to a particular person within the laboratory. Stemming from an appropriate management structure, it is important that each member of the laboratory is adequately trained to perform the areas and activities delegated to them. The level of training required or criteria for correct levels of training for a particular task should be outlined in the tasks' standard operating procedure (SOP).

SOPs provide the core of the day-to-day running of a laboratory and the quality assurance program. They are the laboratory's internal reference manual for any procedure, detailing every relevant step in the procedure, allowing anyone with the relevant training grade to perform the procedure. By nature, scientists often feel that SOPs are not required owing to their technical experience or access to manuals or published literature. However, in practice, SOPs present the procedure in a way that avoids differences in interpretation and therefore circumvents subtle changes in the way methods are performed or equipment is used.

Quality assurance of analytical data is often automated and checks that a range of conditions are optimized. Specifically for NMR analysis, the magnetic field homogeneity is measured by checking the resolution of each spectrum achieved in practice by measuring the width at half height of the calibration signal. If solvent suppression pulse sequences are used for acquisition, the residual signal should not have an effect on the spectrum outside of a given range (spectral ranges of interest). For aqueous solvents, for instance, the residual water peak should not be detected outside a range of 120 Hz from the irradiation point. After the phasing, base-line correction, and calibration routine, an automatic check ensures the quality of data measuring the symmetry of particular peaks, the residual noise at each end of the spectrum, and the position of the calibration signal. At this stage, any anomalous experimental data is often caused by analytical malfunctions, and this can either be checked manually or the sample can set up to be automatically rerun.

Physical or chemical changes to molecules of interest are also common during the storage or preparative process. Chemical or physical processes such as dimerization, polymerization, degradation, or precipitation will result in variance in peak shift,

line widths, or multiplicity. Failure of a sample to compare with a computational simulation of spectral properties implies that either one of these processes has affected the purity of the molecule of interest or the synthesized molecule is of insufficient purity.

Apart from impurity checking, NMR can also be used for stereochemical characterization (see Chapter 16) and chiral analyses. NMR is now often considered quicker and more accurate than chiral high-performance liquid chromatography (HPLC) methods to determine the enantiomeric excess of a synthesis through the addition of a chiral complexing agent or use of a chiral solvent. Moreover, methods also exist for determining absolute stereochemistry of molecules (see Chapter 17).

3.3.4   Data Interpretation Workflow

Classically, the regulatory identification of drugs (ultimately ending up in pharmacopoeias) has not used NMR data. Moreover, until currently, there were very few examples of molecules being identified and properties determined purely by NMR spectroscopy, although many examples exist in the natural products literature. This is largely due to data interpretation workflows still maturing from infancy, and much interpretation is still conducted manually. However, substance identification is becoming much more accurate by comparison to a reference spectrum supplied by the European Directorate of the Quality of Medicines (EDQM) or the internal spectrum of a reference substance.

qNMR is an analytical tool for the quantification of content or purity of organic substances (see Section 3.4.2). If data is acquired using an appropriate NMR sequence and adequate relaxation delays, the area under the curve of an NMR signal is directly proportional to concentration of the substance giving rise to the NMR peak of interest. A reference signal of known concentration can be utilized to calculate the concentration of the compound of interest or impurities, in the case of impurity analysis. The proportionality and linearity of NMR data due to the underlying physical process allow for relatively simply interpretation of standard databases. Basic binding properties of small molecules can also be easily determined by qNMR techniques as peaks arising from small molecules can change markedly in terms of chemical shifts line widths, relaxation times, or diffusion coefficient when bound to large molecules such as proteins.

3.4 QUALITY CONTROL

Preparation of quality control (QC) samples (both internal and external) to a project design helps determine the amount of preparative and analytical variance introduced during a study.[12] Three sample types are useful for these measures, which are a composite study reference (SR) sample, stability QC samples, and external QC samples. A composite SR is simply a reference solution made from equal amounts of each sample within the study. A composite QC is made from the SR in the same way that each individual sample is prepared. A stability QC sample is prepared on a much larger scale with the reagents or buffers necessary and then transferred into multiple tubes or vials for analysis. An external QC is simply produced with samples similar to that in the study of interest, however, obtained completely independently of the study, and prepared similarly to each individual sample. These samples allow for short- and long-term managements of quality assured processes, enabling the monitoring of preparation and analytical procedures.

In pharmaceutical research processes, more emphasis is often put on the QC, not of the analytical process, but rather the QC of the biologically active ingredients and excipients that are fundamentally essential for drug formulation. Methods of QC are regulated in various major governing bodies and are essential for patient safety. NMR spectroscopy is emerging as a modern tool in drug analysis, chosen preferentially to classical HPLC methods as NMR has no problems monitoring very small molecules or molecules of strong polar character. Although pharmacopoeias rarely contain details of measuring ratios of API by NMR, absolute amounts of substances can be determined by qNMR using simple reference substances.

3.4.1 Active and Excipient Ingredients

Examples of the use of qNMR for the identification and quantification of an API are becoming more evident. Furthermore, qNMR is useful for the analysis of excipient and impurities in the presence of the major components of a substance. Furthermore, API compositions can be characterized in the case of multicomponent drugs and the determination of isomeric composition (enantiomeric excess) or the diastereomers can be determined. Finally, the residual signals from solvents used during the synthesis process can be measured.[13] For a fuller description of methods and applications, see Chapter 5.

Fundamentally, qNMR relies on clearly resolved NMR peaks from the nuclei of interest (generally ^{1}H) to enable unambiguous assignment of constituent molecules. Commonly, both 1D and 2D experiments are performed in an effort to enhance signal resolution. Further procedures can be implemented to optimize the separation of signals of components of interest; these include varying the pH of the buffer used or the solvent, altering the temperature at which the experiment is performed, or in more extreme circumstances adding auxiliary reagents such as cyclodextrins or lanthanide shift reagents. If it is a viable option, use of a higher magnetic field strength can generate a spectrum of higher resolution, occasionally resolving signals of interest. The absolute quantification of excipient ingredients or the constituent impurities is discussed in Section 3.4.2.

3.4.2 Impurity Assessment

Assuming that the NMR spectra are acquired with sufficient relaxation delays, the area under an NMR peak is directly proportional to the number of nuclei being observed in the detectable region of solution. Multiple forms of qNMR are then applied depending on the application at hand to assess the amount of excipient or impurity in a drug (Table 3.1).

Assessment of impurities and the purity control of pharmaceutical grade products are common in industry. An example is the widespread measurement of excipient and impurities in amino acids that are used in food additives, and in nutrition, medicine, cosmetics, and agriculture. The absence of the essential chromophore required for UV detection in almost all amino acids makes determination of low-level impurities very difficult. Analysis of amino acid constituents by ^{1}H NMR spectroscopy has proven to be advantageous as not only can small concentrations be easily detected but also amino acid peaks are often well resolved from one another.

Table 3.1.   The types of qNMR for quality assurance and control of drug components

Method	Procedure	Application	Comments
Relative	The concentration ratio of two compounds is measured by comparing the ratios of a peak area from each molecule	Relative qNMR is commonly used for the elucidation of multicomponent drugs and measuring ratios of isomers (diastereomers and enanticmers)	No reference compounds are required and there is the advantage of a quick and easy preparation with only the analyte requiring weighing before solvent addition
Internal standard	The absolute amounts of major constituents of drugs are measured by reference to an internal standard that is added to the sample solution	Internal standards are used for the absolute quantification of constituents of active, excipient, and impurities in drugs	The uncertainty of the method depends on the purity of the reference material, the uncertainties of the weights of standard and sample, and the uncertainties of the measured signal areas. These can be accurately controlled. However, the cost of the technique is higher as more standard/solvent is required. Standards must also not be chemically reactive with any component of the drug
Standard addition	Known amounts of the target analyte are spiked into the sample and the regression of analyte addition amount against analyte peak area is measured	Standard addition can be used similarly to internal referencing; however, it requires no knowledge of the component as it can be a defined chemical structure or a complex mixture	Similar to the internal standard method, this method requires accurate measurement of each component added for precise results
External reference	A standard is either added to a sample in a tube containing a separate sealed capillary or dissolved in a completely separate tube that has been previously calibrated	The reference peak itself can be used for quantification or electronic reference to access in vivo concentrations (ERETIC) can be applied for absolute quantification[14]	Much like internal standard addition, external references need to be measured accurately; however, there is the advantage of never interfering directly with the drug solution. The chemical/binding activity of the reference with the drug solution is therefore not important
Total component	This method measures all constituents of the drug and so from the relative signal intensities, the constituents concentrations can be calculated	Total component analysis is particularly advantageous to applications where the analyte is a fairly simple, high purity compound. It also requires no reference material and so has all the advantageous of relative qNMR	The nonorganic impurities (moisture content, inorganic salts, etc.) need to be determined separately. The method also depends on the complete resolution of all potential impurities and ingredients

3.4.3 Determination of the Origin of Drugs

The assessment of pharmaceuticals has been extended into the analysis of seized recreational drugs. As the isotopic composition of organic molecules is not random but rather depends on the way the organic molecule was synthesized, isotope ratio drug analysis can elucidate important information. Methods that are very closely related to qNMR are employed to measure the isotope ratios of recreational drugs to determine the origin of the molecules as this can become important in criminal investigations.[15,16] The application is based on the wide range of deuterium levels in synthetic ephedrines derived from molasses. Simply by calculating the ratio of ^{1}H peak areas (which can also be used in conjunction with ^{13}C and ^{15}N), similar to the relative qNMR method described in Table 3.1, the origin of ephedrines and so recreational methamphetamines can be determined.

3.5 CONCLUSIONS

Given that NMR spectroscopy is an established platform with many chemical and biomedical applications using both small molecules and macromolecules, it is not surprising that it has found widespread application in pharmaceutical R&D. It is also likely that the pharmaceutical field will continue to invest efforts in conventional and novel imaging technologies, and this is particularly true for molecular imaging applications for which imaging and therapeutic targets are often the same.

RELATED ARTICLES IN EMAGRES

Amino Acids, Peptides and Proteins: Chemical Shifts

Analysis of High-Resolution Solution State Spectra

Biological Macromolecules: NMR Parameters

Biological Macromolecules: Structure Determination in Solution

NMR Spectroscopy of Biofluids, Tissues, and Tissue Extracts

Combinatorial Chemistry

Quantitative Measurements

Spectrometers: A General Overview

Structures of Larger Proteins, Protein-Ligand, and Protein-DNA Complexes by Multidimensional Heteronuclear NMR

Shimming for High-Resolution NMR Spectroscopy

REFERENCES

1. D. LeMaster and F. Richards, *Biochemistry*, 1985, **24**, 7263.

2. D. Waugh, *J. Biomol. NMR*, 1996, **8**, 184.

3. S. Bagby, K. I. Tong, D. Liu, J. R. Alattia, and M. Ikura, *J. Biomol. NMR*, 1997, **10**, 279.

4. C. Lepre and J. Moore, *J. Biomol. NMR*, 1998, **12**, 493.

5. A. Wand, M. Ehrhardt, and P. Flynn, *Proc. Natl. Acad. Sci. U.S.A.*, 1998, **95**, 15299.

6. P. Zhou and G. Wagner, *J. Biomol. NMR*, 2010, **46**, 23.

7. M. Piotto, V. Saudek, and V. Sklenar, *J. Biomol. NMR*, 1992, **2**, 661.

8. B. Meyer and T. Peters, *Angew. Chem. Int. Ed.*, 2003, **42**, 485.

9. A. Viegas, N. F. Brás, N. M. Cerqueira, P. A. Fernandes, J. A. Prates, C. M. Fontes, M. Bruix, M. J. Romão, A. L. Carvalho, M. J. Ramos, A. L. Macedo, and E. J. Cabrita, *FEBS J.*, 2008, **275**, 2524.

10. M. Findeisen, T. Brand, and S. Berger, *Magn. Reson. Chem.*, 2007, **45**, 175.

11. G. Wider and L. Dreier, *J. Am. Chem. Soc.*, 2006, **128**, 2571.

12. M. A. Kamleh, T. M. Ebbels, K. Spagou, P. Masson, and E. J. Want, *Anal. Chem.*, 2012, **84**, 2670.

13. H. Holzgrabe, R. Deubner, C. Schollmayer, and B. Waibel, *J. Pharm. Biomed. Anal.*, 2005, **38**, 806.

14. A. Albers, T. N. Butler, I. Rahwa, N. Bao, K. R. Keshari, M. G. Swanson, and J. Kurhanewicz, *Magn. Reson. Med.*, 2009, **61**, 525.

15. S. Armellin, E. Brenna, S. Frigoli, G. Fronza, C. Fuganti, and D. Mussida, *Anal. Chem.*, 2006, **78**, 3113.

16. P. A. Hays, G. S. Remaud, E. Jamin, and Y. L. Martin, *J. Forensic Sci.*, 2000, **45**, 551.

Chapter 4

^{19}F NMR Spectroscopy: Applications in Pharmaceutical Studies

John C. Lindon and Ian D. Wilson

Computational and Systems Medicine, Department of Surgery and Cancer, Imperial College London, Sir Alexander Fleming Building, South Kensington, London SW7 2AZ, UK

4.1 INTRODUCTION

Fluorine has a nuclear magnetic resonance (NMR)-active nucleus (^{19}F) that is 100% naturally abundant and with a spin quantum number of $^1/_2$. It has the second highest NMR sensitivity of the naturally occurring nuclides, being 83% of the proton (^{1}H) sensitivity. This has resulted in a wide usage and a large number of publications and reviews. In addition, there have been extensive compilations of solution-state ^{19}F NMR chemical shifts and coupling constants in organic, organometallic, and inorganic compounds, these being very useful for confirmation of molecular structures.[1,2] As well as studies on small molecules for structural characterization and impurity profiling there are many studies using ^{19}F NMR spectroscopy in biology including the study of the binding of fluorinated molecules to macromolecules, applications to membrane structure and flexibility, the detection and quantification of drug metabolites in vitro, and the study of the distribution and metabolism of compounds in vivo. For biological systems, the lack of endogenous fluorine-containing compounds means that there is little background interference (except for the detection of fluoride ion which probably arises from drinking water). As a consequence, ^{19}F NMR spectroscopy has found widespread application in areas of research related to pharmaceutical R&D.

There are many pharmaceutical products that contain fluorine (around 60 to our knowledge), including well-known examples such as the antianxiolytic flurazepam, the antidepressant fluoxetine, the antineoplastic 5-fluorouracil, the anesthetic isoflurane, the antifungal fluconazole, the antibacterial flucloxacillin, the lipid-lowering agent atorvastatin, the antimalarial mefloquine, the antipsychotic trifluoperazine, the nonsteroidal antiinflammatory fluribiprofen, and the steroidal antiinflammatory fluticasone.

NMR in Pharmaceutical Sciences. Edited by Jeremy R. Everett, John C. Lindon, Ian D. Wilson, and Robin K. Harris
© 2015 John Wiley & Sons, Ltd. ISBN: 978-1-118-66025-6
Also published in eMagRes (online edition)
DOI: 10.1002/9780470034590.emrstm1479

Here we summarize the main uses of ^{19}F NMR spectroscopy in areas of research relevant to pharmaceutical studies.

4.2 PRACTICAL ASPECTS OF ^{19}F NMR SPECTROSCOPY

Using a 14.1T magnet NMR system, ^{1}H observation is at 600 MHz and ^{19}F observation is at 564 MHz. For small molecule studies, samples for solution study are prepared in the usual manner as for ^{1}H NMR spectroscopy. Degassing the solutions can be important because of the effect of paramagnetic oxygen on the ^{19}F spin relaxation times. A number of substances have been proposed for chemical shift referencing, and earlier an external sample of trifluoroacetic acid (TFA) was commonly used. Now, however, the standard reference compound is internal trichlorofluoromethane, $CFCl_3$, added at a few mole percent and with a fluorine chemical shift (δ_F) taken to be at 0.0 ppm. This compound is very volatile and unreactive, and therefore easy to remove from samples if necessary. At high magnetic fields, its ^{19}F NMR chemical shift is perturbed by the combined isotope effects of the various possible combinations of ^{35}Cl- and ^{37}Cl-containing molecules. This means that at best the signal is broadened and at worst an asymmetric multiplet structure from slightly different chemical shifts can be resolved. However, the International Union of Pure and Applied Chemistry (IUPAC) has approved the use of the 'unified scale', which uses the TMS ^{1}H resonance frequency as a universal NMR reference for chemical shifts of all nuclides, and this overcomes any ambiguity caused by the different shifts of the $CFCl_3$ isotopomers. The current sign convention for fluorine chemical shifts is to assign negative values to low frequency of $CFCl_3$ (to the so-called high field), where almost all organo-fluorine compounds give peaks and positive values to high frequency of $CFCl_3$. Early literature values of chemical shifts generally used the reverse sign convention and therefore care needs to be taken in relating recent data to older data.

In the case of biological and other aqueous samples, $CFCl_3$ cannot be used as an internal reference and then a secondary standard is usually employed. This could be TFA or hexafluorobenzene in a capillary tube as an external reference; the chemical shift of TFA is taken to be -76 ppm from $CFCl_3$ and C_6F_6 is -163 ppm from $CFCl_3$.

The wide spectral frequency widths required to encompass ^{19}F NMR chemical shifts has in the past led to problems in data acquisition, in particular nonuniform excitation and poor digitization. ^{19}F NMR chemical shifts for organofluorine compounds range from around 0 ppm for simple fluorohalomethanes to about -75 ppm and -80 ppm (for CF_3·CH-groups and CF_3·CO-groups respectively), to around -125 ppm for $-CF_2$·CO-moieties, -140 ppm for other $-CF_2$-groups, and -120 to -150 ppm for aromatic fluorines, and finally up to -220 ppm for $-CH_2F$ groups. However the $-SO_2F$ group resonates at $+50$ ppm. Such a spectral range of about 250 ppm or more can mean a spectral width of around 140 000 Hz at an observation frequency of 564 MHz. This requires fast digitization rates because of the Nyquist criterion of sampling at twice the highest frequency, and large numbers of data points to ensure reasonable digital resolution.

The closeness of the ^{1}H and ^{19}F observation frequencies means that careful signal filtering circuits are required for removal of interference on one RF channel from the other. While it is possible to detune a ^{1}H probe to observe ^{19}F, it is preferable to have access to a dedicated ^{19}F-tuned probe. This usually then allows ^{1}H-decoupling to be applied while acquiring ^{19}F NMR spectra.

Spin–spin couplings can have either positive or negative values but the signs have been ignored in this brief overview The geminal spin–spin coupling between ^{1}H and ^{19}F, ^{2}J(HF), generally has values in the range 40–60 Hz (compared to ^{2}J(HH) at about 10–20 Hz) and the vicinal three-bond ^{1}H–^{19}F J couplings, ^{3}J(HF), fall in the range 2–5 Hz, but unlike ^{1}H–^{1}H couplings there is no simple Karplus-type relationship with dihedral angle. Four bond ^{1}H–^{19}F couplings are often observed at up to 5 Hz. For couplings between ^{13}C and ^{19}F the one-bond coupling, ^{1}J(CF), is around 160 Hz but can range up to 250 Hz, while two- and three-bond couplings are in the range 20–50 Hz and around 5 Hz, respectively.

^{19}F–^{19}F spin–spin couplings show a great deal of variation depending on the hybridization of the attached carbons and on the relative configurations of the C–F bonds. Examples include two-bond couplings that can take a wide range of values up to 370 Hz in fluoroalkanes, cis vicinal spin coupling in olefins at 0–60 Hz, with trans values at 100–150 Hz and the two-bond geminal values at 0–110 Hz depending on the other substituents. In phenyl rings, the ortho coupling is around 20–25 Hz, the meta coupling can be anything between 0 and 15 Hz and the para coupling is usually around 5 Hz, but can be much higher, all values again depending on the other ring substituents. Longer-range couplings are often observed and J-couplings have been measured between fluorines many bonds apart. While a through-bond mechanism is recognized for ^{1}H–^{1}H spin couplings, a through-space contribution to the mechanism has been invoked to explain such long range J-coupling constants. One example is the value of 45 Hz for the ^{19}F–^{19}F spin-coupling in 1,12-difluorobenzo[c]phenanthrene, nominally six bonds apart, but where the two fluorine atoms are close in space because of the arrangement of the four aromatic rings.

Fluorine spins relax mainly by dipole–dipole interactions with the nearby protons, but also with a minor contribution from the anisotropy of the fluorine chemical shift. The dipolar interactions are responsible for observed nuclear Overhauser effects (NOEs) on ^{19}F resonances transferred from ^{1}H that provide information about internuclear distances in the same way that ^{1}H–^{1}H NOEs do. Analysis of fluorine relaxation times can also produce quantitative estimates of molecular mobility analogous to similar ^{1}H relaxation studies.

4.3 SMALL MOLECULE STUDIES OF PHARMACEUTICAL INTEREST

Many marketed pharmaceuticals have a fluorine, difluoromethylene, or a trifluoromethyl group incorporated.

These substituents allow the medicinal chemist access to modify electronic and steric properties while probing chemical structure–biological activity relationships. In addition, a fluorine substituent can be useful for blocking a possible site of metabolic transformations on the molecule. The benefits of using fluorine in medicinal chemistry have been reviewed relatively recently.[2]

By way of illustration, Figure 4.1 shows the molecular structure of some well-known drugs with their ^{19}F NMR chemical shifts indicated. These include drugs mentioned elsewhere in this chapter.

Applications fall into several areas. First, there is the use of ^{19}F NMR spectroscopy in support of small molecule structure elucidation, where chemical shifts and J-couplings can be particularly diagnostic.[1] Second, ^{19}F NMR spectroscopy can be used for impurity profiling as each component can give well resolved peaks from the parent compound because of the wide chemical shift range and the sensitive effects

of substituents, geometry and isomerism, including tautomerism, on chemical shifts, and spin–spin coupling constants. Moreover, use of ^{19}F NMR detection in directly coupled high-performance liquid chromatography (HPLC)-NMR can be a valuable tool in complex mixture analysis (see Section 4.6).

The impurity profile of production batches of fluorine-containing drugs can be characterized efficiently using ^{19}F NMR spectroscopy. This yields the number and proportions of impurities in the bulk drug to a level of approximately 0.1 mol% in a few minutes of NMR data acquisition. The approach has been exemplified using a partially purified batch of the steroidal product fluticasone propionate, the impurities in which include a number of dimeric species.[3] In that example, further distinction between the monomer and dimer impurities was achieved through high-resolution chemical shift-resolved NMR measurement of molecular diffusion coefficients (so-called diffusion-ordered spectroscopy, or DOSY) on the intact mixture using ^{19}F NMR spectroscopy, with validation using a standard mixture of authentic materials containing both monomers and dimers.

Finally, the use of solid-state ^{19}F magic-angle-spinning NMR spectroscopy for characterizing crystalline and amorphous forms of drugs should be mentioned. Studies include both pure compounds such as flurbiprofen (Figure 4.1) and drugs in their dosage formulation, such as atorvastatin.

4.4 APPLICATION TO FLUORINE-LABELED MACROMOLECULES

The understanding of drug–receptor interactions can be crucial in pharmaceutical research and ^{1}H, ^{13}C, and ^{15}N multidimensional NMR spectroscopy has been employed widely in determining the structure of proteins. ^{19}F NMR spectroscopy has also found a role in this area in the situations where individual amino acids have been modified by inclusion of one or more fluorine atoms and, from these, proteins have been synthesized using microbial biosynthetic methods leading to proteins containing site-specific fluorine atoms which can be monitored using ^{19}F NMR spectroscopy. As mentioned earlier, the high sensitivity, wide chemical shift dispersion and lack of background makes ^{19}F NMR spectroscopy a good technique for such studies. These ^{19}F NMR

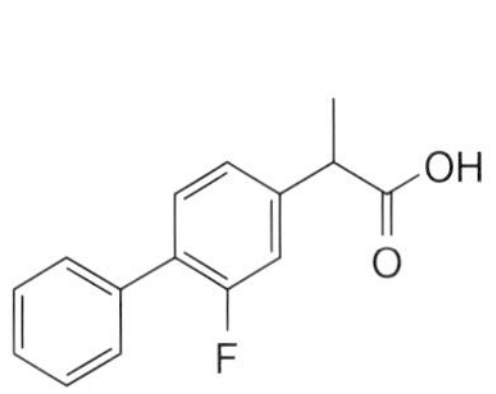

Flurbiprofen $\delta_F = -117.9$ ppm

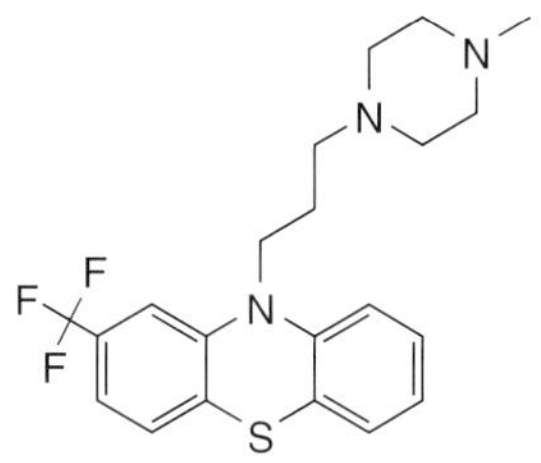

Trifluoperazine $\delta_F = -62.3$ ppm

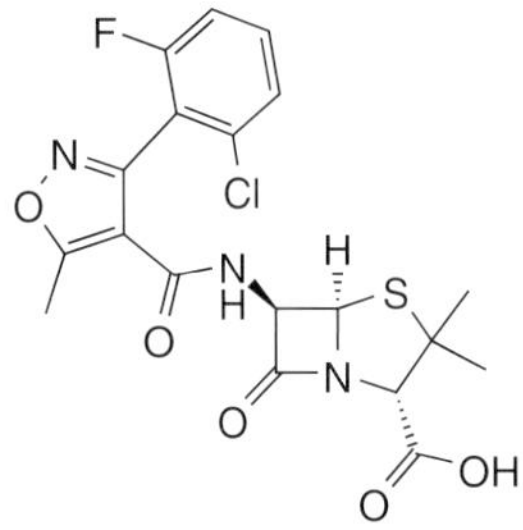

Flucloxacilin $\delta_F = -110.4$ ppm

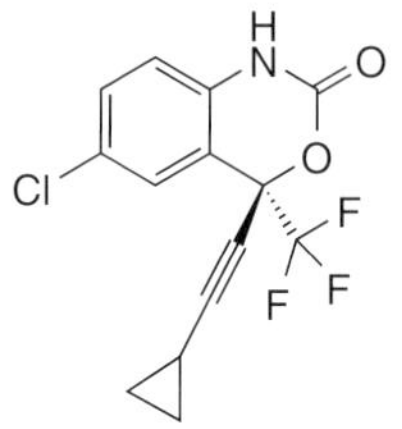

Efavirens $\delta_F = -82.6$ ppm

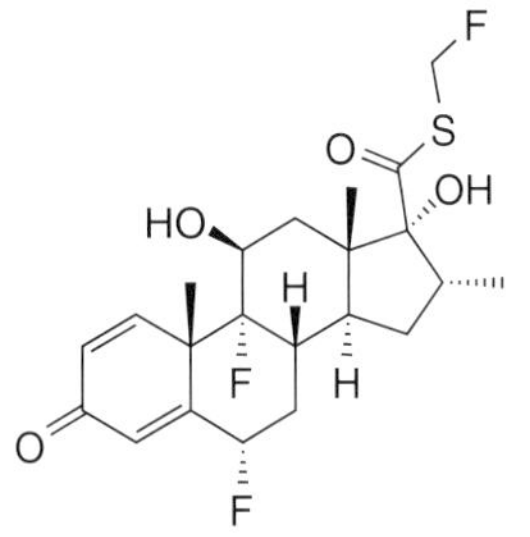

Fluticasone $\delta_F = -164.6$ (CF), -186.7 ppm (CH$_2$F)

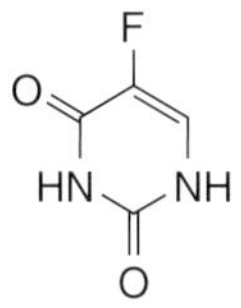

5-Fluorouracil $\delta_F = -169.3$ ppm

Figure 4.1.   The structures of some well-known pharmaceutical products containing fluorine atoms and their ^{19}F NMR chemical shifts (relative to that of CFCl$_3$ set at 0 ppm)

data can yield insight into the local structure of the protein, the protein folding process, and also information on conformational changes, for example, caused by drug binding in the vicinity. One recent example is the use of ^{19}F NMR spectroscopy of modified α-synuclein undergoing fibrillation.[4] This is a 140-amino acid protein implicated in various neurological conditions and which was synthesized with 4-trifluoromethyl-phenylalanine at various positions. Through the use of ^{19}F NMR spectroscopy, the conformational changes near these positions could be monitored when detergent was added to promote fibrillation. In addition, it was possible to study the interaction of α-synuclein with unilamellar vesicles

as model membranes, as the function of the protein is thought to occur through its membrane binding ability.

Other examples are the use of 4-trifluoromethyl-tyrosine in the study of disordered proteins,[5] and 6-fluorotryptophan incorporated into a murine adenosine deaminase to follow the effects of Zn^{2+} binding.[6]

Analogous to fluorine-labeled proteins, fluorine-labeled nucleic acids have also been studied by ^{19}F NMR spectroscopy, using either 2-fluororibose or 5-fluoropyrimide incorporation. Most of the focus has been on DNA duplex and triplex formation,[7] conformational change,[8] and also investigation of the binding of small molecule ligands such as carcinogens.[9] Moreover, RNA has also been studied to characterize secondary structures.[10]

4.5   DRUG SCREENING ACTIVITIES

Clearly with the ^{19}F NMR chemical shift being highly sensitive to environment, the use of ^{19}F NMR spectroscopy in studies of drug–protein interaction has received a great deal of attention. Several separate approaches have been used. The binding of a fluorine-containing drug or other ligand to a protein can be monitored easily by the change in ^{19}F NMR chemical shift on binding if the free and bound molecules are in the fast exchange limit on the NMR time scale. However because of the wide range of ^{19}F chemical shifts, shifts of the free and bound species can be so different such that the exchange rate is in the intermediate or slow regimes, and in the latter case separate chemical shifts are observed and simply measuring the integral of the peak from the unbound moiety gives information on the binding constant. In a similar manner, changes in ^{19}F spin lattice relaxation times on binding have been employed. Dalvit *et al.* have pioneered competitive binding methods that have been used for fragment-based and ligand-based screening (i.e., focusing on ligand NMR spectra). In one implementation, a fluorine-labeled ligand with weak to medium affinity for a target protein is used. On binding, the ^{19}F NMR line width increases because of the protein slow tumbling and the free-bound ligand equilibrium which is in the fast exchange situation for NMR. When a ligand binds to the same site with greater affinity and displaces the fluorine-labeled compound, the NMR peak from the original ligand then returns toward its free solution value and relative binding constants can be deduced.[11] Again, spin–spin relaxation times are also affected by binding and can be used. The use of NMR spectroscopy (including ^{19}F NMR spectroscopy) for probing protein–ligand binding has been comprehensively reviewed by Fielding.[12]

In a different application, for enzyme reactions, an enzyme substrate can be labeled, often with a trifluoromethyl fragment, and ^{19}F NMR spectroscopy is used to detect both the substrate and any enzyme-mediated products.[13]

4.6   APPLICATIONS IN DRUG METABOLISM STUDIES

^{19}F NMR spectroscopy has been widely applied in studies of the metabolic fate of fluorine-containing drugs and xenobiotics with the analysis of biofluids (e.g., urine, bile) providing a relatively sensitive means of obtaining quantitative metabolic profiles of fluorinated compounds and their metabolites. The approach was first described by Everett *et al.*[14] who used ^{19}F NMR spectroscopy to detect metabolites of flucloxacillin (see Figure 4.1 for the molecular structure) in rat urine obtained following administration of the drug. Peaks were observed for flucloxacillin and three metabolites, and there are now numerous published examples on drugs such as the anticancer fluoropyrimidines,[15] fluoroanilines (these are often found as molecular fragments in drugs) using combinations of ^{19}F NMR spectroscopy, HPLC-NMR and HPLC-NMR-MS,[16,17] and haloperidol,[18] to name a few. Moreover, studies have been conducted on systems as diverse as fluorinated gases used as refrigerants[19] and bacterial incorporation of fluoride ion into organic molecules through fluorinase enzymes.[20]

In the recent past, there has been interest in using this approach for the early evaluation of drug metabolism in humans as a result of increasing regulatory guidance concerning the safety evaluation of drug metabolites that are unique to humans, or where exposures to major human metabolites are not sufficiently covered by the preclinical animal toxicology studies. An example of this type of application, where a combination of preparative HPLC and cryoprobe-NMR spectroscopy was used to determine the systemic exposure of volunteers in Phase I clinical studies to a drug and its metabolites enabled novel human plasma metabolites at concentrations as low as $10\,\mathrm{ng\,ml^{-1}}$ to be detected and quantified.[21]

The wide chemical shift range seen with ^{19}F NMR spectroscopy provides exquisite sensitivity to structural changes resulting from biotransformation at sites many bonds removed from the 'reporter' fluorine substituent. Another advantage of using fluorinated substituents as a means of detection and quantification is that, apart from low concentrations of fluoride, the absence of fluorinated endogenous compounds means that detection is highly specific. However, the ^{19}F NMR spectra themselves do not provide a large amount of structural information and metabolite characterization/identification generally requires additional studies such as ^{1}H NMR spectroscopy and mass spectrometry.

Nevertheless, if knowledge of the level of a drug in a biofluid is all that is necessary for monitoring for pharmacokinetic purposes then ^{19}F NMR spectroscopy of fluorinated species is a useful technique. This has been exemplified by the assay of the anti-HIV drug efavirenz in human serum and in pharmaceutical formulations.[22]

A drug metabolism application combining the use of ^{19}F NMR spectroscopy and HPLC-NMR (utilizing both ^{1}H and ^{19}F NMR spectroscopy for detection) was the study of the metabolites of racemic flurbiprofen ([±]-2-(2-fluoro-4-biphenylyl)propionic acid) excreted in human urine (see Figure 4.1 for the molecular structure).[23] The 600 MHz ^{1}H NMR spectra obtained following ingestion of the standard dose of 200 mg of flurbiprofen showed a complex pattern of signals for both endogenous and flurbiprofen-related compounds, limiting structural or quantitative analysis but the ^{1}H-decoupled resolution-enhanced ^{19}F NMR spectra clearly revealed the presence of four major fluorine-containing species and a large number of minor components.

Continuous-flow HPLC-NMR spectroscopy with ^{19}F NMR detection showed four ^{19}F resonances, present in the pseudo-2D contour plot as two pairs at retention times of 30.5 and 36.6 min with those eluting at 30.5 min corresponding to the strongest signals detected by ^{1}F NMR spectroscopy directly on the urine. This is illustrated in Figure 4.2(a). The peaks detected that elute at 36.6 min corresponded to the other major metabolites seen in the whole urine ^{19}F NMR spectrum. A repeat of the separation using stopped-flow ^{1}H NMR spectroscopic detection at 600 MHz provided ^{1}H NMR spectra identifying the peaks as the β-D-glucuronic acid conjugates of the 4′-hydroxy-flurbiprofen (30.5 min) and flurbiprofen acyl glucuronide (36.6 min), as shown in Figure 4.2(b). Both metabolites were found as diastereoisomers because of the conjugation of the β-D-glucuronic acid to the R and S isomers of flurbiprofen/4′-hydroxy-flurbiprofen. Using 'time slicing', that is, repeated NMR data acquisition at short time intervals over the eluting chromatographic peak, it was possible, to obtain a spectrum for one of the diastereoisomers of flurbiprofen-β-D-glucuronide confirming the inhomogeneity of the chromatographic peak. A minor metabolite was identified as 4′-hydroxyflurbiprofen itself by ^{1}H NMR in stop-flow mode.

Acyl glucuronides of the type seen for flurbiprofen are chemically unstable, and reactive, at alkaline pH, where they can undergo both internal acyl migration and hydrolysis to form a complex mixture of isomeric compounds. As these metabolites can react with proteins they have been suggested as a possible cause of adverse drug reactions and both ^{19}F NMR spectroscopy and HPLC-NMR have been used to study the transacylation properties of a range of synthetic β-1-O-acyl-D-glucopyranuronates as models for drugs and to enable the construction of quantitative structure-activity relationship (QSAR) models. For example, ^{19}F NMR spectroscopy of 2-, 3-, and 4-(trifluoromethyl) benzoic acids as models of drug ester glucuronides enabled the overall degradation rate constants of 0.065 h^{-1}, 0.25 h^{-1}, and 0.52 h^{-1} to be obtained for the 2-, 3-, and 4-trifluoromethyl-benzoic acid 1-O-acyl glucuronide isomers respectively. The reasons for the observed differences in reactivity were investigated using computational chemistry methods with differences between the slowly transacylating 2-trifluoromethylbenzoic glucuronide and the much faster 3- and 4-isomers being rationalized from the calculation of the relative bond order of the C–O bonds in the ortho-acid ester intermediates.[24]

The identification of which of the signals in ^{19}F NMR spectra of the mixed transacylated glucuronides correspond to which of the many possible isomers present, was greatly aided by the use of HPLC-NMR spectroscopy employing both ^{19}F and ^{1}H NMR spectroscopy to characterize the eluting peaks.

Another method for studying the in vivo metabolism of fluorine-containing drugs is the use of *Sta*tistical *To*tal *C*orrelation *S*pectroscop*y* (STOCSY),[25] where the correlations that exist between the intensities of spectral features over multiple spectra are exploited to detect intramolecular or intermolecular connectivities. A recent development has been the cross-correlation of heteronuclear NMR spectra (HET-STOCSY), a technique which is applicable to paired spectra following parallel or serial acquisition and this approach with ^{1}H and ^{19}F NMR datasets, recorded in parallel (at 800.32 MHz ^{1}H frequency, 753.05 MHz ^{19}F frequency), has been used to assign the metabolites of the fluorinated antibiotic flucloxacillin in human urine.[26] Intermetabolite correlations were used to aid the interpretation of STOCSY data and assist the structural characterization of drug metabolites by providing information on biotransformation routes. Simply by displaying the correlation coefficient between intensity of the ^{19}F resonance obtained for flucloxacillin (see

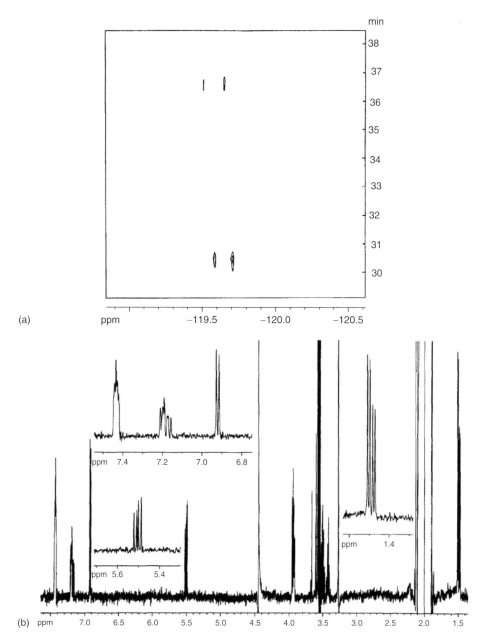

Figure 4.2. (a) Two-dimensional contour plot of an on-flow 470 MHz ^{19}F NMR-detected HPLC chromatogram of the separation of human urine taken after oral administration of a standard dose 200 mg of flurbiprofen. The horizontal axis is the ^{19}F NMR chemical shift and the vertical axis is time. (b) 600 MHz ^{1}H NMR spectrum from an HPLC-NMR experiment measured in stopped-flow mode of the 30.5 min retention time species shown in (a), and corresponding to the β-glucuronic acid conjugate of the metabolite hydroxyflurbiprofen. (Reprinted from J. Pharm. Biomed. Anal., 11, M. Spraul, M. Hofmann, I.D. Wilson, E. Lenz, J.K. Nicholson, and J.C. Lindon, Coupling of HPLC with 19F- and 1H-NMR spectroscopy to investigate the human urinary excretion of flurbiprofen metabolites, 1009–1015, Copyright (1993), with permission from Elsevier.)

Figure 4.1 for the molecular structure) with the values of each data point in the ^{1}H spectrum, all of the ^{1}H NMR peaks for the drug were clearly identified among the thousands of ^{1}H NMR peaks from the endogenous substances (including those of protons up to 13 bonds from the F atom). These data were further explored by performing 2D heteronuclear ^{19}F–^{1}H analysis (HET-STOCSY) and X-nucleus-edited ^{1}H–^{1}H STOCSY analysis (an approach analogous to spectral editing in conventional homonuclear NMR experiments for single samples by magnetization transfer to a hetero-nucleus, e.g., 3D HMQC-TOCSY). By these means, the ^{19}F NMR resonances for all of the major metabolites were identified together with one from a minor metabolite of the drug that was previously unknown. The data were also analyzed for statistical intermetabolite correlations as, in principle, the more distant (in pathway connectivity terms) the two metabolites are in the correlation structure originating from flucloxacillin and its metabolites, the weaker their overall correlation and thus more metabolic steps separate them in the overall pathways of biotransformation. In principle, this provides a means of assessing the likely sequence of biotransformation from parent to drug metabolites with potential for metabolic pathway reconstruction.

4.7 ^{19}F NMR SPECTROSCOPY IN VIVO

The ^{19}F nuclide can be exploited in two ways, by in vivo NMR spectroscopy (MRS) or in vivo magnetic resonance imaging (MRI). The absence of endogenous fluorinated interferences also makes ^{19}F MRI well suited to noninvasive in vivo monitoring of distribution, metabolism, and pharmacokinetics (DMPK) of drugs.[15]

A good illustration of this is provided by studies in humans with the anticancer drug 5-fluorouracil (5-FU) (see Figure 4.1 for the molecular structure). The first MRS work using this drug was conducted over 30 years ago and since then, numerous studies in both animal models and in humans, and in different tissues and biological fluids, including plasma and urine, have been presented.[27] These studies provided improved understanding both of the metabolic fates of the drug and the mechanisms of the cytotoxic activity of the drug. This drug, and several of its metabolites, have been measured in animals and in patients using standard clinical MRI systems. On the basis of the ^{19}F NMR peak intensity, one useful parameter that can

be measured most readily is the tumoral half-life of 5-FU. Patients whose tumoral 5-FU half-life is 20 min or longer are designated as 'trappers', and those whose half-life is less are 'non-trappers'. Trapping of 5-FU in tumors is necessary for therapeutic response, but it is not the only requirement. How modulators of the drug and other agents will affect the half-life has also been investigated. The rationale for the biological processes underlying the fate of 5-FU in humans was illustrated with a 12 compartment model, where several of the steps were discussed and the consequences modulating them were explored. The ^{19}F MRS studies were then extended to other fluoropyrimidines, some of which are pro-drugs of 5-FU, and others that have the fluorine atoms on the ribose ring.[28]

Another important application is the use of the ^{19}F NMR chemical shift of a probe molecule to measure intracellular pH or ion concentration such as for the analyte Ca^{2+} which plays an important role as a second messenger in living cells. This approach has been used for the determination of cytosolic calcium in cells and tissues employing the candidate probe 1,2-bis(o-amino-phenoxy)ethane-N,N,N',N'-tetraacetic acid (BAPTA). Here, the ^{19}F NMR Ca^{2+} indicator is derived from its symmetrically 5,5-substituted difluoro-derivative (FBAPTA), which exhibits a chemical shift response on binding calcium. One of the difficulties in the use of such a probe molecule is the potential lack of penetration of the agent into cells. As the tetracarboxylate does not penetrate the cell, a lipophilic agent, such as acetoxymethyl, is used. Other ^{19}F-bearing ligands have been proposed for ions such as Na^+, Mg^{2+}, Zn^{2+}, and Pb^{2+}.[29]

For ^{19}F detected MRI, perfluorocarbons are often used as probe molecules. They are nontoxic and biologically stable. After in vivo administration, they are not metabolized by the tissue but cleared by circulation and in part vaporized to the air through respiration. They are typically prepared as a nanoparticle emulsion comprising a liquid perfluorocarbon core surrounded by a lipid monolayer.

Applications include monitoring tissue hypoxia by pO_2 mapping as the ^{19}F relaxation time is affected by the paramagnetic oxygen.[30] Since blood-delivered nanoparticles are mainly found in well-vascularized areas not in hypoxic areas, and to overcome this problem, the perfluorocarbon can be injected directly into different tumor regions. Another burgeoning area is the study of cell trafficking, as for example, during stem cell therapy. In the past, in vitro cultured stem cells were incubated with a paramagnetic MRI contrast agent such as iron oxide or Gd-DTPA and the labeled stem cells were detected by ^{1}H MRI contrast studies. In recent years, it was shown that ^{19}F MRI could also be used for cell trafficking in vivo using perfluoropolyether (PFPE) nanoparticles,[31] and this study has been followed up by others. Other applications cover use of ^{19}F MRI to detect cardiac and cerebral ischemia using perfluorocarbon nanoparticles.[32] A related field of using a ^{19}F-labeled probe is functional lung MR imaging that uses fluorine-containing gases such as SF_6, C_2F_6, CF_4, C_3F_8, and liquid perfluorocarbons).

RELATED ARTICLES IN EMAGRES

Applications of ^{19}F-NMR to Oncology

NMR Spectroscopy of Biofluids, Tissues, and Tissue Extracts

Coupling Through Space in Organic Chemistry

Fluorine-19 NMR

Spectroscopic Studies of Animal Tumor Models

HPLC-NMR Spectroscopy

REFERENCES

1. T. S. Everett, in Chemistry of Organic Fluorine Compounds, ed M. Hudlicky, ACS Monograph, 1995, Vol. 187, p. 1037.

2. W. K. Hagmann, *J. Med. Chem.*, 2008, **51**, 4359.

3. N. Mistry, I. M. Ismail, R. D. Farrant, M. Liu, J. K. Nicholson, and J. C. Lindon, *J. Pharm. Biomed. Anal.*, 1999, **19**, 511.

4. G.-F. Wang, C. G. Li, and G. J. Pielak, *ChemBioChem*, 2010, **11**, 1993.

5. C.-G. Li, G.-F. Wang, Y.-Q. Wang, R. Creager-Allen, E. A. Lutz, H. Scronce, K. M. Slade, R. A. S. Ruf, R. A. Mehl, and G. J. Pielak, *J. Am. Chem. Soc.*, 2010, **132**, 321.

6. W.-L. Niu, Q. Shu, Z.-W. Chen, S. Matthews, E. Cera, and C. Frieden, *J. Phys. Chem. B*, 2010, **114**, 16156.

7. K. Tanabe, M. Sigiura, and S. Nishimoto, *Bioorg. Med. Chem.*, 2010, **18**, 6690.

8. F.-T. Liang, S. Meneni, and B. P. Cho, *Chem. Res. Toxicol.*, 2006, **19**, 1040.

9. C. Kreutz, H. Kahlig, R. Konratand, and R. Micura, *Angew. Chem. Int. Ed.*, 2006, **45**, 3450.

10. C. Kreutz, H. Kahlig, R. Konratand, and R. Micura, *J. Am. Chem. Soc.*, 2005, **127**, 11558.

11. C. Dalvit, P. Fagerness, D. A. Hadden, R. Sarver, and B. Stockman, *J. Am. Chem. Soc.*, 2003, **125**, 7696.

12. L. Fielding, *Prog. NMR Spectrosc.*, 2007, **51**, 219.

13. C. Dalvit, E. Ardini, M. Flocco, G. P. Fogliatto, N. Mongelli, and M. Veronesi, *J. Am. Chem. Soc.*, 2003, **125**, 14620.

14. J. R. Everett, K. Jennings, and G. Woodnutt, *J. Pharm. Pharmacol.*, 1985, **37**, 869.

15. M. Malet-Martino, V. Gilard, F. Desmoulin, and R. Martino, *Clin. Chim. Acta*, 2006, **366**, 61.

16. G. B. Scarfe, B. Wright, E. Clayton, S. Taylor, I. D. Wilson, J. C. Lindon, and J. K. Nicholson, *Xenobiotica*, 1998, **28**, 373.

17. K. Kitamura, A. A. Omran, S. Takegami, R. Tanaka, and T. Kitade, *Anal. Bioanal. Chem.*, 2007, **387**, 2843.

18. M. Shamsipur, L. Shafiee-Dastgerdi, Z. Talebpour, and S. Haghgoo, *J. Pharm. Biomed. Anal.*, 2007, **43**, 1116.

19. P. Schuster, R. Bertermann, G. Rusch, and W. Dekant, *Toxicol. Appl. Pharmacol.*, 2010, **244**, 247.

20. D. O'Hagan, C. Schaffrath, S. L. Cobb, J. T. G. Hamilton, and C. D. Murphy, *Nature*, 2002, **416**, 279.

21. G. J. Dear, A. D. Roberts, C. Beaumont, and S. E. North, *J. Chromatogr. B*, 2008, **876**, 182.

22. M. Shamsipur, M. Sarkouhi, J. Hassan, and S. Haghoo, *Afr. J. Pharm. Pharmacol.*, 2011, **5**, 1573.

23. M. Spraul, M. Hofmann, I. D. Wilson, E. Lenz, J. K. Nicholson, and J. C. Lindon, *J. Pharm. Biomed. Anal.*, 1993, **11**, 1009.

24. A. W. Nicholls, K. Akira, J. C. Lindon, R. D. Farrant, I. D. Wilson, J. Harding, D. A. Killick, and J. K. Nicholson, *Chem. Res. Toxicol.*, 1996, **9**, 1414.

25. O. Cloarec, M. E. Dumas, A. Craig, R. H. Barton, J. Trygg, J. Hudson, C. Blanche, D. Gauguier, J. C. Lindon, E. Holmes, and J. K. Nicholson, *Anal. Chem.*, 2005, **77**, 1282.

26. H. C. Keun, T. J. Athersuch, O. Beckonert, Y. Wang, J. Saric, J. P. Shockcor, J. C. Lindon, I. D. Wilson, E. Holmes, and J. K. Nicholson, *Anal. Chem.*, 2008, **80**, 1073.

27. M. J. O'Connell, J. A. Martenson, H. S. Wieand, J. E. Krook, J. S. Macdonald, D. G. Haller, R. J. Mayer, L. L. Gunderson, and T. A. Rich, *N. Engl. J. Med.*, 1994, **331**, 502.

28. W. Wolf, C. A. Presant, and V. Waluch, *Adv. Drug Deliv. Rev.*, 2000, **41**, 55.

29. J. C. Metcalfe, T. R. Hesketh, and G. A. Smith, *Cell Calcium*, 1985, **6**, 183.

30. B. D. Spiess, *J. Appl. Physiol.*, 2009, **106**, 1444.

31. E. T. Ahrens, R. Flores, H. Xu, and P. A. Morel, *Nat. Biotechnol.*, 2005, **23**, 983.

32. U. Flogel, Z. Ding, H. Hardung, S. Jander, G. Reichmann, C. Jacoby, R. Schubert, and J. Schrader, *Circulation*, 2008, **118**, 140.

Chapter 5

Quantitative NMR Spectroscopy in Pharmaceutical R&D

Ulrike Holzgrabe

Institute of Pharmacy and Food Chemistry, University of Würzburg, 97074 Würzburg, Germany

5.1 INTRODUCTION

For some 50 years, NMR spectroscopy has been mainly used for structure elucidation and confirmation for newly synthesized compounds and natural products because constitutional, configurational, and conformational information can be provided by multitudinous two-dimensional techniques, such as COSY, heteronuclear multiple-bond correlation (HMBC), heteronuclear single quantum correlation (HSQC), TOCSY, NOESY, and ROESY experiments, to name the most important only. However, NMR spectroscopy can also be used for quantification purposes because it is a primary method of measurement. Even though

NMR in Pharmaceutical Sciences. Edited by Jeremy R. Everett, John C. Lindon, Ian D. Wilson, and Robin K. Harris
© 2015 John Wiley & Sons, Ltd. ISBN: 978-1-118-66025-6
Also published in eMagRes (online edition)
DOI: 10.1002/9780470034590.emrstm1399

the major pharmacopoeias only rarely make use of the method for quality assessment – it is mainly applied for identification of complex active pharmaceutical ingredients and excipients – it is increasingly used in pharmaceutical industries because it is orthogonal to the chromatographic and electrophoretic techniques that are used compendially. Of note, NMR spectra are able to give additional information about, e.g., new impurities or solvents.

Quantitative nuclear magnetic resonance (qNMR) is one of the major techniques used in very different scientific areas. As already mentioned, the pharmaceutical industries employ NMR spectroscopy for quality and stability assessment of small-molecule drugs, vaccines, peptides, natural products and excipients, as well as drug formulations on the one hand and for metabolic profiling and finger printing of body fluids (metabolomics for disease diagnostic and drug treatment control) on the other. Accordingly, food industries make increasing use of qNMR spectroscopy for product profiling. For example, the wine screener (wine-profiling™) can help to detect fraudulent labeling of origin, grape variety, vintage year, barrique ripening, and dilution using statistical methods. Similar applications are described for, e.g., fruit juices and vegetables.

This article will focus on the basic principles of qNMR to be used in.

- Quality assessment of drugs and excipients
  - Determination of the level of impurities

– Determination of the composition of multicomponent drugs
– Determination of the isomeric composition: the ratio of diastereomers and/or the enantiomeric excess (EE) of chiral drugs by means of chiral additives, e.g., chiral solvating agents or chiral shift reagents
– Evaluation of the content of residual solvents
– Observation of the course of degradation/decomposition of a drug
• Content determination (assay)
• Counterfeit analysis

Impurity profiling and content determination by means of qNMR spectroscopy, using mainly ^{1}H and ^{19}F NMR, do not simply consist of the measurement of a one-dimensional spectrum and the integration of the signals of the main component and the impurities, as people from synthesis departments may think. A lot of parameters have to be measured and considered in order to get precise quantitative results. This will be addressed in this article in depth.

5.2 BASIC PRINCIPLES OF QNMR

qNMR is based on the simple fact that the area under the NMR signal (intensity I) is proportional to the number of nuclei (N) giving rise to this signal.

$$I = k_S \cdot N \qquad (5.1)$$

with k_S being a spectrometer constant. Thus, in comparison to separation methods, no response factor is needed. However, the signals regarded for integration (quantification) have to be well separated from all other signals. Thus, the purity of a signal has to be ensured, e.g., by two-dimensional experiments such as H,H COSY or/and H,C COSY.

5.2.1 Signals to be Considered for Quantification

When analyzing mixtures of compounds, e.g., plant extracts, multicomponent drugs, or impurities in drugs, the signal used for quantification should be unambiguously assigned to an atom of the substance to be analyzed. Moreover, the signals should be as simple as possible. Hence, a singlet in a ^{1}H NMR spectrum is superior to a multiplet. The signal of an exchangeable

proton cannot be used owing to its broadness. Such a signal depends heavily on the sample concentration and temperature, which makes it rather erratic.

The signals of protons can be accompanied by ^{13}C and [29]Si satellites, which may interfere with the signals used for quantification. The spectrum of alanine shown in Figure 5.1(a) demonstrates the structure of the satellites. In order to overcome the problems associated with such interferences, the following experiments can be performed: (i) usage of an instrument of higher or lower field strength that positions the satellites at a different distance in ppm from the signals, (ii) simple subtraction of the integral of those signals from the area of the signal regarded for the quantification should not be used, and (iii) heteronuclear decoupling at the radiofrequency of, for example, ^{13}C nuclei can be performed, resulting in the collapse of the ^{13}C satellites. However, caution has to be exercised, because some decoupling experiments can generate so much heat that the sample is in danger of degrading. The GARP (globally optimized alternating-phase rectangular pulses) sequence may be employed, which provides efficient decoupling without producing too much heat. For details, see Pauli *et al.*[1] On the other hand, the ^{13}C satellites can normally be used for intensity referencing for low-level impurities because of their normal natural abundance of 1.1%. This holds true unless the material has been derived from plant or other unusual sources where isotope depletion can occur. Furthermore, spinning side bands, which are associated with the rotation of the tube in the probe, may interfere with small signals, e.g., of low-level impurities of drugs. Since the introduction of the superconducting magnets that are characterized by very good magnetic field homogeneity B_o and field stability, spinning of the sample is no longer essential. Thus, the spectra can be recorded without spinning, which is normally done in qNMR.

5.2.2 Acquisition and Processing Parameters

Even though equation 5.1 is extremely simple, a number of requirements have to be fulfilled for accurate and precise quantification. With regard to acquisition, the relaxation delay, the digitization, and the pulse sequence design govern the quantification results.

1. A pulse angle of 90° is mostly chosen for acquisition because it results in maximum signal intensity with respect to time of measurement. However, an angle of 30° can reduce the repetition time, but this

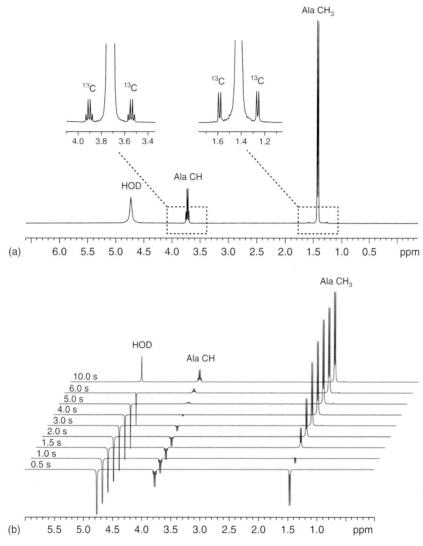

Figure 5.1. (a) 400 MHz ^{1}H NMR spectrum of alanine in D$_2$O; (b) inversion-recovery pulse experiments for T_1 determination (repetition time of the pulse sequence 75 s). (Adapted from Ref. 5. © Elsevier, 2010)

procedure has to be handled with care. The pulse angle must be uniform across the spectrum.

2. The repetition time (also named recycling time) is determined by the longitudinal relaxation time T_1 of the signal used for quantification, and T_1 is described by

$$M_z = M_0 \left(1 - e^{-\tau/T_1}\right) \qquad (5.2)$$

where M_z and M_0 are the magnetization along the z-axis after the waiting (repetition) time τ and at

thermal equilibrium, respectively. The repetition time τ should be five times T_1. In this case, 99.3% of the equilibrium magnetization (signal) is measured. If the repetition time is chosen too short, the quantification will fail. For solutions T_1 is different for each signal in a molecule. Thus, T_1 has to be determined for precise quantification by means of an inversion-recovery pulse sequence for each signal considered in every matrix (caveat: can be different in different solvents). This is exemplified by alanine in Figure 5.1(b): T_1 of the methine proton amounts to 5.2 s and T_1 of the methyl group to 2.2 s. In cases where an internal standard (calibration standard) is employed, T_1 for its signal has to be determined, too. Note that the T_1 of the standard depends on the other compounds in solution (due to dipole–dipole interactions), and vice versa. In particular, symmetrical molecules used as calibration standards are characterized by a very long T_1 (for details see Chapter 3 and Ref. 1).

For nuclei with very long T_1 values, such as ^{13}C, ^{31}P, or ^{29}Si, paramagnetic relaxation reagents [such as $Cr(acac)_3$] can be added for diminution of T_1.

3. Mostly ^{1}H NMR experiments are used for quantification purposes. However, nuclei such as ^{13}C, ^{15}N, or ^{19}F can also be employed, but they may suffer from inherent intensity distortions by the NOE effect created by simultaneous ^{1}H broadband decoupling. This NOE effect on the spectrometer constant k_S can be described by

$$k_S = k_0 \cdot (1 + \eta) \cdot \left(\frac{1 - e^{-\frac{\tau}{T_1}}}{1 - \cos\alpha \cdot e^{-\frac{\tau}{T_1}}} \right) \cdot \sin\alpha$$

(5.3)

where k_0 = constant instrumental factor, η = NOE factor, and α = flip angle of the excitation pulse. The intensity distortion should be not more than 1% for quantification purposes. When using a 90° pulse for excitation, a repetition time between five and seven times T_1 and ^{1}H decoupling only during signal acquisition (to minimize η; inverse gated technique) results in a small distortion.

4. It has to be considered that there is an interrelation between acquisition time t_{aq}, spectral width, dwell time, and sampling rate.[2] In order to avoid truncation of the signals in the time domain (being the FID), the signals should completely decay half way through the acquisition period. In this case, sufficient data points describe the NMR lines [at least 5 (better 10) data points above the half-height of the signal] and the integration does not suffer from artificial distortions.

5. The signal-to-noise (S/N) ratio strongly determines the precision of the integration procedure. A high precision [standard deviation (sdv) < 1%] can be achieved when S/N ratios of >250 : 1 for ^{1}H, >300 : 1 for ^{19}F, and >600 : 1 for ^{31}P are reached. However, in order to achieve very high precision, the S/N should be even higher.

The parameters of the processing have to be chosen as carefully as the acquisition parameters because they sensitively influence the accuracy of the integration.

1. Zero filling (setting additional data points equal to zero before FT) contributes to both the spectral and digital resolutions of the spectrum. However, it should not exceed a factor of two.
2. For the enhancement of the S/N ratio, often exponential multiplication is applied as a weighting function. However, this has to be carefully used because, along with the increase of the S/N, the signals get broader. For a ^{1}H NMR spectrum, the line broadening factor should range between 0.1 and 0.3 Hz.
3. Next, phase, baseline, and drift corrections have to be applied. Even though the software of NMR instruments provides an automatic phase correction, manual phasing is superior to the automatic routine, especially for qNMR purposes. Mathematical functions are available for baseline corrections. Only the baseline should be corrected, not the signal considered for quantification. Sometimes it might be advantageous to correct only the part of the spectrum evaluated.

In order to obtain an accurate integration, the integration limits should be set to a range of 64 times the full width at half height (FWHH) of the signal. Then, it is possible that 99% of signal intensity is obtained (remember, the NMR signal is usually a Lorentzian line). This might be difficult when adjacent signals are very close by. In this case, a compromise has to be made.

A manual correction of the integral has to be avoided and is not necessary when the baseline is optimally corrected.

5.3 SIGNAL SEPARATION/OVERLAP

A prerequisite for qNMR is the clear separation of signals that are considered for integration [or other techniques (see Chapter 3)]. This can be a problem, especially when it comes to the quantification of low-level components of a mixture of compounds or impurities of a drug substance, especially when the components/impurities are structurally closely related. In rare cases, the separation of the partially overlapping signals can be achieved by deconvolution (being algorithms that produce individual signals in the spectrum[3]) or by constrained total lineshape fitting procedures.[4]

Table 5.1. Classification of solvents used for NMR spectroscopy

Polar solvents		Nonpolar solvents	
Protic	Nonprotic	Aromatic	Nonaromatic
D_2O	DMSO-d_6	C_6D_6	$CDCl_3$
CD_3OD	Acctonitrile-d_3	Toluene-d_8	CD_2Cl_2
		Pyridine-d_5	

Even though the position of the signals is defined by the chemical neighborhood of the atoms, a signal can be shifted by the choice of different solvents (or mixture of solvents), by varying sample concentrations and temperatures, by choosing different pH values of the solution (for basic or acidic compounds), or by adding auxiliary reagents such as shift reagents (lanthanides) or cyclodextrins, which may result in signal separation. The different options will be discussed in the following paragraphs.

5.3.1 Choice of Solvent

Variation of the solvent can lead to considerable changes of the chemical shifts of the substance signals, which might be due to complexation of the solvent molecules with the substance to be measured, i.e., hydrogen bonding, the anisotropy of the solvent, van der Waals interactions, and polar effects, or due to an influence on the (self-)association of substance molecules that additionally depends on the sample concentration and temperature (see Chapter 2). Besides the solvent-induced change of the chemical shift, signal dispersion and lineshape can also be affected. Some examples are given in Ref. 5.

Usually, the solvent for dissolving the sample is chosen with regard to the solubility of the substance to be measured. In the case where the solvent should be varied for signal shifting and splitting, the solubility is for sure a limiting factor. Nevertheless, this disadvantage can be overcome using solvent mixtures. Here, the fact has to be considered that different solvents in a mixture may have different vapor pressures. For reproducibility reasons, such mixtures should be measured immediately in order to avoid the evaporation of one component resulting in a change in composition.

The solvents can be classified as polar (subcategories protic and nonprotic) and nonpolar, with the subclasses aromatic and nonaromatic (Table 5.1). Aromatic solvents are known to provide the so-called ASIS (aromatic solvent-induced shift) effect owing to their high diamagnetic anisotropy (ring current

effect) and their tendency to form complexes, especially with carbonyl groups of the analyte. In the days of low-field NMR instruments operating at 60 MHz for protons, ASIS effects were utilized to gain information about the spatial structure of a molecule, high- or low-frequency shifts in the neighborhood of, e.g., a ketone, depending on the position of the ring current of the aromatic solvent moiety, and to separate signals for easier interpretation. The latter fact has been forgotten in the past 20 years because the high- and ultrahigh-field instruments already provide good signal separation, which is further improved by 2D experiments. Nevertheless, the ASIS effect can be extremely useful for signal separation.

As well as aromatic solvents, acetone is able to produce solvent effects owing to its dipole moment and the ability to form hydrogen bonding.

The optimization of signal separation is exemplified in Figure 5.2, which depicts the expansion of the region of H-5 of the four components of the alkaloid codergocrine mesylate, i.e., dihydroergocornine (cor), dihydroergocristine (cr), α-dihydroergocryptine (α), and β-dihydroergocryptine (β). While spectra in DMSO-d_6, $CDCl_3$ and mixtures of both showed only partial separation of the four signals, the spectrum in a mixture of C_6D_6 and DMSO-d_6 (10 : 1) gave a nice separation of signals of H-5, which could be easily integrated.[6] The ratio of components, the overall content, and the relative standard deviation fitted nicely with the high-performance liquid chromatography (HPLC) results that were obtained by means of the official method from the European Pharmacopoea.

However, when using solvent mixtures for the separation of signals it has to be considered that even the purest solvents with a high degree of deuteration still have proton signals and sometimes an HOD signal. All of them can interfere with the signals of interest. This limitation is difficult to overcome, because there is no way to get rid of those signals.

Large water or solvent signals (of nondeuterated solvents) that cause a large distortion of the baseline can be suppressed by different techniques, such as the

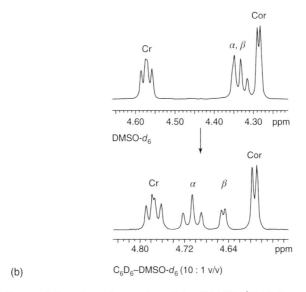

R = Isopropyl (cor)

= Benzyl (cr)

= Isobutyl (α)

= 2-Methyl-propyl (β)

(a)

Sample	Method	Cor (%)	α (%)	Cor (%)	β (%)	Content	
						MW (%)	RSD (%)
Cod-M1	IS	32.4	21.5	31.1	10.5	95.5	0.1
	ES	32.2	21.6	31.2	10.8	95.8	0.5
	ERETIC	32.2	21.8	30.7	10.9	95.6	0.2
	HPLC	33.4	21.9	30.1	10.6	96.0	0.5

(b)

Figure 5.2. (a) Structural formulae and expansion of the 400 MHz ^{1}H NMR spectrum of codergocrine mesylate (region of the proton H-5), being a mixture of dihydroergocornine (cor), dihydroergocristine (cr), α-dihydroergocryptine (α), and β-dihydroergocryptine (β); (b) percentage of the components of codergocrine of a sample, determined using an internal standard, an external standard and the ERETIC method. (Adapted from Ref. 6. © Elsevier, 2010)

gradient-based water suppression by gradient-tailored excitation (WATERGATE), the weak RF irradiation Pre-Sat (presaturation of the solvent signal by a selected soft pulse), or the combination of both as described for Water suppression enhanced through T_1 effects (WET)[2,7]. WET and WATERGATE suppress the solvent signals by a factor of $\sim 10^3$ but may create some small additional signals;

Pre-SAT is advantageous owing to its simplicity and robustness, especially when small molecules are evaluated. However, all available solvent suppression techniques affect the intensity of adjacent signals, which exclude those signals from quantification. PreSAT and the PreSAT180[8] induce fewer baseline distortions than the other methods. In addition, some subtraction methods are available, e.g., the background subtraction exemplified for glucose.[9] Again, a change of the solvent may help to solve the problem.

5.3.2 pH Value of the Solvent

In the cases of bases and carboxylic acids, the pH of the solution plays a decisive role for the chemical shift of the adjacent atoms (for a representative example see norfloxacin, having a basic piperazine moiety and a carboxylic function, at different pH values in Ref. 10) and paves the way to separate signals for quantification.

Besides the fact that signals can be shifted by variation of the pH value of the solution used for measurement, it has to be kept in mind that impurities that are basic or acidic can influence the chemical shifts of the signals, too. The extent depends on the amount of impurities present. This has been impressively shown for the example of alanine, which may contain varying amounts of aspartic acid, malic acid, and glutamic acid. The pH values ranged from 3.8 to 4.9 for impurity concentrations between 2% and 0.1% (Figure 5.3a).[11] In order to overcome such problems, the spectra should be either measured in a medium of higher or lower pH (Figure 5.3b: alanine spectrum in DCl) or in a buffered medium.

In turn, chemical shifts of indicator molecules can be utilized to determine the pH value of a solution, e.g., of a metabolite solution.[12,13] Moreover, the variation of the chemical shift can be used to monitor a titration or to determine pK_A values.

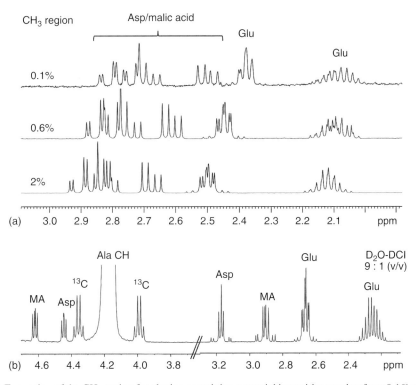

Figure 5.3. (a) Expansion of the CH_2 region for alanine containing potential impurities ranging from 0.1% to 2% (400 MHz, D_2O, 300 K). Asp = aspartic acid; Glu = glutamic acid; MA = malic acid. (b) Spectrum measured in D_2O–DCl (9 : 1)

5.3.3 Concentration Dependence and Temperature

Self-association phenomena caused by noncovalent forces such as hydrogen bonds, aromatic $\pi-\pi$-stacking, or electrostatic interactions induce changes of the chemical shift (depending on concentration and temperature). This phenomenon is well documented for caffeine and aromatic compounds, as well as for drugs such as fluoroquinolones, chlorpromazine, verapamil, mebeverine, and others (for a summary see Ref. 5). As those changes correlate with the sample concentration, they can either be used to determine the concentration of the substance to be analyzed, to monitor the self-aggregation of a compound,[14] or to separate overlapping signals considered for integration. The limitation of these experiments is the solubility of the compound in a given solvent.

The procedure of quantification is demonstrated for ciprofloxacin in the following paragraph: as the association of the molecules depends on the temperature, firstly the temperature of the probe should be calibrated by means of a 4% solution of methanol in CD_3OD for the range of 181 to 300 K, and an 80% glycol solution in DMSO-d_6 for 300 to 380 K. As it takes some time until the probe has adopted the temperature, an equilibration time of $\sim$10 min should be applied.

The dependence of the chemical shift on the temperature is displayed in Figure 5.4(a) and (b) for ciprofloxacin in D_2O at 278, 298, and 308 K. It demonstrates nicely the high-frequency shift of all protons of the quinolone moiety with increasing temperature on the one hand and the sharpening of the signals on the other hand (indicating the dissociation of the self-associated stacking of the ciprofloxacin molecules). The chemical shifts of the piperazine and cyclopropyl protons are almost unaffected by the temperature variation, because they do not experience an interaction of the molecules.

Secondly, concentration-dependent spectra for the range from 1.036 to 10.36 mM ciprofloxacin were measured (Figure 5.4c). As could be seen, with temperature changes the signals of the aromatic protons were shifted more to low frequency than the other protons. This is especially true for H-5 that surpasses the low-frequency shift of H-8 at concentrations higher than 10 mM. Owing to the high $\Delta\delta$ values of H-5, the chemical shift of this proton seems to be appropriate for the determination of the content. The calibration curve given in Figure 5.4(d) shows the good correlation coefficient.

5.3.4 Auxiliary Shift Reagents

As already mentioned, aromatic solvents such as benzene and toluene can shift the signals of a substance to be analyzed if complexation has taken place. Lanthanide shift reagents (LSRs), having a long-standing tradition in NMR spectroscopy, are capable of inducing even larger chemical shifts of the analyte's nuclei owing to the magnetic moment of the unpaired electrons.[15,16] Only lanthanides such as praseodymium, europium, and ytterbium are suitable. The most popular LSR are complexes of 6,6,7,7,8,8-heptafluoro-2,2-dimethyloctane-3,5-dione [Eu(fod)$_3$ and Yt(fod)$_3$] and of 2,2,6,6-tetramethylheptane-3,5-dione [Eu(dmp)$_3$, Pr(dmp)$_3$, and Yt(dmp)$_3$].

In addition, chiral LRSs are available that can be utilized for the determination of the enantiomeric excess (EE) owing to the formation of diastereomeric complexes: tris[3-(trifluoromethylhydroxymethylene)-*d*-camphorato]ytterbium [Yt(tfc)$_3$] and tris{[(heptafluoropropyl)-hydroxymethylene]-*d*-camphorato}europium [Eu(hfc)$_3$]. However, those paramagnetic reagents cause large line broadening that makes the integration of a signal and thus the quantification difficult. This is especially a problem when using high-field instruments because the LSR-induced line broadening is proportional to the square of the magnetic field. Thus, they are nowadays rarely used in qNMR but the LSR methodology has experienced a renaissance in conjunction with MRI contrast agents.[16]

For chiral analysis, organic solvating agents such as α-methylbenzylamine, 1(1-naphthyl)ethylamine, 2,2,2-trifluoro-1-(9-anthranyl)ethanol, crown ethers, calixarenes, or α-, β-, and γ-cyclodextrins and their derivatives are often employed.[17,18] They also form diastereomeric complexes and thus provide signal splitting of the enantiomers, which makes the determination of the EE possible. Alternatively, chiral derivatization agents and metal complexes can be employed for chiral discrimination.[19]

Very recently, the EE determination of a variety of chiral carboxylic acids, terpenoids, and alkaloids using

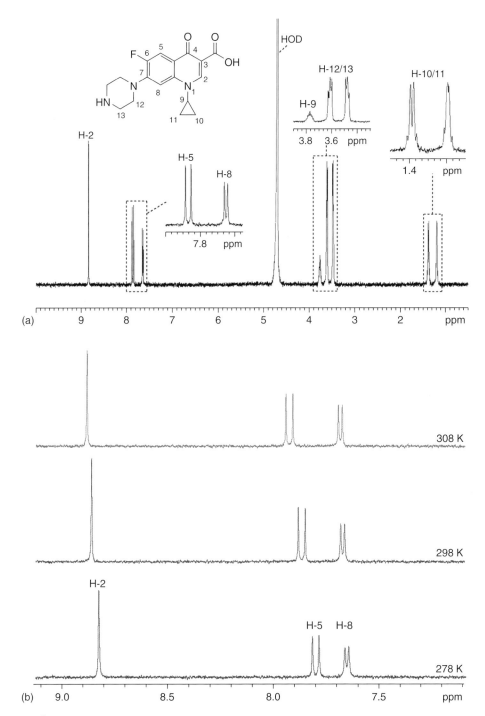

Figure 5.4. (a) ^{1}H NMR spectrum of a 1.875 mM ciprofloxacin in D_2O (400 MHz, 300 K); (b) expansion of the spectra of a 1.039 mM ciprofloxacin hydrochloride solution in D_2O at 278, 298, and 308 K. (see the next page for parts (c) and (d)).

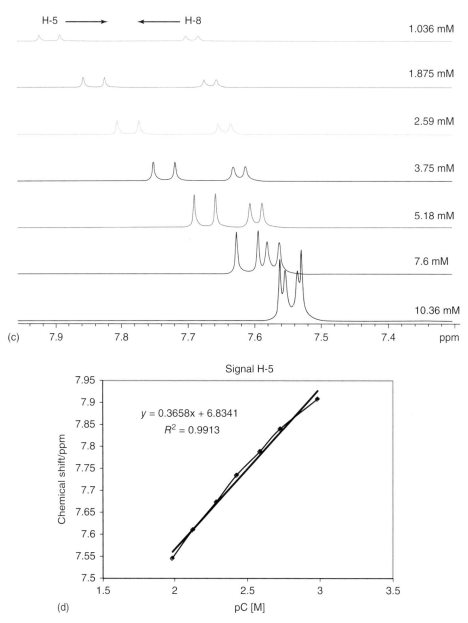

Figure 5.4. (c) expansion of the spectra of ciprofloxacin at different concentrations at 298 K; (d) correlation between the chemical shift of H-5 and the logarithm of the concentration of ciprofloxacin, pC

prochiral solvating agents, i.e., an *N,N'*-disubstituted oxoporphyrinogen host molecule, has been described for the first time.[20] Interestingly, no diastereomeric complexes have been formed. The differentiation was achieved just by the breakdown of the symmetry of the host.

5.3.5 Two-dimensional Nuclear Magnetic Resonance

Instead of shifting the signals for separation, two-dimensional experiments, either homo- or hetero-nuclear, can be employed for quantification.

In this case, the peaks are spread along an additional orthogonal dimension resulting in better separation. However, the separation is at the expense of a long acquisition time and the problem of the complexity of the 2D NMR peak response to this situation. The obtained peak volume depends on many parameters, such as pulse sequence delay, pulse angles, transverse and longitudinal relaxation times, and homonuclear and heteronuclear coupling constants, to name only a few. Even though a lot of techniques have been developed 2D qNMR is left to specialists, currently. A recent summary of methods is given by Giraudeau.[21]

5.4 QUANTIFICATION METHODS

Even though qNMR can be regarded to be a robust, simple, and fast method, a lot of pitfalls are out there which may result in false quantification results. To summarize the procedures described above, the following parameters have to be under control: besides factors that are related to the sample itself, such as sample composition and concentration (self-association effect), temperature, sample volume, tube size, and the solvent (may interact with the analyte), a lot of parameters concerning the instrumental setup have to be controlled for instance apodization of the window function, zero filling, phase, baseline, and drift corrections as well as the relaxation time of the signal considered for quantification. Thus, quantification of an analyte is not 'just the simple measurement' of the 1D NMR spectrum, as already mentioned.

Mostly, the integrated peak areas of pure signals are used for quantification. Alternatively, the signal height can be used, but it needs a normalization of the width at half-height of the internal standard peak (set at, e.g., 3 Hz) through incremental adjustment of Gaussian or line-broadening apodization during the processing of the FID.[22] In addition, for quantification of compounds in complex biological mixtures a curve fitting method (deconvolution) has been described.[23] However, it has to be kept in mind that the peak height is not directly proportional to the number of atoms at the relevant frequency. The widths at half-height of both analyte and standard have to be the same. However, both methods are rarely used because integration is normally far easier. Hence, the integrals are used in the following sections.

5.4.1 Relative Method

The relative method is mainly used for the determination of ratios of components of drugs, of natural products in a mixture, or of synthesized compound mixtures. The molar ratio n_X/n_Y of two compounds X and Y can be calculated simply by using the integrals I of a pair of separated signals of a defined number of nuclei N:

$$\frac{n_X}{n_Y} = \frac{I_X}{I_Y} \cdot \frac{N_Y}{N_X} \tag{5.4}$$

(as K_S is constant during the course of this experiment, it is canceled in the equation). For a valid quantitative determination, the structure of the components should be known. The percentage of a compound X in a mixture of m components is given by:

$$\frac{n_X}{\sum_{i=1}^{m} n_i} = \frac{I_X/N_X}{\sum_{i=1}^{m} I_i/N_i} \cdot 100 \tag{5.5}$$

The solvent signal has to be disregarded. Knowledge of the molar masses of the components is not required in this case, and no reference/calibration standard is necessary. The relative qNMR method is an important tool for quantifying the ratios of isomers, e.g., enantiomers (after addition of chiral agents) and diastereomers (e.g., R/S epimers, and E/Z).

5.4.2 Absolute Method

qNMR offers two different methods for the determination of absolute values, i.e., the normalization method and the use of internal or external standards.

5.4.2.1 Normalization Method

This is called the 100% method because all components of a mixture (components of a drug or the drug including the impurities) are set to 100%. However, some requirements have to be fulfilled: (i) all ingredients of a mixture have to have a nucleus which can be observed (e.g., inorganic impurities, such as KCl, NaCl, or silica gel, do not show up in a ^{1}H NMR spectrum); (ii) the structures of all ingredients have to be known and their spectra unambiguously assigned; and (iii) partially overlapping signals prevent the quantification.

The method is especially useful for the determination of the percentage of a component in a mixture, such as gentamicin consisting of gentamicin C1, C1a, C2, C2a, and C2b in addition to the impurities garamine, sisomicin, deoxystreptamine, and others. The same holds true for the characterization of (structurally modified) polymers.[24]

5.4.2.2 Internal Standard

The concentration of the main component P_X of a solution can be determined using an internal standard, also called a *calibrant*. It can be directly calculated from the following equation:

$$P_X = \frac{I_X}{I_{Std}} \cdot \frac{N_{Std}}{N_X} \cdot \frac{M_X}{M_{Std}} \cdot \frac{m_{Std}}{m} \cdot P_{Std} \qquad (5.6)$$

with N_X and N_{Std} being the numbers of contributing nuclei of the signal X and standard, M_X and M_{Std} the molar masses of analyte and standard, m and m_{Std} the weights of the sample and standard, and P_X and P_{Std} the contents of analyte and standard, respectively. The internal standard method gives the most accurate and precise results in comparison to the external standard and the electronic reference to access in vivo concentrations (ERETIC) methods.[25] However, it may suffer from possible signal overlap.

The uncertainty of this absolute method depends mainly on the following three parameters:

1. The purity of the internal standard and the internal standards have to meet a couple of general conditions: they have to be very pure, nonhygroscopic, nonvolatile (best solid!), inert toward solvent and analyte, stable under typical measurement and storage conditions, soluble in many deuterated solvents, nontoxic, should have a low number of signals (in areas of the spectrum that are normally empty), and last but not least should be available at a reasonable price.

   Many of these properties are related to purity. The purity has to be ensured by chromatographic methods, such as HPLC, or by 2D homo- and hetero-nuclear COSY experiments. This holds true for the initial purity but also for storage over a certain period of time. Thus, storage conditions have to be investigated by means of stress tests.[26,27]

2. The weighing of analyte and standard: In order to reduce the uncertainty of sample weighing, higher amounts of the analyte and the internal standard

should be weighed or the use of stock solutions is recommended, even though a dilution step might be necessary. The internal standard should be easily weighable and the uncertainty can be further decreased by the usage of aluminum pans that are antistatic and decrease humidity absorption.[28]

The ideal molecular masses (M) of an internal standard should be similar to the analyte's M, because when applying low molecular weight compounds (e.g., sodium acetate or dimethylsulfone), only small amounts of the reference substance have to be weighed for comparable signal heights. Of note, standards containing signals of high intensity due to the high number of protons (e.g., *tert*-butyl groups) should also be avoided for the same reason. It goes without saying that the protons chosen for integration should not be exchangeable.

3. The integration: Integration may add a high amount to the uncertainty due to partially manual phase/baseline/drift correction (see Chapter 1) and integration (range) by the operator! However, experienced operating individuals may allow an acceptable inter-individual precision of ±2%.

The search for a single universal standard that can be used in many applications is still ongoing. Nevertheless, a collection of standards is purchasable, among them calcium formate (having an extremely long relaxation time of ~20 s), benzoic acid, 3,5-dinitrobenzoic acid, duroquinone, 1,2,4,5-tetrachloro-3-nitrobenzene, dimethyl sulfone, benzyl benzoate, and maleic acid, to name only a few. The pros and cons of these certified reference materials (CRM in accordance with ISO guidelines) and their traceability are summarized in Ref. 27. In order to conclude the internal standard discussion, one has to keep in mind that one standard (CRM) can be used for multiple applications.

How to minimize the uncertainty using the internal standard method is summarized in Ref. 28.

5.4.2.3 External Standard

In case the analyte ought not to be 'polluted' with the reference standard, a coaxial insert containing the standard in solution can be used. In this case, different solvents can be applied, or a shift reagent can be used (to prevent signal overlap of analyte and standard), or the solutions in the tubes may have a different pH. Moreover, the solution in the insert can be sealed and used for different applications for a long time, as long as the standard is stable.

Alternatively, an analysis of separate but identical tubes for the analyte and the standard can be performed, which is less straightforward.

5.4.2.4 Standard Addition

In case the molecular mass of the analyte is not known, the standard addition method can be applied for the determination of absolute values. Defined amounts of the analyte are stepwise added to the solution to establish a calibration curve that can be used to calculate the content. Even though the method is time consuming (because the samples have to be taken in and out of the spectrometer several times), it has the big advantage of being extremely accurate, because the analyte acts as its own reference/standard. Moreover, it allows for accurate content determination even though the true value is not known.

Typical applications are polymers of varying molecular mass, or the determination of dermatan in unfractionated heparin.[29] More details are given by Kellenbach in Chapter 31.

5.4.2.5 External Calibration

The establishment of a calibration curve by measurement of a series of NMR spectra of samples with different concentrations followed by plotting the obtained signal areas versus the concentration is the same procedure as used for HPLC, ultraviolet (UV) spectrometry, and many other techniques. Having determined the calibration curve, samples of unknown concentration of the analyte can be evaluated.

For the establishment of the calibration curve, the concentration of the analyte has to be known, e.g., by determination using an independent method, such as gravimetry or titration, and a pure standard is needed.

The external method suffers from two disadvantages; it includes the errors from volumetric measurements owing to the preparation of both reference and sample solution, and the variability of the inner diameter of the NMR tubes gives rise to different signal integrals owing to different amounts of material in the tube. The latter can be minimized using high-quality NMR tubes.

If it is intended to determine the analyte in a complex matrix using the calibration curve established with the pure standard, deviation may occur. In this case, the standard addition method (see Section 5.3.2.4) should be preferred.[29]

5.4.2.6 Electronic Reference to Access In Vivo Concentrations (ERETIC)

The ERETIC™ method makes use of an electronic reference signal that can be placed in a free spectral region of the NMR spectrum. This RF signal is fed into the resonance circuit during the acquisition by means of a free coil in the probe. Thus, it needs extra hardware along with the manipulation of the NMR instrument.

Besides the prevention of signal overlap, further advantages are the full control of the size and phase of the signal. However, when it comes to the evaluation of small molecules the accuracy of the method is inferior to the aforementioned methods.

Further details and applications can be found in Refs. 25 and 30.

5.4.2.7 Pulse Length-based Concentration Determination (PULCON)

As already mention in Chapter 3, the use of two different tubes has some drawbacks. This is because proportionality between the signal area and the concentration can be sample-dependent. This can be avoided by using the pulse length-based concentration determination (PULCON) method that is based on the principle of reciprocity saying that the NMR signal intensity is inversely proportional to the 90° pulse duration. Hence, the difference of sensitivity between the samples can be corrected by means of the pulse duration. Even though a couple of improvements have been published in recent years, the method is more or less restricted to the protein field. However, use of the ERETIC 2 method, which is based on PULCON, may be a good alternative to the internal standard methods, e.g., in natural product evaluation.[31]

Further information can be retrieved from Ref. 30.

5.5 APPLICATIONS

Even though the international pharmacopoeias do not make much use of qNMR spectroscopy, qNMR is widely applied in the pharmaceutical and food industries. Quality evaluation of drugs and excipients can take advantage of this method. This is described in Chapters 28 and 30.

5.5.1 Solid Pharmaceutical Systems

In addition to qNMR in solution, solid-state nuclear magnetic resonance (SSNMR) can also be employed for quantification.[32] SSNMR is a nondestructive and noninvasive method.[33] It provides detailed information about structure and dynamics in the solid state for both crystalline and amorphous solids. For example, complexes consisting of cyclodextrins and a drug can be structurally characterized. It is also applicable to heterogeneous solids; hence, it is appropriate for quantification of an active pharmaceutical ingredient in a tablet or other solid formulation. Manufacturing processes can be monitored for crystallization and during milling. Furthermore, the dynamics of a system, i.e., the mobility of atoms and molecules in a lattice, can be observed, and it is possible to distinguish between polymorphs (and solvates) and to quantify them (see Chapter 21).[34] While a lot of potential applications have been reported, SSNMR has a couple of drawbacks: (i) the spectrometers are rather expensive and the sensitivity, while benefitting by the absence of solvent volume, is generally somewhat lower than that for NMR in solution; (ii) the signals suffer from line broadening owing to homonuclear ^{1}H dipolar coupling. Therefore, ^{13}C, ^{15}N, ^{19}F, and ^{31}P are the most observed nuclei and high-power ^{1}H decoupling is applied to narrow the signals. Moreover, the molecules can take all possible orientations with equal probability in a powder. The resulting line broadening (and broadening caused by heteronuclear dipolar coupling) can be overcome by magic-angle spinning (MAS). Sensitivity enhancement can be obtained by cross-polarization (CP). Very high-speed MAS (>60 kHz) can even sharpen proton spectra. For quantification in SSNMR, the ERETIC method can be used, but it is vital to check the dependence of signal intensities on CP time and on the various relaxation times.

The basics of quantitative SSNMR and a collection of applications can be found in Refs. 35, 36 and in the handbook on 'NMR Crystallography'.[37]

5.6 CONCLUSION

The sensitivity problem that might be a result of impurity profiling can be overcome by ultrahigh-field instrumentation (600–1000 MHz), and/or special probes, such as the cryoprobes or microprobes. Thus, when applying ^{1}H NMR, limits of detection of 10^{-10} mol can be reached.

In addition, time is always an issue in pharmaceutical industries. Nowadays, most of the hardware enables automated sample handling, including preparation, tube filling, and sample changing, or flow injection techniques (see Chapter 15). This is often accompanied by automated data evaluation. Thus, even though NMR spectroscopy is far more sophisticated than chromatographic methods, it is as automated and as fast as the separation techniques but provides addition structural information.

ACKNOWLEDGMENTS

Thanks are due to Johannes Wiest and Dr. Tanja Beyer for measuring spectra.

REFERENCES

1. G. F. Pauli, B. U. Jaki, and D. C. Lankin, *J. Nat. Prod.*, 2005, **68**, 133.

2. H. Günther, NMR Spectroscopy – Basic Principles, Concepts, and Application in Chemistry, 3rd edn, Wiley-VCH: Weinheim, 2013, Chapter 8.

3. E. Alvarado, Practical guide for quantitative 1D NMR integration, 2010. http://www.umich.edu/~chemnmr/docs/Quantitative_NMR.pdf cited MestreNova Research global spectral deconvolution (GSD); http://www.acdlabs.com/resources/freeware/nmr_proc/index.php

4. P. Soininen, J. Haarala, J. Vepsäläinen, M. Niemitz, and R. Laatikainen, *Anal. Chim. Acta*, 2005, **542**, 178.

5. U. Holzgrabe, *Prog. NMR Spectrosc.*, 2010, **57**, 229.

6. T. Beyer, C. Schollmayer, and U. Holzgrabe, *J. Pharm. Biomed. Anal.*, 2010, **52**, 51.

7. G. Zheng and W. S. Price, *Prog. NMR Spectrosc.*, 2010, **56**, 267.

8. H. Mo and D. Raftery, *J. Magn. Reson.*, 2008, **190**, 1.

9. T. Ye, C. Zheng, S. Zhang, G. A. Nagana Gowda, O. Vitek, and D. Raftery, *Anal. Chem.*, 2012, **84**, 994.

10. S. K. Branch and U. Holzgrabe, *Magn. Reson. Chem.*, 1994, **32**, 192.

11. U. Holzgrabe, C.-J. Nap, T. Beyer, and S. Almeling, *J. Sep. Sci.*, 2010, **33**, 2402.

12. G. Orgován and B. Noszál, *J. Pharm. Biomed. Anal.*, 2011, **54**, 958.

13. T. Tynkkynen, M. Tiainen, P. Soininen, and R. Laatikainen, *Anal. Chim. Acta*, 2009, **648**, 105.

14. S. R. LaPlante, N. Aubry, G. Bolger, P. Bonneau, R. Carson, R. Coulombe, C. Sturino, and P. L. Beaulieu, *J. Med. Chem.*, 2013, **56**, 7073.

15. A. F. Cockerill, G. L. O. Davies, R. C. Harden, and D. M. Rackham, *Chem. Rev.*, 1973, **73**, 583.

16. C. F. G. C. Geraldes, in The Rare Earth Elements, ed D. A. Atwood, Wiley, 2012, 501.

17. W. H. Pirkle and D. J. Hoover, in Topic in Stereochemistry, eds N. L. Allinger, E. L. Eliel and S. H. Wilen, Wiley: Hoboken, 2007, Chap. NMR áchiral solvating agents, 13.

18. G. Uccello-Baretta and F. Balzano, in Differentiation of Enantiomers II, *Topics in Current Chemistry*, ed V. Schurigt, Springer: Switzerland, 2013, vol. 341, Chap. Chiral NMR solvating additives for differentiation of enantiomers, 69.

19. T. J. Wenzel, in Differentiation of Enantiomers II, *Topics in Current Chemistry*, ed. V. Schurigt, Springer: Switzerland, 2013, Vol. **341**, Chap. Chiral derivatization agents, macrocycles, metal complexes, and liquid crystals for enantiomer differentiation, 1.

20. J. Labuta, S. Ishihara, T. Sikorsky, Z. Futera, A. Shundo, L. Hanykova, J. V. Burda, K. Ariga, and J. P. Hill, *Nat. Commun.*, 2013, **4**, 2188. DOI: 10.1038/ncomms3188.

21. P. Giraudeau, *Magn. Reson. Chem.*, 2014, **52**, 259.

22. P. A. Hays and R. A. Thompson, *Magn. Reson. Chem.*, 2009, **47**, 819.

23. D. J. Crockford, H. C. Keun, L. M. Smith, E. Holmes, and J. K. Nicholson, *Anal. Chem.*, 2005, **77**, 4556.

24. B. W. K. Diehl, F. Malz, and U. Holzgrabe, *Spectrosc. Eur.*, 2007, **19**, 15.

25. C. H. Cullen, G. J. Ray, and C. M. Szabo, *Magn. Reson. Chem.*, 2013, **51**, 705.

26. T. Rundlöf, M. Mathiasson, S. Beriroglu, B. Hakkarainen, T. Bowden, and T. Arvidsson, *J. Pharm. Biomed. Anal.*, 2010, **52**, 654.

27. M. Weber, C. Hellriegel, A. Rueck, J. Wuethrich, and P. Jenks, *J. Pharm. Biomed. Anal.*, 2014, **93**, 102.

28. T. Schoenberger, *Anal. Bioanal. Chem.*, 2012, **403**, 247.

29. T. Beyer, B. Diehl, G. Randel, E. Humpfer, H. Schäfer, M. Spraul, C. Schollmayer, and U. Holzgrabe, *J. Pharm. Biomed. Anal.*, 2008, **48**, 13.

30. P. Giraudeau, I. Tea, G. S. Remaud, and S. Akoka, *J. Pharm. Biomed. Anal.*, 2014, **93**, 3.

31. O. Frank, J. K. Kreissl, A. Daschner, and T. Hofmann, *J. Agric. Food Chem.*, 2014, **62**, 2506.

32. R. K. Harris, *Analyst*, 1985, **110**, 649.

33. D. C. Apperley, R. K. Harris, and P. Hodgkinson, Solid-state NMR: Basic Principles and Practice, Momentum Press: New York, 2012.

34. T. Virtane and S. L. Maunu, *Int. J. Pharm.*, 2010, **394**, 18.

35. R. T. Berendt, D. M. Sperger, P. K. Isbester, and E. J. Munson, *Trends Anal. Chem.*, 2006, **25**, 977.

36. K. Paradowska and I. Wawer, *J. Pharm. Biomed. Anal.*, 2014, **93**, 27.

37. R. K. Harris, R. E. Wasylishen, and M. J. Duer (eds), NMR Crystallography, John Wiley: Chichester, 2009.

Chapter 6

High-throughput NMR in Pharmaceutical R&D

John C. Hollerton

GlaxoSmithKline R&D, Gunnels Wood Road, Stevenage, Herts SG1 2NY, UK

6.1 INTRODUCTION

The term *high-throughput NMR* is difficult to define and will mean different things in different contexts. This chapter is focused on the key components of the NMR analysis process and uses some examples of what could be considered to be 'high-throughput' analyses to exemplify the considerations needed to achieve enhanced throughput.

The introduction of high-throughput screening (HTS) has dramatically increased the number of compounds that pharmaceutical companies have in their screening collections. These compounds come from a variety of sources both internal and external and lead to a challenge in ensuring that they are both pure and structurally correct. Many companies have adopted the use of high-performance liquid chromatography-mass spectrometry (HPLC-MS (LCMS)) as a primary analysis tool[1] owing to its speed and simplicity of data analysis. In particular, the advent of ultra-high pressure (UHPLC)[2] systems has allowed very rapid sample analysis in around 2 min per sample. There are inevitable limitations with LCMS in both purity and identity confirmation. Purity is normally determined using the ultraviolet (UV) absorbance and this depends on the response factors of the various components of the sample. Entities without a chromophore are invisible. Some implementations use a secondary detector such as the evaporative light scattering (ELS) detector but this also has sample-dependent response factors, particularly around the sample's volatility. The search for the universal detector continues for HPLC, although what we really need is a universal detector that does not detect the mobile phase!

High-throughput identity determination by LCMS is normally limited to identification of the parent molecular ion. Even if generation of this ion were 100% successful, it does not distinguish between isobaric compounds. Normally systems use mass spectrometers capable of measuring nominal mass (for cost and simplicity) but even

NMR in Pharmaceutical Sciences. Edited by Jeremy R. Everett, John C. Lindon, Ian D. Wilson, and Robin K. Harris
© 2015 John Wiley & Sons, Ltd. ISBN: 978-1-118-66025-6
Also published in eMagRes (online edition)
DOI: 10.1002/9780470034590.emrstm1390

high-resolution systems would not be able to distinguish regiochemistry unless fragmentation is considered and then data analysis becomes a more complex process.

NMR offers considerable advantages; in that, it is a truly quantitative detector (provided the experiments are set up correctly) and that it offers much richer structural information. The main disadvantages are that it is relatively slow, insensitive, and produces complex data.

There are other areas that may need high-throughput NMR (e.g., fragment-based drug design and structure–activity relationships (SAR) by NMR[3]). It is not possible to cover all of these topics; however, this chapter will look at three different example workflows. There are many other workflows but they can all be characterized in terms of their overall process and the components that support that process. It is hoped that some of the considerations for these workflows may be applicable in other cases not covered here.

6.2 OVERALL PROCESS VIEW

If we look at the whole process (not just the data acquisition), achieving high throughput relies on optimizing all of the components of that workflow (Figure 6.1). Simply performing fast data acquisition does not necessarily address the high-throughput requirements of the process. We can consider the process to consist of five parts: sample preparation, sample introduction, system preparation, data acquisition, and data processing/analysis. All of these parts are interrelated and choices made at one part of the process may affect the other parts. While for any individual sample, the process is linear, it is possible to parallelize some of the process components to achieve higher overall throughput. For example, sample preparation can take place in parallel to the other steps. Likewise, data processing and analysis is a parallel process. The components that must operate sequentially are the middle three of sample introduction, system preparation, and data acquisition.

6.2.1 Sample Preparation

How the sample is prepared has an impact on all of the downstream processes. In the example workflows, it can be seen that the choice of solvent may impact how data acquisition is performed and may also affect the data analysis step at the end of the process. A full understanding of the purpose of the analysis is a key to selection of all of steps of the process.

6.2.2 Sample Introduction

Flow or tube? Traditionally, NMR systems have been designed to work with NMR tubes but flowing samples into the probe has been possible for many years.[4] Tube introduction is simple and discrete, whereas flow can be complex. Examples of extremely high-throughput NMR have been demonstrated using flow processes – in particular, microflow systems using capillaries and microcoil probes.[5] The trade-off is the complexity of robust control of microfluidic systems and potentially compromising the quality of the NMR data at the end of the process. Despite these difficulties, sample throughput rates of over 500 samples per day have been achieved.[6] More conventional flow volumes are offered by Agilent (VAST) and Bruker (BEST) using a Gilson liquid handler and a flow probe (or a probe adapted to flow).

6.2.3 System Preparation

Before data acquisition, a number of preparatory steps are needed to ensure that the data quality is acceptable. These may be broken down into the following steps.

- Temperature equilibration
- Probe tuning and matching
- Shimming
- Locking

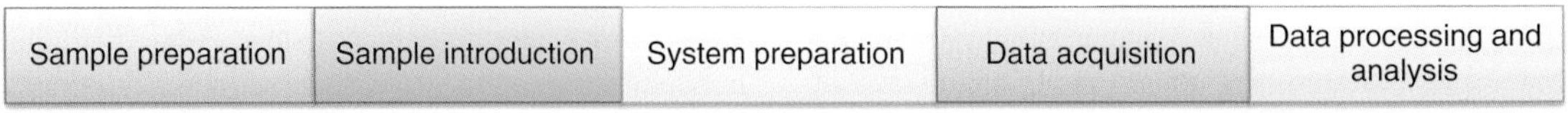

Sample preparation	Sample introduction	System preparation	Data acquisition	Data processing and analysis

Figure 6.1. Overall process for NMR workflow.

- Receiver gain setting
- Steady-state equilibration

Many of these steps can be skipped depending on the desired outcome of the experiment and the type of data acquisition needed. For example, if all of the samples are very similar (in the same solvent and similar conductivities), then it is possible to skip probe tuning and even shimming. If the desired experiment is a short, simple 1D proton experiment, then locking may be unnecessary as well as temperature equilibration, as modern instruments show good stability in our experience. For simple 1D experiments, steady-state spin equilibrium (often achieved with the use of 'dummy scans') may also be omitted. If samples are of a reasonable concentration for the probe/magnet combination, then a fixed receiver gain will produce little degradation of signal to noise (in-house observations).

The purpose of skipping these steps is that each one takes time and it is often the case that the sample changing and system preparation steps take longer than the actual data acquisition time.

6.2.4 Data Acquisition

Obviously, the choice of data acquisition experiments is driven by the information required at the end of the data analysis process. It is also driven by the medium in which the sample is dissolved. If the process uses protonated solvents, then some form of solvent suppression is probably required.[7] The length of this part of the process is often the focus of optimization efforts although it can be worthwhile looking elsewhere in the process for saving time.

For simple 1D proton experiments, data acquisition is relatively quick as long as there is sufficient material. If quantitative results are needed (e.g., if trying to quantify solution concentrations of screening solutions), care must be taken in ensuring that there is a sufficient relaxation delay to allow true quantification.[8]

If more complex experiments are needed, particularly multidimensional experiments, there are a number of developments that allow this data to be acquired more rapidly. Once again, choice of the right experiment is driven by the downstream needs of the data analysis step at the end of the process.

For all experiments in a high-throughput mode, there are often compromises that have to be made in order to have sufficient throughput. This is not a problem as long as everyone who uses this data is aware of the compromises.

6.2.5 Data Processing and Analysis

The product of the process is the output of the data analysis, whether this is a concentration figure, a likelihood of the correct structure, or a measure (such as protein binding). The result must be fit-for-purpose and this drives the choices in all of the previous steps of the process.

6.2.5.1 Processing

For a 1D proton experiment, the simple processes of zero-filling, apodization, Fourier transformation, phase correction, baseline correction, and integration are fraught with difficulties when performed under automation. Despite many years of automatic processing, phase correction remains unreliable,[9] especially in the case of samples with very broad peaks or poor baselines. Likewise, integration and baseline correction are affected by unreliability. These inadequacies are fundamental limitations in the reliability of the automated NMR analysis process. Recent developments have shown ways that the inadequacies of the Fourier transform process can be avoided by direct analysis of the time-domain data. This approach 'CRAFT'[10] shows some promise in reliable, accurate quantification by NMR.

6.2.5.2 Analysis

Quantitative measurements require identification of known multiplets in order to correctly identify the number of protons that the multiplet represents as well as a good quality integral. For samples with no reference spectra, multiplet identification requires some form of spectral prediction. The integral itself may be extracted by measuring the area under the peaks or, more accurately, by performing high-quality peak fitting. The advantage of the latter is that it is not subject to influence by poor baseline.

If the outcome desired is structural confirmation, then some form of automated structural verification software is needed if the numbers are high. Once again, this outcome drives the choices in earlier parts of the process. Getting a reliable result requires good quality data and this will affect what time-saving steps are taken.

6.3 WORKFLOW – PURITY AND IDENTITY OF SOLID SAMPLES

6.3.1 Overall Process and Needs

This is the typical workflow for samples acquired from an external organization or in the case of performing some quality control on in-house solid samples. There are normally several milligrams of material available, enough for at least 1 mg to be used for NMR analysis. Because we wish to verify that the data fits the proposed chemical structure, we need data that is capable of being used for this purpose. It is entirely possible to acquire data on many hundreds of samples per day but it is unlikely that there is the resource to look at all of this data. This means that we need some form of automated system to interpret the data and this then guides the selection of each of the parts of the process.

6.3.2 Sample Preparation

When starting from solid (i.e., not dissolved), there is a freedom to select the best solvent. In practice, this is often d6-DMSO as it is capable of dissolving most pharmaceutical entities. In theory, it is possible to select the solvent based on the proposed structure but this leads to greater complexity in the sample preparation step and process complexity often works against high-throughput methods. The use of a deuterated solvent also enables higher quality data acquisition as the sample can be locked and shimmed using deuterium gradient shimming. The dynamic range of the instrument is not compromised by using deuterium and there is no need for solvent suppression to be used. If the sample numbers are very great, then it is possible to use a conventional liquid-handling robot to dispense the dimethyl sulfoxide (DMSO) into the sample vials (or plates). It is important to ensure that the samples are fully dissolved. This can be checked in a number of different ways (automatic and manual); the challenge comes from samples that do not fully dissolve. In this case the sample is either filtered (which gives rise to potentially incorrect sampling) or it is taken out of the automated system and dissolved manually with

sufficient solvent to take up the whole solid. These solubility issues can severely impact the efficiency of the sample preparation process.

6.3.3 Sample Introduction

Many people have used flow systems to load samples into the NMR probe. This approach was particularly popular in the 1990s as it was one of the few ways of automatic sample introduction from the 96-well plate format samples. A typical flow system consists of a liquid handler and a flow probe (Figure 6.2).

The liquid handler injects a certain volume of the solution into a tube that leads to the NMR probe. The slug of material is pushed along the tube using a 'push-solvent', which can be the same as the dissolution solvent or another immiscible liquid. There are a number of challenges with this approach; in that, the slug must be positioned accurately in the active volume of the probe and must not mix too much with the push-solvent. Numerous technical methods have been used to make this as reliable as possible but it remains a complex problem to get a sample into the probe. An approach of using an immiscible solvent and queuing up 'sample-trains'[6] allow for a much faster and efficient way of loading samples into a microcoil probe but this limits the experiments possible as there is only a small amount of material in the probe.

Developments by Bruker in recent years offer an alternative approach with the combination of their modified Gilson liquid handler (to transfer the solutions to short NMR tubes) and the SampleJet sample changer

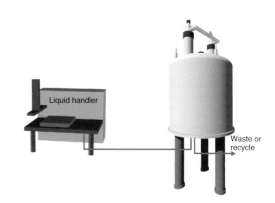

Figure 6.2. Example of a flow-inject system.

that handles short NMR tubes in the 96-well plate format directly. This leaves the scientist in a position where they can choose sample tubes with the ideal volume to support their sample needs. We have used 1, 1.7, and 3 mm tubes in this mode. There are no complexities of sample positioning with this approach and samples may be reanalyzed (or further experiments performed) by reloading the sample into the magnet. If necessary, it is possible to reclaim the samples once analyzed.

6.3.4 System Preparation

Because a flow cell does not move in and out of the magnet and maintains its geometric relationship to the coils, it is often possible to skip shimming, locking, and probe tuning. This can make for a very efficient process as long as the sample introduction is not too time-consuming. In the case of tube introduction, the best quality data will require some shimming although probe tuning may be unnecessary.

6.3.5 Data Acquisition

With a sample in an NMR tube, it is possible to perform almost any NMR experiment on that sample. In this workflow, the requirement is to check a sample's identity and purity. This demands that we acquire a 1D proton experiment as a minimum. If we are to perform some sort of automated analysis (see the next section), we also benefit from an HSQC (heteronuclear single quantum coherence). For a 1 mg sample, it is possible to acquire both 1D proton and DEPT-edited HSQC experiments in around 10 min on a cryoprobe. This allows us to acquire data on around 140 samples per day. If accurate quantification is required, this slows down the process as we need a longer relaxation delay for the 1D proton experiment. Quantification can be performed using an absolute intensity approach such as QUANTAS.[11] For increased acquisition speed, the proton experiment may be acquired in a single scan (eliminating the need for a long relaxation delay) and nonuniform sampling (NUS) approaches can decrease the time taken to acquire the HSQC data. Under these conditions, it is possible to acquire a full dataset in around 5 min, bringing the number of samples per day to about 280.

6.3.6 Data Analysis

A rate >200 samples per day is beyond the capacity of most groups performing manual analysis unless they are very well resourced. Work on automated structure verification (ASV) has been going on over the last 20 years (or longer). There are now commercial tools that are designed to do this. Most will work with a 1D proton alone but the reliability of the analysis is considerably improved by the addition of a DEPT-edited HSQC.[12] These systems often work by producing a score for how well the experimental data fits the calculated data. This attempts to give a probability that the sample has the proposed chemical structure. Often the data is not sufficient to prove that the structure is what is proposed; however, it is possible to be certain that the data does not fit the proposed chemical structure. For example, the absence of a large singlet somewhere around 1δ for a compound having a tBu group would be a strong indication that it was not right. Because the vast majority of samples are correct, identification of incorrect structures is more important than identification of correct structures. Bruker, ACD Labs, and Mestrelab all have software to perform this analysis.

High-throughput data analysis requires information systems that streamline and automate the process for getting the proposed chemical structure and NMR data into the application carrying out the analysis. This can be achieved through the use of chemical structure and data files (e.g., the ubiquitous 'SD File') or, more conveniently, through links to in-house informatics systems.

6.4 WORKFLOW – PURITY AND IDENTITY OF SCREENING SOLUTIONS

6.4.1 Overall Process and Needs

Screening solution stores are often highly automated due to the large numbers of samples held in them. The solutions are often in DMSO and can be held in tubes or deep-well plates. The overall 'unit of currency' is the microtiter plate that comes in various sizes (96-, 384-, 1536-well plates are common sizes). For most NMR work, 96-well plates are most popular as they hold sufficient volume for most sample introduction techniques. The 96-well plate is also the standard unit (or 'form-factor') that is accepted by the SampleJet sample changer from Bruker. Sample concentration and volume direct the possibilities for data acquisition. Often these will limit options for experiments in a reasonable amount of time. The information needed from the process may be varied. A common question is the concentration of the sample and how it varies from its nominal concentration. In this case, care must be taken in the sample preparation and introduction processes to ensure that these do not affect the concentration in an uncontrolled way.

6.4.2 Sample Preparation

In the case of screening collection liquid samples, we are often limited in terms of solution volume and this may demand the choice of either microcoil flow systems or small tubes (1 mm or 1.7 mm o.d.). Sample preparation may be nonexistent in the case of solutions although they are often in protonated solvents (DMSO being popular). If spectrometer lock is needed, introductions of a small amount of deuterated solvent can be sufficient to achieve this. If an addition of deuterated solvent is required, then it must be done in a controlled way if quantification is required at the end of the process. Screening solutions are almost always wet and this must be taken into account with the choice of data acquisition conditions.

6.4.3 Sample Introduction

Samples can be introduced to the NMR system in the same ways as the previous approach (i.e., flow-inject or tubes). The volume of solution and its concentration will drive the best approach. In our laboratories, we tend to use tube introduction where possible as this has been the most reliable and robust method for us.

6.4.4 System Preparation

If a flow system is used then there is often no need to perform many of the system preparation steps (as in the last workflow). If the samples are very wet and/or in nondeuterated solvent, then there may be a need for solvent suppression in which case it may be necessary to perform a 'scout scan' to identify the strong

signals and allow some form of solvent suppression to be performed on them. In this case, selection of the correct receiver gain is important to get the best signal to noise from the sample.

6.4.5 Data Acquisition

Driven by the available volume and concentration (often a few microliters of 10 mM samples), a simple 1D proton experiment may be all that is possible in the available time. The sample is likely to contain large amounts of protonated DMSO and water, forcing the use of solvent suppression experiments such as 'WET'[13] to decrease the intensity of the DMSO and water signal. For specific samples, it is also possible to acquire longer experiments (such as HSQC) but this is often not possible for weak samples in a high-throughput operation.

6.4.6 Data Analysis

The quality of the data and the limited experiments possible may lead to limited data analysis in a high-throughput system. The question of concentration can be readily addressed using an external standard technique (such as QUANTAS) but requires identification of a known signal in the proton spectrum. There are various approaches available to do this which rely on spectral analysis alone (such as identification of the largest common integral). These approaches are often wrong if there is no single proton integral or if impurities give misleading results. A more reliable approach is to use a structurally directed approach to multiplet analysis (similar to ASV). This approach can often identify a single proton (or a multiplet with a known number of protons) to allow quantification.

6.5 WORKFLOW – FRAGMENT-BASED DRUG DISCOVERY

6.5.1 Overall Process and Needs

Fragment-based drug discovery screens small molecule 'fragments' to explore larger areas of chemical space more rapidly than is possible with larger molecules. NMR is often used as part of the process to identify weakly binding fragments to a protein of interest. The protein and a number of potential ligands are mixed together and an NMR experiment is performed to determine which (if any) ligands bind to the protein. The most common experiments to do this are saturation transfer difference (STD)[14] and waterLOGSY.[15] The example chosen here is that of STD.

6.5.2 Sample Preparation

A number of fragments will be mixed with the protein of interest. Several are observed at the same time for speed and also to minimize protein usage. This mixing can be performed in a liquid handler and then transferred into NMR tubes. Choice of suitable fragments for mixing is determined by their ^{1}H NMR dissimilarity. Fragments with considerable spectral overlap are avoided to ease deconvolution of the resultant data. Some proteins are unstable at room temperature, so the use of a cooled NMR sample changer can be advantageous. Samples are normally run cold to improve the dynamics of the STD effect, so keeping them close to the acquisition temperature speeds up data acquisition (less time required to thermally equilibrate the sample in the magnet).

6.5.3 Sample Introduction

In this example, sample introduction is using tubes and a conventional sample changer.

6.5.4 System Preparation

Low temperature control is important to achieve good quality data for STD, as this can help stabilize the protein as well as improve the dynamics of the saturation transfer.[16] Many of the other components of system preparation can be configured once for a batch of samples owing to the similarity between consecutive samples.

6.5.5 Data Acquisition

The STD experiment is a difference experiment. An 'on-resonance' experiment (irradiating the protein signals) and an 'off-resonance' experiment are performed and subtracted one from the other.

6.5.6 Data Analysis

This simple difference experiment will show signals for compounds that are weakly bound to the protein and no signals for those that are not bound. Because there are several potential ligands in the sample, it is necessary to be able to determine which of the components show signs of binding. Often this is done by running spectra of all of the fragments independently and using software to display the result of the experiment against a synthesis of all of the spectra of the components of that mixture. This allows easy identification of the fragment(s) that show binding.

6.6 COMMON THEMES

It is important to identify the rate-limiting step in any high-throughput process. Optimizing one step may make no difference to the overall process if there is a slower step that limits throughput. Sometimes, optimization of one part of the process can help free up a resource for other processes. For example, dramatically speeding up acquisition time may free up an instrument to perform other tasks.

6.7 GETTING FASTER

There are a number of developments that have improved the speed of the process. It is worth reiterating that speeding up one part of the process on its own may have little effect on the overall process if there are other, much slower, parts of the process that remain and cannot be performed in parallel. An obvious way of speeding up the process is to replicate systems. This is very effective but has significant cost associated with it in terms of hardware, people to run/maintain/fix that hardware, and the building space needed to house it.

6.7.1 Sample Preparation

There are many sample preparation systems available and these are capable of producing around one hundred samples per day (depending on the sample preparation required). In the case of sample preparation from solid, the dynamics and challenges of sample dissolution may make it difficult to speed up the process hugely.

6.7.2 Sample Introduction

Many modern sample changers are much faster than their predecessors. There is still room for further advances in this area, with sample change times in the order of 20–30 s, but even doubling the speed of the system would make little impact on the overall process unless all the other components are improved.

6.7.3 System Preparation

As has been seen earlier, many of the steps of system preparation can be omitted for a number of workflows. Where locking and shimming are needed, modern NMR systems achieve this rapidly through modern lock circuitry and the use of deuterium gradient shimming.

6.7.4 Data Acquisition

Much work has been carried out in recent years to try to improve the speed of data acquisition. With the advent of cooled probes, 1D proton experiments can be carried out in a single scan for most samples, leaving little room to improve speed. The concept of multiplexing the acquisition using parallel microcoil probes[17] has been investigated, together with selective excitation and chemical shift imaging approaches, but these are not in mainstream use today.

Improvements in acquisition times for 2D experiments have been considerable. The use of gradient experiments has eliminated the need for a phase cycle and so single transient per increment experiments are now possible. Further improvements have come with sparse sampling methodologies[18] such as NUS, which enable the indirect dimension to be reconstructed from a partial sampling. Ultra-fast 2D experiments such as single-shot 2D experiments may improve things further but they lack the signal to noise for many samples and may require other technologies, such as dynamic nuclear polarization (DNP) using parahydrogen[19] to get the sensitivity to a useable level.

6.7.5 Data Analysis

The large amounts of data produced by high-throughput NMR analysis are a challenge to interpret. For some workflows, it is possible to automatically analyze the data (e.g., fragment based screening) but for others this proves to be a challenge. Certainly, anything that aims to use the rich structural information from NMR spectra will rely on either human interpretation or the use of modern software tools to assist in the process. ASV shows promise and has been implemented in some places[20] but still has considerable limitations compared with trained NMR spectroscopists.

6.8 SUMMARY

'High-throughput NMR' is a relative term and is hard to define. Achieving a fast throughput is the result of optimizing all of the steps in the process from sample preparation through to information generation. While extremely high throughput NMR is possible (hundreds of samples per day), this is often at the expense of data quality so a clear view of the desired outcome is needed before trying to implement such a system.

REFERENCES

1. S.J. Lane, D.S. Eggleston, K.A. Brinded, J.C. Hollerton, N.L. Taylor, S.A. Readshaw, *Drug Discov. Today*, 2006, **11**, 267.

2. J. E. MacNair, K. C. Lewis, and J. W. Jorgenson, *Anal. Chem.*, 1997, **69**, 983.

3. S. B. Shuker, P. J. Hadjuk, R. P. Meadows, and S. W. Fesik, *Science*, 1996, **5292**, 1531.

4. P. A. Keifer, S. H. Smallcombe, E. H. Williams, K. E. Salomon, G. Mendez, J. Belletire, and C. D. Moore, *J. Comb. Chem.*, 2000, **2**, 151.

5. R. Subramanian, W. P. Kelley, P. D. Floyd, Z. J. Tan, A. G. Webb, and J. V. Sweedler, *Anal. Chem.*, 1999, **71**, 5335.

6. R. A. Kautz, W. K. Goetzinger, and B. L. Karger, *J. Comb. Chem.*, 2005, **7**, 14.

7. G. Zheng and W. S. Price, *Prog. Nucl. Magn. Reson. Spectrosc.*, 2010, **56**, 267.

8. G. F. Pauli, B. U. Jaki, and D. C. Lankin, *J. Nat. Prod.*, 2005, **68**, 133.

9. H. de Brouwer, *J. Magn. Reson.*, 2009, **201**, 230.

10. K. Krishnamurthy, *Magn. Reson. Chem.*, 2013, **51**, 821.

11. R. D. Farrant, J. C. Hollerton, S. M. Lynn, S. Provera, P. J. Sidebottom, and R. J. Upton, *Magn. Reson. Chem.*, 2010, **48**, 753.

12. S. S. Golotvin, E. Vodopianov, R. Pol, A. J. Williams, R. D. Rutkowske, and T. D. Spitzer, *Magn. Reson. Chem.*, 2007, **45**, 803.

13. R. J. Ogg, P. B. Kingsley, and J. S. Taylor, *J. Magn. Reson. B*, 1994, **104**, 1.

14. M. Mayer and B. Mayer, *Angew. Chem. Int. Ed.*, 1999, **38**, 1784.

15. C. Dalvit, P. Pevarello, M. Tato, M. Veronesi, A. Vulpetti, and M. Sundstrom, *J. Biomol. NMR*, 2000, **18**, 65.

16. P. Sledz. C. Abell, A. Ciulli, NMR of Biomolecules: Towards Mechanistic Systems Biology, Chapter 15: Ligand-observed NMR in fragment-based approaches (2012) (ISBN: 9783527328505).

17. J. A. Norcross, C. T. Milling, D. L. Olson, D. Xu, A. Audrieth, R. Albrecht, K. Ruan, J. Likos, C. Jones, and T. L. Peck, *Anal. Chem.*, 2010, **82**, 7227.

18. Z. Zhang, P. E. Smith, and L. Frydman, *J. Chem. Phys.*, 2014, **141**, 194201.

19. R. W. Adams, J. A. Aguilar, K. D. Atkinson, M. J. Cowley, P. I. Elliott, S. B. Duckett, G. G. Green, I. G. Khazal, J. Lopez-Serrano, and D. C. Williamson, *Science*, 2009, **323**, 1708.

20. P. Keyes, M. Messinger, and G. Hernandez, *Am. Lab.*, 2012, **44**, 26.

Chapter 7

Multivariate Data Analysis Methods for NMR-based Metabolic Phenotyping in Pharmaceutical and Clinical Research

Kirill A. Veselkov, James S. McKenzie, and Jeremy K. Nicholson

Computational and Systems Medicine, Department of Surgery and Cancer, Faculty of Medicine, Imperial College London, London SW7 2AZ, UK

7.1 INTRODUCTION

NMR spectroscopy is an exceptionally powerful technique for the analysis of metabolites and other small molecules. It is a diverse analytical technique, in as much as it can probe a range of nuclides commonly found in metabolites (e.g., ^{1}H, ^{13}C, ^{31}P, and ^{15}N). While a number of different NMR experiments (e.g., two-dimensional homonuclear and heteronuclear,

diffusion ordered, solid state) have been shown to enhance metabolite analysis, it is ^{1}H NMR (with the greater sensitivity of the ^{1}H nuclide) that is most ubiquitous in the field of pharmaceutical and clinical research.[1,2]

Two-dimensional NMR spectroscopy can be performed either as homonuclear or heteronuclear correlation experiments. COSY is a ^{1}H–^{1}H homonuclear technique that provides structural information regarding proton connectivity and the enhanced signal dispersion across two dimensions, and information content make it a useful tool for the structural elucidation of unknown metabolites. Heteronuclear correlation NMR experiments, such as heteronuclear single quantum coherence (HSQC) spectroscopy pairs the ^{1}H and ^{13}C nuclides via direct one-bond coupling. As with COSY, this provides structural information and reduces spectral overlap. The relatively poor sensitivity of ^{13}C (and other nuclides, e.g., ^{15}N) is overcome through the transfer of magnetization between it and its neighboring ^{1}H nuclide. The acquisition of 2D spectra is more time consuming than the standard 1D ^{1}H NMR experiment, and 2D spectra are not routinely collected for each sample. Instead, representative specimen spectra are typically used as part of structural assignment to complement a ^{1}H NMR analysis.[1]

High (upward of 600 MHz)-frequency ^{1}H NMR allows for the detection of micromolar concentrations

NMR in Pharmaceutical Sciences. Edited by Jeremy R. Everett, John C. Lindon, Ian D. Wilson, and Robin K. Harris
© 2015 John Wiley & Sons, Ltd. ISBN: 978-1-118-66025-6
Also published in eMagRes (online edition)
DOI: 10.1002/9780470034590.emrstm1407

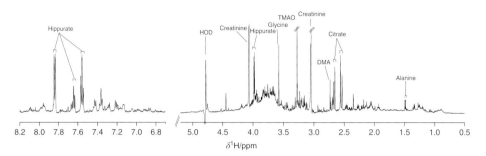

Figure 7.1.   A ^{1}H NMR spectrum of urine acquired using high-field 600 MHz NMR spectroscopy

of proton-containing metabolites, whose chemical signals are typically highly reproducible and unique. The chemical signature of a complex biological mixture ('metabolic phenotype'), such as that obtained from analysis of biofluids, consists of overlapping signals of hundreds to thousands of distinct chemical entities influenced by genes, drug treatment, gut microbiota activity, or other environmental factors (Figure 7.1). Comparison of samples allows for the identification of chemically distinct observations, and the chemicals responsible for such sample-to-sample differentiation (i.e., biomarkers) can be subsequently identified and may be used for phenotypic prediction, e.g., the presence or indeed severity of a disease. The assignment of spectral features and observed interrelationships between molecules can provide a greater understanding of underlying metabolic pathways.[3,4] This so-called systems biology approach can be used to place temporal changes in the metabolome (i.e., a complete set of small-molecule chemicals in the biological sample) in the context of protein and gene expression. Metabolic phenotyping experiments are typically carried out using the rapid and reproducible ^{1}H NMR, and the high-throughput nature of modern instrumentation is capable of handling the large quantity of samples potentially available to researchers.

In comparison to other widely used analytical techniques, the sample preparation required for liquid-state NMR samples is typically minimal. The advent of high-resolution MAS NMR allows solid samples to be analyzed intact. MAS NMR is widely applied in clinical settings, notably for excised samples following (or indeed during) surgical intervention.[5]

Metabolic phenotyping employing NMR is applied across a range of pharmaceutical applications, such as for disease diagnoses/prognoses, and personalized

healthcare studies gauging the effect of xenobiotic molecules (e.g., drug candidates) on metabolic expression levels.[2] A metabonomics study may involve the collection of dozens or hundreds of individual biological samples and the acquisition of NMR or liquid chromatography–mass spectrometry (LC-MS) data on each of them. Regardless of the specific application, each analysis needs to be complemented with suitable spectral preprocessing and multivariate statistical methods. The former is necessary to correct a range of instrumental and experimental heterogeneities, and the latter to extract pertinent information regarding sample-to-sample differences.[3,4] Both of these topics are elaborated later.

7.2   RAW ANALYTICAL SIGNAL PROCESSING

NMR-based metabolic phenotyping is generally highly reproducible and robust. However, the recovery of biological information is complicated by instrumental imperfections: noise, lineshape variations, and phase and baseline distortions all contribute to quantitative errors. Thus, NMR data sets often require the application of preprocessing before statistical interrogation.[6,7]

Spectroscopic noise can be reduced by the application of weighting functions to the FID. For example, exponential window functions enhance the signal-to-noise ratio (SNR), but at the expense of spectral line broadening, which may adversely affect the observation of smaller peaks in complex mixtures such as biofluids. Alternatively, more sophisticated wavelet-based approaches have been proposed.[8] Without suitable identification and correction, an imperfect spectral baseline is likely to adversely affect subsequent multivariate analyses. Systematic trends, where present, need to be removed in an unbiased manner; however, hypercorrection may adversely affect low intensity signals and significant validation of methods should be performed to ensure that additional bias ('overfitting') is not introduced. A suitable ^{1}H NMR baseline has been identified as one having no trend toward nonzero intensity values in noise-only regions. Typical (nonmanual) correction methods involve fitting polynomial curves, and numerous such examples, generally applicable to any technique, have been published in the literature.[9]

Additional challenges in the analysis of biochemical data sets can arise owing to the natural, variable dilution of samples, uncontrolled variation in peak position across spectra (also referred to as positional noise), and heteroscedastic noise structure.[10,11] These effects induce unwanted systematic variation, which can compromise results and interpretation of statistical techniques, e.g., statistical total correlation spectroscopy (STOCSY), principal component analysis (PCA), and partial least squares (PLS).[12,13] For example, without adequate correction, the various treatment groups may potentially be discriminated on the basis of irrelevant variation in peak positions[14] rather than from true changes in an organism's metabolic composition. Correlation between metabolites can also arise because of confounding variable dilution of samples rather than due to specific metabolic changes of interest such as biologically relevant pathway activity. Such difficulties are often observed in ^{1}H NMR spectroscopic studies of urine, which varies in pH, ionic strength, metal ion concentrations, and osmolality, and also has a large dynamic range of multiply overlapped species. Correcting variation in peak position and variable dilution across sample profiles is therefore necessary for improved information recovery.[15,16]

Sample dilution induces a uniform change in concentrations of all molecules in a sample, scaling all corresponding spectral intensities by the same factor. Changes induced by internal or external perturbations, in contrast, result in relatively few overall concentration changes typically across a small subset of molecules. This is due to the modular-like architecture of metabolic pathways, which are organized into a group of spatially isolated or chemically specific biological components that are tailored for a specific biological function. From a computational point of view, metabolite peaks affected purely by a dilution

will exhibit the same fold changes between two spectra, and this is used in the median fold change normalization method. Widely applied alternatives involve either division of a spectrum by its creatinine peak area (creatinine area normalization) or by its area over all peaks (total area normalization). The former method is usually used in a conventional clinical chemistry setting; however, these methods are not generally robust as their assumption that the concentration (of either creatinine or of all metabolites) exhibits little or no change across samples is not generally fulfilled in metabonomic studies.[17] Nonlinear normalization methods (e.g., quantile) that use different scaling factors across signal intensity range have also been shown to be nonrobust for metabolic datasets.[18]

Various analytical strategies have been developed in order to correct for variations in peak position across multiple spectra over high-throughput technologies such as NMR, chromatography, and MS.[10] The general aim of these alignment methods is to group signals attributable to the same chemical species. For example, while the measured $-CH_2-$ shift of ethanol is likely to occur at marginally distinct ppm values across multiple observations, alignment is needed to match like resonances. While in simple spectra this is a trivial matter, it becomes nontrivial for spectra of complex mixtures involving many overlapping resonances. The original correction method applied was 'binning', in which small segments, or bins, of a fixed spectral width (classically $\delta = 0.04$ ppm), or variable size, are defined and integrated. This approach is relatively coarse, and while it can deliver good classification information, this is often at the expense of spectral resolution for biomarker recovery. The reduction from, for example, 65 536 (a full data point approach for an NMR spectrum of '64 K' points) to just 256 variables via binning leads to difficulties in the analysis of small peaks and interpretation of statistical models. In order to overcome such limitations, spectral profiles can instead be aligned at high resolution. Peak positions of sample profiles are generally corrected with respect to reference profile(s), although some algorithms make use of a model peak.[10] The most widely applied peak alignment methods include recursive segment-wise peak alignment (RSPA)[16] and correlation optimized warping (COW).[19] The advantages and limitations of peak alignment algorithms have been recently reviewed in the context of metabolic phenotyping studies.[10]

In addition, NMR peak intensities of biological samples are subject to noise from various sources

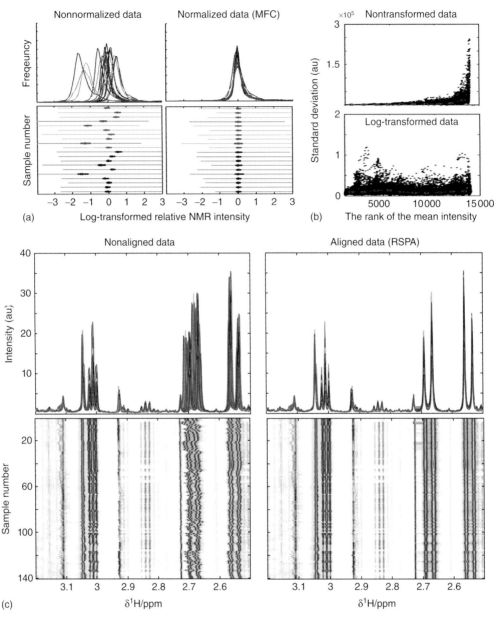

Figure 7.2. Performance diagnostics for various preprocessing algorithms. (a) Box plot summary statistics of sample peak intensity distributions of 20 ^{1}H NMR urine profiles from male (*red*) and female (*black*) Wistar rats via box plots. After successful normalization, the log ratios of nondifferentially abundant metabolite peaks should have a relatively small spread around zero across samples. (b) Standard deviation as a function of the rank of mean signal intensity of a typical ^{1}H NMR urine profile when subject to repeated measurements. The effect of log-based variance stabilizing transformation is seen by comparison of the two plots. After successful application of variance stabilizing transformation, the running median of the standard deviation should approximate a horizontal line with minor fluctuations but no overall trend. (c) Nonaligned ^{1}H NMR urine profiles (left) and profiles processed by recursive segment-wise peak alignment (RSPA, right), where lower plots represent peak intensity positions. The successful peak alignment should transform signals so that peaks arising from the same metabolite have the same spectral position across samples

that typically manifests as heteroscedastic noise structure.[11,20,21] Consequently, metabolites with higher peak intensities are progressively more variable when measured repeatedly. This violation of constant variance across the measurement range imposes a serious challenge when standard statistical techniques are applied.[18] Several variable scaling methods (e.g., unit-variance or Pareto) are usually applied to reduce the influence of the increased intensity variability between large and small peaks. More recently, the application of approximate log-based variance stabilization transformations has been validated to stabilize the technical variance across the intensity range. It has been demonstrated that appropriate variance stabilizing transformation can greatly benefit metabolic information recovery from NMR biological datasets when widely applied chemometrics methods are used.[11]

High spectral overlap in ¹H NMR spectra of complex mixtures essentially precludes the extraction of peak areas for individual analytes. However, several approaches have been recently developed for (semi-)automated metabolite deconvolution and quantification from NMR spectra via spectral templates of pure compounds. These approaches include the proprietary Chenomx NMR deconvolution suite (http://www.chenomx.com) and the freely available Bayesian (BATMAN) deconvolution method.[22]

Generally, the choice of NMR preprocessing algorithms should be accompanied by adequate qualitative or quantitative diagnostics. For example, the performance of the normalization methods can be diagnosed by use of summary statistics of sample peak intensity distributions via box plots (Figure 7.2a). If it is assumed that up to a quarter of peak intensities are asymmetrically increased or decreased owing to biological effects, the log ratios of nondifferentially abundant metabolite peaks should have a comparably small spread around zero across sample and median zero. The successful performance of variance stabilizing transformation can be assessed by variance/mean diagnostics plots, which show the standard deviation as a function of the rank of the mean of peak intensity for replicate measurements (Figure 7.2b). In the absence of heteroscedastic noise structure, the moving median of the standard deviation should approximate a horizontal line with minor fluctuations but no overall trend. The success of any peak alignment algorithm can be assessed by use of peak position maps

to ensure that all peaks are matched between samples (Figure 7.2c).

7.3 EXPLORATORY ANALYSIS OF ¹H NMR-BASED METABOLIC PHENOTYPES

Once data have been adequately preprocessed, they can be analyzed to address specific biological or clinical questions. The applicability of specific computational methods is critically dependent on biological objectives. Furthermore, the structure and complexity of ¹H NMR data sets across studies must also be considered.

Metabolic profiling studies result in an *m* by *n* data matrix of *m* observations and *n* variables. The influence of myriad intrinsic and extrinsic factors on the functional integrity of an organism, as well as intramolecular ¹H nuclide interactions, results in complex interrelationships between both spectral observations and variables. As a result of such relationships between NMR spectral variables, statistical analyses require a multivariate perspective, as univariate analyses act independently of all other variables. A variety of multivariate statistical and machine learning methods have been developed and applied to the analysis of ¹H NMR datasets.[2–4,23] Correlation analysis has been widely used to explore correlations (or inferred networks) among metabolic variables[3,14] with a number of excellent reviews addressing the issue of the origin of metabolic correlations.[24,25] This has provided important information on metabolite regulation in health and disease, as pathological and physiological processes alter the typically observed behavior in metabolic networks.

Peaks attributable to the same molecule are stoichiometrically linked and exhibit strong correlations across multiple samples; this has resulted in the development of a new approach termed *STOCSY*. This approach is used to identify multiple NMR peaks deriving from the same molecule in a complex mixture and, hence, enhances molecular identification (Figure 7.3).[14,26] Many hours of experimental time and resources are required to extract similar information using conventional two-dimensional NMR experiments such as total correlation spectroscopy (TOCSY) and COSY. Various extensions of STOCSY have been introduced and reviewed for improved

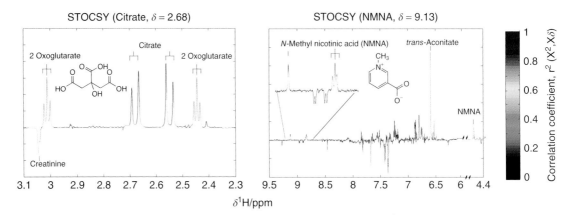

Figure 7.3.   1D STOCSY analysis to identify intensity variables correlated to the chemical shifts $\delta = 2.68$ (the high-frequency citrate doublet resonance) and $\delta = 9.13$. Upper (lower) sections of the plot indicate variables positively (negatively) covarying with a variable at defined chemical shift with color code proportional to the strength of statistical relationships between variables

biomarker identification and recovery of biologically related pathway connectivities between molecules.[27]

A number of multivariate approaches such as matrix factorization and (bi-)clustering algorithms have also been developed and applied for exploratory analyses of large data sets.[13] Matrix factorization approaches take advantage of inter-related spectral variables, i.e., changes in metabolite concentrations, in order to extract a small number of latent components or factors that approximately summarize the measured data with minimal information loss. This usually provides a more efficient representation of the relationship between variables, ignoring components largely attributable to noise. These approaches include PCA, independent component analysis (ICA),[28] and nonnegative matrix factorization (NMF).[29] All involve the linear decomposition of a data matrix into factor scores and loadings, yet differ due to the mathematical constraints imposed on the resultant factors. NMF enforces a nonnegative constraint on factors, whereas orthogonality or statistical independence is respectively enforced on components in PCA and ICA. Factor scores in these methods effectively summarize relationships between observations, which may be related to, for example, physiological traits, gender, and dose- or time-dependent changes. A subset of metabolites related to these factor scores can be identified using the corresponding factor loadings. Of these methods, PCA is routinely applied for ^{1}H NMR metabolic studies, while relatively few applications to date have involved NMF and ICA.

The applications of NMF and PCA are exemplified for the evaluation of related changes in metabolic outcomes, obtained from ^{1}H NMR spectra of rat urine samples collected 2 days after hydrazine exposure (Figure 7.4). The scores of the first factor in NMF are indicative of dose-dependent phenotypic changes due to hydrazine, which are clearly distinguished from the control specimens. The subset of metabolites related to the changes in the factor scores is represented in the first factor loadings. Certain metabolites, e.g., taurine, 2-aminoadipate, and N-acetyl-citrulline, increase in concentration in response to hydrazine toxicity. Such multiple related changes in metabolite concentrations in biofluids usually signal the disruption of the status quo in metabolic pathway activity.

The control samples are separated from the hydrazine-induced effects in the second factor scores, but not in a dose-dependent manner, as there is no distinction between the low- and high-dose cases for this factor. The loadings for this factor indicate the metabolites, e.g., citrate, 2-oxoglutarate, *trans*-aconitate, and hippurate, with higher concentrations in biofluid profiles of control animals. It can be seen in Figure 7.4 that factor scores and loadings are both nonnegative in the NMF method.

In PCA, the first principal component (PC) mainly summarizes changes in the biofluid composition caused by hydrazine toxicity, accounting for the single largest amount of variation in the data set (72%). Because of a lack of nonnegative constraints, the first PC loadings include positive and negative values, incorporating information of positive and negative covariance

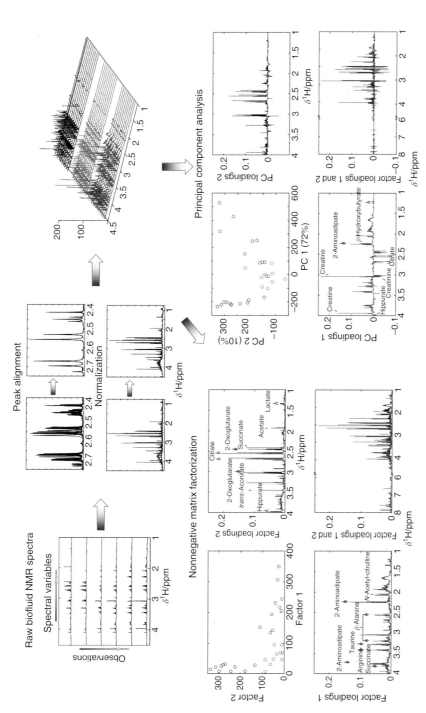

Figure 7.4. Evaluation of the interrelated changes in metabolic phenotypes, obtained from ¹H NMR spectra of rat urine 48 h posttreatment, in response to hydrazine toxicity. Factor scores summarize relationships between phenotypic observations related to dose-dependent hydrazine-induced changes in metabolic phenotypes. A subset of metabolites related to these factor scores can be identified using the corresponding factor loadings. Control, low-, and high-dose groups are color-coded

of the metabolites with the hydrazine-induced effects. This suggests that the metabolites – taurine, creatinine, and 2-aminoadipate – increase in concentration, while concentrations of 2-oxoglutarate, succinate, and citrate decrease in response to hydrazine toxicity (Figure 7.4, PC loadings). These results are somewhat similar to NMF. However, the dose-related phenotypic changes in the PC scores are less clear. This is because the metabolites not responding in a dose-dependent manner, as evaluated in the second NMF factor (e.g., citrate and succinate), contribute to the first PC. These examples illustrate that changes in metabolic phenotypes can be effectively represented using a few underlying factors (components) and that the application of several methods is of particular benefit to understand and cross-validate the results.

The components (or factors) of the matrix factorization methods discussed earlier are derived as a linear superposition of all spectral (original) variables with typically nonzero factor loadings. Difficulties often arise in the interpretation of factor scores, leading to the contribution of a large number of noninformative variables in a statistical model. Moreover, disruption of an organism's functional activity usually results in changes to other metabolites owing to the complex nature of metabolic networks. Hence, additional sparse constraints imposed on factor loadings that minimize the influence of uninformative (noisy or unrelated) variables have been shown to be of particular benefit in facilitating the interpretation of statistical models. This leads to exploring a subset of highly related variables across a subset of sample observations – similar to the objective of biclustering approaches.[30]

Cluster analyses typically calculate the pair-wise distances between variables or observations.[31] There are various methods available to calculate the distance matrix (e.g., Euclidean, Chebyshev, Mahalanobis), and provided that the choice of metric is suitable, objects that cluster closer together exhibit similar behavior compared to more distant clusters. Conventional clustering approaches such as *k*-means, self-organizing maps, and hierarchical clustering involve grouping biological observations or changes in metabolite concentrations using, respectively, all peaks in biofluid profiles or metabolite changes across a complete set of observations.[30,31] An exciting application of hierarchical clustering to ^{1}H NMR spectra of urine samples has demonstrated the grouping of human metabolic phenotypes according to diverse populations.[32] Such change in metabolic levels across populations is affected by a variety of genetic, environmental, and lifestyle factors.

Pathological processes can, however, induce specific and related metabolic changes only in a subset of molecules across a subset of samples. For example, specific and related metabolic marker changes effectively characterize the dose-related hydrazine effects. This information can be lost when grouping is performed on the basis of an overall metabolic profile or changes in metabolite concentrations across a complete data set.[30] Biclustering approaches have been recently introduced to overcome these limitations by identifying a subset of highly correlated variables (changes in metabolite concentrations) across a subset of observations (^{1}H NMR spectra of biofluid samples). These methods are of great potential for characterizing large and heterogeneous biofluid data sets, although all clustering approaches face a general problem of result validation.

7.4 PREDICTIVE ANALYSIS OF ^{1}H NMR-BASED METABOLIC PHENOTYPES

A critical objective of NMR-based metabonomic studies is the establishment of predictive (e.g., mathematical, rule-based) models capable of relating changes in biofluid NMR spectra to single or multiple biological response variables, e.g., the presence or severity of drug toxicity.[2,13] These models are derived empirically using a defined subset of 'training' data, which incorporates information on both explanatory (spectral) variables and associated response variable(s). A complementary test subset is used to monitor and assess the predictive accuracy of the model to demonstrate its capability to generalize to unseen samples. Once a model has been established, its predictive capacity and generality needs to be validated with respect to another independent set of observations. The prediction of continuous response variable(s), e.g., disease severity, is referred to as regression, while the term *classification* is used for categorical predictor variable(s). Numerous approaches have been developed and routinely applied for both regression and classification.[33]

The choice of method(s) usually depends on the following considerations: (i) predictive accuracy, (ii) model interpretability, (iii) robustness to noise and missing values, and (iv) computational efficiency and scalability to large data sets.[33] The optimal predictive accuracy of models is typically overlooked in favor of the interpretative capacity.[13] This refers to the recovery of a subset of spectral variables (biomarkers) that are strongly associated with specific biological variance or class, in addition to interclass similarities or differences (or fuzziness) between metabolic phenotypes. Pertinent biomarkers provide insights into the biological function of an organism in terms of health and disease and are of potential diagnostic and prognostic values.

Supervised linear multivariate methods such as PLS have powerful interpretative capabilities; consequently, they are widely applied. PLS involves linear decomposition of a data matrix into orthogonal components derived to maximize the covariance between descriptor (spectral) and associated response variables. The PLS components consist of scores and loadings and are interpretable in a similar way to PCA components. The PLS algorithm has been recently combined with orthogonal signal correction (the combination being called *O-PLS* (*orthogonal projection to latent structures*)[34]), which removes information from descriptor variables that are not correlated to response, e.g., baseline distortions for spectroscopic data sets. This reduces the number of predictive components, allowing optimal linear predictions that substantially improve the interpretability of PLS models.[13] O-PLS is currently the most widely used method for biomarker recovery in metabolic phenotyping studies. O-PLS discriminant analysis is exemplified in Figure 7.5 for the evaluation of spectral patterns of metabolites that are strongly associated with hydrazine toxicity. The O-PLS loadings represent urine metabolites positively or negatively covarying with the hydrazine class variation, and the color coding relates the metabolite correlations with the discrimination of classes. The O-PLS scores show the relationship between sample observations: a complete discrimination of samples of hydrazine-treated and control animals (Figure 7.5).

In the case of discriminant analysis, it has been recently shown that PLS-based discriminant components are derived by the maximization of between-class variance, i.e., the difference between class means (Table 7.1).[35] A more mathematically eloquent mode of discriminant analysis is to maximize the difference between class means, while simultaneously minimizing within-class variability. This is the objective of linear discriminant analysis (LDA), which maximizes the ratio of between- versus within-class variance. LDA cannot be directly applied in circumstances

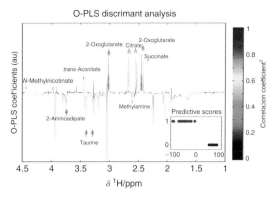

Figure 7.5. Supervised analysis of metabolic phenotypes obtained from 600 MHz ^{1}H spectra of Sprague–Dawley rat urine collected 48 h following hydrazine exposure. The main figure shows the O-PLS-DA coefficients over the low frequency region of the spectrum. They are colored according to their correlation with the predictive scores. These scores are shown (inset) highlighting the successful discrimination of hydrazine-dosed (red) from control (blue) samples

where the number of variables exceeds the number of samples, as is the case with the NMR datasets. PCA has been commonly applied as a preprocessing step before LDA (PCA-LDA) to mitigate this problem. However, a problem arises here with respect to selection of the optimal number of components. The use of a modified recursive maximum margin criterion (RMMC, Table 7.1) has been recently proposed to improve class-specific recovery of molecular ion patterns, while simultaneously avoiding arbitrary selection of the number of PCs before discriminant analysis.

If optimal predictive accuracy (of drug toxicity, for example) is required, the above-mentioned linear methods have been extended to nonlinear cases using 'kernel' methods (e.g., kernel-based PLS, or kernel-based MMC) and shown to increase predictive accuracy. In brief, the central idea is to reformulate algorithms using only pair-wise products ('similarities') between spectra. Kernel-based support vector machines (SVM) and artificial neural networks (ANN) have been also successfully applied for classification.[33] In addition, the classification method termed *classification of unknowns by density superposition* (*CLOUDS*)[36] has been developed to address specific challenges of kidney and liver toxicity predictions in preclinical drug toxicity screening. It is a probabilistic classification method with the advantage of providing information about the fuzziness

Table 7.1. The relationships of widely applied supervised and unsupervised dimensionality reduction techniques in metabolic phenotyping studies via eigenvalue decomposition

Method	Eigenvalue decomposition	Components derivation
PCA	$[S_t]w = \lambda w$	Maximizes overall dataset variation (S_t), without specific regard to between-class differences
PLS	$[S_b]w = \lambda w$	Maximizes between-class variation (difference in class means S_b), disregarding within-class variability
LDA	$[S_w^{-1}S_b]w = \lambda w$	Maximizes ratio of between- versus within-class variation (S_w), with the condition that within-class variance is nonsingular (i.e., number of samples $\gg$ number of variables)
RMMC	$[S_b - S_w]w = \lambda w$	Maximizes the difference of between-class and within-class variation

and uniqueness of classes and also classification quality. CLOUDS has been evaluated with respect to hepatotoxicity and nephrotoxicity prediction, using NMR spectroscopy of urine samples ($n = 12\,935$) from laboratory rats subjected to a total of 80 toxin treatments. The proportion of correctly classified animals in the disease states was, respectively, 67% and 41% for hepatotoxicity and nephrotoxicity with corresponding sensitivity (the proportion of correctly classified healthy animals) of 77% and 100%. This currently constitutes the largest validation of metabolic phenotyping in preclinical drug toxicity screening.

7.5 TIME-COURSE ANALYSIS OF NMR-BASED METABOLIC PHENOTYPES

Biological systems dynamically respond and adapt to environmental changes in order to maintain functional activity. The analysis of time-resolved metabolic data has, therefore, proven to be useful in deciphering the dynamic behavior of an organism in response to stimuli, including complex regulation mechanisms, development of dysfunction, and the capacity to

adapt or recover.[37] Depending on the objective, the time-series methods can be classified into three major categories: (i) explorative trajectory analysis, (ii) differential trajectory analysis, and (iii) predictive modeling from time-series datasets. Method selection is largely dependent on the experimental setup and available datasets, including (uniform or nonuniform) type of sampling rates, missing values, number of sampling points, and, group sizes. The number of sampling points in NMR time-series metabolic studies is usually limited (typically less than 10).

Explorative trajectory or trajectory cluster analysis	Aims to identify metabolic patterns exhibiting statistically significant time-related changes and to group similar metabolic time-related patterns together
Differential trajectory analysis	Its purpose is the identification of differential trajectories of metabolic patterns in relation to outcome variable(s), e.g., good or bad responders. It can be combined with trajectory cluster analysis to identify groups of diagnostic or prognostic markers exhibiting similar time-related behavior
Prediction (classification) from time-series metabolic datasets	Its goal is to improve prediction of outcome variable(s) by incorporation of time-related phenotypic measurements

7.5.1 Explorative Trajectory Analysis

PCA-based metabolic trajectory analysis has been successfully applied to characterize interrelated metabolic changes associated with various stages of disease and toxicity development (onset, progression, and recovery).[38] The time-related metabolic effects are represented in terms of changes in PC scores, while PC loadings are used to identify metabolic patterns co-related with those changes. Other matrix factorization tools, e.g., NMF, can be used instead of PCA to derive metabolic trajectories.

The simultaneous influence of several factors (e.g., time progression and disease development) results in various overlapping sources of factorial variation in metabolic phenotypes. These are related to normal metabolic flux, disease (or toxicity) development, and subject-specific behavior. This leads to nonoptimal PCA models as PCs are extracted by maximizing the overall variation: this is not exclusively time related. For example, PCs will reflect difference in subject responses rather than time-related treatment effects if the former accounts for the major variation. Integrated analysis of variance and principal component analysis (ANOVA-PCA or ASCA) has been recently introduced to separate confounding phenotypic variation into additive contributions related to factors varying in accordance with experimental design. As the confounding variation is separated, the transparency and interpretability of PCA models is substantially improved, leading to enhanced information recovery.[39]

PLS- and O-PLS-based batch processing has been described in the literature as a tool for modeling time-related metabolic variation.[13,38,40] Here, (O-)PLS

components are derived by maximization of the metabolic variation that is linearly related to time. The PLS scores represent time-related metabolic effects, while metabolic patterns related with those changes are identified in the PLS loadings. However, this results in suboptimal models, as biological processes usually develop nonlinearly through time. A piece-wise multivariate modeling method based on O-PLS has been proposed for the evaluation of local metabolic changes over time (between consecutive time points) with the advantage of predicting different stages of time-related events in the development of toxicity or disease processes. A variety of clustering approaches has been also introduced to group metabolites exhibiting similar time-related behavior.[41]

7.5.2 Differential Trajectory Analysis

Several methods based on generalized linear and nonlinear mixed effect models have been recently

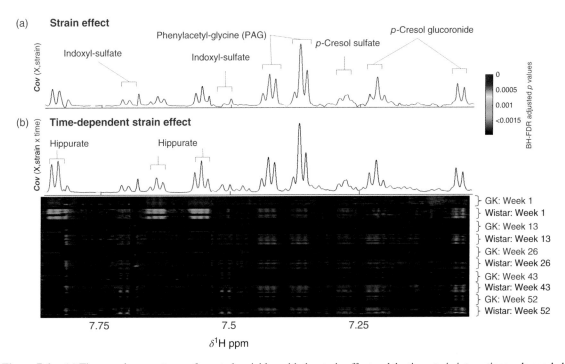

Figure 7.6. (a) The covariance patterns of spectral variables with the strain effect and the time strain interactions color-coded by the related FDR (false discovery rate) adjusted *p* values. (b) Heat map shows metabolite level ratios calculated for each ^{1}H NMR signal intensity value in a given sample relative to the average signal intensity across all GK profiles. The data are colored red for a relative increase in metabolite levels (upregulation) and green for a relative decrease in metabolite levels (downregulation)

developed to test the statistical significance of time-related changes in metabolic phenotypes, derived from ^{1}H NMR spectra of biofluid samples.[42,43] These methods take into account a time-dependent (correlation) structure for a variety of intra- and intersubject variations and assume that time-related changes in metabolic profiles are not necessarily linear. An intuitive example of the application of generalized linear models to recover metabolic signatures associated with risk of developing type-2 diabetes in Goto–Kakizaki rats is shown in Figure 7.6. The heat map represents the difference in metabolite levels between Goto–Kakizaki and Wistar control rats with the red and green colors indicating, respectively, a relative increase and a relative decrease in metabolite levels. It shows that comparatively lower levels of phenylacetyl-glycine, 4-cresolsulfate, or 4-cresol-glucoronide are observed in Wistar control rats consistently over time. This results in statistically significant strain effects and nonsignificant time-dependent strain interactions, i.e., the group trajectories are shifted and parallel to each other. In turn, relatively lower levels of hippurate in Wistar rats are found only at week 1 leading to the statistical significant time-dependent strain interactions, i.e., differential trajectories between animal strains.

7.5.3 Classification from Time-series NMR Datasets

The majority of classification and prediction algorithms developed thus far rely on static metabolic datasets. In many cases, the outcome prediction (e.g., drug-response prediction) can be improved by considering the time-related changes in metabolic phenotypes, as diseases or drug responses (onset, progression, and recovery) develop over time.[2,44] Several approaches based on Hidden Markov Models and dynamic system kernel SVM have been recently introduced for classification from time-series high-throughput biological datasets, which are yet to be applied to NMR time-series datasets. From these, an interesting approach proposed by Borgwardt *et al.*[45] involves building an individual model for each patient of metabolic time-related changes using linear dynamic systems. This model assumes that a few latent trajectories drive the patient dynamics. A variety of model-based metrics can be used to derive distances between patients. Once calculated, a variety of distance-based classifiers (including SVM

and *k*-nearest neighbors) can be applied to predict outcomes.

7.6 CONCLUSIONS

We have outlined various chemoinformatics strategies that maximize disease- and pharmacologically relevant molecular information recovery from ^{1}H NMR spectra of biological samples. In broad terms, these strategies involve (i) raw analytical signal preprocessing for improved information recovery, (ii) multivariate statistical explorative and predictive analysis of NMR biological spectra, and (iii) time-course analysis. While various methods have been presented, there is no single best bioinformatics workflow; the interrogation of datasets must be tailored according to the design of the experiment and its objectives. The current NMR data processing pipelines rely on a heterogeneous array of bioinformatics packages that are poorly optimized and only suitable for specialists in the field. The future lies in the creation of integrated bioinformatics solutions that can streamline interrogation of highly complex metabolic datasets to address a variety of biologically and clinical relevant questions in a user-friendly manner.

RELATED ARTICLES IN EMAGRES

Metabonomics: NMR Techniques

Software Tools for NMR Metabolomics

Plant Metabolomics

Recent Progress in Clinical Magnetic Resonance Spectroscopy

NMR Quantitative Analysis of Complex Mixtures

REFERENCES

1. J. Lindon, E. Holmes, and J. Nicholson, *Pharm. Res.*, 2006, **23**, 1075.

2. J. K. Nicholson, J. Connelly, J. C. Lindon and E. Holmes, *Nat. Rev. Drug Discov.*, 2002, **1**, 153.

3. W. Weckwerth and K. Morgenthal, *Drug Discov. Today*, 2005, **10**, 1551.

4. D. S. Wishart, *Methods Mol. Biol.*, 2010, **593**, 283.

5. R. Mirnezami, B. Jiménez, J. V. Li, J. M. Kinross, K. Veselkov, R. D. Goldin, E. Holmes, J. K. Nicholson, and A. Darzi, *Ann. Surg.*, 2014, **259**, 1138.

6. U. L. Gunther, C. Ludwig, and H. Ruterjans, *J. Magn. Reson.*, 2000, **145**, 201.

7. J. C. Hoch and A. Stern, NMR Data Processing, Wiley-Liss: New York, 1996.

8. N. Trbovic, F. Dancea, T. Langer, and N. Günther, *J. Magn. Reson.*, 2005, **173**, 280.

9. H. Witjes, W. J. Melssen, H. J. in't Zandt, M. van der Graaf, A. Heerschap, and L. M. Buydens *J. Magn. Reson.*, 2000, **144**, 35.

10. T. N. Vu and K. Laukens, *Metabolites*, 2013, **3**, 259.

11. H. M. Parsons, C. Ludwig, U. L. Günther, and M. R. Viant, *BMC Bioinform.*, 2007, **8**, 234.

12. O. Cloarec, M. E. Dumar, A. Craig, R. H. Barton, J. Trygg, J. Hubson, C. Blancher, D. Gauguier, J. C. Lindon, E. Holmes, and O. Cloarec, *Anal. Chem.*, 2005, **77**, 1282.

13. J. Trygg, E. Holmes, and T. Lundstedt, *J. Proteome Res.*, 2007, **6**, 469.

14. O. Cloarec, M. E. Dumas, J. Trygg, A. Craig, R. H. Barton, J. C. Lindon, J. Nicholson, and E. Holmes, *Anal. Chem.*, 2005, **77**, 517.

15. J. Forshed, I. Schuppe-Koistinen, and S. P. Jacobsson, *Anal. Chim. Acta*, 2003, **487**, 189.

16. K. A. Veselkov, J. C. Lindon, T. M. Ebbels, D. Crockford, V. V. Volykin, E. Holmes, D. B. Davies, and J. K. Nicholson, *Anal. Chem.*, 2009, **81**, 56.

17. F. Dieterle, A. Ross, G. Schlotterbeck, and H. Senn, *Anal. Chem.*, 2006, **78**, 4281.

18. K. A. Veselkov, L. K. Vingara, P. Masson, S. L. Robinette, E. Want, J. V. Li, R. H. Barton, C. Boursier-Neyret, B. Walther, T. M. Ebbels, I. Pelczer, E. Holmes, J. C. Lindon, and J. K. Nicholson, *Anal. Chem.*, 2011, **83**, 5864.

19. G. Tomasi, F.van den Berg, and C. Andersson, *J. Chemometr.*, 2004, **18**, 231.

20. S. Zhang, C. Zheng, I. R. Lanza, K. Sreekumaran Nair, D. Raftery, and O. Vitek, *Anal. Chem.*, 2009, **81**, 6080.

21. T. K. Karakach, P. D. Wentzell, and J. A. Walter, *Anal. Chim. Acta*, 2009, **636**, 163.

22. J. Hao, M. Liebeke, W. Astle, M. De Iorio, J. G. Bundy, T. M. D. Ebbels, *Nat. Protoc.*, 2014, **9**, 1416.

23. V. Shulaev, *Brief. Bioinform.*, 2006, **7**, 128.

24. D. Camacho, A. de la Fuente, and P. Mendes, *Metabolomics*, 2005, **1**, 53.

25. R. Steuer, *Brief. Bioinform.*, 2006, **7**, 151.

26. E. Holmes, O. Cloarec, and J. K. Nicholson, *J. Proteome Res.*, 2006, **5**, 1313.

27. S. L. Robinette, J. C. Lindon, and J. K. Nicholson, *Anal. Chem.*, 2013, **85**, 5297.

28. C. Pierre, *Signal Process.*, 1994, **36**, 287.

29. D. D. Lee and H. S. Seung, *Nature*, 1999, **401**, 788.

30. S. C. Madeira and A. L. Oliveira, *Comput. Biol. Bioinform.*, 2004, **1**, 24.

31. X. Rui and D.Wunsch II, *IEEE Trans. Neural Netw.*, 2005, **16**, 645.

32. E. Holmes, R. L. Loo, J. Stamler, M. Bictash, I. K. S. Yap, Q. Chan, T. Ebbels, M. De Iorio, I. J. Brown, K. A. Veselkov, M. L. Daviglus, H. Kesteloot, H. Ueshima, L. Zhao, J. K. Nicholson, and P. Elliott, *Nature*, 2008, **453**, 396.

33. S. B. Kotsiantis, In *Proceedings of the 2007 conference on Emerging Artificial Intelligence Applications in Computer Engineering: Real Word AI Systems with Applications in eHealth, HCI, Information Retrieval and Pervasive Technologies*, 2007, 3–24.

34. J. Trygg and S. Wold, *J. Chemometr.*, 2002, **16**, 119.

35. K. A. Veselkov, R. Mirnezami, N. Strittmatter, R. D. Goldin, J. Kinross, A. V. M. Speller, T. Abramov, E. A. Jones, A. Darzi, E. Holmes, J. K. Nicholson, Z. Takats, *Proc. Natl. Acad. Sci. U. S. A.*, 2014, **111**, 1216.

36. T. Ebbels, H. Keun, O. Beckonert, H. Antti, M. Bollard, E. Holmes, J. Lindon, J. Nicholson, *Anal. Chim. Acta*, 2003, **490**, 109.

37. K. A. Veselkov, V. I. Pahomov, J. C. Lindon, V. S. Volynkin, D. Crockford, G. S. Osipenko, D. B. Davies, R. H. Barton, J. -W. Bang, E. Holmes, J. K. Nicholson, *J. Proteome Res.*, 2010, **9**, 3537.

38. J. Azmi, J. L. Griffin, H. Antti, R. F. Shore, E. Johansson, J. K. Nicholson, E. Holmes, *Analyst*, 2002, **127**, 271.

39. A. K. Smilde, J. J Jansen, H. C. J. Hoefsloot, R. -J. A. N. Lamers, J. van der Greef, M. E. Timmerman, *Bioinformatics*, 2005, **21**, 3043.

40. S. Wold, M. Sjostrom, and L. Eriksson, *Chemom. Intell. Lab. Syst.*, 2001, **58**, 109.

41. T. Warren Liao, *Pattern Recogn.*, 2005, **38**, 1857.

42. M. Berk, T. Ebbels, and G. Montana, *Bioinformatics*, 2011, **27**, 1979.

43. Y. Mei, S. B. Kim, and K.-L. Tsui, *Expert Syst. Appl.*, 2009, **36**, 4703.

44. J. K. Nicholson, J. C. Lindon, and E. Holmes, *Xenobiotica*, 1999, **29**, 1181.

45. K. M. Borgwardt, S. V. Vishwanathan, and H. P. Kriegel, *Pac. Symp. Biocomput.*, 2006, 547.

PART C
Idea to Lead

Chapter 8

The Role of NMR in Target Identification and Validation for Pharmaceutical R&D

Krishna Saxena[1,2] and Harald Schwalbe[1,2]

[1]*Institute for Organic Chemistry and Chemical Biology, Center for Biomolecular Magnetic Resonance, Johann Wolfgang Goethe-University Frankfurt, Max-von-Laue-Str. 7, Frankfurt am Main, D-60438, Germany*
[2]*German Cancer Consortium (DKTK), Im Neuenheimer Feld 280, 69120 Heidelberg, Germany*

8.1 INTRODUCTION: WHAT IS DRUG DISCOVERY?

Drug discovery is the process by which potential new candidate medications (lead compounds or biologics) are identified for the treatment of specific diseases. Traditionally, small molecules (chemical compounds) are seen as drugs whereas biologics represent large molecular entities (e.g., monoclonal antibodies, cytokines, tissue growth factors, and therapeutic proteins) produced in living cells. Today, small molecule drugs are also referred to as *New Chemical Entities* (*NCE*) and biomolecular drugs (biologics) as *New Biological Entities* (*NBE*). Two goals of biomedical research have to be achieved by the drug discovery process: (i) to understand the origin and causality of a human disease, its progression and prognosis and (ii) to identify safe and effective therapeutics that can ameliorate the disease state.[1]

Drug discovery can generally be divided into three stages: target identification and validation, lead identification and lead optimization to select a drug candidate (Figure 8.1). After this discovery process, the so-called drug development phase starts with preclinical research (microorganisms and animals), clinical trials (on humans), regulatory approval of the drug to the market and sales, and finally leads from the clinics back to the laboratory by a so-called reverse translation process. In modern drug discovery, translation is required to happen in both directions: Forward translation, or translating from preclinical studies to patient studies, and reverse translation, in which patient data from clinical studies are used to improve the early drug discovery. Whereas the first steps of drug discovery (target identification and validation) are often carried out at universities and nonprofit institutions, the later stages (lead identification and optimization), and the drug development are traditionally conducted by pharmaceutical industry.

NMR in Pharmaceutical Sciences. Edited by Jeremy R. Everett, John C. Lindon, Ian D. Wilson, and Robin K. Harris
© 2015 John Wiley & Sons, Ltd. ISBN: 978-1-118-66025-6
Also published in eMagRes (online edition)
DOI: 10.1002/9780470034590.emrstm1429

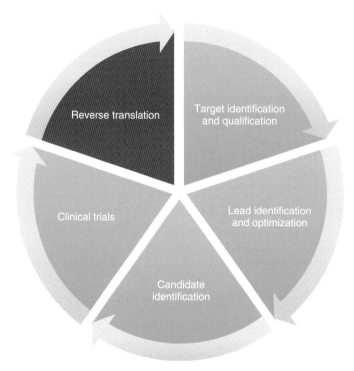

Figure 8.1. After the drug discovery (target identification and qualification, lead identification and optimization, and candidate identification) the drug development starts. The reverse translation process identifies new disease states (e.g., after acquisition of resistance) from clinic back to the laboratory.

The drug discovery process involves a wide range of scientific disciplines, including biology, chemistry, and pharmacology. Despite its multidisciplinary scientific character, drug discovery and development remains a complex very unpredictable process with a high failure rate. In 2010, the research and development cost for a single New Molecular Entity (NME is a drug that contains an active moiety that has never been approved by the FDA (US Food and Drug Administration) or marketed in the United States of America) was around 1.8 billion US dollars.[2] The first goal of modern pharmaceutical research and development today is to identify the biological origin of disease or malfunction, and the potential targets for intervention. In this chapter, we will describe the complex process of target identification and validation and the potential role of NMR.

8.2   DRUG TARGETS

Selecting the right target to work on is the most important decision pharmaceutical companies are facing.[3]

A 'drug target' is a broad term that can be used for any naturally existing cellular or molecular structure (proteins, DNA, and RNA) involved in the pathology of interest which is the underlying motivation for a drug discovery program. A validated drug target needs to be efficacious, specific, safe to target, meet clinical and commercial needs and, above all, be 'druggable'.[4] A druggable target is able to bind its putative drug molecule with high specificity (and affinity) and thereby modulate the activity of the target, leading to a beneficial therapeutic effect in the targeted disease population. Examples for very promising disease-linked targets with a 'druggability gap' are the mutated RAS proteins or transcription factors such as c-Myc. These targets are currently regarded as extremely challenging to be targeted by small-molecule compounds.

However, the phrases 'drug target' or 'validated drug target' are controversial in the pharmaceutical industry. In contrast to the huge amount of data about the multiple in vitro and in vivo approaches to validate a drug target, the precise meaning of a validated drug

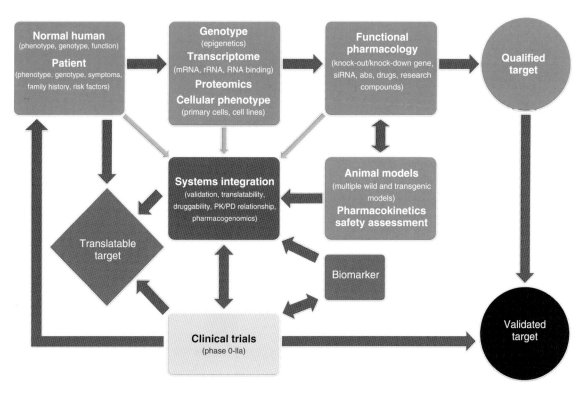

Figure 8.2.   Translatable, qualified and validated target in drug discovery. Different activities are required to progress an identified translatable target to a qualified target, finally leading to a validated target. (Adapted from Ref. 1)

target is unclear. Actually, the term *validated drug target* is often used throughout the whole drug discovery process, that is, it is confused with the activity of qualifying a drug target: an iterative, hierarchical process (Figure 8.2) which builds confidence in the drug target.[1] This semantic definition issue is not a meaningless discussion treated in text books; the consequence of this confusion leads to the erroneous assumption that a drug target could be definitively validated preclinically or in the early stages of clinical trials.

A recent analysis by Andersen *et al.* using the Drug-Bank database identified 435 effect-mediating drug targets in the human genome, which are modulated by 989 unique drugs (FDA-approved) (Figure 8.3), through 2242 drug target interactions.[5]

Antihypertensive drugs, followed by antineoplastic and antiinflammatory agents, represent the most common indication for the drugs in the analysis of Andersen, as shown in Figure 8.3. 193 proteins (44% of the human drug targets) are receptors, and 82 (19%) of these are G protein-coupled receptors (GPCRs). In

the overall data set (Figure 8.3), ~36% of drugs target GPCRs (antihypertensive, antiallergic therapeutic indication) ~29% enzymes (anti-inflammatory, antineoplastic), and ~15% transporter proteins (antihypertensive, antiarrhythmia).[6]

In the past, researchers had a tendency to work on commonly known drug target proteins, often identified in the literature by academic groups, amenable to low-throughput analysis. Thus, a majority of successful drug discovery projects have targeted a rather small number of protein classes that were available to pharmaceutical development. Besides GPCRS, other favored protein classes include ion channels, kinases, and nuclear receptors.

8.3   TARGET IDENTIFICATION AND VALIDATION

Target identification and validation are the first key stages in the drug discovery pipeline. Selecting the

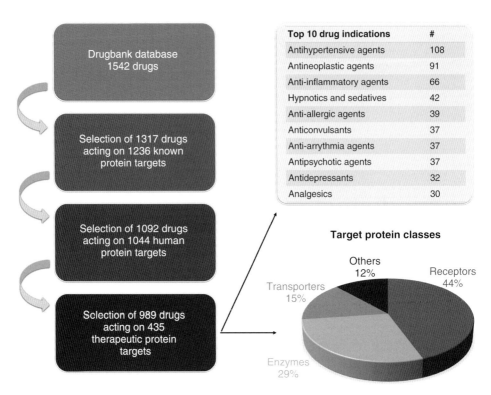

Figure 8.3. Drugs and drug targets (left) and classification of the currently utilized drug targets that are encoded by the human genome and the most common indications. (Adapted from Ref. 6)

right target to work on is of utmost importance as choosing a wrong target wastes considerable resources.[7] However, before we move toward these two processes, we will give a short overview of how drugs were discovered in the past and how they are currently discovered.

8.3.1 Classical Pharmacology, Systems-Centric, Holistic Approach

Historically, therapeutic drugs were discovered by isolation, identification, and derivatization of active ingredients from traditional medicinal plants. Nature was the most important source for drugs, with knowledge of toxic or medicinal properties often long predating knowledge of precise target mechanism.[8] A prominent example for such a case is aspirin: The active ingredient of aspirin was first discovered from the bark of the willow tree in 1763 and was used to cure fever long before modern pharmacology was established. Classical

pharmacology (forward pharmacology) was a process which investigated the effects of substances or natural products in intact cells or whole organisms to identify compounds that show a desired therapeutic effect. As this approach is not based on knowledge of the drug target and/or its signaling pathways and only observes the physiological effects in complex biological systems (like an animal), it is also called the *phenotypic, system-centered,* or *holistic drug discovery approach* (see Figure 8.4, left). A major advantage of this holistic strategy lies in the identification of candidate compounds which work in a more biologically relevant context than approaches dealing with purified proteins. In addition, this approach measures cellular function without imposing preconceived notions of the relevant targets and offers the possibility of discovering new therapeutic targets. Therefore, this strategy was and still remains the basis for the discovery of many highly successful drugs.

However, many beneficial and prominent drugs (Penicillin) have been discovered in an incidental way,

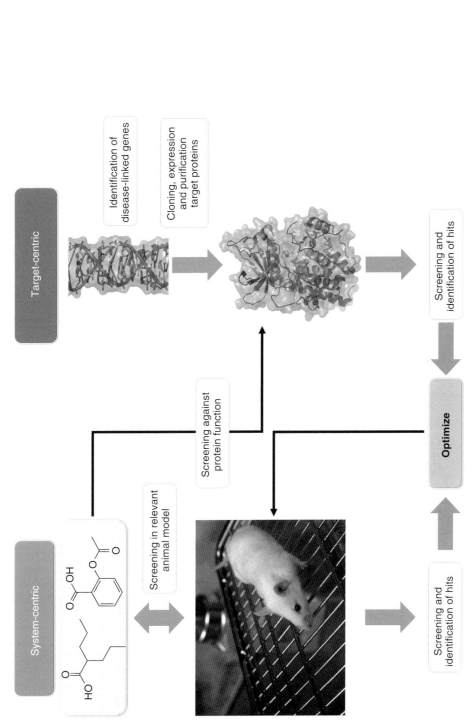

Figure 8.4. Two different concepts for drug discovery. The systems-centric (holistic) approach is shown on the left and the target-centric (reductionist) on the right. (Adapted from Ref. 9. Image of the animal model: Reproduced from File:Lab mouse mg 3158.jpg, 2008, http://commons. wikimedia.org/wiki/File:Lab_mouse_mg_3158.jpg © Rama 2008, distributed under the Creative Commons Attribution-Share Alike 2.0 France license)

mostly during scientific research studies in academic or pharmaceutical laboratories. Such unplanned accidental discoveries are nowadays referred to or summarized as *drug serendipity* and it is hard to define if the discovery was made in the context of a rational drug design or a totally blind approach.

8.3.2 Target (reductionist) Centric Approach, Rationale

Paul Ehrlich (1854–1915) delivered the basis of modern pharmacology by advancing the idea that a chemical substance could selectively affect physiological processes in the human body. Today, it is clear that the specific molecular interactions of a drug with its target link structure to function in such a way that a therapeutically effective and safe response can be provided. According to his concept, called the *magic bullet* (term for an ideal therapeutic agent), pure compounds – instead of crude extracts – were screened for biological activity in different systems.

Significant progress in modern molecular genetics and high-throughput approaches to synthetic chemistry have revolutionized the way that drug discovery is conducted today.[9] Sequencing of the human genome in combination with modern recombinant technology for gene cloning and protein expression enabled the availability of almost any protein for scientific investigations.[9] Therefore, a paradigm shift from the system-centered pharmacology has taken place to a reductionist approach to drug discovery.[9] This reductionist approach, shown in Figure 8.4 on the right, focusses on isolated biological targets based on a rational hypothesis that the specific modulation of a target by an agonist or antagonist in vitro may induce a therapeutic benefit – independent of the requirement for the compound to be useful in vivo.[9] Once a suitable target for the relevant disease has been identified, cloning and heterologous expression of the target protein in suitable hosts enables screening in high throughput assays using large compound, specific protein family tailored or fragment-based libraries (Figure 8.4). In addition, structure-based tools including modern drug discovery methods such as NMR, X-ray crystallography, and computational modeling and screening may be used to aid lead identification and optimization of the target.

8.3.3 Target Discovery and Validation

Before any new drug can be discovered, research has to be performed to identify and understand the underlying cause of the disease or malfunction to be treated. Therefore, it has to be investigated how the genes are altered, how that influences the proteins they encode and how those affected proteins interact with each other leading to a diseased state.

The sequencing of the human genome identified 21 000 protein-encoding genes, and genetics has provided powerful biological insights, allowing characterization of protein function by manipulating specific genes. So, a key task in biomedical research is no longer target identification (as by now, in theory all targets have been identified), but target validation.[10]

However, it is becoming increasingly evident that the complexity of biological systems lies with the currently available scientific tools at the level of the proteins, and that genomics alone will not be sufficient to understand the effects of drugs in the human body. Practically every medicine used today is interacting with its protein targets and either stimulating or inhibiting its function, resulting in a therapeutic effect. Therefore, the analysis of proteins (including protein–protein interactions (PPIs), protein–nucleic acid, and protein–ligand interactions) and their post-translational modifications will be of paramount importance to target discovery.

In the past, the human proteome data appeared too complex and too big to be exploited for drug discovery: How can we find out which proteins are deregulated in the diseased state? Which proteins do we need to modulate to achieve cure?

As proteins are translated from genes, the sequencing of the human genome has made it easier to study genes than the bigger and more complicated protein world. One general approach to identify new drug targets involves comparing the genes, proteins and the transcriptome of healthy individuals with those of patients with the disease. The differences between these maps (gene sequence and gene expression) can help to generate hypotheses as to which genes, protein(s), mutations, or lack thereof cause the diseased state. Typically, proteins that are highly expressed in diseased tissues, but have low expression rates in healthy tissues, will be selected as potential drug targets for therapy.

Target identification and validation is an iterative process, characterized by the use of many different methods. Besides classical methods of cellular and molecular biology, new techniques of target identification are becoming increasingly important. These new applied techniques can be divided into four different technology platforms.

1. Genomics
2. Bioinformatics
3. Transcriptomics
4. Proteomics

The diversity in methods is shown in Figure 8.5.

Besides the publication of the human genome sequence and many infectious disease genomes, cancer genome sequencing identified important and frequently mutated oncogenes for which addiction has been demonstrated and against which drugs were successfully developed.[7] A recent prominent example for such a successful development is the drug Vemurafenib which targets the BRAF V600E mutation in melanoma patients.

After having established that the selected target is important in the biology of the disease, the second important step is to show that modulation of the target will lead to the desired therapeutic effect. This process, called *drug target validation*, generates data demonstrating the relevance of a drug target in a diseased state. Such validation experiments range from in vitro tools through the use of whole animal models, to modulation of a desired target in disease patients.[4] Frequently, target validation is performed by suppressing the expression of the targeted protein(s) by RNAi and/or a suitable chemical compound and/or antibodies. In general, antisense oligonucleotide-mediated effects are reversible, in contrast to gene knockout approaches, but RNAi often shows undesired off-target effects. Excellent target validation can be achieved by the application of monoclonal antibodies as they interact with a larger region of the target molecule, mediating higher affinity and specificity to the target, which results in fewer off-target effects than from small-molecule drugs. On the other hand, the application of antibodies is restricted to protein target classes acting on the cell surface because antibodies are not yet able to cross the cell membrane.[4]

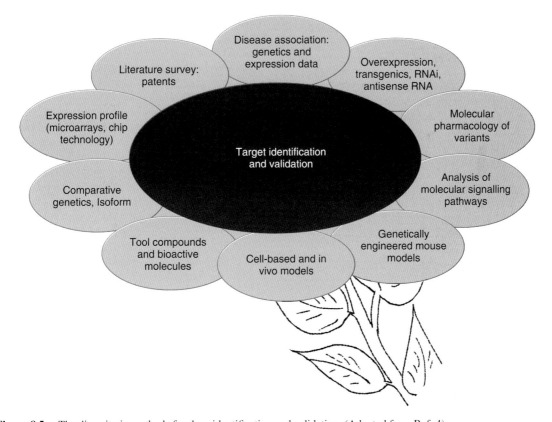

Figure 8.5. The diversity in methods for drug identification and validation. (Adapted from Ref. 4)

Transgenic animals serve as attractive validation tools as these experiments involve whole organisms and allow observation of phenotypic endpoints to clarify the functional consequence of gene manipulation.[4] Unfortunately, the robustness and reproducibility of such validation experiments published in the literature appear to be problematic, so that in-house laboratories of pharmaceutical companies are not able to confirm the literature data. In almost two-thirds of the published projects from literature, there were inconsistencies between published data and in-house data that either considerably prolonged the duration of the target validation process or, in most cases, resulted in termination of the projects.[11] Therefore, literature data on potential drug targets should be viewed with caution, and underline the importance of independent and confirmatory validation studies for pharmaceutical companies and academia before larger investments are made for a costly drug discovery program.

8.4   APPLICATION OF NMR IN TARGET IDENTIFICATION AND VALIDATION

The two different drug discovery approaches – reductionist target-based screening and systemic phenotype screening (Figure 8.4) – have advantages and disadvantages, but today a combination of both approaches is used in drug discovery programs running in pharmaceutical industry.

Independently of any drug discovery approach chosen, identification and understanding of the molecular mechanism of action (MMOA) of the drug with its target has been a key factor contributing to the success of drug programs in pharmaceutical industry.[12] NMR, as a versatile biophysical technique with wide applicability, should be involved in these processes. A major disadvantage of the reductionist approach compared to the phenotypic one is that a drug identified on the basis of a specific molecular hypothesis may not be relevant to the disease pathogenesis. However, the more modern target-based approach generally has the strength of applying molecular and chemical knowledge to investigate specific molecular hypotheses and the ability to apply, as well as small-molecule screening strategies, biologically based approaches, such as identification of monoclonal antibodies.[12] In fact, biomolecular drugs are of utmost importance in pharmaceutical industry (for example, 8 out of the top 20 biopharmaceuticals in 2012 were biomolecular

drugs) but the use of NMR in antibody development in pharmaceutical companies is limited so far. Recently developed NMR methods allow the generation of structural and hydrodynamic profiles of unlabeled antibodies.[13] However, although NMR spectroscopy is widely used as an analytical tool, the application of NMR techniques for larger molecules in the biopharmaceutical industry has been hampered by the perception that these approaches are too complex or insensitive.

Generally, NMR spectroscopy is now a well-established technique to elucidate the structure, interaction, and dynamics of molecules in solution and to guide structure-based lead discovery approaches. The more a particular compound has been designed, optimized, and understood in its binding to the target, the less likely it will bind to other proteins and cause undesired effects. This is rational and indisputable. In most pharmaceutical companies, NMR detection and characterization of molecular interactions between a target and a small compound are considered to be more attractive than solving protein structures by NMR. NMR structure determination is regarded as time-consuming and costly in comparison to X-ray diffraction, with the limitation that NMR is not able to deliver high-throughput structure data of many drug targets. However, this view of NMR being an expensive redundant structural biology technique besides X-ray diffraction has changed in the last years, with the realization that a significant percentage (20–40%) of proteins are only solvable by NMR.[14]

In addition, NMR-based drug interaction assays neither need to be developed in a specific way nor require knowledge of the function of the protein. Therefore, it can be easily analyzed if a compound interacts with the suggested drug target, providing a simple robust approach to validate the chosen drug candidate at any stage of the drug discovery process, and also in the target identification and validation process.

However, what will be the future role and application of NMR in this early drug discovery phase? So far, NMR is beginning to support the target identification and validation phase. Both processes are dominated by a multitude of different approaches and new technologies. The process of target identification cannot generally be solved by a single method but rather by analytical integration of multiple, complementary approaches.[8]

8.4.1   Metabolomics

The key for understanding disease causality for developing successful drugs is the ability to correctly diagnose both the disease and its response to treatment. Recently, NMR has moved beyond its traditional role to a new field called *Metabolomics*. Here, NMR studies biological samples (cell suspension, biofluids such as urine and blood) to monitor noninvasively the effects of candidate drugs on the metabolome. Currently, the main focus of NMR metabolomics is the identification of novel biomarkers on the basis of classification of healthy and diseased biological samples. Biomarkers generally serve as quantitatively measurable reporters for a biological or pathogenic state and therefore play an important role in disease diagnostics, the prediction of disease progression, and/or monitoring of the response to a treatment.[15] Interestingly, biomarkers may also represent drug targets, opening new avenues for NMR to be applied to the target identification process.

Unlike genomics or proteomics samples, metabolomics samples are not static and will change with time and depend strongly on the user operations.[16] Therefore, one major challenge in the field of metabolomics is of sample measurement protocols defining a gold standard for all scientific measurements.[16] This criterion seems to be required for all applications used as routine components in pharmaceutical industry processes. In contrast to toxicity studies, metabolomics application in target identification is in its infancy. Currently, NMR can detect low micromolar analyte concentrations, but is approximately five or more orders of magnitude less sensitive than MS (mass spectrometry), which can detect low to sub-femtomole levels. Together, NMR and MS will synergistically enable comprehensive profiling of the metabolome of cells and tissues.

8.4.2   Difficult Targets Not Accessible by Conventional Drug Discovery

Moreover, in recent years of drug discovery a clear trend has been emerging with the realization that many of the most attractive drug targets are not so easily accessible and have a high complexity, so that conventional drug discovery does not lead to a successful drug. In contrast to the historically investigated drug targets (enzymes or single domains of individual proteins shown in Figure 8.6a), these novel targets (nonenzymes or macromolecular complexes shown in Figure 8.6b) have proven to be challenging in drug discovery as these proteins use PPIs to function as an organizing, regulating signaling complex.[17] Identification of inhibitors of PPI has a huge therapeutic potential, but discovering small molecule drugs that modulate PPI has proven to be a challenging endeavor in drug discovery.

Structural biology appears to be central to successfully target PPI.[18] In a collaboration of Sanofi with the University of Frankfurt, the biophysical mechanism of action of SSR128129E (SSR), the first small molecule multi-FGFR (fibroblast-growth-factor-receptor) inhibitor could be elucidated by NMR.[19] Most drugs targeting the extracellular domains of RTK (receptor tyrosine kinases) are traditionally antibodies, but small chemical compounds, acting extracellularly, which are capable of inhibiting RTK signaling, have not been described so far. SSR allosterically modulates the PPI between FGFR and its hormone FGF. Interestingly, the D3 domain, the binding domain of SSR, exists in a molten globular state in solution. Intrinsically disordered proteins (IDPs) (Figure 8.6b) play crucial roles in many important cellular processes such as signaling or transcription and therefore represent attractive therapeutic targets for several diseases. Given their inherent structural flexibility, X-ray crystallography is not applicable to study these proteins. Moreover, the considerable flexibility of IDPs poses a challenge for drug discovery approaches. NMR spectroscopy offers unique opportunities for structural and dynamic studies of IDPs

Another area of highly complex and frustrating drug discovery, leading to high attrition rates, is the development of cancer drugs. Owing to its high complexity and its high importance, cancer drug discovery is accompanied with the most advanced efforts in patient registration and patient stratification activities.

Most notably, in the cancer drug discovery and development process the one-size-fits-all approach (cytotoxic chemotherapy) changed to a personalized medicine strategy that focusses on molecularly targeted drugs emphasizing the structural biology role (NMR and X-ray) in drug design.[7] Modern drug discovery has to be translational, incorporating early biomedical research findings into later stages of development, and requires permanent

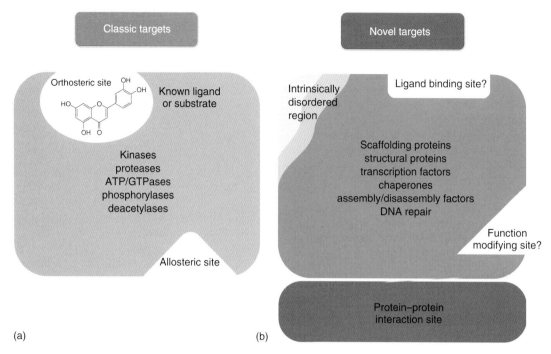

Figure 8.6. (a, b) Classic and novel targets in pharmaceutical drug discovery programs. Whereas classic targets consist of enzymes with well-defined active sites for orthosteric inhibitors, novel targets lack these well-defined binding pockets only offering protein–protein-interaction sites. (Adapted from Ref. 17)

updating of new results in the progression of the development process. This process does not occur in a linear way, as it progresses in both directions, leading to circle of the drug discovery and development process by the reverse translation (Figure 8.1) process.

The presence of drug-acquired resistance mechanism in cancer patients documents the need for close collaboration between the target identification process and its drug-induced change occurring in the patients. Structural elucidation of the binding mechanism of drugs enables the understanding of cellular signaling and is the basis for rational drug design. Even for a prototype molecule of a structure-based drug development program such as Gleevec, important features of the famous drug (imatinib) binding to its target have been recently discovered by NMR, revealing new options against drug resistance.[20]

So far, understanding of the interactions of oncogenic tyrosine kinase (like c-Abl) with small molecule inhibitors is largely based on X-ray diffraction. This methodology presents snapshots of molecules and may not always reflect physiologically relevant conformations and dynamics.[20] Recently, NMR has revealed that various c-Abl complexes with allosteric (targeting the Abl myristoyl pocket) and ATP-competitive inhibitors show a dynamic equilibrium between several c-Abl conformations that is distinctly modulated by both types of inhibitors. Binding of the ATP-site inhibitor imatinib leads to an unexpected open conformation of the multidomain SH3-SH2-kinase c-Abl core and the combination of imatinib with the allosteric inhibitor GNF-5 restores the closed, inactivated state of the kinase. These NMR data provided detailed insights for the poorly understood combined effect of the two inhibitor types, which are able to overcome drug resistance, opening new possibilities for the development of new drugs.

In summary, NMR spectroscopy is undoubtedly destined to play a more prominent role in the target identification (metabolomics) and validation processes in pharmaceutical industry. Especially the application of NMR for 'difficult' discovery programs (PPI and cancer) is highly recommended for successful drugs.

ACKNOWLEDGMENTS

This work was supported by grants from the German Cancer Consortium (DKTK, Heidelberg).

REFERENCES

1. K. Mullane, R. J. Winquist, and M. Williams, *Biochem. Pharmacol.*, 2014, **87**, 189.

2. S. M. Paul, D. S. Mytelka, C. T. Dunwiddie, C. C. Persinger, B. H. Munos, S. R. Lindborg, and A. L. Schacht, *Nat. Rev. Drug Discov.*, 2010, **9**, 203.

3. M. E. Bunnage, *Nat. Chem. Biol.*, 2011, **7**, 335.

4. J. P. Hughes, S. Rees, S. B. Kalindjian, and K. L. Philpott, *Br. J. Pharmacol.*, 2011, **162**, 1239.

5. M. Rask-Andersen, M. S. Almen, and H. B. Schioth, *Nat. Rev. Drug Discov.*, 2011, **10**, 579.

6. S. Hoelder, P. A. Clarke, and P. Workman, *Mol. Oncol.*, 2012, **6**, 155.

7. M. Schenone, V. Dancik, B. K. Wagner, and P. A. Clemons, *Nat. Chem. Biol.*, 2013, **9**, 232.

8. P. J. Hajduk, *Mol. Interv.*, 2006, **6**, 266.

9. M. Wehling, *Nat. Rev. Drug Discov.*, 2009, **8**, 541.

10. F. Prinz, T. Schlange, and K. Asadullah, *Nat. Rev. Drug Discov.*, 2011, **10**, 712.

11. D. C. Swinney and J. Anthony, *Nat. Rev. Drug Discov.*, 2011, **10**, 507.

12. L. Poppe, J. B. Jordan, K. Lawson, M. Jerums, I. Apostol, and P. D. Schnier, *Anal. Chem.*, 2013, **85**, 9623.

13. R. Powers, *Expert Opin. Drug Discov.*, 2009, **4**, 1077.

14. D. C. Anderson and K. Kodukula, *Biochem. Pharmacol.*, 2014, **87**, 172.

15. R. Powers, *J. Med. Chem.*, 2014, **57**, 5860.

16. L. N. Makley and J. E. Gestwicki, *Chem. Biol. Drug Des.*, 2013, **81**, 22.

17. H. Jubb, A. P. Higueruelo, A. Winter, and T. L. Blundell, *Trends Pharmacol. Sci.*, 2012, **33**, 241.

18. C. Herbert, U. Schieborr, K. Saxena, J. Juraszek, F. De Smet, C. Alcouffe, M. Bianciotto, G. Saladino, D. Sibrac, D. Kudlinzki, S. Sreeramulu, A. Brown, P. Rigon, J. P. Herault, G. Lassalle, T. L. Blundell, F. Rousseau, A. Gils, J. Schymkowitz, P. Tompa, J. M. Herbert, P. Carmeliet, F. L. Gervasio, H. Schwalbe, and F. Bono, *Cancer Cell*, 2013, **23**, 489.

19. L. Skora, J. Mestan, D. Fabbro, W. Jahnke, and S. Grzesiek, *Proc. Natl. Acad. Sci. U. S. A.*, 2013, **110**, E4437.

20. M. Vogtherr, K. Saxena, S. Hoelder, S. Grimme, M. Betz, U. Schieborr, B. Pescatore, M. Robin, L. Delarbre, T. Langer, K. U. Wendt, and H. Schwalbe, *Angew. Chem. Int. Ed. Engl.*, 2006, **45**, 993.

Chapter 9

High-resolution MAS NMR of Tissues and Cells

Leo L. Cheng

Massachusetts General Hospital, Harvard Medical School, Pathology, MGH 149, 13th Street Charlestown, Boston, MA 02129, USA

9.1 INTRODUCTION

For chemists and physicists from the late 1950s onward, the rapid mechanical spinning of solid samples at the ***magic angle*** (θ) (where $3\cos^2\theta - 1 = 0$, $\theta = 54.74°$) relative to an applied permanent magnetic field has provided a well known and powerful solid-state NMR line-narrowing technique – magic angle spinning (MAS). Before 1996, however, no application of this solid-state approach to measurement of proton NMR signals from biological tissue and/or cell samples was attempted or discovered.[1,2] The 1996 advent of this application extended the use of solid-state MAS technology to analyses of semi-solid samples, such as intact tissues, through a technique now termed *high-resolution MAS (HRMAS) NMR*. With the discovery of HRMAS, the landscape

of biomedical NMR investigations, and particularly of proton NMR studies, dramatically expanded.

Beginning in the 1970s, the unique potential of NMR to reflect biological and metabolic states in human disease, along with the rich, natural abundance of protons, prompted applications of NMR to tissue analysis. Before HRMAS, however, proton NMR measurements were obtained by performing conventional liquid-state NMR techniques either on intact tissue and cell samples, resulting in broad-line spectra, or on solutions obtained by chemical extractions from these semi-solid samples.

Each of these approaches achieved important results; however, both were fraught with disadvantages. Extraction procedures, performed with acids or other chemical solvents on mechanically homogenized tissues or cells, were able to produce aqueous solutions that satisfied liquid-state NMR's requirement that only negligible interactions should occur among solute molecules, thereby permitting high-resolution spectra to be obtained. In fact, the very extraction that denatured cellular enzymes and other proteins (producing the welcome effect of freezing metabolism and sample metabolic states) also destroyed biological membranes. Overall, the result of these processes was to liberate into solution metabolites that had previously been protein-bound or sequestered within such organelles as mitochondria. However, concern that some metabolites would not completely undergo extraction could

NMR in Pharmaceutical Sciences. Edited by Jeremy R. Everett, John C. Lindon, Ian D. Wilson, and Robin K. Harris
© 2015 John Wiley & Sons, Ltd. ISBN: 978-1-118-66025-6

cast an extract's composition into doubt: such a composition might not reflect concentrations of metabolites present in the cytosol, or under in vivo conditions.

Moreover, the tissue homogenization process rendered subsequent identification of heterogeneous pathological features impossible in all samples that were not already of a homogeneous nature (e.g., cultured cells from cell-lines or some tumor xenografts in animal models). This obfuscation of heterogeneity introduced a particularly critical weakness for all analyses involving human oncology.

In an intact tissue approach, cell or tissue samples placed within tubes of the type used for solution NMR are analyzed through conventional NMR procedures for aqueous solutions to permit subsequent pathology evaluations. These samples, maintained in conditions more closely resembling in vivo states than extract solutions, can be studied at higher magnetic field strengths, and so achieve better spectral resolution, than is possible with in vivo studies. However, the spectra produced by this approach are generally characterized by broad spectral lines that not only make it impossible to identify individual metabolites, but that render accurate estimations of resonance intensities a challenge. The urgent need for a reliable means to bypass these limitations so as to obtain reliable NMR measurements motivated the development of the HRMAS technique.

HRMAS permitted high-resolution NMR spectra to be observed directly from intact tissue and cell samples, while preserving their architectural structures for evaluation and quantification, after completion of the NMR measurements, by histology. Figure 9.1 illustrates the effects of HRMAS on the proton NMR measurements of a human prostate specimen of ~10 mg.

9.2 THE DEVELOPMENTS

9.2.1 Proton NMR: From Solid State MAS to Biological Tissue HRMAS

Among NMR observable nuclei, the proton's high natural abundance has granted it a prominent place in biomedical NMR. This popularity was further secured as technical developments in MRI scanners increasingly rendered commonplace the clinical implementation of imaging, as well as in vivo NMR spectroscopy (MRS). However, a further result of the proton's high natural abundance was the strength of interactions among proton nuclei, termed *couplings* or *solid-effects* (e.g., dipolar coupling, chemical shift anisotropy).

The importance of each of these proton attributes can be grasped as follows. Were we to view any spectroscopy as a histogram depicting an entity-of-interest's population distributions (i.e., solutes in a solution, or proton nuclei in a solution for proton NMR), each entity in an aqueous solution, whether solute or proton nucleus, could be considered to exist in the independently identical state of any other identical entity. This correspondence would exist owing to the 'long-distance' separating the one entity in the aqueous solution from the other. These entities' identical states result in their narrow distribution in the histogram, and produce a spectrum characterized by sharp peaks spiked at the values of the distributions. Added to the 'long-distances' between proton nuclei in an aqueous solution are isotropic and rapid molecular tumbling motions that further impede their couplings. From these collective effects, observation of the desirable, highly resolved, spectroscopic lines was achieved.

The harmony of proton nuclei in aqueous solutions, however, is disrupted in nonliquid materials, such as biological tissues, due to the presence of chemical and physical heterogeneities. In such materials, the natural abundance of protons leads to proton nuclei being packed more closely together, where they exhibited far stronger couplings, or interactions, much greater interdependence, and far more limited, even truncated, tumbling motions than in aqueous solutions. In tissue and other semi-solid samples, slow and restricted molecular motions result in regional environmental and magnetic susceptibility differences, non-completely averaged dipolar couplings, and residual chemical shift anisotropy, as well as shortening of proton spin–spin relaxation times, all contributing to anisotropic and isotropic line-broadenings. In other words, through these 'solid-effects' each proton nucleus occupies a somewhat different physical environment, as compared with its neighboring, chemically identical nucleus. This difference results in a broadening of the distribution pattern for the nuclei of a specific chemical structure, thereby producing the broad-line proton NMR for these materials of semi-solids.

Generally speaking, distribution patterns for proton in solid states are broad, with the width of proton NMR peaks for solids >50 kHz. Therefore, line narrowing by the mechanically achievable MAS rate alone is often insufficient. An example of a solid-state proton NMR spectrum under moderately fast MAS is illustrated in

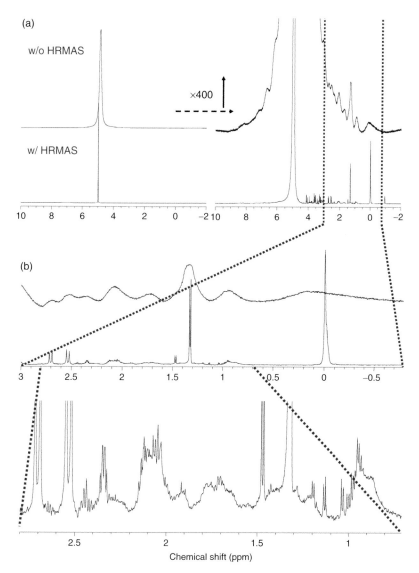

(a)

w/o HRMAS

×400

w/ HRMAS

Figure 9.1. An illustration of HRMAS proton NMR spectra of a human prostate tissue sample measured at 600 MHz; (a) without and with HRMAS and (b) expansions of the result with HRMAS. With 400 times vertical expansion from the top left to the top right spectra, cellular metabolite resonances are clearly visible under the condition of HRMAS even without water suppression. Red dashed lines delineate spectral regions that are further expanded

Figure 9.2(a). The figure shows the measures for crystalline alanine powder, but presents no clear separation among different functional groups in the molecule.

Pondering the physical characteristics of aqueous solutions, wherein proton nuclei are far apart, made it reasonable to expect that couplings, or interactions, among proton nuclei might be greatly reduced by increasing internucleus distances. Through the same crystalline alanine powder, but after a uniform exchange of its protons with deuterons, it was possible to achieve a drastically improved spectral resolution. This proton NMR spectrum, measured from the residual protons in perdeuterated alanine and shown in Figure 9.2(b),[3] was the first high-resolution proton

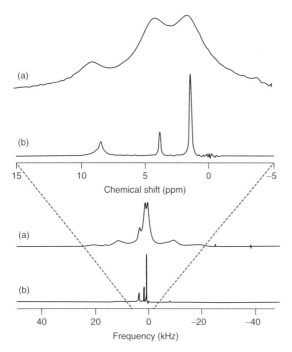

Figure 9.2. Proton NMR spectra of (a) natural abundance alanine crystalline powder measured with MAS rate of 10 kHz, and (b) perdeuterated alanine crystalline powder with MAS rate of 9 kHz, at room temperature at 400 MHz. (Cheng, Griffin, Herzfeld, unpublished data in PhD Thesis, 'Solid State NMR Studies of Proton Diffusion,' Brandeis University, 1993)

NMR of crystalline alanine ever observed. This study also provided instrumental guidance for the successful development of HRMAS.

The examples shown in Figure 9.2 nicely illustrate the need for MAS for effective line narrowing. This is concisely summarized in Fukushima and Roeder's classic *Experimental Pulse NMR, A Nuts and Bolts Approach*[4]: '(MAS) is only successful when the rate of rotation is rapid in comparison to the linewidth (in hertz) that you are attempting to reduce.' In the above-mentioned examples, the static linewidth ($\sim$50 kHz) of natural abundance alanine was much greater than the spinning rate ($\sim$9 kHz), and so resulted in a relatively poor resolution even with MAS. By contrast, perdeuterated alanine shows the relative clarity of resolution attainable when the static linewidth has been significantly reduced from its naturally abundant state. These results suggested that if, by some mechanism, the natural abundance of proton NMR linewidth for a sample of interest could be brought within the range

of MAS rates, spinning the sample at the 'magic angle' would result in a greatly improved proton NMR spectral resolution. This revelation was of paramount importance to the discovery of tissue HRMAS proton NMR.

In 1996, the first observation of the MAS effect on biological tissue was recorded for rat metastatic lymph nodes. After suppressing tissue water signals, the natural proton NMR linewidth of the metabolites was observed to be on the order of 2–3 kHz, primarily contributed by bulk magnetic susceptibility broadening and within the spinning capabilities of MAS. The application of MAS resulted in highly resolved proton NMR spectra of intact tissue never previously observed. From these spectra of normal and malignant tissues, the identification and quantitation of metabolite changes associated with malignant transitions in lymph nodes were achieved.[1] This methodology of applying MAS beyond its initial design for solid-state NMR was termed *HRMAS*, to differentiate it from the original MAS NMR used for solid-state samples.

The first application of HRMAS in a human study was soon reported (1997) for the analysis of post-mortem human brain tissue from a patient with a neurodegenerative disease. In addition to quantifying brain metabolites from different cortical regions, the study confirmed correlations between the populations of surviving neurons and the metabolic concentrations of *N*-acetylaspartate (NAA), a proposed neuronal marker, as measured within adjacent tissue samples from these regions.[2]

The ultimate strength of HRMAS in terms of its compatibility with histopathological analysis was discovered later by Dr. D. C. Anthony, then a neuropathologist at Boston's Children Hospital, in a study of human brain tumors. This study demonstrated that, despite the earlier perception that tissue pathological architectures would be completely destroyed by the centrifugal force of MAS on tissues during NMR measurement, pathology evaluations of brain tissue could be carried out with samples after the HRMAS NMR analysis.[5]

9.2.2 Methodologies and Advances

The experimental procedures for intact tissue HRMAS are extremely straightforward. Either fresh or previously frozen tissue samples can be measured directly; further preparation is not required. Procedures for intact tissue studies with HRMAS proton NMR have been detailed in earlier publications.[6,7] Since that publication, other technical details, discussed later, have emerged.

9.2.2.1 Sample Handling

Although fresh tissue samples, such as biopsy cores, may be washed briefly with deuterated water (D_2O) to remove blood and foreign materials related to procedures at the time of harvesting, samples should not be immersed in any liquid medium during the transfer of fresh tissues or when storing frozen tissues. Avoiding such immersion will help prevent the loss of cellular metabolites into the medium, which can be severe. For instance, immersion of a 10 mg block of human prostate specimen in D_2O for 10 min could result in complete depletion of NMR measurable metabolites.

To prevent dehydration of tissue samples of limited volume, particularly for those with a large surface-to-volume ratio, such as needle biopsy cores, a simple apparatus may be constructed with daily laboratory materials. One example of such an apparatus, constructed for transferring a human prostate biopsy core, is illustrated in Figure 9.3. Depicted in this figure are two eppendorf tubes (0.5 ml for PCR, polymerase chain reaction, and 1.5 ml regular). Inside the 1.5 ml tube, the PCR tube (without cap) is situated on top of a D_2O saturated cotton ball. This design allows the moisture provided by the D_2O to keep the biopsy core from dehydrating, while the inner tube prevents the loss of metabolites from the core to the medium. This type of apparatus may also be used for snap-freezing of tissue specimens, a tactic used to prevent tissue from directly contacting the liquid nitrogen.

When working with frozen samples, if the size of the tissue block exceeds the size required for HRMAS NMR analysis (for instance, about 10–12 mg if a Bruker 4 mm rotor and a spherical cavity insert are used to place the sample in the middle of the receiving coils so as to minimize shimming, as well as the effects of the magnetic field inhomogeneity), samples must be sectioned on a frozen surface to avoid multiple freezing and thawing. Borrowing the routine practice of clinical neuropathology for brain slice preparation, a thin metal plate, covered with gauze and placed on top of dry ice, should be prepared as a surface for any sectioning operation.

If a systematic study is planned, the tissue type studied should be individually evaluated for specifying the potential effects of both freezing the tissue and long-term storage on the metabolites to be measured.

To apply the HRMAS technique to the study of cells, where the cell supply is not usually of concern (between 2×10^6 and 50×10^6 are required to obtain a spectrum), cells can be washed, centrifuged, and directly transferred into NMR rotors. The transfer of cells into NMR rotors can present a challenge if cell supply is limited. However, tests have shown that with even ~100 000 HL-60 (human promyelocytic leukemia) cells, reasonable spectra can be measured within <10 min of HRMAS proton NMR analysis, although the pellet formed by such a low number of cells can hardly be visible with the naked eye. In such cases, it may prove useful to pack the pellet directly inside the rotor by transferring a cell suspension, centrifuging, and removing the supernatant medium in situ. If such a procedure is employed, every caution should be exercised to safeguard the integrity of the cells in the pellet and to ensure that cellular metabolites will not be washed away with the removal of the medium.

9.2.2.2 HRMAS Proton NMR

After following the routine procedures of MAS adjustment (as specified by the manufacturer) and probe shimming (for tissue measurements, this may be achieved according to the measured splitting of the lactate doublet at 1.33 ppm), the experimental conditions required for tissue and cell HRMAS, (such as temperature, spinning rates, etc.) will depend on the type of tissue to be studied, metabolites of interest, and whether histopathology is scheduled after the NMR measurements.

Figure 9.3. An example of the apparatus constructed for transferring a human prostate biopsy core

Generally speaking, a low temperature (e.g., 4 °C) will reduce the potential for tissue metabolic degradation during the measurement, while a slow spinning rate better preserves tissue histological architecture. When the tissue type can endure a strong centrifugal force (skin provides a notable example), or where histopathology is of no concern, spinning at the frequency of 6×10^{-6} of the NMR magnet field strength will ensure that all the spinning side-bands on proton NMR spectra with the RF centered on the water resonance (4.7 ppm) will be located to the outside of the metabolite spectral regions (0–10 ppm). However, if experimental realities do not permit this approach, multiple schemes for slow spinning HRMAS may be investigated and implemented.

Any NMR pulse sequence may be applied to tissue and cell HRMAS NMR in order to address such concerns as rotor-synchronization, if necessary. However, most potential clinical applications of intact tissue analysis will employ one dimensional HRMAS proton NMR, acquired with or without water suppression and with a rotor-synchronized CPMG sequence of the determined T2 filter effect. This approach can achieve a flat spectral baseline by filtering out undesired broad resonances from such macromolecules as proteins and cell membranes, as well as the probe background. However, to elucidate and interpret these one dimensional observations, multidimensional proton homonuclear and heteronuclear NMR of intact tissues and cells may also be carried out. As a general caution, the longer the time needed to conduct multidimensional experiments, the greater the need to investigate potential metabolic and architectural degradations of the tissues under study.

It is generally safe to assume that alterations in physiologically and pathologically related cellular metabolites will not appear simply as the presence or absence of peaks, but rather will present continuous changes in intensity throughout disease development and progression. Thus, the estimation and quantification of metabolites is extremely important. Metabolite concentrations may be estimated from HRMAS proton MR spectra by using the intensity of tissue water signals (without water suppression), or by using an external standard permanently attached to the inside of the rotor or to the rotor inserts. Reports have shown a trend toward applying ERETIC (Electronic REference To access In vivo Concentrations) to HRMAS NMR.

To process 1D HRMAS proton NMR spectra, commercial software can be purchased either from the manufacturer of the spectrometer or from independent suppliers; programs can also be developed in-house. In any of these cases, if estimations of cellular metabolite concentrations are attempted, spectra in the frequency domain must be curve-fitted.

The traditional 'binning' approach for the processing of broad-line tissue NMR spectra involves dividing the entire spectrum into equal buckets, usually in 0.01 ppm intervals; it was widely used in the pre-HRMAS era and is still practiced today. Although the binning method is simple and requires significantly less data-processing time, it often produces data sets with variables much larger than subjects/cases, generating an apparently tranquil data landscape in which it is very easy to fall into false-positives.

Regardless of the approach employed, the high resolution obtained through HRMAS measurement makes it necessary to consider accurate spectral calibrations according to known metabolites, along with spectral alignment among the evaluated sample sets.

9.2.2.3   Histopathology

Owing to the known pathological heterogeneity existing within a single lesion of human disease, a quantitative examination of tissue histopathology is extremely important, particularly when studying intact tissue from human malignancies. For instance, histological analysis following an HRMAS NMR study of 199 human prostate specimens from 82 biopsy-proven cancer patients identified cancer cells in only 20 of the specimens – the other 179 prostate tissues appeared benign at pathology.[8]

It is also important to recognize that, unlike the numerical and continuous data generated by HRMAS NMR, routine clinical histopathology data may depict only the presence or absence of a certain pathology, and so be inadequate for quantitative research studies. Serial-sections, throughout the entire NMR measured specimen, along with the use of computer image-analysis programs, may be necessary to estimate the volume percentages of pathologies. The frequency of serial sections required should be determined by the distribution of pathological changes due to the medical condition under study.

9.2.2.4   Statistical Data Analysis

The improved spectral resolution afforded by HRMAS allows resonance peaks to be identified and assigned to certain cellular metabolites. Earlier tissue or cell studies were dedicated to the confirmation and discovery of metabolite correlations with pathologies; these primarily made use of linear regression analysis or differentiations between clinical conditions according to analysis of variances (ANOVA). As long as a Bonferroni correction is applied to cases in which multiple comparisons are performed, these approaches remain statistically valid.

As the power of metabolomics over each of these individual metabolites became more widely recognized, 'profile' (either metabolic or metabolomic) became the term of choice in presenting tissue or cell HRMAS NMR data. In pursuing metabolomic profiles, spectral data from cases studied are either binned or integrated into a large data matrix in which the number of variables is often greater than the number of cases. Such datasets often then go through dimension-reduction processes, such as principal component analysis (PCA) or factor analysis (FA), and are subjected to classifier discovery through, for example, canonical correlation (CANCOR) analysis.

Here again, great caution must be exercised to guard against false positives. Although in the past it was considered appropriate to study samples from a reasonably large patient population and find classifiers that would differentiate disease conditions, we now know that such results for a 'training cohort' alone should not be reported as facts until they have been verified with an independent 'testing cohort'.

The most trustworthy test of metabolomic profiles obtained from the training cohort is to apply the coefficients for the construction of these profiles from the training cohort directly to the testing cohort, and then to evaluate the sensitivity and specificity of these results with receiver operating characteristic (ROC) curves. From these ROC curves, the thresholds may be defined and further tested in the clinic for a specific clinical question. The more complicated the statistical procedure needed to reach publishable results of statistical significance, the less likely is it that these results might be directly applicable and implementable in the clinic. While it is completely valid to analyze a cohort of data by using the commonly used 'leave-one-out' approach (an approach which may indeed provide valuable information as it maximizes the statistical power available for that cohort), the results obtained will present little information from which to construct and test thresholds of clinical utility.

9.3 APPLICATIONS

Over the past 18 years since the discovery and development of HRMAS proton NMR for intact tissue analysis, this methodology has provided the driving force for many studies designed to investigate human physiology and pathology, animal models of human diseases, and cells through the evaluation of intact specimens. According to our survey of PubMed (http://www.ncbi.nlm.gov/pubmed), some 100 institutions around the world have contributed reports based on HRMAS studies, with the number of publications appearing since 1996 quadratically increasing, as shown in Figure 9.4.

9.3.1 Animal and Cell-line Studies

After the initial HRMAS study of metastatic lymph nodes from a rat tumor model, the first decade of the applications of HRMAS to animal studies were focused on either to determine how tissues from certain organs would appear under spinning conditions,

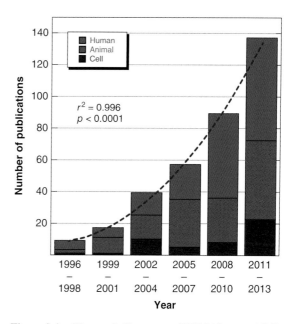

Figure 9.4. The trend of increases of HRMAS proton NMR studies on animal models, cell-lines, and human subjects published annually between 1996 and 2013 shows a quadratic relationship

or to measure the toxicity of particular agents or conditions. Most of these reports described experiments conducted with a limited number of animals, and were presented as 'proof-of-concept' that high-resolution proton NMR spectra could be obtained with HRMAS for the studied type of tissue. Such explorations did not achieve lasting advances within the field, but they provided the determination of utility and limits that typically accompanies technological innovation.

Since 2005, HRMAS technology for routine intact specimen measurements has been mature, and the need for validating its effectiveness has vanished. Thus, the experimental work reported for much of the past 10 years has probed the fundamentals of science that can be revealed by applying HRMAS NMR as the primary means of investigation for a range of animal models of human malignant disease. It has also been applied to a spectrum of such nonmalignant diseases as nonalcoholic fatty liver diseases,[9] brain metabolic changes for neuroAIDS in SIV-infected macaques, rat hippocampus and cortices at different ages, animal models of Alzheimer's disease, acute lung injuries, long-term alcohol consumption, early myocardial infarction, experimental sepsis, schizophrenia, and posttraumatic stress disorder (PTSD), and the effects of ecstasy abuse. Such less-investigated areas as the metabolism of the eye for evaluation of cataract models and after UVB radiation have also made use of HRMAS methodology.

HRMAS investigations of malignancy have evolved from the identification of tumor-associated metabolic changes to the evaluation of therapy responses, including measurements of phospholipid metabolic responses to therapy with quantitative ^{31}P HRMAS NMR.[10] The most noticeable, recent trend of such studies of malignancies has been their combination with molecular biology measurements, such as receptors and gene expression microarrays.[11] HRMAS NMR data have further been analyzed and compared with results obtained from other analytic methods, including HPLC and mass spectrometry.[12]

Reviewing the studies of cell-lines reported in the past also reveals a similar transition from demonstrating the possibility that high-resolution NMR spectra of cells could be measured to utilize the technique to investigate the effects of therapies on metabolite changes in cell-lines. In addition, HRMAS has been applied to interrogate metabolic pathways with the assistance of ^{13}C-labeled substrates using proton and ^{13}C NMR.[13]

9.3.2 Human Studies

The applications of HRMAS NMR to investigations of human disease have dramatically increased over the past 10 years. On the basis of the previously mentioned literature search, a summary of trends within human studies utilizing HRMAS and published between 1996 and 2013 is presented in Figure 9.5. The trends presented in this figure clearly indicate the heavily skewed application of HRMAS toward human malignancy studies, particularly for those areas where in vivo MRI/MRSI (magnetic resonance spectroscopic imaging) are most widely applied, such as brain tumors, prostate, and breast cancers. These studies show a diverse spectrum of attempts to evaluate human malignancies; however, their overall aims, as we have observed, have recently shifted from the identification of cancer metabolite markers for use as independent measures permitting cancer detection, diagnosis, and patient prognosis, to attempts at establishing clinical protocols that can function as adjunct measures for assisting current histopathology in clinical decision-making. Rather than summarizing the results of these investigations for specific cancers,[14,15] we highlight the emerging concepts fostered by the studies.

9.3.2.1 Bridging In Vivo MRSI with Pathology

Studying samples through HRMAS methodology requires that tissues be removed by invasive procedures, such as biopsy or surgery. Were HRMAS NMR analysis only able to differentiate tissue pathologies already categorized by histopathology, its utility would be miniscule; histopathology can readily perform this task accurately and with ease. Thus, for HRMAS analysis to be clinically useful, it must either contribute critical information about the disease that cannot be assessed by current pathology, such as tumor biology, or provide a means of direct noninvasive testing, capable of providing clinically relevant information without recourse to invasive procedures. The potential of HRMAS to reach beyond the capacity of current pathology will be discussed in the section titled 'Advancing Beyond the Current Capabilities of Clinical Pathology'; here we will only examine how HRMAS analysis can provide a bridge connecting *in vivo* observations with histopathology.

A major motivation for analyzing tissue specimens of human disease has been the attempt to better understand in vivo observations, especially in our age of rapid technical developments in the clinical use of MR

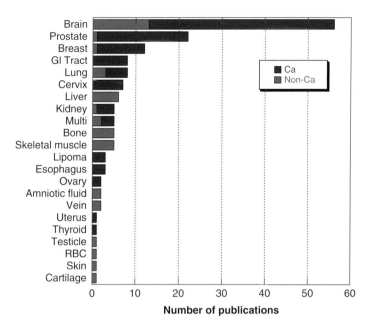

Figure 9.5. Distributions of HRMAS proton NMR studies on human subjects, for malignant (Ca) and nonmalignant (non-Ca) diseases of different organs between 1996 and 2013 - RBC: red blood cell, GI: gastrointestinal.

scanners. As ex vivo NMR analyses generally are conducted at higher field strengths than in vivo scans, resonance peaks can be better resolved and identified. By taking advantage of the compatibility of histological analysis after HRMAS NMR measurement, the resonance peaks quantified through HRMAS NMR can be interrogated according to tissue pathology, permitting profiles of the ex vivo spectra to be overlaid with in vivo spectral patterns so as to confirm histopathological interpretations of in vivo findings. The use of this concept has been reported for studies of human brain tumors[16] and of uterine cervical,[17] breast,[18] and prostate cancers.[19]

9.3.2.2 Advancing Beyond the Current Capabilities of Clinical Pathology

The clinical implementations of new cancer detection paradigms, for example, with such tools as mammography for breast cancer and blood serum prostate specific antigen (PSA) for prostate cancer, has provided access to care for a much larger number of early cancer-stage, asymptomatic patients, whose conditions formerly were discovered only accidentally. However, at present cancer clinics mainly treat patients

based on pathological evidence collected from previous, symptomatic patients. The challenge presented by patients with early-stage disease to present-day cell morphology-based pathology urgently necessitates innovation within pathology.

Tissue metabolic status, quantifiable with HRMAS NMR, is considered an ideal candidate through which to accomplish such innovation, owing to its potential to subcategorize existing pathology criteria. This use of HRMAS has been applied to characterize brain tumor subgroups within meningiomas, oligodendrogliomas, and gliomas,[20–22] to predict estrogen receptor (ER) and lymph node involvement status for breast cancer,[23] to identify tumor aggressiveness and recurrence potential for prostate cancer patients,[24,25] colorectal cancer,[26] and to indicate chronic liver disease from biopsy.[27–29]

Since 2010, we have witnessed an increasing trend of utilizing HRMAS proton NMR for measurements of biopsy samples aimed at the establishment of protocols to enhance clinical real-time capability for intraoperative consultations. Active clinical research studies categorizing the malignant potential of brain tumors has been reported, as have studies in breast cancer aimed at predicting responses to neoadjuvant chemotherapies.

Before tissue metabolic measurements can be considered for diagnosis, and particularly for prognosis, one must undertake rigorous tests, including randomly selected training and testing cohorts, as these will be absolutely essential to minimizing the potential of false positives. Increasingly, evidence supporting the implementation of such crucial rigor has been documented in research reports.

9.3.2.3 Analyzing Biological Samples of Limited Supply

Applications of HRMAS NMR in studies of biological fluids, such as those of blood serum and amniotic fluid, have been reported. HRMAS can assist in reducing spectral broadening caused by the bulk magnetic susceptibility and the presence of macromolecules, which are found in all fluids. For instance, with human blood serum samples, Figure 9.6 shows that the spectral result obtained with HRMAS from a 10 μl sample presented more clearly the metabolic information compared to the result from using more than 200 μl using the conventional solution NMR approach.

The above-mentioned example illustrated one of the advantages of utilizing HRMAS in measuring biological specimens, that is, its requirement for a relatively small sample amount (<10 mg of intact tissues, <10 μl of blood serum, or 100 000 cells). While this advantage has represented a drastic (more than one order of magnitude) reduction in the sample volume/mass from that required for conventional solution NMR to study either intact samples or chemical extractions, this sample size is still magnitudes greater than the ideal 'single cell' measurement. To increase measurement sensitivity, a number of NMR approaches may be considered, including the now well appreciated methodology based on the concept of dynamic nuclear polarization (DNP). However, where such drastic new hardware installations as the microwave sources and low temperature apparatus required for DNP are lacking, a recent clever proposal has yielded a brightened, new horizon for tissue and cell HRMAS NMR. This approach, termed *magic angle coil spinning*, or *MACS*, has the potential to be implemented in every laboratory interested in carrying out HRMAS NMR measurements for samples of nanoliter (nl) scale with minimal hardware modification.[30] Realizing that other than targeting the intrinsic nuclear spin limitation as attempted by DNP to improve NMR measurements, the NMR sensitivity may be optimized through the design of the close fitting between the RF receiving coil and the microscopic

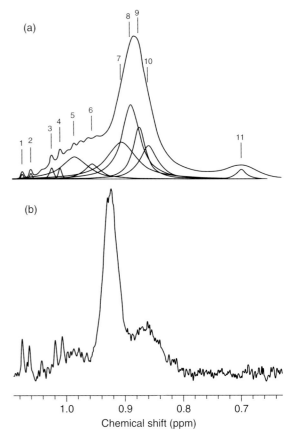

Figure 9.6. Comparison of human blood serum NMR spectra (a) in literature measured using conventional liquid state method at 500 MHz, where it was reported that Lorentzian lines obtained are labeled from 1 to 11. Components 1–4 arise from the methyl function of valine, component 5 from methyl groups of proteins, components 6–10 from the lipoproteins, and component 11 from carbon 17 of cholesterol,[35] with (b) obtained with HRMAS methodology at 600 MHz

samples (a.k.a. optimizing the coil filling factor). The proposed MACS methodology includes the building of a tuned micro-coil circuit immediately surrounding samples of microscopic scale, and introduces both the micro-coil and the sample into a standard HRMAS rotor and a commercial NMR probe.

On the basis of the concept of inductive coupling between the micro-coil in the rotor and the standard HRMAS probe coil, the RF pulse and NMR signals can be wirelessly transmitted between the micro-coil and the probe coil under the spinning condition. With the optimization of microscopic coil filling factors, it has been shown that the normal sample size for regular HRMAS in the order of microliter scale can be further reduced to the nanoliter scale with the assistance of high-resolution MACS, or HRMACS. This advancement is of critical importance for overcoming pathological heterogeneity in NMR studies of human diseases.[31]

It is evident that better characterizations of human diseases with intact tissue NMR, such as that of various human malignancies, rely to a great degree on the quantification of NMR measurable metabolites of different pathological features. Unfortunately, at present, while micro-dissections of these pathological features, such as cancer glands, are possible under a microscope, the resulting amounts of specimens are too scant to be measured, even with the HRMAS methodology. With maturation, HRMACS NMR holds a great potential to help in the evolution of intact tissue NMR for clinical utilizations and applications.

9.3.2.4 Developing Metabolomics and Metabolomic Imaging

Another rapidly evolving frontier, and one closely associated with the development of HRMAS NMR for intact tissue analysis, is the establishment of NMR-based metabolomics. The most active area of metabolomics lies in its application within studies of human malignancies, often referred as *oncometabolomics*, which is now defined as ' ... study of the global variations of metabolites and (the) measurement of (their) global profiles ... from various known metabolic pathways under the influence of oncological developments and progressions'.[32]

Numerous reports, which should be examined with caution, identify metabol(om)ic profiles for various types of disease. However, studies of tissue HRMAS NMR findings in conjunction with pathology have indicated the potential existence of metabolomic field effects – that is, the existence of malignancy-associated metabolomic changes in histologically benign tissue regions surrounding cancer cells.

Similar to our earlier discussion of HRMAS NMR applications to serum analysis, the metabolomic analysis of tissue-serum pairs from lung cancer patients has shown the potential existence of serum metabolomic profiles able to distinguish different cancer types, as well as to differentiate between samples from cancer patients and healthy control subjects. Such profiles have the potential eventually to be used to provide blood tests for lung cancer screening and patient triage.[26,33]

Metabolomic imaging has been performed by the co-registration of prostate cancer metabolomic profiles, obtained from tissue analysis, to measures obtained by localized, multivoxel MRS throughout the entire prostate.[34] After further developments in both imaging technology and the characterization of metabolic profiles it should be possible to produce three dimensional maps of the distributions of various diseases from a single dataset of in vivo measurements. Such metabolomic maps would be calculated through data processing paradigms according to the metabolomic profile of each individual disease.

9.4 CONCLUSIONS

After its discovery in 1996, HRMAS proton NMR has met with widespread adoption for analyses of intact tissues and cells of various human and animal model origins. While it may remain too early to predict the degree of its clinical impact, recent biopsy studies have demonstrated the potential of this technique for clinical implementation, particularly given the exploding rate of HRMAS investigations of human diseases witnessed in literature. Whether by identifying metabolomic pathology to supplement morphologic pathology, or assisting with in vivo disease characterization, HRMAS is likely to develop into a tool of clinical importance. However, many clinical topics must now be rigorously investigated to identify the true and accurate extent of the clinical capabilities of HRMAS NMR.

RELATED ARTICLES IN EMAGRES

Magic Angle Spinning

Tissue NMR Ex Vivo

Tissue and Cell Extracts MRS

Magnetic Susceptibility and High Resolution NMR of Liquids and Solids

Brain Neoplasms Studied by MRI

Multiparametric Magnetic Resonance Imaging of Prostate Cancer at 3T

Kidney, Prostate, Testicle, and Uterus of Subjects Studied by MRS

REFERENCES

1. L. L. Cheng, C. L. Lean, A. Bogdanova, S. C. Wright Jr, J. L. Ackerman, T. J. Brady, and L. Garrido, *Magn. Reson. Med.*, 1996, **36**, 653.

2. L. L. Cheng, M. J. Ma, L. Becerra, T. Ptak, I. Tracey, A. Lackner, and R. G. Gonzalez, *Proc. Natl. Acad. Sci. U. S. A.*, 1997, **94**, 6408.

3. L. Zheng, K. W. Fishbein, R. G. Griffin, and J. Herzfeld, *J. Am. Chem. Soc.*, 1993, **115**, 6254.

4. E. Fukushima and S. B. W. Roeder, *Experimental Pulse NMR – A Nuts and Bolts Approach*, Perseus Books: Reading, Massachusetts, 1981, 281.

5. L. L. Cheng, D. C. Anthony, A. R. Comite, P. M. Black, A. A. Tzika, and R. G. Gonzalez, *Neuro Oncol.*, 2000, **2**, 87.

6. L. L. Cheng, M. A. Burns, and C. Lean, in *Modern Magnetic Resonance*, ed G. A. Webb, Springer: Netherlands, 2006, 1037.

7. O. Beckonert, M. Coen, H. C. Keun, Y. Wang, T. M. D. Ebbels, E. Holmes, J. C. Lindon, and J. K. Nicholson, *Nat. Protoc.*, 2010, **5**, 1019.

8. L. L. Cheng, M. A. Burns, J. L. Taylor, W. He, E. F. Halpern, W. S. McDougal, and C.-L. Wu, *Cancer Res.*, 2005, **65**, 3030.

9. J. F. Cobbold, Q. M. Anstee, R. D. Goldin, H. R. Williams, H. C. Matthews, B. V. North, N. Absalom, H. C. Thomas, M. R. Thursz, R. D. Cox, S. D. Taylor-Robinson, and I. J. Cox, *Clin. Sci. (Lond.)*, 2009, **116**, 403.

10. M. Esmaeili, T. F. Bathen, O. Engebraten, G. M. Maelandsmo, I. S. Gribbestad, and S. A. Moestue, *Magn. Reson. Med.*, 2014, **71**, 1973.

11. M. T. Grinde, S. A. Moestue, E. Borgan, O. Risa, O. Engebraaten, and I. S. Gribbestad, *NMR Biomed.*, 2011, **24**, 1243.

12. J. L. Izquierdo-Garcia, S. Naz, N. Nin, Y. Rojas, M. Erazo, L. Martinez-Caro, A. Garcia, M.de Paula, P. Fernandez-Segoviano, C. Casals, A. Esteban, J. Ruiz-Cabello, C. Barbas, and J. A. Lorente, *Anesthesiology*, 2014, **120**, 694.

13. E. Maes, C. Mille, X. Trivelli, G. Janbon, D. Poulain, and Y. Guerardel, *J. Biochem.*, 2009, **145**, 413.

14. E. A. Decelle and L. L. Cheng, *NMR Biomed.*, 2014, **27**, 90.

15. S. Moestue, B. Sitter, T. F. Bathen, M. B. Tessem, and I. S. Gribbestad, *Curr. Top. Med. Chem.*, 2011, **11**, 2.

16. A. Elkhaled, L. E. Jalbert, J. J. Phillips, H. A. Yoshihara, R. Parvataneni, R. Srinivasan, G. Bourne, M. S. Berger, S. M. Chang, S. Cha, and S. J. Nelson, *Sci. Transl. Med.*, 2012, **4**, 116ra115.

17. M. M. Mahon, I. J. Cox, R. Dina, W. P. Soutter, G. A. McIndoe, and A. D. Williams, *J. Magn. Reson. Imaging*, 2004, **19**, 356.

18. D. W. Klomp, B. L. van de Bank, A. Raaijmakers, M. A. Korteweg, C. Possanzini, V. O. Boer, C. A. van de Berg, M. A. van de Bosch, and P. R. Luijten, *NMR Biomed.*, 2011, **24**, 1337.

19. K. M. Selnaes, I. S. Gribbestad, H. Bertilsson, A. Wright, A. Angelsen, A. Heerschap, and M. B. Tessem, *NMR Biomed.*, 2012, **26**, 600.

20. D. Monleon, J. M. Morales, A. Gonzalez-Segura, J. M. Gonzalez-Darder, R. Gil-Benso, M. Cerda-Nicolas, and C. Lopez-Gines, *Cancer Res.*, 2010, **70**, 8426.

21. G. Erb, K. Elbayed, M. Piotto, J. Raya, A. Neuville, M. Mohr, D. Maitrot, P. Kehrli, and I. J. Namer, *Magn. Reson. Med.*, 2008, **59**, 959.

22. A. Elkhaled, L. Jalbert, A. Constantin, H. A. Yoshihara, J. J. Phillips, A. M. Molinaro, S. M. Chang, and S. J. Nelson, *NMR Biomed.*, 2014, **27**, 578.

23. G. F. Giskeodegard, S. Lundgren, B. Sitter, H. E. Fjosne, G. Postma, L. M. Buydens, I. S. Gribbestad, and T. F. Bathen, *NMR Biomed.*, 2012, **25**, 1271.

24. G. F. Giskeodegard, H. Bertilsson, K. M. Selnaes, A. J. Wright, T. F. Bathen, T. Viset, J. Halgunset, A. Angelsen, I.S. Gribbestad, and M. B. Tessem. *PLoS One* 2013, **8**(4), e62375.

25. A. Maxeiner, C. B. Adkins, Y. Zhang, M. Taupitz, E. F. Halpern, W. S. McDougal, C. L. Wu, and L. L. Cheng, *Prostate*, 2010, **70**, 710.

26. R. Mirnezami, B. Jimenez, J. V. Li, J. M. Kinross, K. Veselkov, R. D. Goldin, E. Holmes, J. K. Nicholson, and A. Darzi, *Ann. Surg.*, 2014, **259**, 1138.

27. S. Cacciatore, X. Hu, C. Viertler, M. Kap, G. A. Bernhardt, H. J. Mischinger, P. Riegman, K. Zatloukal, C. Luchinat, and P. Turano, *J. Proteome Res.*, 2013, **12**, 5723.

28. J. F. Cobbold, I. J. Cox, A. S. Brown, H. R. Williams, R. D. Goldin, H. C. Thomas, M. R. Thursz, and S. D. Taylor-Robinson, *Hepatol Res.*, 2012, **42**, 714.

29. B. Martinez-Granados, J. M. Morales, J. M. Rodrigo, J. Del Olmo, M. A. Serra, A. Ferrandez, B. Celda, and D. Monleon, *Int. J. Mol. Med.*, 2011, **27**, 111.

30. D. Sakellarious, G. Le Goff, and J. F. Jacquinot, *Nature*, 2007, **447**, 694.

31. A. Wong, B. Jimenez, X. Li, E. Holmes, J. K. Nicholson, J. C. Lindon, and D. Sakellarious, *Anal. Chem.*, 2012, **84**, 3843.

32. L. L. Cheng and U. Pohl, in *The Handbook of Metabonomics and Metabolomics*, eds J. C. Lindon, J. K. Nicholls and E. Holmes, Elsevier: Amsterdam, 2007, 345.

33. K. W. Jordan, C. B. Adkins, L. Su, E. F. Halpern, E. J. Mark, D. C. Christiani, and L. L. Cheng, *Lung Cancer*, 2010, **68**, 44.

34. C. L. Wu, K. W. Jordan, E. M. Ratai, J. Shen, C. B. Adkins, E. M. DeFeo, B. G. Jenkins, L. Ying, W. S. McDougal, and L. L. Cheng, *Sci. Transl. Med.*, 2010, **2**, 16ra18.

35. L. Le Moyec, P. Valensi, J. C. Charniot, E. Hantz, and J. P. Albertini, *NMR Biomed.*, 2005, **18**, 421.

Chapter 10

NMR Studies of Inborn Errors of Metabolism

Sarantos Kostidis[1] and Emmanuel Mikros[2]

[1] Center for Proteomics and Metabolomics, Leiden University Medical Center, P.O. Box 9600, 2300RC Leiden, The Netherlands

[2] Faculty of Pharmacy, University of Athens Panepistimiopolis, 15771 Zografou, Greece

10.1 INTRODUCTION

Inborn errors of metabolism (IEM) form an expanding group of genetic disorders that are caused by an inherited alteration in an enzyme or a functional protein resulting to the interruption of a metabolic pathway.[1] In most cases, this alteration is a result of the abnormal function of the respective genes encoding those enzymes. In the typical IEM case, the reduced activity of specific enzymes in a pathway of the intermediate metabolism results either to the accumulation of upstream substrates or the shortage of downstream products. Some of the accumulated upstream substrates may be further metabolized by alternative secondary pathways resulting in unusual metabolites. Pathogenesis usually derives from toxic activity of the accumulated metabolites, causing a variety of symptoms. In

more severe cases, this can lead to coma or even death if not timely detected. Initial clinical symptoms can occur at any age from prenatal development to adulthood, and interactions with the environment and more importantly nutrition determine the individual patient phenotype.[2,3]

IEM manifestation is extremely diverse, within the same disease, a single genotype can be manifested by more than one phenotype, thus making IEM classification difficult. Approaches include categorization according to the affected organ or organelle, those resulting in acute intoxication, and those comprising energetic process deficiencies[4]; however, there is not a universal classification system.[2] The accumulated metabolites are certainly of great diagnostic importance, occupying a significant role in the pathophysiology of the disease while, very often, classification is related to the detection of those metabolites. The following classification involving both small molecules and organelles is quite frequent: aminoacidemias, organic acidurias, carbohydrate disorders, fatty acid oxidation disorders, purine/pyrimidine metabolism disorders, lysosomal disorders, peroxisomal disorders, mitochondrial disorders, and others.

Although all IEM are considered individually as rare disorders, their collective burden is considerable. The occurrence of each IEM mainly depends on the geographical and ethnic composition of the population.[5] The overall number is expected to increase owing to

NMR in Pharmaceutical Sciences. Edited by Jeremy R. Everett, John C. Lindon, Ian D. Wilson, and Robin K. Harris
© 2015 John Wiley & Sons, Ltd. ISBN: 978-1-118-66025-6
Also published in eMagRes (online edition)
DOI: 10.1002/9780470034590.emrstm1400

the progress in molecular biology until all variants of the enzymes and transporters that are involved in the homeostatic mechanisms of the human are fully identified.[2] Unraveling the human genome will enable the discovery of the etiology and pathophysiology of newly recognized disorders with a range of clinical findings not yet related to any defective gene. Moreover, the development of diagnostic methods and the expansion of neonatal screening will facilitate improved awareness and discovery of new diseases.

The IEM are a serious public health concern, as a delayed diagnosis can lead to intensely long treatment. Thus, early diagnosis and prompt therapy are critical for preventing development of severe symptoms and minimizing the likelihood of premature death. Neonatal control (Newborn Screening) is one of the best and most important programs in current preventive medicine. Detecting specific metabolic and endocrine disorders in the neonatal population is an important health care aspect organized in the Western world mainly for phenylketonuria (PKU), congenital hypothyroidism, and transferase deficient galactosemia, and most states test for hemoglobinopathies such as sickle cell disease using as the sample collection method the newborn's dried blood spots (DBS) on paper Guthrie cards.

Recent advances in tandem mass spectrometry have permitted evaluation in a time- and cost-effective manner of a number of metabolites from small samples. The methods involve screening of various body fluids for organic acids (in urine), or amino acids (AAs) and fatty-acid-conjugation products (acylcarnitines in serum/plasma). This approach has reduced the times required for diagnosis improving the management of both classical and complex metabolic disorders. However, further progress to improve diagnosis and personalized treatment of more complex phenotypes is needed and the development of new methods is imperative both for understanding and managing IEM.

One method of analysis that shows great potential is NMR spectroscopy, which was already used back in 1985 to diagnose various types of organic acidurias.[6] In contrast to other techniques, NMR spectroscopy captures the metabolic profile by detecting both known and unknown metabolites in a nontargeted way. The rapid and extremely reproducible, simultaneous detection of a large number of known and unknown chemical substances over an extended concentration range renders NMR a versatile tool for newborn screening. Modern spectrometers are nowadays capable of analyzing biological samples such as

urine, plasma, serum, and cerebrospinal fluid (CSF) with high-throughput automatic routines, coupled with powerful data analysis methods. The holistic nature of the method allows for the detection of metabolites, which are not determined routinely by other techniques such as mass spectrometry (MS).

The main limitation of NMR is the inherent low method sensitivity. The detection limit for body fluid NMR spectroscopy is in the micromolar range and depends on the field strength, probe technology, proton chemical equivalence, and signal overlapping. Apart from the ex vivo study of ^{1}H NMR spectroscopy, magnetic resonance spectroscopy (MRS) has been also utilized for the in vivo determination of intracellular concentration for key metabolites in the disease site. Both approaches have been reviewed in detail for the period up to 2008.[6-9] In this article, we highlight methodological issues, provide routine application examples, and present some outstanding recent conceptual applications of NMR spectroscopy in the field of diagnosis and management of IEM.

10.2 BODY FLUID NMR

10.2.1 Body Fluids for NMR-based IEM Diagnosis

Urine is the most commonly used body fluid for IEM screening as it is easily collected in large quantities, allows for multiple sampling per day, and sample preparation for NMR analysis is minimal. More than 2000 small molecular weight metabolites (MW < 1500 Da) are present in urine and can potentially be identified by a single ^{1}H NMR experiment.[10,11] However, practice has shown that per sample, some tens (up to ∼200) of compounds can be annotated and quantified owing to either the extensive overlap of signals in the proton spectrum or the inherent sensitivity of NMR. An important characteristic of urine composition is the high variability of metabolites levels, which hinders the definition of 'normal' physiological ranges for healthy human. In order to overcome this, large population of samples need to be analyzed and form the 'normal condition' dataset, which in turn can be the basis for rapid screening and discovery of abnormal spectral data.

Blood plasma and serum are also used for IEM diagnosis by NMR, albeit less frequently than urine owing to the more invasive sampling method. A technique that makes blood collection less invasive and thus more appropriate for neonates makes a pinprick puncture in one heel, collects a minimal quantity of blood on filter papers, and stores it as DBS.[12] The rich content of serum/plasma in various lipids (including triglycerides, phospholipids, and cholesterol esters) bounds into lipoprotein particles, and offers the possibility to monitor changes in the organization of lipids within lipoprotein particle density subclasses using several NMR techniques.[13] This is particularly useful if, for example, suspicions of inherited defects of lipid metabolism exist.

CSF sampling is far more invasive than blood and urine and it is less common in the field of IEM screening. However, as the composition of CSF is dependent on metabolite production/consumption rates in the brain, it is of particular interest for screening of disorders related to the central nervous system (CNS).[8,9,14]

Typically, about 1 ml of body fluid is required for a conventional NMR system that uses 5 mm probes. However, this quantity can be reduced down to a few tens of microliters using microprobe technology. The 1 mm microprobes require about 10 to 20 µl of NMR sample, which is particularly useful for studies focusing on DBS or CSF analysis.

Sampling techniques for IEM screening should be considered as a very critical step, as they can severely affect the NMR analysis results if proper treatment of samples is not followed. Specifically in the case of IEM screening, it is common practice that sampling of material and NMR analyses are often conducted by more than one site (hospitals, clinics, and NMR laboratories) and may last for several years, especially in cases of patient follow-up or large cohort studies. In order to minimize any form of additional variation to the already complex content of body fluids, standard operating protocols (SOPs) should be agreed and followed for the whole duration of a study.[10] There is extensive research in sampling methods, mainly aiming for metabolomics studies and a brief overview of them adjusted for NMR-based IEM diagnostic screening is presented in the next section.

10.2.2 Sample Collection, Storage, and Preparation

10.2.2.1 Urine

Urine is preferably collected in the morning after 8–12 h of fasting unless sampling of multiple time points per day is required. In the case of neonates, urine collection bags are commonly employed. If long-term storage is intended, addition of a preservative (sodium azide, 0.01–0.1%) is recommended before storage at $-80\,°C$.[10,11] Fresh urine contains human cells, bacteria, fungi, and noncellular components and additional steps such as filtration or mild centrifugation will further improve sample stability.[15] Urine sample processing should occur within a short time. Samples left at $4\,°C$ for an extended period of time and without the use of any preservative will more likely exhibit changes in their NMR spectra owing to bacterial and enzymatic activity.[10,11,15] Sample preparation for NMR includes centrifugation at $4\,°C$ and addition of a buffer to minimize pH variation among the samples. Human urine has a broad range of pH, from ~5 to 8, and contains a variable amount of ionic species (Na^+, Ca^{2+}, K^+, and Mg^{2+}) that can affect the individual resonance ranges. In general, the use of phosphate buffer (pH 7.0–7.4, 0.3–1 M) in H_2O/D_2O is preferred.[11] However, excellent work by Wevers and coworkers of IEM diagnosed by NMR using human urine has been conducted at pH 2.5.[8,9] A reference compound is also added to the buffer for chemical shift calibration and quantitation and for urine, 4,4-dimethyl-4-silapentane-1-sulfonic acid (DSS) or 3-trimethylsilylpropionic acid (TSP) with the methylene groups deuterated are commonly used.

10.2.2.2 Blood

The important factors that should be controlled to avoid biased results from blood sampling are clotting times, freeze/thaw cycles, and temperature.[11,13] It is recommended to process blood into serum immediately and not more than 2 h after collection at $4\,°C$[15] as prolonged periods at room temperature may induce sample degradation that mainly affects glucose, lactate, and pyruvate resonances. Serum separation is achieved by allowing the sample to coagulate at room temperature for ~30 min followed by centrifugation at $15–25\,°C$. For plasma separation, blood is collected directly into tubes containing an anticoagulant agent such as lithium heparin, ethylene diamine tetraacetic acid (EDTA), and sodium citrate or oxalate, and subsequently, centrifuged at $4\,°C$ before storage.[11,13] From the NMR point of view, lithium heparin is preferable as use of EDTA and citrate results in intense peaks in the proton spectrum. Both plasma and serum should be subsequently stored at $-80\,°C$.

Before NMR analysis, samples are mixed with phosphate buffer (pH 7.0–7.4, H_2O/D_2O solution), including a reference compound (TSP, DSS). In addition, filtration or the extraction of serum/plasma followed by drying may also be used in order to further remove proteins.[11,13]

Processing of DBS samples requires an extraction step to collect the blood content from the filter papers. The extraction with D_2O or methanol-d_4 has been proposed with the first being more suitable for NMR as only one broad signal of residual water has to be suppressed in the proton spectrum.[12] Owing to the limited quantities of blood stored on the filters (~10–20 μl blood per spot), it is common to cut more spots to produce one blood NMR sample. Clearly, the selected volume for extraction and NMR sample preparation is dependent on the NMR hardware availability. Microprobes and cryogenically cooled probes (cryoprobes) as well as systems of higher field (600 MHz and up) are more suitable for analysis of DBS. Moreover, care has to be taken regarding the times and the temperature of DBS storage as spots stored more than a week at ambient temperature are prone to changes in their metabolites content. It is recommended to store the DBS below −20 °C, preferably at −80 °C until NMR analysis to avoid sample degradation.

10.2.2.3 Cerebrospinal Fluid (CSF)

Recommendations for CSF handling include centrifugation immediately after collection to remove erythrocytes and white blood cells, and storage at −80 °C.[14] Deproteinization by filtration is not always followed as the protein content in CSF is significantly less than plasma and serum. For NMR sample preparation, phosphate buffer in D_2O (pH 7–7.4) is added, including an internal standard (TSP, DSS) and sodium azide. Processing of CSF for NMR sample preparation should take place over a short time as an increase of pH might occur when the sample is left standing at room temperature or even when stored at −20 °C. This phenomenon has been attributed to CO_2 evaporation.[14]

10.2.3 NMR Measurements of Body Fluids for IEM Diagnosis

The selection of NMR experiments that may be used for IEM screening depends on the study specific purpose, the kind of collected samples, and the hardware availability.[16] Almost always, the fast, simple to set up, and fully automated 1D ^{1}H NMR experiments are employed for a robust initial screening. About 60–100 samples can be measured within 24 h and this number is increased even more if a flow-injection system is available. A drawback of the 1D ^{1}H NMR of body fluids is that the spectra are very crowded and peaks are overlapped within a narrow region of ~10–12 ppm, leading to uncertainty in detection or quantitation of several metabolites. This problem can be overcome with homo- or heteronuclear 2D NMR techniques, which provide a much better resolution and allow for the assignment of more compounds. On the other hand, 2D NMR techniques require longer experimental times and this hampers their use for high-throughput screening. Exceptions are the 2D [1]H J-resolved (Jres) experiment, which is nowadays routinely used in body fluids NMR, and some recent developments in 2D NMR techniques, which are briefly overviewed in the following paragraphs.

Higher observation frequencies (^{1}H, 750–1000 MHz) obviously provide better spectral resolution and sensitivity, but the high capital cost of such instruments also increases the cost of such NMR assays. Site-comparison studies with replicates measured in several laboratories on the same or similar field-strength spectrometers resulted in reproducible metabolic profiles.[10,11] A compromise between the cost, the speed, and sensitivity factors together indicates that NMR systems of 500 or 600 MHz provide an excellent choice for IEM screening.

One common characteristic of urine, serum, plasma, and CSF is that they all contain water. Therefore, it is necessary to apply solvent suppression techniques in order to increase the dynamic range of the method and minimize the water resonance in the NMR spectrum. Several pulse sequences exist for this purpose, including presaturation, NOESY-presaturation, water suppression through gradient-tailored excitation (WATERGATE), water suppression enhanced through T_1 effects (WET), and excitation sculpting. Among them, the NOESY-presaturation method is preferred for body fluids as it requires little optimization and produces excellent quantitative results.[11,17]

In the following paragraphs, we provide an overview of the NMR experiments, which have been or could be used for the detection of IEM, with emphasis given to high-throughput screening techniques and developments.

10.2.3.1 One-dimensional ^1H Nuclear Magnetic Resonance

The best way to perform fast and quantitative NMR analysis of any kind of biofluid is to use the 1D ^1H NOESY-presaturation experiment. The pulse program employs the first increment of a 2D NOESY pulse sequence with water irradiation during the relaxation delay (RD) and the mixing time (τ_m) and has the form (RD)-90°-τ_1-90°-τ_m-90°-AQ. An τ_m of 100 ms is usually applied for all biofluids and the 90° pulse can be automatically calibrated for each sample. Figure 10.1 shows one typical 1D ^1H spectrum of urine from a neonate, employing the 1D NOESY-presaturation experiment.

Owing to the presence of macromolecules (MW > 1500) mainly in serum and plasma and less in CSF, the 1D NOESY spectra of these fluids are dominated by several broad signals, which induce difficulties in the quantification of overlapped sharp resonances. Special NMR editing techniques are available and employed for simplification of the spectra. The Carr–Purcell–Meiboom–Gill (CPMG) spin-echo experiment with presaturation utilizes a transverse relaxation (T2) filter that removes or attenuates the broad resonances of molecules having short T2 relaxation times. In contrast to the 1D NOESY experiment, the CPMG pulse sequence can only be used for semi-quantitative analysis because of dissimilarities in signal attenuation of compounds exhibiting different T2 relaxation times. On the other hand, a diffusion-edited experiment takes advantage of the distinctive translational diffusion rates of molecules in solution based on their size and shape and efficiently eliminates the sharp signals of small molecules leaving only the broad resonances of macromolecules. Combination of the CPMG and the diffusion-edited experiments allow for the holistic monitoring of both small molecular weight metabolites and the lipoprotein profile within just ~4.5 min per sample on a 600 MHz NMR using a 5 mm cryoprobe.

10.2.3.2 Two-dimensional Nuclear Magnetic Resonance

The 2D Jres NMR technique is routinely used in the analysis of body fluids together with the aforementioned 1D experiments. It separates J-couplings and chemical shifts in the 2D plane and provides a simplified proton decoupled projection spectrum (p-Jres) with only singlet peaks (Figure 10.2a and b). The Jres experiment can be acquired within a few minutes and consists together with 1D NOESY the standard high-throughput set of measurements for all biofluids.

Other useful and common homonuclear (^1H–^1H) 2D NMR techniques are the correlation experiments COSY and TOCSY.[13,16] They are used to gain structural information on spin systems belonging to the same compound and assist the unambiguous annotation of spectra. Owing to their prolonged experimental times, COSY and TOCSY are acquired for selected spectra only and often in cases of targeted analyses.

Considering the very low natural abundance of ^{13}C (~1.1%), heteronuclear NMR methods rely on initial polarization and detection of the sensitive nuclei (e.g., ^1H) via transfer of magnetization through the insensitive ones (^{13}C, ^{15}N). The ^1H–^{13}C heteronuclear single quantum correlation (HSQC) experiment is mainly used from this category of experiments as it contains valuable structural information well resolved in the ^{13}C dimension, which minimizes peak identification ambiguities. Figure 10.3 shows an HSQC spectrum of urine from a newborn. In this particular example, the distortionless enhancement through polarization transfer (DEPT) sequence is implemented in the HSQC experiment, providing additional information by differentiating between primary, secondary, and tertiary carbons.

Developments in the pulse sequences of HSQC have improved its sensitivity and reduced the experimental times; however, it has not yet been applied in routine high-throughput NMR screening for IEM diagnostic purposes. Shanaiah *et al.*[18] presented an interesting approach that uses the ^1H–^{13}C HSQC for spectral annotation and quantitation of AAs related to IEM. They employed a fast chemical derivatization step to enhance the ^{13}C NMR sensitivity using a ^{13}C enriched acetic anhydride in aqueous medium, which induced acetylation of AAs and other compounds. This way, they managed to quantify acetylated metabolites present in the ^1H–^{13}C HSQC spectra of serum and urine on a time scale that allows routine NMR screening.

There is increasing interest in the area of developing fast quantitative multidimensional NMR techniques for the comprehensive analysis of complex mixtures.[19] Efforts are made for the reduction

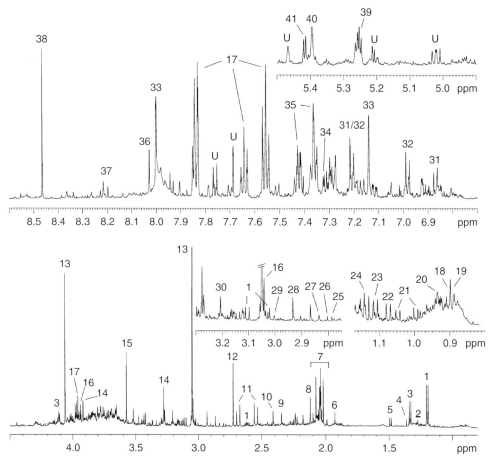

Figure 10.1. Regions of a 600 MHz 1D ^{1}H NMR spectrum of urine from a neonate depicting the assignment of the most notable resonances. 2D NMR techniques are needed for the unknown peaks. Annotations: 3-aminoisobutyric acid, 1; 3-hydroxyisovaleric acid, 2; lactic acid, 3; 2-hydroxyisobutyric acid, 4; alanine, 5; acetic acid, 6; acetyls, 7; dimethylsulfide, 8; pyruvic acid, 9; succinic acid, 10; citric acid, 11; dimethylamine, 12; creatinine, 13; betaine, 14; glycine, 15; creatine, 16; hippuric acid, 17; isocaproic acid, 18; isovaleric acid, 19; leucine, 20; valine, 21; methylglutaric acid, 22; 2-methylsuccinic acid, 23; propionylglycine, 24; ephedrine, 25; succinimide, 26; methylguanidine, 27; dimethylglycine, 28; ketoglutaric acid, 29; carnitine, 30; tyrosine, 31; chlorotyrosine, 32; histidine, 33; gentisic acid, 34; phenylalanine, 35; methylxanthine, 36; hypoxanthine, 37; formic acid, 38; glucuronic acid, 39; allantoin, 40; sucrose, 41; unknown, U

of time increments along the indirect dimension (t_1), which is the limiting factor for high-throughput applications, without sacrificing resolution and sensitivity. One of such methods, covariance processing, uses the spin evolution times as the determining factor of variation and provides the same resolution among the direct and indirect dimensions independent of the number of t_1 increments. An alternative technique, termed *doubly indirect covariance processing*, combines 2D ^{1}H–^{1}H COSY and ^{1}H–^{13}C

HSQC spectra and linear algebraic tools to construct a high-resolution ^{13}C–^{13}C correlation spectrum.[19] Other significant advances in the area of fast multidimensional NMR include the reduction of interscan delay with methods such as the band-selective optimized flip-angle short-transient (SOFAST) and the band-selective excitation short-transient (BEST), acceleration by sharing adjacent polarization (ASAP), and small recovery times (SMART).[20] In addition, techniques applying nonuniform sampling (NUS)

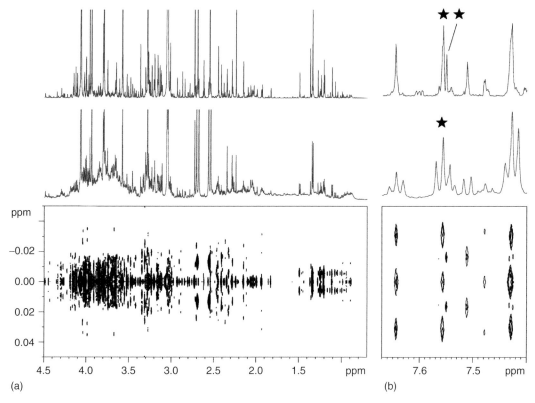

Figure 10.2. (a) Region of a 600 MHz 2D J-resolved NMR spectrum of urine and the corresponding 1D ^{1}H (red) and p-Jres (blue) spectra. The crowded and severely overlapped region of 3–4.2 ppm in the 1D ^{1}H is better resolved in the p-Jres. (b) Expanded spectral window of the 2D Jres. The triplet of hippuric acid indicated with a star on the normal 1D ^{1}H spectrum (red) is overlapped with a doublet from an unknown metabolite as it is shown in both 2D Jres and 1D p-Jres (blue) spectra

and ultra-fast (UF) 2D NMR principles have also been proposed.[21] Although, not yet used for IEM diagnosis, these developments clearly show that fast multidimensional NMR is becoming a realistic goal and may be used in the arsenal of NMR techniques for holistic and accurate screening in the near future.

10.2.4 NMR Data Analysis

Processing of acquired NMR data of biofluids is nowadays performed with automatic routines, which include zero filling and line broadening, chemical shift referencing, phasing, and baseline correction. Two main approaches exist for the analysis of NMR data acquired for disease diagnostic purposes. The

first approach, so-called untargeted profiling, treats the full spectral data of a sample as a metabolic fingerprint and tries to identify differences or similarities across many samples through multivariate statistical processing.[10,16] Typically, the processed spectra are divided into equidistant or variable size integrals (bins) to form a numeric matrix of N rows (samples) and K columns (variables, bins), which in turn is analyzed using chemometric techniques such as principal components analysis (PCA), projections to latent structures – discriminant analysis (PLSDA), and soft independent modeling of class analogy (SIMCA), among others.[10,14,16] An advantage of this approach is that it is very fast as no peak assignment is needed and allows for the robust identification of atypical fingerprints. The resulted discriminating features (integrals), if any, are then further explored

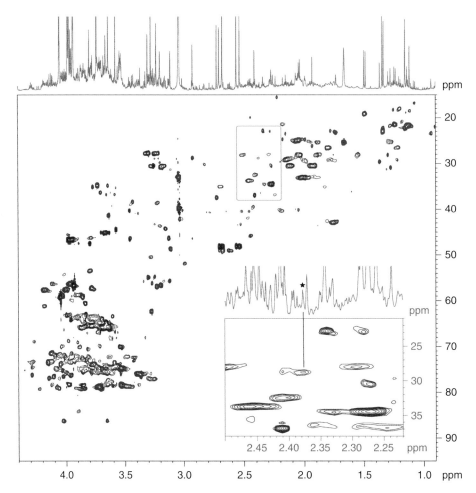

Figure 10.3. Region of a 600 MHz ^{1}H–^{13}C 2D edited HSQC spectrum of urine. Implementation of the DEPT135 sequence within the HSQC experiment allows for the discrimination of primary and tertiary carbons (blue peaks) from the secondary ones (red peaks). In this particular case, the expanded region indicates the presence of oxalacetic acid (CH$_2$; negative peak) rather than the commonly found in urine pyruvic acid (CH$_3$). The protons (and attached carbons) of these two compounds share the same chemical shift at pH 7, but can be correctly annotated with this version of HSQC

with several methods, including 2D NMR, in order to identify the corresponding metabolites. As the full spectral window is encountered, this approach is sensitive to experimental variations, such as improper phasing, baseline artifacts, and drifting of resonances due to matrix effects (pH, ionic strength). Several techniques have been developed to deal with these problems such as the adaptive binning and the adaptive intelligent binning, which produce bins of variable size based on the actual positions and lineshapes of resonances, and peak alignment routines that correct

peak drifting.[10] In addition, mathematical manipulations are applied to correct the baseline such as polynomial subtraction, cubic spline, and penalized parametric smoothing on the frequency domain or backward linear prediction that acts on the time domain.[17]

The second method to analyze NMR data of complex mixtures is the targeted profiling approach, in which metabolite resonances are identified and quantified and their concentrations are recorded. This approach uses classical quantitative NMR techniques and benefits from the fact that each metabolite has a unique fingerprint in the spectrum provided that pH and temperature control have occurred during acquisition of data. Despite the complexity of the spectra, the use of reference compounds data acquired under the same conditions makes the assignment and quantitation process less time consuming, often performed automatically.[10,11,16,17] Many NMR laboratories have developed their own in-house spectral libraries and automation routines, or use commercial ones accompanied with dedicated software for peak identification and quantitation (MestReNova, Chenomx, and Bruker's AMIX with BBIOREFCODE-2.0 database). There are also some freely available online databases (HMDB, BMRB, MetaboLights), which can be used for manual annotation; however, in this case, the conditions under which the data has been acquired may be different or sometimes poorly described. Quantitation of metabolites is performed either through single peak integration and comparison with the integral of an internal standard (TSP, DSS) or by peak deconvolution based on a reference compound library. Considering the Lorentzian nature of NMR resonances their tails are usually overlapped with neighboring peaks (especially in crowded spectra of biofluids) and this makes single integration vulnerable to produce biased results. For this reason, signal deconvolution is currently preferred for the quantitative analysis of body fluids.[17] It is based on fitting reference peak shapes of the identified molecule to the selected resonance in the mixture spectra and subsequently extracting the fitted peak for quantitation. An alternative method for deconvolution that does not require reference compounds spectral libraries is to compare the deconvoluted resonance with that of an internal standard of known concentration (e.g., TSP).

Often, in cases of NMR screening of urine for IEM diagnosis, a relative quantification approach is used and metabolite levels are expressed with regard to the creatinine quantity in the sample (mmol of compound/mol of creatinine).[8,9] This approach is based on the correlation of creatinine's excretion rate to the overall body weight and that for healthy humans, there is little metabolic variance in excretory levels. Independently of the method for quantitation, parameters such as the relaxation times of individual protons, the J-couplings, and the signal-to-noise ratio of peaks should be considered for accurate results. For a comprehensive overview of the aspects of quantitative NMR in the field of body fluids analysis, we suggest the review of Barding

et al.[17] In the following sections, we present some examples of NMR-based screening for diagnosis of IEM in which the aforementioned principles are used.

10.3 APPLICATIONS OF NMR IN IEM DIAGNOSIS

10.3.1 Selected Cases

More than 90 IEM can be diagnosed utilizing NMR spectra of body fluids including most organic acidurias, aminoacidurias, purine, and pyrimidine metabolism disorders. A comprehensive and very useful and detailed handbook has been published by Wevers and coworkers, including the characteristic parts of urine NMR spectra of detectable disease along with a short description.[9]

The major advantage of NMR is the possibility to simultaneously detect more than 200 metabolites in a urine ^{1}H NMR spectrum and about 100 in a CSF sample. Notably, among these metabolites, several cannot be routinely determined by MS. Characteristic examples of these metabolites are 1-methylnicotinamide, allantoin, betaine, choline, creatine, dimethylamine, dimethylglycine, *myo*-inositol, urocanic acid, and trimethylamine (TMA) in urine; 2-hydroxyisovaleric, 3-hydroxyisovaleric, acetoacetic acid, citric acid, and creatine in plasma; and 2-hydroxyisovaleric acid, 3-hydroxybutyric acid, 3-hydroxyisovaleric acid, creatine, glycolic acid, *myo*-inositol, and *N*-acetylneuraminic acid in CSF.[8]

In the next four examples, the diagnosis of well-known IEM by the use of NMR will be concisely presented, highlighting in two of those cases the advantages of untargeted detection.

10.3.1.1 Propionic Aciduria

Propionic aciduria is one of the first IEM for which an NMR-based diagnosis was described.[6] That disease is due to a deficiency of the mitochondrial biotin-dependent enzyme propionyl-CoA carboxylase and results in the accumulation of propionyl-CoA and subsequent excretion of a number of abnormal propionyl-derived metabolites. Methylcitrate is formed by competition of propionyl-CoA with acetyl-CoA for oxaloacetate and it is regarded as a

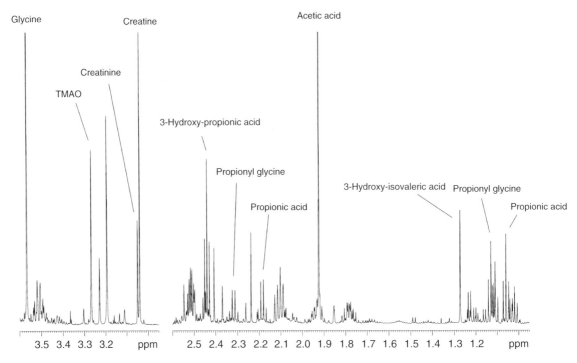

Figure 10.4. 600 MHz ^{1}H 1D NMR spectrum of urine from a newborn diagnosed with propionic aciduria. Characteristic metabolites of this disorder are simultaneously detected with a single 1D NMR experiment

characteristic diagnostic metabolite for this disorder. However, methylcitrate is usually not detected in urine NMR spectra possibly owing to binding. Characteristics in NMR are the signals of propionyl metabolites: 3-hydroxypropionate, propionylglycine, and tiglylglycine (an isoleucine catabolite). Meanwhile, in the NMR spectrum presented in Figure 10.4 of a propionic aciduria sample, propionic acid and relatively high concentrations of glycine, creatine, acetic acid and 3-hydroxyisovaleric acid are also detected.

10.3.1.2 Methylmalonic Aciduria

Methylmalonic aciduria results from the deficiency of the adenosylcobalamin-dependent mitochondrial enzyme methylmalonyl-CoA mutase that converts methylmalonate to succinate. Perturbations of the cofactor vitamin B_{12} of the enzyme can also lead to disease. Increased signals of methylmalonic acid and 3-hydroxypropionic acid are observed in the urinary ^{1}H-NMR spectrum.[9]

In the spectrum presented in Figure 10.5, intense signals of methylmalonic acid are observed; however, it is important that the presence of urocanic acid can be detected simultaneously. Urocanic acid is not routinely detected by MS screening and its presence may constitute evidence of the existence of another metabolic disease, urocanic aciduria. Urocanic acid is a histidine metabolism product transformed through urocanase (hydratase enzyme of urocanic acid) to glutamic acid. Urocanate accumulation suggests deficiency of this enzyme. In this case, the untargeted NMR analysis provides evidence for diagnosis that would otherwise be easily neglected. In the spectrum, the high concentrations of trimethylamine N-oxide (TMAO), carnitine, and acetylcarnitine that can be also observed should be taken into account for final diagnosis and personalized treatment of the patient.

10.3.1.3 Type II Prolinemia

Type II prolinemia is due to the deficiency of pyrroline-5-carboxylate (P5C) dehydrogenase.

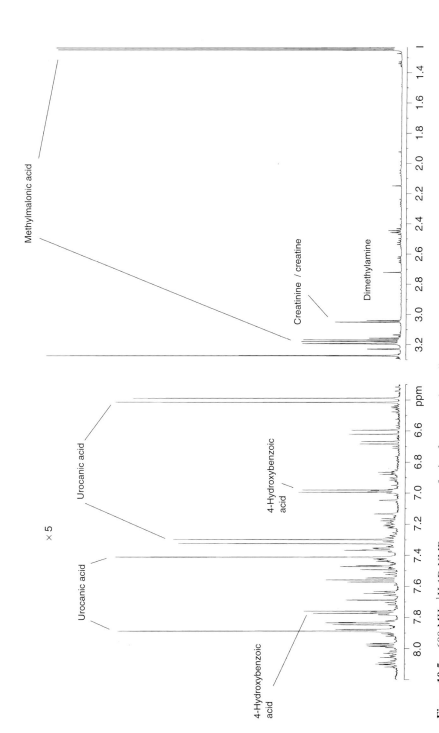

Figure 10.5. 600 MHz ^{1}H 1D NMR spectrum of urine from a newborn diagnosed with methylmalonic aciduria. Except for the methylmalonic acid, high levels of urocanic acid and 4-hydroxybenzoic acid are also detected in the same spectrum. These metabolites were not detected with MS-based targeted screening

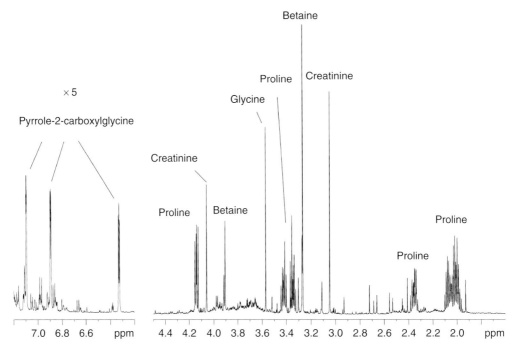

Figure 10.6. 600 MHz ^{1}H 1D NMR spectrum of urine from a newborn diagnosed with type II prolinemia. This disease is characterized by increased levels of pyrrole-2-carboxylglycine, as well as for proline. A single 2 min ^{1}H NMR experiment is able to detect both markers

The characteristic metabolite of this disease is pyrrole-2-carboxylglycine, the accumulation of which differentiates it from prolinemia type I (in both disorders proline also accumulates). This metabolite is not detectable in routine MS screening while is easily detected by NMR.[9]

In the NMR spectrum in Figure 10.6, major differences from a typical urine metabolic profile can be observed concerning mainly proline and pyrrole-2-carboxyglycine; however, other metabolite levels such as betaine and hippurate are outside the normal range. Betaine is a metabolite usually not detected by routine screening. In that case, NMR-based studies of this IEM highlighted that neonates are characterized by increased betaine levels decreasing with age.[8]

10.3.1.4 Trimethylaminuria

A last example where NMR spectroscopy provides a rapid and unique method for diagnosis of IEM is trimethylaminuria or 'fish odor syndrome'. TMA is produced by bacterial activity from dietary choline, it

is normally oxidized in the liver to the corresponding N-oxide (TMAO) and both are secreted in urine. Fish odor syndrome is related with the reduced activity or expression of flavin-containing monooxygenase (FMO3) and this could be genetic (the most frequent), acquired from viral infections, or transient related with increased consumption of choline derivatives.[8]

Patients with trimethylaminuria have the smell of rotten fish, but they are usually not aware of it and thus they face psychological and social isolation problems. NMR determination of trimethylaminuria can be rapid without any derivatization or sample preparation, providing detection and quantification of both TMA and TMAO along with all other metabolites that, notably, cannot be easily detected through routine conventional screening.

10.3.2 Novel Diseases and New Applications

By exploiting the NMR capabilities of holistic inspection of the metabolism, novel IEM have been identified while its application to known diseases provided new valuable insights. The novel inborn errors of metabolism characterized by NMR are dimethylglycine dehydrogenase deficiency, ureidopropionase deficiency, and a defect in polyol metabolism, the ribose-5-phosphate isomerase deficiency. The metabolites associated with these diseases are dimethylglycine, ureidopropionic acid, ureidoisobutyric acid, and arabitol and ribitol molecules, which are not detected by routine screening techniques.[8] All three have been reviewed in detail hence we decided to make reference in this review to more recent studies on genetic disorders for which NMR had never been applied to the past.

10.3.2.1 Genome-wide Association Study Combined with NMR-based Metabolome Profiling

In a recent NMR-based genome-wide association study (GWAS) by Suhre *et al.*,[20] a number of interesting findings concerning metabolic traits related to the detoxification machinery and kidney functionality of the human were reported. In this study, the authors used simple one-dimensional urine spectra in an effort to identify distinct and clearly differentiated metabolic phenotypes [genetically determined metabotypes (GDMs)], which could assist in rationalizing the metabolic basis for a number of diseases and pathological conditions related mainly to the kidney. For this purpose, a large number of urine samples were used in combination with replication studies, including a 5-year follow-up, reaching a total of 2893 points. More specifically, an initial set of 862 male samples were obtained and analyzed by ^{1}H NMR spectroscopy using a 400 MHz instrument, and the concentration of 59 metabolites and 1661 ratios between metabolite concentrations were examined for their possible associations with 645 249 autosomal single-nucleotide polymorphisms (SNPs). Among the five genetic loci demonstrating statistically significant associations with the studied SNPs, the most interesting finding concerned the gene encoding the enzyme alanine-glyoxylate aminotransferase-2 (AGXT2). The encoded enzyme is in turn a mitochondrial aminotransferase that is primarily expressed in the kidney and regulates the catabolism of metabolite β-aminoisobutyrate (BAIB) by catalyzing its reaction with pyruvate to afford 2-methyl-3-oxopropanoate and alanine. Given that high excretion levels of BAIB are strongly correlated to the pathological state called

hyper-β-aminoisobutyric aciduria, the authors suggested that this particular gene identified through the aforementioned NMR-based GWAS, which is notably the first published study of this kind using urine, can be considered as the genetic basis of this metabolic disease.

10.3.2.2 Following Up and Directing Disease Treatment

A very interesting application of NMR-based metabolomics in clinical practice is described in the study of Legido-Quigley *et al.*,[22] where monitoring of the metabolome of a single patient diagnosed with ornithine transcarbamylase (OTC) deficiency assisted rationalization as well as successful implementation of the therapy consisting of hepatocyte and subsequently liver transplantation. In this case, NMR was used as a real-time versatile tool for measuring metabolites that were considered critical for determining the course of treatment progress and response to therapy. Continuous monitoring of urine and plasma metabolite profiles over the 350-day treatment period was shown to be valuable in assisting fast and efficient decision-making in clinical practice with respect to necessary therapy readjustments or fine-tuning. Moreover, the NMR determination of high ammonia levels, the major metabolic abnormality derived from the unpaired urea cycle due to OTC, contributed to the successful implementation of liver transplantation instead of continuing hepatocyte transfusion by the time this approach would no longer be capable to sustain a therapeutic effect to the OTC patient.

10.3.2.3 Respiratory Chain Diseases Fingerprint

An additional case where an NMR metabolite measurement approach was applied with success for deriving an IEM biomarker signature can be found in the work of Smuts *et al.*,[23] which is, to our knowledge, the first contribution of NMR-based metabolic profiling to the study of respiratory chain deficiencies (RCDs). The RCDs comprise a range of relatively underexplored conditions yet are highly challenging in terms of correct diagnosis mitochondrial abnormalities. The authors present a combined NMR- and MS-based analysis of 101 pathological urine samples that concludes to the establishment of a group of 13 metabolites as a basic biosignature, rather than a single biomarker, for efficient RCD detection. Considering that accurate diagnosis of the different RCDs can be performed by only using highly invasive techniques, the possible application of a NMR-based profiling diagnostic test for prescreening could be of high clinical value, especially for infant and children patients.

10.3.2.4 Glycogen Storage Disease Type 1a

Duarte *et al.*[24] describe an additional case where biofluid metabolite analysis using NMR succeeded in identifying a possibly useful metabotype of the pathological condition named glycogen storage disease type 1a (GSD1a). In this first NMR study of that specific IEM impairing the liver ability to release free glucose, spectral analysis of five pathological and five healthy samples was focused on the methyl region of plasma lipoproteins. The result was a quite clear metabolite pattern capable of discriminating GSD1a individuals, with the notable exception of creatine and creatinine levels that were found to contribute relatively inconclusively to the model. Although not a full metabolomics study due to the small sample size, this example further demonstrates the wide applicability range of NMR as a valid diagnostic tool for several underexplored IEM cases.

10.3.2.5 Use of Multivariate Modeling

All advantages previously mentioned, i.e., simultaneous and untargeted detection, rapid and extremely reproducible measurements, and minimal sample preparation, render NMR as the method of choice to be combined with multivariate chemometric methods of data analysis under the metabonomics approach as described by Nicholson *et al.*[25] The possibility however to expand NMR applications to routine IEM diagnosis still requires long steps toward automation of detection and quantification and, more importantly, generation of universal databases. Although physiological values for metabolite concentrations do exist, it is evident that specifically in newborn screening it is important to define the influence of different factors such as population genetics, nutrition, drug treatment, and environment in general. Fluctuations of the neonate metabolism have been fairly heavily studied,[26] revealing at least six metabolites varying within the first days of the newborn life such as *N*-methylnicotinamide (NAD$^+$ pathway), formate, hippurate, betaine (kidney development), taurine (neuronal development), and

bile acids (hepatic clearance). In this aspect, multivariate statistics is of great importance to the generation of population models. Early application of the NMR-based multivariate modeling approach to IEM suggested that diagnosis of different diseases is feasible through the generation of adequate models for both urine and blood-spot-extract analysis.[11,27] The authors focused on two specific IEM, namely PKU and maple syrup urine disease (MSUD), and created a model utilizing simple multivariate statistical analysis such as PCA and PLSDA, which could efficiently discriminate pathological from healthy samples. The major limitation of this NMR-based method is related to the limit of detection of metabolites derived by extraction of blood spots, as this is inherently defined in that spectroscopic application.

More recently, a very interesting example of implementing the previously described methodologies in an automated manner is described in the study of Aygen *et al.*,[28] concerning screening of newborns for IEM in Turkey, a country showing high prevalence of diseases of this kind. In this study, the authors aimed at creating a database of ~1000 samples and subsequently a model that could serve as a routine and automatic tool for diagnosing congenital metabolic diseases. The untargeted PCA approach was successful at predicting the small number of confirmed pathological samples that were included in the total population. More importantly, the study succeeded in showing the importance of atypical profiles in such massive screening efforts, as the percentage of such samples was found to be markedly high, reaching almost 20% eventually posing obstacles to statistical analysis, especially using the untargeted approach.

A rather distinct approach that involves NMR-based metabolomics aiming at the detection of metabolic diseases is the very recently introduced statistical health monitoring (SHM) method by Engel *et al.*[29] In this particular methodology, attention is principally focused on healthy representatives of a population. Using metabolome data derived from healthy individuals along with sophisticated multivariate statistical analysis, SHM can be used to create a 'healthy metabolome profile' at a high degree of precision. Having such a consistent model describing the healthy metabolites landscape, the technique can be subsequently used for discriminating pathological samples by simply identifying abnormal metabolite concentrations or patterns after comparison with the healthy profile model. The main advantage of the method is that construction of a SHM model is exclusively based on data derived from healthy samples and thus it is absolutely not disease dependent.

New developments are expected to render NMR spectroscopy as a suitable method for the diagnosis and management of IEM. Automated data acquisition and more specifically self-locking, shimming, field and temperature stability, and so on, as well as deconvolution and automatic quantification software for analysis, and finally adequate spectral databanks of standard metabolites would definitely be a step forward. However, the most important prerequisite in this aspect is the generation of large databanks of physiological subject metabolism, capable to sustain creation of models accounting for population-specific characteristics as well as environmental factors. Diagnosis and treatment of IEM is one of the most evident examples of diseases needing personalized management. The untargeted character of NMR-based profiling provides an invaluable tool complementing the arsenal needed for the medicine of the future.

RELATED ARTICLES IN EMAGRES

Analysis of High-Resolution Solution State Spectra

NMR Spectroscopy of Biofluids, Tissues, and Tissue Extracts

Two-Dimensional *J*-Resolved Spectroscopy

Metabonomics: NMR Techniques

Inherited Disease Studies by NMR

REFERENCES

1. The Online Metabolic and Molecular Bases of Inherited Disease. 2011 D. Valle, Editor-in-Chief, The McGraw-Hill Companies, Inc. http://ommbid.mhmedical.com (last update of online version) New York, USA.

2. B. Lanpher, N. Brunetti-Pierri, and B. Lee, *Nat. Rev. Genet.*, 2006, **7**, 449.

3. K. de Meer, in Encyclopedia of Food Sciences and Nutrition, 2nd edn, ed. B. Caballero, Academic Press: Oxford, 2003. DOI:10.1016/B0-12-227055-X/00628-3.

4. J.-M. Saudubray, I. Desguerre, F. Sedel, and C. Charpentier, in Inborn Metabolic Diseases Diagnosis and Treatment, 4th edn, eds J. Fernandes, J.-M. Saudubray, G. van den Berghe, and J. H. Walter, Springer: Heidelberg, 2006.

5. A. Gupta, in Anthropology Today: Trends, Scope and Applications, eds V. Bhasin and M. K. Bhasin, Kamla-Raj Enterprises, 2007, Vol. 3.

6. R. A. Iles, *Curr. Med. Chem.*, 2008, **15**, 15.

7. R. A. Iles, in Encyclopedia of Magnetic Resonance, John Wiley and Sons Ltd: Chichester, West Sussex, UK, 2010. DOI: 10.1002/9780470034590.emrstm1106.

8. U. Engelke, M. Oostendorp, and R. Wevers, in The Handbook of Metabonomics and Metabolomics, eds J. C. Lindon, J. K. Nicholson, and E. Holmes, Elsevier: Amsterdam, 2007, chapter 13.

9. U. Engelke, S. Moolenaar, S. Hoenderop, E. Morava, M. van der Graaf, A. Heerschap, and R. Wevers, Handbook of ^{1}H-NMR Spectroscopy in Inborn Errors of Metabolism: Body Fluid NMR Spectrum and in Vivo MR Spectroscopy, 2nd edn, SPS Verlagsgesellschaft mbH: Heilbronn, 2007, 143.

10. A.-H. M. Emwas, R. M. Salek, J. L. Griffin, and J. Merzaban, *Metabolomics*, 2013, **9**, 1048.

11. J. S. McKenzie, J. A. Donarski, J. C. Wilson, and A. J. Charlton, *Prog. NMR Spectrosc.*, 2011, **59**, 336.

12. M. A. Constantinou, E. Papakonstantinou, D. Benaki, M. Spraul, K. Shulpis, M. A. Koupparis, and E. Mikros, *Anal. Chim. Acta*, 2005, **511**, 303.

13. R. Mallol, M. A. Rodriguez, J. Brezmes, L. Masana, and X. Correig, *Prog. NMR Spectrosc.*, 2013, **70**, 1.

14. A. Smolinska, L. Blanchet, L. M. C. Buydens, and S. S. Wijmenga, *Anal. Chim. Acta*, 2012, **750**, 82.

15. P. Bernini, I. Bertini, C. Luchinat, P. Nincheri, S. Staderini, and P. Turano, *J. Biomol. NMR*, 2011, **49**, 231.

16. J. C. Lindon and J. K. Nicholson, *Annu. Rev. Anal. Chem.*, 2008, **1**, 45.

17. G. A. Barding Jr, R. Salditos, and C. K. Larive, *Anal. Bioanal. Chem.*, 2012, **404**, 1165.

18. N. Shanaiah, M. A. Desilva, G. A. N. Gowda, M. A. Raftery, B. E. Hainline, and D. Raftery, *Proc. Natl. Acad. Sci. U.S.A.*, 2007, **104**, 11540.

19. K. Bingol and R. Brüschweiler, *Anal. Chem.*, 2014, **86**, 47.

20. K. Suhre, H. Wallaschofski, J. Raffler, N. Friedrich, R. Haring, K. Michael, C. Wasner, A. Krebs, F. Kronenberg, D. Chang, C. Meisinger, H.-E. Wichmann, W. Hoffmann, H. Völzke, U. Völker, A. Teumer, R. Biffar, T. Kocher, S. B. Felix, T. Illig, H. K. Kroemer, C. Gieger, W. Römisch-Margl, and M. Nauck, *Nat. Genet.*, 2011, **43**, 565.

21. A. L. Guennec, P. Giraudeau, and S. Caldarelli, *Anal. Chem.*, 2014, **86**, 5946.

22. C. Legido-Quigley, O. Cloarec, D. A. Parker, G. M. Murphy, E. Holmes, J. C. Lindon, J. K. Nicholson, R. R. Mitry, H. Vilca-Melendez, M. Rela, A. Dhawan, and N. Heaton, *Bioanalysis*, 2009, **1**, 1527.

23. I. Smuts, F. H. van der Westhuizen, R. Louw, L. J. Mienie, U. F. H. Engelke, R. A. Wevers, S. Mason, G. Koekemoer, and C. J. Reinecke, *Metabolomics*, 2013, **9**, 379.

24. I. F. Duarte, B. J. Goodfellow, A. Barros, J. G. Jones, C. Barosa, L. Diogo, P. Garcia, and A. M. Gil, *NMR Biomed.*, 2007, **20**, 401.

25. J. K. Nicholson, J. C. Lindon, and E. Holmes, *Xenobiotica*, 1999, **29**, 1181.

26. S. Trump, S. Laudi, N. Unruh, R. Goelz, and D. Leibfritz, *Magn. Reson. Mater. Phys.*, 2006, **19**, 305.

27. M. A. Constantinou, E. Papakonstantinou, M. Spraul, S. Sevastiadou, C. Costalos, M. A. Koupparis, K. Shulpis, A. Tsantili-Kakoulidou, and E. Mikros, *Anal. Chim. Acta*, 2005, **542**, 169.

28. S. Aygen, U. Duerr, P. Hegele, J. Kunig, M. Spraul, H. Schaefer, D. Krings, C. Cannet, F. Fang, B. Schuetz, S. F. H. Buelbuel, H. I. Aydin, S. U. Sari, R. Örs, R. Atalan, O. Tuncer JIMD Reports 2014, SSIEM and Springer, Berlin, Heidelberg, 2014, DOI 10.1007/8904_2014_326S.

29. J. Engel, L. Blanchet, U. F. H. Engelke, R. A. Wevers, and L. M. C. Buydens, *PLoS One*, 2014, **9**, e92452.

Chapter 11

NMR-based Structure Confirmation of Hits and Leads in Pharmaceutical R&D

Philip J. Sidebottom

GlaxoSmithKline R&D, Chemical Sciences – Analytical Chemistry, Gunnels Wood Road, Stevenage, Herts SG1 2NY, UK

11.1 INTRODUCTION

During the early stages of a small-molecule drug discovery program, a sample is classified as a hit when it shows reproducible activity in the primary biological screen. In this context, even molecules with molecular weights of around a thousand would be considered small. However, the majority of projects are targeting an orally administered drug. As a result, in most cases, owing to the need for effective drug absorption, the

NMR in Pharmaceutical Sciences. Edited by Jeremy R. Everett, John C. Lindon, Ian D. Wilson, and Robin K. Harris.
© 2015 John Wiley & Sons, Ltd. ISBN: 978-1-118-66025-6
Also published in eMagRes (online edition)
DOI: 10.1002/9780470034590.emrstm1401

desirable molecular weight for hits and leads is limited to a few hundred.[1] The exact definition of a lead will vary depending on the institution and the specific project. Typically, it will encompass a pure compound of known structure having an appropriate level of activity in the primary screen. Some evidence of selectivity particularly with respect to closely related targets may also be needed. Increasingly, a clean profile in certain safety-related screens, evidence of achievable pharmacokinetic parameters, and a range of desirable physicochemical properties, such as a minimum level of aqueous solubility, are being included in the definition. Also important is that it should be part of a readily accessible series of related active compounds where variation in biological activity can be related to changes in structure. This ensures that the medicinal chemists have the best chance of success during the lead optimization phase that comes next.

This article will explore the questions of why the structure confirmation of hits and leads is needed and what the benefits are of doing so. It will also address the issues of how and when this work should be carried out. Finally, an example will be given of just how badly things can go wrong if this small but important step is neglected.

11.2 WHY IS IT NECESSARY?

There are a number of different approaches to the discovery of hits and leads. In some, such as the natural products paradigm, the structure elucidation of the active compound is an inherent part of the process. In others, for example, high-throughput screening, the structure of any hit is supposedly known. Methods such as virtual screening and fragment-based drug discovery can fall into either camp. This will depend on whether the resulting active compounds have been specifically synthesized for the program or were drawn from some internal or external collection. By definition, structure confirmation is carried out in cases where the structure is supposedly known. An examination of the size and origins of the compound collections from which hits and leads are plucked shows us why this is an important and necessary step.

With the exception of fragment screening sets, company screening collections are generally large as this improves the chances of success. Even a small screening set will contain many tens of thousands of compounds and the largest collections are numbered in millions. These collections have been put together over the past 50 years or more. As well as internally generated samples they will generally also include samples from a range of external academic and commercial sources. The opportunities for mistakes to occur are numerous. They range from a wrong initial structure elucidation to sample handling mistakes or simple drawing errors when computerizing the structure. It is thus inevitable that some of the supposedly known structures will be incorrect. Considerable resources are needed to establish and maintain the quality of such a screening collection.[2] The upfront structure confirmation by NMR of such a large number of compounds, other than at the point of synthesis, is highly impractical and financially prohibitive. Checking those that turn out to be of significant interest is essential.

All the same issues are likely to apply to commercially available screening sets. Even when a supplier indicates that NMR was used in the quality control of its compounds, this will rarely extend beyond a quick check of the 1D proton spectrum. A full NMR analysis of the structure of any important, externally acquired, compound should always be carried out.

11.3 WHEN SHOULD THE STRUCTURE CONFIRMATION BE CARRIED OUT?

While initial screening may produce a large number of hits, the resources available to pursue them are always limited. Inevitably, these resources will be focused on only the most promising samples. These will have been identified from the results of secondary screening and other measurements made on the initial hits. It is at this point, when the resynthesis of a hit or some close analogs is being contemplated, that carrying out the structure confirmation of a hit has value. Before being designated as a lead, the structure of any compound will need to be confirmed. However, in many cases, this will have been done when the compound was synthesized and characterized as part of the hit to lead process.

11.4 WHAT ARE THE BENEFITS?

In most cases, hit and lead compounds are shown to have the correct structures. It is tempting to assume that in these cases the carrying out of structure confirmation provides little benefit. This is not the case. The NMR data from the structure confirmation are useful during resynthesis as they make it easy to establish that the correct compound has been made. Unfortunately, from time to time, the resynthesized sample of a hit is inactive. Work on the inactive compound can be promptly terminated because of this structural confidence. Pursuing the source of the activity in the original sample is unlikely to be fruitful. Had any impurities above trace levels been observed by NMR or liquid chromatography (LC)/mass spectrometry (MS), the sample should already have been purified to remove them. It should then have been retested, and unless found still to be active, it should have been removed from the list of hits.

When the original structure is not confirmed, it usually proves possible to establish the correct one. If the revised structure has any undesirable features, such as a reactive center, further work on it will be abandoned; otherwise, effort will be redirected toward the new correct structure. In either case, resources are not wasted on designing and carrying out the synthesis of the wrong material. When an historic synthetic route allows the production of new active material, structure confirmation is even more vital. Without it, any error may remain undetected for a long time. In such a case, the exact cost of failing to confirm the structure will depend on who uncovers the error and when they do so.

11.5 HOW IS STRUCTURE CONFIRMATION CARRIED OUT?

11.5.1 The Basics

Typical hits and leads are small organic molecules with molecular weights in the 250–450 range. Most will contain at least one heterocyclic ring, either saturated or aromatic. The majority will have one or more nitrogen atoms. They may contain one or more chiral centers. The structure confirmation of most such molecules is readily accomplished using a combination of 1D and 2D NMR data in conjunction with molecular weight information from MS.

The solvent of choice is DMSO-d_6. Solubility is essentially assured because nearly all biological screening is carried out using dimethyl sulfoxide (DMSO) solutions. Unlike some other commonly used NMR solvents, it also has the advantage of giving access to NMR data from exchangeable protons (NH, OH, and SH). In addition, the high boiling point of DMSO allows NMR data to be obtained at elevated temperatures if necessary.

In most cases, in addition to 1D proton and carbon data, one-bond (HSQC or HMQC) and long-range (HMBC) ^1H–^{13}C correlation spectra and ^1H–^1H coupling data (COSY) will provide the information needed. If stereochemistry needs to be confirmed, then ^1H–^1H NOE data will often be required, and for some structures, long-range ^1H–^{15}N correlation data will also be helpful.

If a state of the art 600 MHz instrument, equipped with a 5 mm cryoprobe, is available, then these data can be acquired in a few hours using a sample prepared from only a milligram of solid. If a cryoprobe is not available, then the data collection for such a sample will require a weekend of instrument time. On the other hand, if there is access to a microcryoprobe (1.7 mm), then samples in the 100 µg range can be utilized successfully.[3]

A detailed discussion of how the basic set of NMR experiments, as listed earlier, is interpreted will not be given here. Such discussions are at the heart of modern NMR textbooks focused on the structure elucidation of organic compounds.[4,5] In most cases, this analysis enables a full set of assignments and details of the connectivities between the 1D proton and carbon signals to be obtained. The final stage of a structure confirmation involves ensuring that the heteroatoms are located as expected. This is usually straightforward, especially when the number of such atoms is limited. It will

generally depend on establishing that the local chemical shifts are as expected. The prediction of chemical shifts is thus vital to the process. Even experienced spectroscopists will occasionally find themselves unsure of what to expect so we need to consider what to do in these circumstances.

11.5.2 Chemical Shift Prediction

Methods to assist with the prediction of chemical shifts have been in use since the early days of NMR. In simple cases, long established rule-based systems may provide the answer.[6] For more complex cases, it is now usual practice to turn to one of several commercial computer software packages.[7,8] These utilize databases containing chemical shift information from a very large number of compounds. In general, they work very well but they are not infallible. The nature of the compounds used to make any key predictions should always be checked. If data for good model compounds are not being used because they are not in the database, then the prediction will be unreliable. In such cases, NMR data for model compounds obtained directly from the literature may be helpful. For this latter approach, the Reaxys® chemical literature database[9] is particularly useful. It enables the fast location of references containing NMR data by nucleus and solvent in conjunction with chemical structure. Finally, the possibility of calculating the chemical shifts directly using ab initio methods exists.[10] Currently, these methods are rarely used during structure confirmation. Recent advances in the speed and ease of use of the necessary software suggest that they may play a bigger role in future.

11.5.3 Challenging Samples

While the structure confirmation of most samples is routine, from time to time a more challenging sample will be encountered. In this section, some of the possible problems will be highlighted and potential solutions presented. Despite these, it may remain impossible to confirm the structure fully by NMR. In such cases, it will be necessary to decide if the available NMR data, while not sufficient to prove the structure, are consistent with it. If possible, at this point, it is good practice to establish the molecular formula using high-resolution mass spectrometry. When all the available data are consistent with the structure, interest in the sample may continue without full structure confirmation. This will then be revisited when circumstances improve. For example, more material may become available or a new active analog lacking the problematic features may be synthesized. On the other hand, if the problems preventing structure confirmation by NMR appear to be fundamental to a particular series of compounds, that series may be abandoned. This is particularly likely if competing series are trouble free. If all the initially available data do not fit the structure and no obvious alternative structure is apparent, then work on the sample is also very likely to be terminated.

11.5.3.1 The Sample Is Not Available in Solid Form

In some cases, the only available material for structure confirmation will be the residual DMSO solution used for biological screening. DMSO is hygroscopic so these solutions invariably contain a significant level of water. Typically the concentration will be 5–10 mM or less. Using them directly to obtain NMR data will thus require the suppression of the water peak as well as the DMSO and its ^{13}C satellites. Fortunately, solvent suppression methods, originally developed for the acetonitrile/water mixtures frequently encountered in HPLC/NMR, are available.[11–13] Alternatively, in cases where the structure does not contain any aliphatic protons, it will be possible simply to avoid the solvent signals. This can be achieved by reducing the proton spectrum width and exploiting the excellent digital filters incorporated into modern instruments. For compounds with aliphatic protons, there is always the possibility that one or more key solute signals will be removed along with those from the solvent. This is certain to complicate the structure confirmation. In practice, it is usually much simpler to avoid all these solvent suppression issues by removing the DMSO. A sample in 0.5 ml of DMSO can be blown down overnight to obtain the solid using a stream of dry nitrogen.

11.5.3.2 There Is Not Enough Material

The amount of material needed to confirm the structure of a compound by NMR will depend on the instrumentation available. However good this is, there will come a point at which all the data needed to complete the task cannot be obtained. Acquiring the 1D ^{13}C NMR spectrum is usually the most demanding. In many cases, it

is possible to obtain the ^{13}C chemical shift information from the more sensitive, proton-detected, 2D experiments such as HSQC and HMBC. In such cases, it is possible to complete the structure confirmation without a 1D ^{13}C spectrum. When the required data cannot be obtained using the initially available material, all is not necessarily lost. It may be possible to obtain more material either using the same synthetic route or by acquisition from the original source.

11.5.3.3 Proton-poor Molecules

For an organic molecule, the smaller the ratio of the number of protons to heavy atoms (P/HA), the more difficult it is to determine its structure using NMR spectroscopy. This is because the correlations observed in HMBC spectra are the key to joining structural fragments together correctly, and these correlations are not generally observed over more than three bonds. As the P/HA ratio falls, it becomes increasingly likely that one or more carbon or nitrogen atoms in the structure is located more than three bonds away from a proton. The absence of long-range correlations may then make it impossible to position such remote atoms unambiguously within a structure. When confirming the structure of a proton-poor molecule, the lack of long-range correlation information may well leave only chemical shift information from the 1D carbon spectrum to work with. Even this will be more difficult to obtain as carbon atoms that are remote from protons generally give rise to the weakest signals because they have the longest relaxation times. However, when enough material is available and the compound is sufficiently soluble, other less-sensitive NMR experiments can be employed. For example, the 1,n-ADEQUATE experiment that makes use of ^{13}C–^{13}C coupling may provide the extra data needed to confirm the structure of a proton-poor molecule.[14]

11.5.3.4 The Sample Gives Poor Quality Spectra

Even when a pure sample fully dissolves to give a clear solution, there is no guarantee that good quality spectra with sharp signals will be obtained. If some of the signals for part of the molecule appear broad, in either the 1D proton or carbon spectrum or both, then confirming the structure may be very difficult. Such selective broadening is nearly always the result of some sort of exchange process among conformers, tautomers, or protonated and unprotonated forms. In order for a signal to appear broad, the rate of exchange must be close to the separation, in Hz, between the positions of that signal in each of the two exchanging forms. If the rate of exchange can be increased so that it becomes fast compared to the separation, then a sharp signal will be observed. A common way to achieve this is to reacquire the spectra at an elevated temperature, typically >100 °C, when using DMSO-d_6 as the solvent. However, this approach often cannot be used effectively with a cryoprobe because most of them have an upper operating limit of 60 °C for the sample temperature. Exchange between tautomers may well occur more quickly if a protic solvent such as CD_3OD is used instead of the aprotic DMSO-d_6. Where partial protonation of the compound is thought to be the cause of the broad lines, then adding acid or base to obtain either the fully protonated or fully deprotonated state may enable sharp spectra to be obtained.

11.5.3.5 The Structure Contains One or More Chiral Centers

Where the hit or lead structure contains one or more chiral centers, the question of the absolute stereochemistry arises. This question is rarely addressed for hits produced by chemical synthesis. This is because these are usually assumed to be racemic. In such cases, an early objective in the hit to lead process is to obtain activity data on the individual enantiomers. To this end, initial synthetic efforts will be directed toward making the racemate. Usually, the separate enantiomers can then be obtained by chiral chromatography. When a single enantiomer is under consideration as a possible lead, establishing or confirming its absolute stereochemistry becomes a high priority.

The use of NMR for the determination of absolute stereochemistry dates back many decades, and many different methods have been published.[15] In general, each method is only applicable to a narrow range of compounds and often involves the synthesis of derivatives and/or the use of unusual chiral auxiliaries. These limitations and complications count against NMR. As a result, other techniques such as vibrational circular dichroism (VCD)[16] or X-ray crystallography[17] are the methods of choice for the determination of absolute stereochemistry of leads.

11.5.4 Computer Assistance

Recently, computer programs to carry out automatic structure verification (ASV) have been developed. In order to determine if they have a role to play in the structure confirmation of hits and leads, a closer look at how they work is needed. These ASV programs have been designed to meet the structure confirmation challenges presented by parallel synthetic methods. Large libraries of compounds can now be produced and basic NMR data collected.[18] However, the manual review of such large amounts of data is normally beyond the resources available. Thus, ASV programs usually use a limited set of quickly obtained NMR data (e.g., ^1H or ^1H and HSQC). They work by comparing the observed signal intensities, multiplicities, and chemical shifts with those predicted for the proposed structure and then return a pass/uncertain/fail answer. There is always going to be some difference between the observed and predicted data. How these differences are evaluated and the sample then classified is one of the key parts of the software. At what point is the fit good enough for a pass or so bad that the sample fails? If these criteria are set wrongly, then some false positive and/or false negative results will be produced. Alternatively, a large number of samples may end up in the uncertain category. No single solution exists as long as the prediction accuracy is compound dependent. Thus, applications have been focused in two areas. Firstly, following optimization of the settings, ASV can be used to check large sets of compounds with similar structures. Secondly, with more diverse sets of compounds, it can provide a way of picking up the gross errors while accepting that some wrong compounds may get through. ASV programs continue to improve and ways to overcome some of the shortcomings continue to appear.[19,20] However, ASV is probably not yet the tool needed to confirm the structures of a small number of diverse structures with a high level of confidence.

A better idea might be to employ a computer-assisted structure elucidation approach. These programs use all the available NMR data and the molecular formula to generate structural fragments consistent with the data. All possible structures resulting from combining the fragments and having the correct molecular formula are then generated. Some programs can make use of lists of good and bad substructures to speed up this process. Finally, the chemical shifts are predicted for each generated structure. The structures are then ranked by how well their predicted data match the experimental values. If the original structure is confirmed by the NMR data, then it should emerge clearly at the top of the list. If it does not, then a list of one or more alternative structures that need to be considered will be available.

11.6   HOW BAD CAN IT GET?

The story of TIC10, also known as ONC201, has been well publicized.[21,22] It serves as a salutary reminder of the risks that are mitigated by the structure confirmation of hits and leads.

TIC10 was part of a set of screening compounds available free of charge from the American National Cancer Institute (NCI). The structure of this sample was listed as (**1**). Screening of this set at Pennsylvania State University showed that TIC10 had interesting anticancer activity.[23] A patent was filed, followed by a licensing agreement with Oncoceutics, who by early in 2014 were planning an initial clinical trial. Smooth progress was possible because further active material was synthesized using the route that had been used to make TIC10 in the first place. This appears in a lapsed 1973 German patent.

1

The first sign of trouble occurred when a group from the Scripps Research Institute became interested in the compound.[24] Unaware of the German patent they devised their own synthesis of (**1**). This produced inactive material that differed from a reference sample of TIC10 obtained from NCI. They then established that the correct structure of TIC10 is (**2**), an isomer of (**1**). Interestingly, they also report that a commercial sample of TIC10 proved to be yet another isomer, (**3**). According to reports, a patent covering the revised structure, (**2**), has been applied for by Scripps and licensed exclusively to Sorrento Therapeutics.

2

3

The complicated intellectual property position now surrounding TIC10 could so easily have been avoided. Carrying out a proper structure confirmation at an early stage was clearly indicated. The fact that the listed structure (**1**) dated from a time before any 2D NMR methods existed is a clear indicator of an increased risk that it could be wrong.

11.7   CONCLUSION

The timely use of NMR for the structural confirmation of hits and leads will always be beneficial. At the simplest level, it provides confidence and a set of NMR assignments that will be useful to the project as it progresses. Occasionally problems will be discovered that result in the project progressing a revised structure or terminating work with that particular material.

RELATED ARTICLES IN EMAGRES

Analysis of High-Resolution Solution State Spectra

Combinatorial Chemistry

REFERENCES

1.   M. P. Gleeson, A. Hersey, D. Montanari, and J. Overington, *Nat. Rev. Drug Discovery*, 2011, **10**, 197.

2.   S. J. Lane, D. S. Eggleston, K. A. Brinded, J. C. Hollerton, N. L. Taylor, and S. A. Readshaw, *Drug Discov. Today*, 2006, **11**, 267.

3.   B. D. Hilton and G. E. Martin, *J. Nat. Prod.*, 2010, **73**, 1465.

4.   S. A. Richards and J. C. Hollerton, Essential Practical NMR for Organic Chemistry, Wiley: Chichester, 2011.

5.   J. H. Simpson, Organic Structure Determination Using 2-D NMR Spectroscopy: A Problem Based Approach, 2nd edn, Academic Press: San Diago, 2012.

6. E. Pretsch, P. Buhlmann, and M. Badertscher, Structure Determination of Organic Compounds, 4th edn, Springer: Berlin, 2009.

7. Advanced Chemistry Development, Inc. (ACD/Labs), Product Details Available via their Home Page; http://acdlabs.com (Date accessed, September 27, 2014).

8. Mestrelab Research, Product Details Available Via their Home Page; http://mestrelab.com (Date accessed, September 27, 2014).

9. Reaxys® is owned by Reed Elsevier Properties SA. For product details see http://elsevier.com/online-tools/reaxys (Date accessed, September 27, 2014).

10. M. W. Lodewyk, M. R. Siebert, and D. J. Tantillo, *Chem. Rev.*, 2012, **122**, 1839.

11. C. Dalvit, G. Shapiro, J.-M. Bohlen, and T. Parella, *Magn. Reson. Chem.*, 1999, **37**, 7.

12. S. H. Smallcombe, S. L. Patt, and P. A. Keifer, *J. Magn. Reson., Ser. A*, 1995, **117**, 259.

13. D. Neuhaus, I. M. Ismail, and C.-W. Chung, *J. Magn. Reson., Ser. A*, 1996, **118**, 256.

14. M. M. Senior, R. T. Williamson, and G. E. Martin, *J. Nat. Prod.*, 2013, **76**, 2088.

15. T. J. Wenzel and C. D. Chisholm, *Chirality*, 2011, **23**, 190.

16. S. S. Wesolowski and D. E. Pivonka, *Bioorg. Med. Chem. Lett.*, 2013, **23**, 4019.

17. J. R. Deschamps, *Life Sci.*, 2010, **86**, 585.

18. J. Hollerton, *Encyclopedia of Magnetic Resonance*, 2015, **4**, 289. DOI: 10.1002/9780470034590. emrstm1390.

19. S. S. Golotvin, R. Pol, R. R. Sasaki, A. Nikitina, and P. Keyes, *Magn. Reson. Chem.*, 2012, **50**, 429.

20. C. Cobas, F. Seoane, E. Vaz, M. A. Bernstein, S. Dominguez, M. Perez, and S. Sykora, *Magn. Reson. Chem.*, 2013, **51**, 649.

21. S. Borman, *Chem. Eng. News*, 2014, **92** (Issue 21; May 26), 7.

22. S. Borman, *Chem. Eng. News*, 2014, **92** (Issue 23; June 9), 32.

23. J. E. Allen, G. Krigsfeld, P. A. Mayes, L. Patel, D. T. Dicker, A. S. Patel, N. G. Dolloff, E. Messaris, K. A. Scata, W. Wang, J.-Y. Zhou, G. S. Wu, and W. S. El-Deiry, *Sci. Transl. Med.*, 2013, **5**, 171ra17.

24. N. T. Jacob, J. W. Lockner, V. V. Kravchenko, and K. D. Janda, *Angew. Chem. Int. Ed.*, 2014, **53**, 6628.

Chapter 12

Fragment-based Drug Design Using NMR Methods

Leonor Puchades-Carrasco and Antonio Pineda-Lucena

Structural Biochemistry Laboratory, Advanced Therapies Program, Centro de Investigación Príncipe Felipe, Eduardo Primo Yúfera 3, 46012-Valencia, Spain

NMR in Pharmaceutical Sciences. Edited by Jeremy R. Everett, John C. Lindon, Ian D. Wilson, and Robin K. Harris
© 2015 John Wiley & Sons, Ltd. ISBN: 978-1-118-66025-6
Also published in eMagRes (online edition)
DOI: 10.1002/9780470034590.emrstm1405

12.1 INTRODUCTION

Fragment-based drug design (FBDD) has significantly developed over recent years and is now recognized as an alternative and complementary technology to more traditional screening approaches, such as high-throughput screening (HTS). This technique, currently predominant in pharmaceutical research, has already led to the development of several clinical candidates.[1] The approach relies on the consecutive screening of low-molecular weight (LMW) compounds to identify fragments that are then developed and converted into lead compounds. The technique consists of two well-defined phases: identification of fragment hits and their conversion into leads. For the consecutive incorporation of each fragment, structural and biophysical information about the interaction between the fragment and the specific drug target is used and this leads to final compounds with higher affinity and selectivity. Fragment library design, fragment-screening methods, and prioritization of fragment hits are key factors in the fragment hit identification phase.

The affinity of small fragments for their target is generally weak and, thus, requires the application of sensitive biophysical techniques for its evaluation. Several of these techniques have been successfully applied in the FBDD field, and they provide very useful information about the existence of interactions between a fragment and a drug target. Among them, NMR and

X-ray crystallography are undoubtedly the most powerful tools in both the discovery and optimization of fragment hits into leads. In particular, NMR presents some clear advantages over other biophysical techniques because of its sensitivity in the identification of very weak intermolecular interactions (binding constants in the mM range) and low propensity to the appearance of false positives.[2]

12.2 FRAGMENT-BASED DRUG DESIGN VS HIGH-THROUGHPUT SCREENING

HTS has been extensively used in the pharmaceutical industry to search for new chemical compounds that might serve as lead structures for medicinal chemists. However, the HTS strategy is complex, expensive, and has, like any other method, a number of drawbacks that limit the possibility of finding new compounds. Therefore, the pharmaceutical industry is permanently exploring innovative technologies that might complement and/or improve the difficult process of discovering novel drug candidates. In this context, FBDD aims to evolve new tight-binding molecules in a step-by-step approach, which contrasts with the one-step procedure used by HTS. As its name indicates, FBDD involves the selection, screening, and optimization of simple small molecules with much lower molecular weights than the complex full-size molecules typically found in HTS compound collections. The use of small molecules offers several appealing features compared with the use of classical compound libraries (Table 12.1). Thus, despite their low potency, fragments are usually more efficient binders because, given their lower complexity, they have higher binding energy per unit mass than HTS hits. The probability that small fragments match the binding site of a target protein is also higher than that of complex molecules, and the subsequent assembly of low affinity fragments can generate hits with improved target affinity over independent fragments. As a result of this fragment combination, FBDD molecules are more likely to exhibit greater ligand efficiency than the molecules found by conventional screening of compound libraries.[2]

Table 12.1. FBDD vs HTS

FBDD	HTS
Screening of 10^2–10^3 compounds	Screening of 10^5–10^6 compounds
Low complexity of the tested compounds	Fully assembled compounds
Biophysical techniques	Biochemical assays
High coverage of the chemical space	Proportionally low coverage chemical space
Step-by-step approach, design	One-step approach, modification
Hit range μM to mM	Hit range nM to μM

12.3 FBDD APPROACHES

Given the moderate to poor affinities expected in early phases of drug discovery projects that rely on FBDD, specialized and highly sensitive techniques are required to screen fragments within the affinity 10 mM to 100 μM range. A wide variety of biophysical techniques is available that can be applied to FBDD. However in practice, they are commonly combined to exploit the synergistic information that each provides. Traditionally, these biophysical techniques have been used as secondary assays in the confirmation and profiling of the hits obtained from biochemical or cell-based screening and, until very recently, their utility for primary screening has been hampered by low throughput and relatively high protein consumption. However, this situation has changed in the last few years with the introduction of technological advances and improved instrumentation. Biophysical techniques often require highly specialized equipment, personnel with specific expertise, a supporting informatics structure and access to moderate amounts of purified protein. Fragment molecules also need to be soluble at the high concentrations used for screening. Apart from NMR, several biophysical techniques are now widely used in the pharmaceutical industry for FBDD-based hit identification.

12.3.1 Surface Plasmon Resonance

Surface plasmon resonance (SPR) is a phenomenon that occurs when light is reflected off thin metal films to which target molecules are immobilized and addressed by ligands in a mobile phase. In the commonest setup, SPR is more widely used in a mode wherein the protein is captured and immobilized on a chip, and small-molecule ligands are flowed over the immobilized protein with interacting ligands detectable through a change in the refractive index. Thus, it is possible to screen thousands of molecules sequentially using the same surface before protein renewal is required. In SPR, the binding of a fragment to a protein is observed as a change in the mass of the molecular system attached to a chip surface. SPR presents high sensitivity, with fragments as small as 100 Da being detectable. Yet, because of minor changes in molecular weight, fragment binding detection becomes challenging if the immobilized receptor is larger than approximately 50 kDa. The major challenge in SPR entails achieving a robust immobilization of either target or fragment to the chip surface without affecting binding properties. If appropriate immobilization is achieved, SPR is a very powerful technique because it uses relatively small amounts of protein. If the protein itself is attached to the surface, then measurement is direct and can provide kinetics of binding and a K_d value. Although SPR is normally used in drug discovery to determine the affinity of small-molecule–protein interactions, it can also provide information about stoichiometry, reversibility, and about changes in compound behavior over a range of concentrations.[2]

12.3.2 Mass Spectrometry

Mass spectrometry (MS) has played numerous supporting roles in the drug development process for many years, including assessment of compound purity, quantification of absorption, distribution, metabolism and excretion, as well as compound-specific pharmacokinetic analyses. More recently, MS has been widely used as a screening technology in FBDD as it enables the sensitive label-free detection of LMW modulators and accurate detection of chemical modifications of biomolecules. Targets of unknown biological function, or targets with a known function for which the development of a functional assay with an optical or radiometric readout would be tedious or even impossible, can be successfully addressed by the application of MS-based identification of chemical binders. Electrospray ionization (ESI) and matrix-assisted laser desorption ionization (MALDI) are the most widely used ionization techniques to identify chemical compounds. These techniques allow the direct identification of different complexes as the mass of each molecule serves as the intrinsic detection label. In addition to exploiting the

'*x*-axis' of the mass spectrum (i.e., the mass-to-charge ratio, *m/z*), the '*y*-axis' of the mass spectrum (i.e., abundance/intensity) provides information about affinity and specificity. Therefore, the combined information from the *x* and *y* axes can be used to rapidly determine which compounds from a mixture bind to which targets and with what relative affinity.[2]

12.3.3 X-ray

Traditionally, X-ray crystallography has focused more on the lead optimization phase of the drug discovery process than on HTS. It is well-established that the 3D structure of a target–ligand complex provides very detailed information of the interaction at an atomic level, which is of much help in the design and chemical evolution needed during the optimization of initial hits to increase their affinity. However, its importance in drug discovery has augmented lately, and nowadays, together with NMR, it is one of the main biophysical tools used in FBDD. One of the major benefits of X-ray diffraction is its high sensitivity in binding affinity terms as it is able to detect weak binders that other methods miss. This is particularly important in the field of FBDD as small molecules are not expected to bind tightly to the target. Hence, it is much more sensitive than other biochemical assays and has the advantage of providing structural information for structure-based drug design.

One of the commonest methods of carrying out fragment screening by X-ray crystallography consists in soaking the already crystallized protein with solutions of fragments either individually or as mixtures (cocktails) of molecules at high concentrations (25–100 mM). However, finding suitable crystallization conditions and subsequent protocol optimization to obtain good diffraction single crystals are considered some of the biggest hurdles to overcome during the crystallization of a protein. However in the last few years, major improvements in the field have been made, and one of the main advances is the implementation of robotics in the production of crystals to find optimal crystallization conditions. This automation can also be used later for screening fragments against the biomolecular target. Other good advances made stem from the collection of the diffraction data in both the X-ray generator and automation of the collection and subsequent data processing. Technologically speaking, significant improvements have been in the available synchrotrons and in laboratory X-ray generators, which now offer more intense X-rays and more sensitive X-ray detectors to allow much faster data collection. With the introduction of automation in sample changers, the process of screening a batch of crystals has been greatly facilitated. This process has also been improved with the development of new software, which integrates all the different data collection steps.[2]

of the advantages that this approach offers include no limit to the size of the protein or ligands used, provision of K_i information, and exclusive reporting of functionally relevant molecules. However, high ligand and dimethyl sulfoxide (DMSO) concentrations can often interfere with the assay by creating high false positive rates, and no information about the binding mode is available.

12.3.4 Biochemical Assays

Screening fragments by biophysical methods, such as those already described, does not reveal whether a fragment found to bind to a target protein has an effect on normal protein function or not. Thus, it is usual in all FBDD programs, where screening has been performed by a biophysical technique, to confirm whether the binding of a fragment to a target protein has any functional relevance by performing a biochemical assay at high concentration. Then, the biochemical assay results are used to drive the medicinal chemistry optimization process along with structural information on fragment binding to the target protein revealed by protein NMR or X-ray crystallography. An alternative fragment-screening strategy is to directly screen by biochemical assays. Indeed, there are a number of reports on the successful application of high concentration bioassays for screening fragment libraries, in which a variety of different assay protocols have been used, and which have demonstrated the applicability of bioassay methods for screening fragments. Some

12.4 FRAGMENT LIBRARY

The two prerequisites for a successful fragment-based NMR screening campaign are a robust assay system capable of detecting and quantifying weak binding (approximately mM) and a library of compounds that fulfill certain requirements.[3] Compound libraries can be generic if compared to target or directed using protein structural information or known ligands.[3] The commonest fragment libraries (Table 12.2) are diverse sets of compounds that have been selected according to specific criteria, such as maximal diversity and/or physicochemical properties, and have then been filtered to remove those containing functional groups associated with chemical reactivity, toxicity, and false positives.[4]

As the size of chemical space increases exponentially with molecular size, statistically speaking, covering a chemical space is more manageable with fragments than with drug-sized molecules.[5] In fact, most libraries for FBDD are restricted to 1000–10 000 compounds,[6] a number that is three or two orders of

Table 12.2. Commercial fragment libraries

Company supplier	Number of fragments	Characteristics[a]
Asinex	4500	RO3 compounds
ChemBridge Corporation	5000	RO3 compounds
Edelris	1900	Expanding to '3D' fragments
Enamine	1190	RO3 compounds
InFarmatik	198	Diverse fragments
Iota Pharmaceuticals	1500	Designed for FBDD
Key Organics	6335	RO3 compounds
Life Chemicals	22 000	MW < 300 Da; clogP < 3
Maybridge	1000	RO3 compounds
Otava	3800	RO3 compounds
Prestwick Chemical	2800	RO3 compounds
Pryxis/Merachem	317	RO3 compounds
Zenobia Therapeutics	352	Small fragments

[a]RO3: 'rule-of-three', MW < 300 Da, clogP $\leq$ 3, up to three hydrogen bond acceptors and up to three hydrogen bond donors.

magnitude smaller than most corporate HTS compound libraries.[7] However, this comparatively small number of compounds does not compromise the success of fragment-screening campaigns as a fragment library captures a substantially larger chemical diversity space than conventional HTS.[7] Still there are over 700 000 fragments that are commercially available and it is estimated that representing the biologically relevant space would require a library of minimally over 32 000 fragments. In this context, it has been proposed that the combination of empirical and computational fragment screenings could enable the discovery of unexpected chemotypes and the capture of the chemotype holes missing in the fragment library.[6]

Physicochemical property restraints are crucial for the development of any fragment library and represent the first filter to remove compounds with little likelihood of being an attractive hit. These include properties such as topological polar surface area (PSA), molecular weight, and calculated octanol/water partition coefficient (clogP).[8] Other factors to be considered include aqueous solubility, novelty, chemical and metabolic stability, availability of analogs, and vendor reliability.[4] Particular emphasis has to be placed on solubility as the concentration of the ligand in the screen is typically between $100 \, \mu M$ and $1 \, mM$, but can easily be higher. These ligand concentrations must be attained in a protein buffer containing typically no more than 5% DMSO-d6[8] as this compound is commonly used as a cosolvent in fragment-screening experiments because it is a very good solvent for a range of organic molecules, is nonvolatile, and is miscible with water.[9] Molecular size also plays a key role in the design of fragment libraries as, according to Hahn's model of molecular complexity, an increased probability of detection binding to a target protein can be expected when small, low-complex molecular fragments are screened with high sensitivity instead of full-sized ligands with low sensitivity.[10] Historically, it has been found that libraries containing fragments with a molecular weight range of 150–300 Da allow sufficient opportunity for further optimization of the potency and physicochemical properties of the identified leads.[5] The lower limit, of approximately 150 Da, offers less chance of a fragment reorienting on the target protein upon preparation. Smaller, less complex fragments will only contain single rings with small substituents that, therefore, are more likely to bind in multiple orientations.[11]

Finally, although NMR-based methods are among the most sensitive techniques, they are intrinsically slower than other screening techniques. Thus, to increase throughput in NMR-based screening, cocktails of fragments are often used instead of single fragments.[12] This procedure drastically reduces protein consumption and data collection time but requires more library preparation. Mixtures are designed to avoid possible chemical reactions between fragments and to minimize the overlap between the NMR resonances of the components. In general, mixtures are prepared by considering the minimum separation between peaks (0.04 ppm) and the number of compounds per mixture (5–10).[4]

12.5 TARGET DRUGGABILITY

Despite all the genomic and proteomic efforts made, a minimal number of targets are addressed each year with new drugs. Failures in late-stage drug development are costly and significant cost savings can be achieved by improving the target selection process. Accordingly, the ability to accurately predict the probable success of a screening campaign may directly affect target attrition and can make the drug discovery process more efficient by allowing prioritization of targets.[13]

In general terms, the druggability concept refers to the likelihood of finding orally bioavailable small molecules that bind to a particular target in a disease-modifying way. However, it is difficult to assess the intrinsic druggability of a protein, unless there are other known compounds that act on that particular target. In recent years, it explains why this term has been slightly downgraded to describe a target's ability to bind small molecules with high affinity, a property that some authors tend to describe as ligandability.[14] Either way, any tool capable of distinguishing protein targets that are able to bind ligands would be of much interest to the pharmaceutical industry.

Hajduk *et al.*[15] suggested that experimental fragment-based screening is a good indicator of target druggability. These authors observed a strong correlation between the experimental NMR hit rate and the ability to bind drug-like high-affinity ligands at a particular binding site. They derived a linear regression model that includes terms for polar and apolar surface areas, surface complexity, and pocket dimensions, which is capable of predicting experimental hit rates with a statistically significant prediction value. Other authors extended Hadjuk *et al.*'s findings by analyzing the ligandability of targets based on hit rates, affinities, and diversity of hits.[14] Using this approach, it was found that when targets were nondruggable in the fragment-based screen, HTS also failed to produce sufficient high-quality hits to enter a hit-to-lead program. On the contrary, when targets were deemed druggable (with a medium to high druggability score), HTS proved successful in the vast majority of cases.[14] Taken together, these findings highlight the potential of FBDD-based NMR to identify and characterize hot spots on target surfaces and its ability to confidently predict the probability that high-affinity, drug-like leads can be identified for a particular target.

12.6 TARGET-BASED VS LIGAND-BASED FBDD NMR METHODS

The NMR-based screening field has evolved rapidly since the first report appeared in 1996 describing the use of NMR spectroscopy to screen for potential drug molecules. Over the last several years, several novel approaches have been introduced and have found widespread application in both pharmaceutical and academic research settings.[16] NMR methods can be broadly categorized into target- and ligand-based approaches, depending on whether the signals from either the drug target or the ligand are detected to characterize intermolecular interaction (Figure 12.1). All these methods have advantages and limitations and can provide information about the ligand–target interaction with various levels of detail, including the determination of ligand affinities and potencies, or their binding site and binding mode, when bound to the drug target.[17]

Target-based methods observe and compare the NMR parameters of target molecule resonances in the presence and absence of compound mixtures. By identifying the perturbations of assigned protein resonances, ligands are not only identified, but their binding sites are also localized. The site-specific characterization of binding suggests strategies for

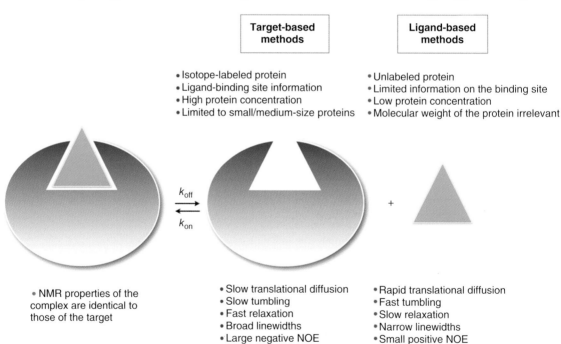

Figure 12.1. The NMR properties of a fragment hit dramatically change upon binding to the target resembling those of the free target. The modifications include alterations in the translational diffusion, tumbling, relaxation, linewidth, and nature of the NOEs that are exploited in a number of target- and ligand-based NMR experiments. Both approaches present advantages and disadvantages regarding their application to labeled/unlabeled proteins, the information provided about the binding site, the amount of protein required for the experiments, and the molecular weight of the target

fragment-based lead generation, in which the lower affinity molecular fragments binding to distinct sub-sites can be linked or prepared to yield higher affinity compounds.[16] Despite their obvious advantages, they require large amounts of isotope-labeled drug target, as well as knowledge of the 3D structure and NMR assignments of active site residues to reveal active site binders. In practical terms, target-based methods are limited to a subset of drug targets (MW < approximately 40–60 kDa) that give good quality NMR spectra and do not aggregate at relatively high concentrations (approximately 25–100 μM) in aqueous NMR buffer.[17]

Ligand-based methods rely on the monitoring of change in some NMR parameters of the ligand upon its binding to the protein.[17] This approach renders the molecular weight of the target irrelevant. In fact, the most powerful ligand-based approaches become more sensitive when dealing with large targets.[16] Although some details about the ligand-binding epitope may be obtained, ligand-based approaches do not reveal the ligand-binding site on the drug target.

Clearly, if target-based methods are possible (low MW, efficient expression, available 3D structure, and NMR assignments), then the potentially higher information content obtainable makes them methods of choice. However, given the scarcity of LMW drug targets and the complexity and time required to perform 3D structural studies, the applicability of ligand-based screening is generally broader and places less demands on other research disciplines and infrastructures.[16]

12.7 PROTEIN TARGET PRODUCTION

In target-based methods, the availability of an efficient protein expression protocol that allows the production of an isotope-labeled protein is particularly important. Although it is possible to use the isolated domain of the protein target for screening, it is generally preferable to use a full-length target. The functional/structural viability of the protein target should also be confirmed by showing activity in a biochemical assay or by binding a known ligand with expected affinity. The nature and relevance of post-translational modifications (e.g., phosphorylation, glycosylation, and acetylation) should be known, as should the need for cofactors. There is generally no problem with screening in the presence of cofactors (e.g., ATP, NAD, NMN, Mg^{2+}), even if they give rise to a binding signal, provided that there is no direct interference with the fragments

or components of the buffer (e.g., Mg^{2+} precipitating with phosphate).[4]

Target-based NMR screening works best if the drug target is relatively small and can be produced cost-effectively in a uniformly isotope-labeled form at high yields. For larger proteins, ^{13}C-labeling and/or deuteration may be required. Site- or amino acid-selective labeling can extend the size of the proteins that can be analyzed. Selective ^{13}C-labeling of the methyl groups of valine, leucine, and isoleucine has proved especially advantageous.[18] These labeling schemes are usually achieved using *Escherichia coli* expression systems because of the advantages of easy handling, rapid growth, high-level protein production, and low-cost isotope labeling. When post-translational modifications are important for the biological activity of the protein, isotope labeling in non-*E. coli* prokaryotic and eukaryotic cells is required to produce active proteins that are useful for target-based NMR screening. Thus, target production can be carried out in yeast, insect cells, or mammalian cells.[17] In contrast, ligand-based NMR methods do not require isotope-labeled material, work better for larger proteins, require less material, and do not need high-quality protein NMR spectra. Thus, these methods are more broadly applicable, and a wider range of expression hosts can be considered for protein production.

Finally, choice of NMR screening buffer can be critical for a successful NMR screening campaign. In general, the buffer must be selected to promote a folded, monomeric protein with relatively high solubility and long-term stability.

12.8 NMR-BASED SCREENING EXPERIMENTS

NMR screening can be defined as the identification of small-molecule ligands for macromolecular targets through the observation of a change in an NMR parameter that occurs upon their interaction. As previously mentioned, NMR screening methods can be divided into those that detect interactions by the observation of either macromolecule NMR parameters or small-molecule NMR parameters.[19]

Target-based methods are based on the use of chemical shift changes to screen low-affinity ligands in combination with structural information to direct fragment optimization approaches leading to an enhancement of the binding affinity. In the first step, the fragment library is screened to identify molecules that bind to the protein. Binding is detected by comparing the 2D ^{1}H–^{15}N heteronuclear single quantum coherence (HSQC) spectra of the isotopically ^{15}N-labeled target protein in the absence and presence of ligands in order to elucidate ligand-induced changes in a chemical shift (Figure 12.2a). Binding constants for the identified ligands can be determined by monitoring

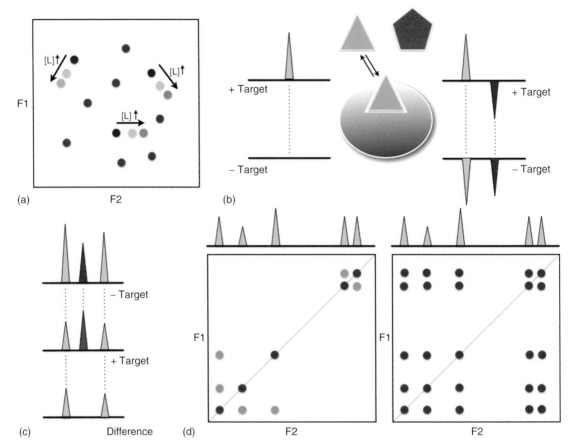

Figure 12.2. Selection of 1D and 2D NMR experiments used in FBDD. (a) Target-based method relying on the ^{1}H–^{15}N HSQC experiment. The chemical shifts of the amino acids directly involved in the interaction (blue) undergo changes in the presence of increasing concentrations of the fragment hit (light and dark green) as opposed to other amino acids that do not interact with the ligand (red). (b) Ligand-based methods relying on the excitation of the target–fragment complex. Left, the STD experiment achieves the selective excitation of the target through the application of a train of radiofrequencies over a frequency window that only contains resonances of the target. A fragment binding to the target (green) is identified in this experiment by the presence of a positive signal in the presence of the target. Right, in the waterLOGSY experiment, the excitation of the target–fragment complex is obtained by a selective perturbation of the water. In this case, fragments binding to the target (green) are characterized by positive signals or by a reduction in the negative signals in the presence of the target, as opposed to nonbinders (red) that do not experience that modification. (c) Ligand-based methods relying on the alteration of the relaxation or diffusion properties of the fragment upon binding to the target. These experiments require the acquisition of the spectra in the absence (upper panel) and presence (middle panel) of the target. In both cases, the difference (absence-presence) spectrum (lower panel) reveals the existence of fragment hits. (d) Ligand-based methods relying on trNOEs. In these experiments, intramolecular NOEs of the potential ligands are observed via 2D NOESY spectra in the absence (left panel) and presence (right panel) of the target. Fragment hits (green signals in the trace 1D NMR experiments) are identified by NOE cross peaks that have changed sign (blue to red) in the presence of the target

shift changes in accordance with the ligand concentration. If the sample contains only one sort of potential ligand molecule, the binding ligand can be identified. If a mixture has been used, the active compound has to be identified by deconvolution.[20] Despite the wealth of information generated from these methods, in the crowded spectra and resonance assignment of large monomeric protein targets, bottlenecks remain that continue to limit the utility of target-based NMR methods. However, advances in isotope-labeling strategies show real promise for expanding the general applicability of these approaches.[16]

Ligand-based methods predominantly use 1D NMR spectra and are, therefore, comparably fast and allow higher throughput in screening. For small molecules, the choice of NMR parameters that can be exploited for the identification of the fragments that interact with the protein target is more diverse.[19] The detection of ligand signals provides the opportunity to screen mixtures of fragments without the need for deconvolution, as long as the signals of the ligands do not overlap. Moreover, the amount of target required for the screening process is smaller and there are fewer restrictions on the properties of the target. The target is not frequently labeled isotopically and its molecular weight is practically unlimited. Some information about the binding epitope and the binding mode can be extracted from ligand-based methods.[18]

In ligand-based screening (Figure 12.2b–d), the differences in NMR properties between small and large molecules form the basis of NMR experiments. When a small ligand binds to a large target, it adopts the properties of the target to the extent that it is dependent on the residence time of the binding event,[18] and the choice of physical mechanisms manifested in measurable NMR parameters includes longitudinal, transverse, DQ relaxation, diffusion coefficients, and intermolecular and intramolecular magnetization transfer. The latter includes transferred nuclear Overhauser effect (trNOE), NOE pumping and reverse NOE pumping, saturation transfer, and waterLOGSY (water–ligand observed via gradient spectroscopy) experiments.[19]

It is well-established that NOEs are extremely useful for determining the 3D structure of molecules in solution. When ligand molecules bind to receptor proteins, NOEs undergo drastic changes, which lead to the observation of trNOEs. These changes are the basis of a variety of experimental schemes that are designed to detect and characterize binding events. The observation of trNOEs relies on different properties of free and bound molecules. LMW or medium-molecular-weight molecules (MW < 1–2 kDa) have a short correlation time, τ_c, and consequently, such molecules exhibit positive NOEs, no NOEs, or very small negative NOEs, depending on their molecular weight, shape, and field strength. Large molecules, however, exhibit strongly negative NOEs. When a small molecule (fragment) is bound to a large-molecular-weight protein (the target protein), it behaves as if it formed part of the large molecule and adopts the corresponding NOE behavior; that is, it shows strong negative NOEs, the so-called trNOEs. These trNOEs reflect the bound conformation of the ligand. The binding of a ligand to a receptor protein can, thus, be easily distinguished by looking at the sign and size of the observed NOEs.[20]

Saturation transfer difference (STD) and water-LOGSY both take advantage of the fact that the intermolecular NOE transfer is strongly negative. Nonequilibrium magnetization is created on the target–ligand complex, and the subsequent alteration of the free ligand signal intensities is monitored. Two principal implementations are imaginable. Either all resonances (target, binding, and nonbinding ligands) are perturbed simultaneously, followed by selective suppression of some signals, or excitation itself is already selective. Selective suppression, for instance, can be implemented by a diffusion filter, eliminating signals of small molecules, or by a relaxation element to filter out target resonances.[18]

The STD is the difference of two experiments. In the first experiment, the 'on-resonance' experiment, target proton magnetization is selectively saturated via a train of selective radiofrequency pulses. The radiofrequency train is applied to a frequency window that contains receptor resonances but, for which, ligand resonances are absent (e.g., 0.0 to −1.0 ppm for proteins). Saturation propagates from the selected receptor protons to other receptor protons via the network of intramolecular ^1H–^1H cross-relaxation pathways. This spin diffusion process is quite efficient owing to the typically large molecular weight of the receptor. Saturation is transferred to binding compounds via intermolecular ^1H–^1H cross-relaxation at the ligand–target interface. Small molecules then dissociate back into solution, where the saturated state persists because of their small free-state R1 values. At the same time, a 'fresher' unsaturated ligand exchanges on and off the receptor, while saturation energy continues to enter the system through the sustained application of radiofrequencies, thus increasing the population of saturated free ligands. A reference experiment (the 'off-resonance' experiment) is then recorded, which applies the identical radiofrequency train far off-resonance so that no NMR resonances are perturbed. 'On-resonance' and 'off-resonance' experiments are recorded in an interleaved manner and subtracted. The resulting difference spectrum yields only those resonances that have experienced saturation, namely target and binding compound resonances. As they are present at a minimal concentration, target resonances are not usually visible and, if this were the case, can be eliminated by R2 relaxation filtering before detection. The result is a simple 1D ^{1}H spectrum that reveals only the ligand signals that bind to the receptor.[16]

Like STD, the waterLOGSY technique relies on the excitation of the receptor–ligand complex through a selective radiofrequency pulse scheme. However, waterLOGSY achieves this indirectly by the selective perturbation of bulk water magnetization as opposed to the direct perturbation of receptor magnetization. The intended transfer of magnetization is water–receptor–ligand, which may occur via a number of mechanisms. All of them, however, allow binding compounds to pick up the bulk water inversion while residing at the receptor binding site. Ligands then dissociate into free solution where, analogous to the case of the STD experiment, their perturbed magnetization state is maintained owing to their small free-state R1 values. The lower these R1 values are, the more time is available for ligands to complex with the receptor. They receive the inversion transfer from the target–ligand complex, and then dissociate back into free solution, where they add to the growing pool of spin-inverted ligands.

Distinguishing binding from nonbinding compounds in the waterLOGSY experiment differs slightly from STD and is achieved through the observation of the differential cross-relaxation properties of these ligands with water. In the waterLOGSY experiment, binders and nonbinders display peak intensities of opposite signs that, thus, provide an easy means to discriminate between them.[16]

12.9 ^{19}F NMR SCREENING

While ligand-based ^{1}H NMR fragment-screening methods are well documented and are widely used, the favorable properties of fluorine NMR spectroscopy

in FBDD have only been recently introduced.[13,21] Although ^{19}F NMR has similar sensitivity to ^{1}H NMR, ligand-based ^{19}F NMR experiments have the limitation of their applicability only to the screening of chemical libraries of fluorinated molecules.[21]

Perhaps, the most popular experiment based on the spectroscopy properties of the ligand molecules whose structure contains fluorine is the so-called FAXS (fluorine chemical shift anisotropy and exchange for screening). This experiment utilizes a spy molecule that contains either a CF or CF3 moiety and monitors changes in the transverse relaxation rate of the ^{19}F resonances in the presence of a mixture of test compounds. This approach offers some unique advantages: high sensitivity and 100% natural abundance of the ^{19}F NMR active isotope; protonated solvents, buffers, or detergents do not interfere with measurements; no overlap.[21]

Another interesting application of ^{19}F NMR is the possibility of performing NMR-based functional screening. Classically, NMR has been extensively used for characterizing the product or products of an enzymatic reaction and for gaining insight into the kinetics of the reaction. A high substrate concentration is necessary for these studies because of the low sensitivity of the NMR technique. The required high concentration is a major hurdle to overcome to utilize this approach for functional screening purposes in general and for FBDD in particular. One way to overcome this limitation is the use of ^{19}F NMR as the detection method. Prototypical fluorine-based biochemical screening experiments are collectively known as FABS (*fluorine atoms for biochemical screening*) and individually as *n-FABS*, where *n* is the number of fluorine atoms used for detection.[21]

Together, ligand- and substrate-based fluorine NMR approaches are powerful tools for performing screening against a protein target and for determining the K_{d}, IC50, and K_{i} values for the identified hits. NMR-based screening techniques FAXS and FABS offer some important advantages in terms of their sensitivity, reliability, reproducibility robustness, and a few interferences.[21]

More recently, ^{19}F NMR-based fragment screening has also been demonstrated as a key tool for the rapid development of structure–activity relationships (SAR) on the hit-to-lead path, and also as a quick and efficient means to assess target druggability.[13]

12.10 HIT VALIDATION BY NMR AND VALIDATION OF NMR FRAGMENT HITS

NMR techniques for the detection and characterization of protein–ligand interactions can also be exploited for the validation and characterization of HTS hits. The outcome of HTS is hits, not leads. To convert hits into leads, the compounds must be investigated very closely. Binding HTS hits to the target protein is usually validated by independent biochemical and/or biophysical methods. NMR is an independent binding assay and can be used to validate the binding of HTS hits to the target protein.

Experience shows that many HTS hits actually do not bind to the target protein. This is sometimes due to a not completely validated assay or to the complex nature of some HTS assays. In other cases, it is due to the poor quality of protein employed for HTS. Cell-based assays or coupled assays such as ELISA are necessarily complex and contain several other proteins apart from the target protein. Assay response is positive if any of the proteins conveying the signal is inhibited and not necessarily the target protein. In other cases, compounds interfere with the detection method; for example, fluorescent compounds can cause artifacts in fluorescent-based assays.[22]

On the other hand, validation is required for the fragment hits identified by comparing the STD or water-LOGSY spectra of mixtures with individual ^{1}H NMR reference spectra. It requires manually examining the spectrum of every hit. Some apparent hits may be subtraction artifacts due to instrument instability. Others may be due to a spillover from irradiation of the protein or water. This problem can be solved by using an alternative irradiation frequency.

Reliability of hits can improve by combining the results from different experiments, each with different sources of artifacts. A comparison is often made between the STD and waterLOGSY experiments, with and without an added competitor. The most robust hits appear in both experiments, while those that bind, but show no change, upon the addition of a competitor are classified as 'binding noncompetitively'. It has also been observed that a small number of fragments appear as hits in nearly every screen against a wide variety of targets. It is noteworthy that some of them contain scaffolds that have previously been identified as 'privileged' or 'preferred' motifs for protein binding.[23]

Finally, fragment hits should also be validated using 'orthogonal' methods (i.e., based on different read-outs and prone to different sources of interferences), such as a combination of NMR methods with biophysical methods (SPR, isothermal calorimetry (ITC), etc.) or biochemical methods.

12.11 NONSPECIFIC BINDING

A basic challenge in ligand-based fragment screening is distinguishing the spectral signatures of bona fide binding from those stemming from nonspecific binding. In this context, it is important to define and distinguish 'nonspecific' from 'low-affinity' binders. Low-affinity binders are ligands with K_d values over 10 μM and in fast exchange on the chemical shift time scale. Low-affinity interactions may, indeed, be specific for well-defined sites on the receptor. Nonspecific binders may also be low-affinity binders. The criterion for specificity for these low-affinity binders is that they bind preferentially to either the targeted active site or another site that directly modulates target activity (e.g., an allosteric interaction). In contrast, nonspecific binders usually bind to receptor surface regions that have no direct effect on target activity. These interactions result from general interactions of the ligand with hydrophobic patches on the protein surface.[16]

A preferred method to expose nonspecific binding for ligand-based approaches is to test for displacement effects upon addition of a known specific and competitive binder. To reduce the risk of nonspecific binding, it is helpful to work at lower target and ligand concentrations. It is also worthwhile to consider alternative expressed forms of the target for screening because working with full-length proteins might present fewer exposed hydrophobic patches than truncated proteins.[16]

12.12 FRAGMENT PRIORITIZATION

FBDD compared to HTS offers several advantages. Among them, lower structural complexity results in a higher hit rate as smaller ligands bind with greater inherent efficiencies. In FBDD, a set of fragments is identified that binds to the desired target. As fragments are small by design, their binding affinities are low and, thus, optimization is necessary.

The combination of both effects (increased hit rate and modest initial affinities) renders the selection of fragment hits necessary for follow-up at the outcome of any fragment-screening campaign. Various criteria have been introduced to perform this evaluation.[24] Two of the most popular ones are the application of the ligand efficiency (LE) and ligand lipophilicity efficiency (LLE) concepts.[17] The weak binders identified in the FBDD process may be a good starting point for lead generation if they exhibit good LE and LLE.[17]

LE is simply the binding free energy for a ligand divided by its molecular size. Potency can be defined just as well as pIC_{50}, pK_i, or ΔG. LE is the useful metric for measuring the impact of the addition of more molecular bulk on activity. The molecules that achieve a given potency with fewer heavy atoms are, by definition, more efficient. LLE is a measure of minimally acceptable lipophilicity per unit of in vitro potency. Chemists will have more freedom to prepare LMW and high ligand efficiency hits before reaching unacceptable limits of molecular weight and complexity, which often lead to compounds that exhibit unacceptable solubility, absorption, and permeability properties. Similarly, fragments with good LLE allow more room to increase lipophilicity during lead optimization without obtaining an unfavorable physical profile for the drug candidate.[17] Other parameters for quantitating the relative potency of fragment-screening hits include the percentage efficiency index (PEI), the binding efficiency index (BEI), obtained by dividing percentage inhibition at a given concentration, and pK_i, respectively, by molecular weight. The surface-binding efficiency index (SEI) is also obtained by dividing pK_i by the PSA. Finally, an alternative metric called *fit quality* (*FQ*) attempts to address the empirical discovery that small ligands have an inherently higher LE value than large ligands.[25] FQ is a scaled ligand efficiency parameter that centers optimal binders near 1.0 by scaling raw ligand efficiencies, thus facilitating the identification of ligands with exceptional properties.[24]

In addition to all these chemical parameters, when considering the interaction between a fragment and a target and the possibility of progressing original hits into leads, a number of thermodynamic considerations should also be taken into account. Affinity, or binding energy, comprises two components: enthalpy and entropy. Some advantages of starting with enthalpically driven leads in which binding arises from specific molecular interactions, such as hydrogen bonds, salt bridges, and van der Waals interactions, have been recently proposed; in contrast, entropically driven binding generally arises from nonspecific hydrophobic interactions. The contribution of each component is usually assessed by isothermal calorimetry and is evaluated in conjunction with a detailed structural model of the binding interaction obtained by X-ray crystallography and/or NMR. Thermodynamic analyses and structural information are an excellent support to guide synthetic efforts.[17]

Finally, the tractability of a hit from chemical preparation should be judged by medicinal chemists, who possess the expert knowledge needed to assess the possibilities for preparing a hit with substituents, or recasting a chemotype into an isostere. From a pharmaceutical viewpoint, chemical novelty should also be evaluated, especially if the target has been extensively studied by other research groups.

12.13 HIT-TO-LEAD

Although initially the fragment hits chosen for follow-up tend to be only weak binders, their ligand efficiency is typically quite high. This means that the number of atoms involved in the desired interaction with the drug target is usually high for fragment hits.[17] To turn an NMR hit into a lead, its affinity must be increased to a low micromolar or better to a submicromolar range. Structural information about the binding mode of the fragment hit usually plays a critical role in achieving this goal.

Follow-up strategies for fragment hits depend strongly on the nature and characteristics of the drug target and fragment hits. For more challenging targets, structural data are critical for efficient fragment hit-to-lead optimization, whereas this may not necessarily be the case for other targets with deep, well-defined active sites.[17]

The first optimization process step usually includes the exploration of the SAR around the primary hit by testing, as much as possible, close analogs of the primary hit, and measuring its affinity. This may result in an optimized hit with a potency of two or three orders of magnitude higher than the original hit. Next, as much structural information as possible on the interaction between the optimized hit and the target protein should be collected. Ideally, the structure of the target protein complexed to the NMR hit should be determined by either X-ray crystallography or NMR spectroscopy.[22]

After collecting this structural information, the NMR hit can be further followed up by adopting different strategies (Figure 12.3). The combination strategy entails combining the molecular fragments

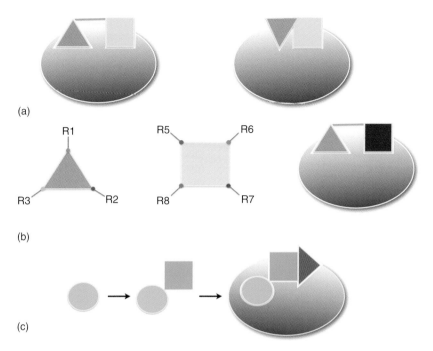

Figure 12.3. FBDD strategies for fragment optimization. (a) Combination strategies are based on the linking (left) or merging (right) of fragment hits binding to two different subsites. (b) Variation strategies rely on the design of combinatorial libraries around the primary fragment hits (left, middle) or in the application of optimization approaches (right) to improve the properties of a particular fragment hit. (c) Elaboration strategies pursue the progressive building of a lead through the rational evolution of fragment hits based on structural information

that have been demonstrated to bind individually to the target. This approach takes full advantage of the fact that many drugs are modular, with groups that bind to distinct subpockets within an active site. When multiple, weakly binding fragments are combined into a single, more complex molecule, which comes into contact with the same set of subsites, potency can be substantially enhanced. Fragments may be combined by either linking them together or merging chemical features from multiple fragments in one molecule. In a linked-fragment approach, a second ligand is identified by second-site screening, which binds simultaneously and in the vicinity to the NMR hit (the 'first ligand'). Linking both ligands then results in a significant increase in binding affinity if the linker is well designed. In the merged-fragment approach, information about binding epitopes from several ligands is used to construct a merged ligand that exploits the binding interactions from two or more distinct ligands. These ligands cannot be linked as they bind at overlapping binding sites. Both the linked-fragment and merged-fragment approaches

rely heavily on the structural characterization of the binding of ligands to the target protein.[22] In the variation strategy, systematic modifications are made to individual portions of a primary screening hit. In the simplest approach, substructure- and similarity-based searches of compound databases are carried out to find analogs of primary screening hits. More sophisticated approaches require the synthesis of compounds. As starting hits already contain numerous connected functional groups, it is much easier to synthesize follow-up compounds than it is with the fragment linking method; groups are already oriented correctly and structural information is not required. The variation strategy has been implemented most effectively in two ways by (i) designing directed combinatorial libraries and (ii) fragment optimization. The combinatorial library design requires compounds that contain either a central scaffold with multiple positions for varying substituents (so that the SAR of each one can be explored separately) or two scaffolds connected by a linker, which is amenable to combinatorial chemistry, and allows scaffolds to be readily replaced or modified.

Fragment optimization is also used to improve existing lead compounds that suffer from unacceptable characteristics, such as low solubility, lack of novelty, poor bioavailability, toxicity, or metabolic instability. A portion of the lead molecule is chosen for the selective replacement with fragments that bind to the same subsite on the protein but possess superior properties. Finally, the elaboration strategy consists in systematically building upon primary screening hits to make more complex molecules. The analogs that contain functional groups or ring systems capable of making additional interactions with the target are selected. Therefore, potency increases without disrupting the binding of the core scaffold. The SAR from each generation of analogs is used to bias the selection of compounds for the next, thus the process converges rapidly to more potent inhibitors. Structural information about the bound leads is often used to direct lead optimization.[16]

12.14   LINKER DESIGN

Once two ligands that bind simultaneously and in the same vicinity have been identified, the goal is to generally chemically link them to enhance their binding affinity, which is expected to be higher owing to enthalpic or entropic considerations. Thus, the binding enthalpy of a linked compound is, in the first approach, the sum of the binding enthalpies of both individual compounds. The binding affinity is, therefore, using the same approximation, the product of the affinities of each individual fragment. However, the linker generally constrains the linked compound, so that the individual fragments cannot make optimal interactions, and the affinity diminishes. Also, the linker itself generally has interactions with the target, which can be favorable or unfavorable, and will further influence the affinity of the linked compound. Entropic considerations take into account that only a binary complex, instead of a ternary complex, needs to be formed with a linked compound. This reduces the entropic cost of complex formation.

The most direct linker design guidelines are offered if the 3D structure of the ternary complex, formed by the target protein and both the first-site and second-site binders, is available. In principle, both X-ray crystallography and NMR can be used for the 3D structure determination of the complex. In practical terms, crystallization of protein targets with weakly binding ligands

tends to be difficult and NMR may be preferred. However, for most therapeutically relevant targets, determination by NMR of its 3D structure is practically not feasible. Very few protein targets pursued by the pharmaceutical industry are amenable for a complete 3D structure determination by NMR. For large, soluble, and well-expressing proteins, deuteration and selective ^1H, ^{13}C labeling offer an appealing option to obtain highly specific information on the ligand-binding mode. However, with unlabeled proteins, essential information for linking first-site and second-site binders can also be deduced by NMR in favorable cases. For example, the identification of interligand trNOEs between protons of the first-site ligand and protons from the second-site ligand can provide sound guidance for chemistry to choose the correct attachment points and linker length. This strategy is known as *SAR-by-NOE*, in analogy to the term *SAR-by-NMR*.[26]

12.15   PROTEIN–LIGAND AFFINITIES

By monitoring biologically relevant protein–ligand interactions, NMR can provide a means to validate these discovery leads and to optimize the drug discovery process. NMR-based screens use a combination of approaches to detect a protein–ligand interaction.[27]

Thus, when an NMR target-based screening method is used, affinities can be determined for $K_d >$ [target] by the titration of hits. When ligand-based methods are used, binding affinities can be measured directly using initial STD rates or by the titration of linewidths or a waterLOGSY/STD signal and indirectly via changes in the binding signal from a competing ligand of known affinity.[4] However, these methods generally work only for weakly or moderately binding ligands and not for tightly binding ligands.[28] More recently, several NMR competition-based methods have been proposed to extend the limit of K_d determination to the high-affinity (nanomolar and subnanomolar) range, although no absolute affinities, but only relative affinities of two competitive ligands, can be determined.[28]

In general, the relatively low throughput of current NMR screens and their high demand of sample requirements generally make it impractical to collect complete binding curves to measure the affinity of each compound in a large, diverse chemical library. As a result, NMR ligand screens are typically limited to identifying candidates that bind to a protein and do not provide any estimate of binding affinity.[27]

12.16 PROTEIN–LIGAND STRUCTURE DETERMINATION

Hit-to-lead optimization in FBDD heavily relies on knowledge of the 3D structure of the complexes between the target and fragment hits. Given the limited time available for research projects in the pharmaceutical environment, obtaining a rapid protein structure in early drug discovery project stages is essential. Despite the prevalent use of X-ray crystallography in industry for determining protein structures, NMR and X-ray should be viewed as complementary techniques.[29]

A classical NMR approach for the structure determination of target–fragment complexes includes 2D and 3D spectra acquisition. In a 600 MHz spectrometer equipped with a cryogenic probe, these experiments can be acquired in 1 week for proteins under 25–30 kDa at concentrations over 0.5 mM. This time can be significantly shortened if the target concentration is high enough or by using nonuniform sampling (NUS) methods. Continuous collection of data on a single sample reduces spectral variations and results in improved data analyses. Intermolecular distance constraints for target–fragment complexes are determined from another experiment, 3D F1-^{13}C/^{15}N-filtered, F3-^{13}C-edited NOESY–HSQC, performed on a sample of a ^{15}N/^{13}C-labeled target saturated with an unlabeled fragment. This experiment detects only the NOEs arising between the protons directly bound to ^{13}C (target) and the protons bound to any non-^{15}N or ^{13}C atom (fragment), while suppressing all other cross peaks by isotope filtering and editing. The constraints generated from the filtered NOESY spectrum can, therefore, be unambiguously assigned to protons on the small-molecule ligand. If the ^{13}C/^{15}N-labeled ligand is available, a 3D F1-^{13}C/^{15}N-filtered, F3-^{13}C-edited NOESY spectrum can be collected on a sample containing an unlabeled target and labeled fragment to thereby provide a complementary data set that further facilitates atom assignments for intermolecular NOEs.[30] These strategies are also usually combined with advanced labeling schemes and NMR techniques, which include deuterium labeling, selective residue labeling, selective methyl labeling, and TROSY-based experiments. Low-resolution structures may also be obtained from a minimal NOE data set combined with other structural constraints, such as residual dipolar couplings (RDCs) and pseudo-contact shifts.[29]

NMR chemical shift perturbations (CSPs) are often used to identify fragment binding sites and are based on the clustering of residues on the target surface that incur CSPs. High-quality target–fragment complex models can be obtained rapidly by combining traditional in situ ligand docking software with experimental NMR CSPs. Specifically, CSPs can be used to define a three-dimensional (3D) search grid for docking programs, such as AutoDock. AutoDockFilter (ADF) is then used to select the best conformers based on consistency with experimental CSPs. ADF uses an NMR energy function based on the magnitude of CSPs and the proximity of the ligand to amino acid residues with a CSP. Similarly, HADDOCK takes advantage of the relationship between CSPs and ligand binding to generate target–fragment complexes by ambiguous interaction constraints (AIRs) to refine these structures. An advantage of the HADDOCK approach is the fact that the complex structure is the result of direct refinement against experimental CSPs.[29]

Other methods, which do not require target resonance assignments, are based on trNOE experiments, which have long since been applied to determine internal conformations of weakly bound fragments. Long-range paramagnetic distance constraints from site-directed spin labeling have also been explored computationally as a possible approach. Furthermore, some of the ligand-based screening experiments already described can be combined with other experimental approaches to determine target–fragment complex structures. Thus, a method based on STD and residue type-specific labeling has been proposed (SOS-NMR).[31] This method utilizes STD NMR spectroscopy performed on a ligand complexed to a series of target samples that have been deuterated everywhere, except for specific amino acid types. In this way, the amino acid composition of the ligand-binding site can be defined and, given the three-dimensional structure of the protein target, the three-dimensional structure of the protein–ligand complex can be determined.[32] A method called *SALMON* (*solvent accessibility, ligand binding, and mapping of ligand orientation by NMR spectroscopy*), based on the waterLOGSY experiment, has also been proposed to determine the orientation of a fragment bound to a target by mapping its solvent accessibility.[33]

12.17 CLINICAL CANDIDATES ORIGINATING FROM FBDD

Over the last few years, a number of biotechnology companies that have focused on FBDD have

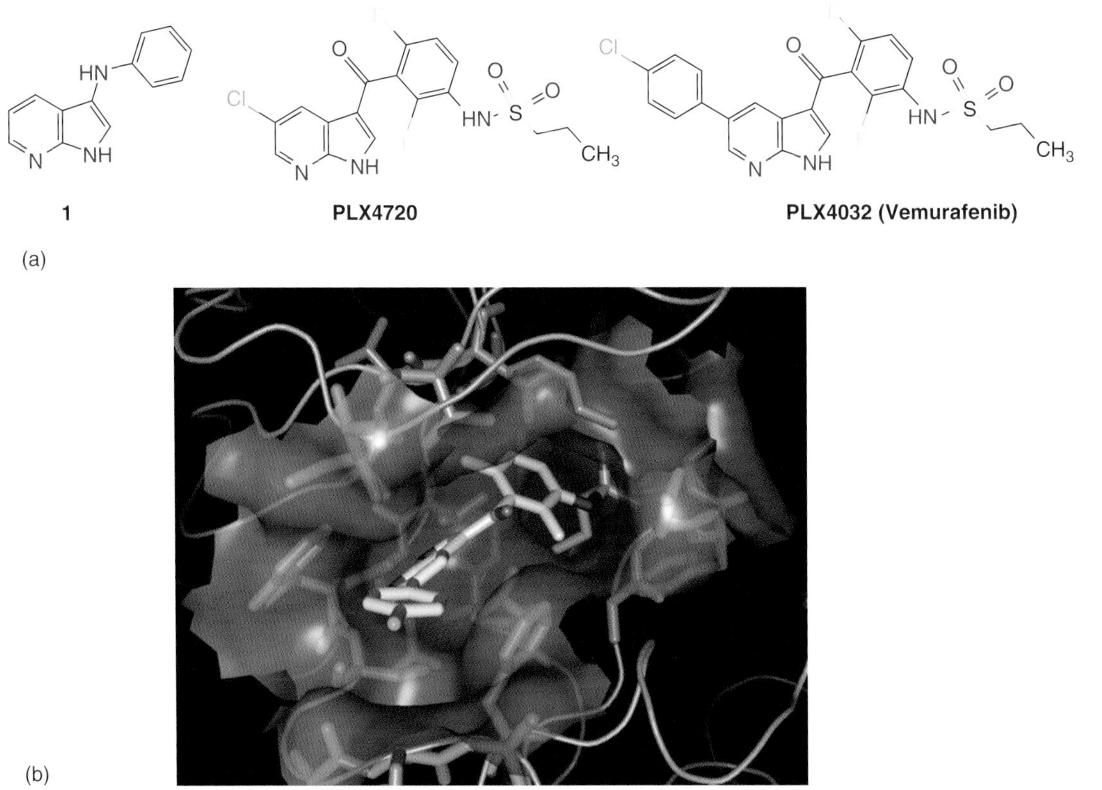

(a)

(b)

Figure 12.4. First FBDD drug approved. (a) Structures of individual compounds leading to the discovery of the first marketed drug based on FBDD approaches. Compound **1** (3-aminophenyl-7-azaindole) led to the discovery of PLX4720 and eventually to PLX4032 (Vemurafenib), a relevant drug in the treatment of metastatic melanoma. (b) Structural details of the interaction between PLX4032 and its target, B-RAF (V600E)

emerged, including Astex (X-ray, NMR, ITC), Sunesis (MS), Plexxikon (X-ray), Evotec (SPR, NMR, X-ray), Carmot Therapeutics (X-ray), Vernalis (NMR, SPR, ITC, X-ray), ZoBio (NMR, SPR), and so on. These companies, and some other big pharmaceutical companies (Abbott, Novartis, AstraZeneca, Hoffmann-La Roche, etc.), have shown the potential of this technology for advancing fragments into clinical settings.[5]

A recent report[34] has described several clinical-stage compounds originating from fragment-based screening campaigns. These molecules cover different therapeutic areas (Alzheimer's disease, solid tumors, multiple myeloma, leukemia, etc.) and are the result of internal research programs conducted in different biotechnology and pharmaceutical companies. Interestingly, FBDD approaches have already resulted in a marketed drug (Figure 12.4), Vemurafenib

(PLX4032), a compoundy that binds to mutant BRAF kinase, a pharmaceutically relevant target in metastatic melanoma. Several other compounds are now in clinical development phase I/II/III and are expected to deliver new drugs in forthcoming years.

RELATED ARTICLES IN EMAGRES

Understanding the Role of Conformational Dynamics in Protein-Ligand Interactions Using NMR Relaxation Methods

REFERENCES

1.   M. Congreve, G. Chessari, D. Tisi, and A. J. Woodhead, *J. Med. Chem.*, 2008, **51**, 3661.

2. R. Gozalbes, R. J. Carbajo, and A. Pineda-Lucena, *Curr. Med. Chem.*, 2010, **17**, 1769.

3. N. Blomberg, D. A. Cosgrove, P. W. Kenny, and K. Kolmodin, *J. Comput. Aided Mol. Des.*, 2009, **23**, 513.

4. C. A. Lepre, *Methods Enzymol.*, 2011, **493**, 219.

5. G. Chessari and A. J. Woodhead, *Drug Discov. Today*, 2009, **14**, 668.

6. S. Barelier, O. Eidam, I. Fish, J. Hollander, F. Figaroa, R. Nachane, J. J. Irwin, B. K. Shoichet, and G. Siegal, *ACS Chem. Biol.*, 2014, **9**, 1528.

7. P. J. Hajduk and J. Greer, *Nat. Rev. Drug Discov.*, 2007, **6**, 211.

8. J. F. Bower and A. Pannifer, *Curr. Pharm. Des.*, 2012, **18**, 4685.

9. E. H. Mashalidis, P. Śledź, S. Lang, and C. Abell, *Nat. Protoc.*, 2013, **8**, 2309.

10. A. Schuffenhauer, S. Ruedisser, A. L. Marzinzik, W. Jahnke, M. Blommers, P. Selzer, and E. Jacoby, *Curr. Top. Med. Chem.*, 2005, **5**, 751.

11. G. Siegal, E. Ab, and J. Schultz, *Drug Discov. Today*, 2007, **12**, 1032.

12. X. Arroyo, M. Goldflam, M. Feliz, I. Belda, and E. Giralt, *PLoS One*, 2013, **8**, e58571.

13. J. B. Jordan, L. Poppe, X. Xia, A. C. Cheng, Y. Sun, K. Michelsen, H. Eastwood, P. D. Schnier, T. Nixey, and W. Zhong, *J. Med. Chem.*, 2012, **55**, 678.

14. F. N. B. Edfeldt, R. H. Folmer, and A. L. Breeze, *Drug Discov. Today*, 2011, **16**, 284.

15. P. J. Hajduk, J. R. Huth, and S. W. Fesik, *J. Med. Chem.*, 2005, **48**, 2518.

16. C. A. Lepre, J. M. Moore, and J. W. Peng, *Chem. Rev.*, 2004, **104**, 3641.

17. H. L. Eaton and D. F. Wyss, *Methods Enzymol.*, 2011, **493**, 447.

18. J. Klages, M. Coles, and H. Kessler, *Analyst*, 2007, **132**, 693.

19. B. J. Stockman and C. Dalvit, *Prog. Nucl. Magn. Reson. Spectrosc.*, 2002, **41**, 187.

20. B. Meyer and T. Peters, *Angew. Chem. Int. Ed. Engl.*, 2003, **42**, 864.

21. C. Dalvit, *Prog. Nucl. Magn. Reson. Spectrosc.*, 2007, **51**, 243.

22. H. Widmer and W. Jahnke, *Cell. Mol. Life Sci.*, 2004, **61**, 580.

23. S. Barelier, J. Pons, K. Gehring, J.-M. Lancelin, and I. Krimm, *J. Med. Chem.*, 2010, **53**, 5256.

24. S. D. Bembenek, B. A. Tounge, and C. H. Reynolds, *Drug Discov. Today*, 2009, **14**, 278.

25. M. Orita, K. Ohno, and T. Niimi, *Drug Discov. Today*, 2009, **14**, 321.

26. W. Jahnke, A. Flörsheimer, M. J. J. Blommers, C. G. Paris, J. Heim, C. M. Nalin, and L. B. Perez, *Curr. Top. Med. Chem.*, 2003, **3**, 69.

27. M. D. Shortridge, D. S. Hage, G. S. Harbison, and R. Powers, *J. Comb. Chem.*, 2008, **10**, 948.

28. X. Zhang, A. Sänger, R. Hemmig, and W. Jahnke, *Angew. Chem. Int. Ed. Engl.*, 2009, **48**, 6691.

29. R. Powers, *Expert Opin. Drug Discov.*, 2009, **4**, 1077.

30. J. J. Ziarek, F. C. Peterson, B. L. Lytle, and B. F. Volkman, *Methods Enzymol.*, 2011, **493**, 241.

31. K. L. Constantine, M. E. Davis, W. J. Metzler, L. Mueller, and B. L. Claus, *J. Am. Chem. Soc.*, 2006, **128**, 7252.

32. P. J. Hajduk, J. C. Mack, E. T. Olejniczak, C. Park, P. J. Dandliker, and B. A. Beutel, *J. Am. Chem. Soc.*, 2004, **126**, 2390.

33. C. Ludwig, P. J. Michiels, X. Wu, K. L. Kavanagh, E. Pilka, A. Jansson, U. Oppermann, and U. L. Günther, *J. Med. Chem.*, 2008, **51**, 1.

34. M. Baker, *Nat. Rev. Drug Discov.*, 2013, **12**, 5.

Chapter 13

Hit Discovery from Natural Products in Pharmaceutical R&D

Olivia Corcoran

Medicines Research Group, School of Health, Sport and Bioscience, University of East London, London E15 4LZ, UK

13.1 INTRODUCTION

An ongoing debate to justify natural product (NP) drug discovery centers on the fact that between 1981 and 2002 in the United States, roughly half of the 877 small-molecule New Chemical Entities (NCEs)

NMR in Pharmaceutical Sciences. Edited by Jeremy R. Everett, John C. Lindon, Ian D. Wilson, and Robin K. Harris
© 2015 John Wiley & Sons, Ltd. ISBN: 978-1-118-66025-6
Also published in eMagRes (online edition)
DOI: 10.1002/9780470034590.emrstm1432

were NPs, semisynthetic analogs or synthetic compounds based on NP pharmacophores.[1] An enduring inspiration for chemically diverse new pharmaceuticals and nutraceuticals remains the evolving NP scaffold. By definition, NPs are compounds produced by living systems and the secondary metabolites are those which give particular species their characteristic features. These metabolites include alkaloids, phenylpropanoids, polyketides, and terpenoids, among others. Dereplication is therefore a key concept in NP hit discovery and can be defined as the process of testing sample mixtures that are active in screening (hits) in order to differentiate the novel compounds from active substances that have already been studied.

The World Health Organization recently estimated that over 80% of the world's population depends on plants for healthcare. However, of approximately 400 000 flowering species on our planet, some 80 000 will have become extinct by 2050, largely due to anthropogenic activity. The ^{1}H and ^{13}C NMR spectroscopic analysis of organic compounds from valuable natural resources continues to play a key role in the development of medicinal chemistry.[2–4] Beyond this, we are only now starting to understand the important role that NMR sciences in fact play in conserving our natural resources. Despite a decline in the NP sector since 2002, with 31 NP or NP-derived drug approvals in phase III or in registration by the end of 2013, a steady stream of approvals has been predicted for at least the next 5 years.[5] With this comes anticipated future issues because only five new drug pharmacophores discovered in the last 15 years are currently being evaluated in clinical trials.

For pharmaceutical purposes, access to diverse drug scaffolds discovered from plant, microbial, marine, reptile, insect, and animal resources is often perceived as high risk. A major problem is the typically low yield of active constituents from raw materials: 4 tons of dried leaves from *Catharanthus roseus* yields just 1 kg of vinarelbine, the active ingredient in the antimitotic drug Navelbine, that is used to treat cancer. Where ethical and economic considerations feature, plants and microbes are often genetically manipulated to sustainably engineer virus, antibody, enzyme, and small molecule products. These approaches can often provide cost-effective synthetic intermediates not feasible by agricultural means. The nondestructive nature of NMR spectroscopy also provides a critical advantage over mass spectrometry (MS) as the first line analytical tool when faced with characterizing microgram amounts of novel bioactive compounds. Given higher demands from decreasing amounts of precious, chemically diverse, and preferably intellectual property-rich raw material, samples initially used for NMR spectroscopy can later be tested for bioactivity.

Sustainability being a key economic driver, biotechnology can often improve yields of target bioactives beyond 1%. Background knowledge of crude drug including the conservation status, agricultural history, botanical authentication, origins, harvest and treatment of material, and pharmacopeia monographs relates to the sciences of ethnobotany, pharmacognosy, and herbal medicine.[6,7] Beyond the systematic screening of NP resources from soil and marine environments, and the biodiverse regions of South Africa and South America targeted by the National Cancer Institute programs of the 1960s, ethnobotany remains a valuable underpinning discipline. Arguably, indigenous knowledge, both documented and oral, can inform the systematic review of the scientific and economic botany literature to focus drug discovery programs on those plants with a higher likelihood of potent hits.

NMR spectroscopy and MS are key mining technologies for dereplication in NP drug discovery. These analytical tools service similar objectives despite issues of sensitivity (NMR spectroscopy) and ion suppression from interferences as well as contra-sustainability (MS being both sample and analyte destructive). The main historical functions that relate NMR spectroscopy to NP hit discovery have

been (i) stereochemical elucidation of isolated, pure compounds (see Sections 13.2, 13.3, and 13.5), (ii) characterization of pharmacophore-receptor binding (other chapters herein) (iii) screening of NP extracts (see Section 13.4), and (iv) multiply hyphenated NMR spectroscopic technologies (see Section 13.6). Where other chapters elucidate generic theoretical and practical aspects of ^{1}H and ^{13}C NMR spectroscopy spanning applications (i)–(iv), the focus here is on advances in NMR spectroscopy related to accelerating hit discovery from natural resources, mainly plants.

13.2 PRACTICAL ASPECTS OF NMR SPECTROSCOPIC EXPERIMENTS ON NATURAL PRODUCTS

This section considers sample preparation for NMR spectroscopy in light of the wide diversity of NP chemistry. The main requirements are for stable, homogenous material devoid of paramagnetics and suitably soluble in deuterated solvents. Unlike routine NMR spectroscopic acquisition of aqueous biological fluids, NP samples cover crude solvent extracts and semi-purified fractions, as well as the final isolated pure compounds. A wider range of deuterated solvents are thus necessary to dissolve NPs of varying polarity: D_2O, methanol-d_4, chloroform-d, pyridine-d_5, and DMSO-d_6. As much of classical NMR theory and practice depends on the available amount and purity of sample, these facts govern the suitable preparation of NP samples for a given NMR system.

Classical extraction protocols are based on partitioning sciences and are specific to the various classes of secondary metabolites. These protocols include liquid–liquid partitioning, steam distillation for oils, soxhlet extraction, and solid phase extraction (SPE). Newer technologies using microwave extractions, ionic liquids, HILIC (hydrophobic interaction liquid chromatography), and SPE based on molecularly imprinted polymers are among those methods reviewed elsewhere.[8] Secondary metabolites may be selectively extracted based on class chemistries such as alkaloids, flavonoids, and steroids. These are known to vary between different plant tissues, the bulk of plant material being made up of primary metabolism and structural features. Thus it is important to dereplicate at an early stage to avoid saturating NMR samples with common plant constituents. This also focuses 2D experiments on novel NPs. A hexane solvent extraction step is often the first step in defatting root, stem, and bark samples. Pigments including chlorophyll from leaf extracts can degrade chromatographic column performance and interfere with HPLC-UV detection. Ubiquitous tannins from barks and stems can mask aromatic spectral regions. Such constituents must be removed from samples before NMR spectroscopic data acquisition and to avoid false positives in bioassay after analysis.[8]

In a busy pharmaceutical department, the onus is generally on the customer to present a suitably concentrated and purified sample to the open access NMR system. To efficiently generate the common 1D ^{1}H and ^{13}C NMR spectroscopic datasets, NMR scientists generally favor a minimum amount for their available system configuration of magnet strength (typically 300–600 MHz ^{1}H NMR observe systems), probe specification, and autosampler capacity. Often between 2 and 10 mg dissolved in 0.6–1 ml of deuterated solvent is enough for rapid 1D ^{1}H NMR spectral data to check a reaction product. However, 10–50 mg may be necessary for full structure elucidation of a purified, isolated NP requiring the full armory of 1D (^{1}H, ^{13}C, and ^{15}N), 2D (gCOSY, gTOCSY, NOESY, HSQC, HMBC, and HNMBC) datasets.[2] The structural analysis of mass-limited samples in the microgram range typical of NP isolates remains challenging. It is mainly 500 MHz and higher field systems equipped with microprobe and cryoprobe configurations that reliably handle the sub 25 μg amounts for the full suite of NMR spectroscopic experiments to be generated in less than a day.[9]

Although the probe specifications for homo- and hetero-nuclear 2D NMR experiments vary from probes that require standard 5 mm tubes (500–600 μl sample volume) down to the newer microprobe diameters of 1.7–1 mm (30 and 5 μl sample volumes), most open access academic NMR spectroscopy systems tend to use room temperature static 5 mm selective (^{1}H, ^{13}C) inverse probes with pulsed field gradients.[2] The advantages in going from a 5 to 1.7 mm probe include halving sample amounts, reducing expensive deuterated solvent costs and minimizing solvent impurity signals. Ultimately, the same signal-to-noise ratio in the same number of scans is possible on half the mass. As yet, the higher cost of cryogenic probes and platforms often precludes their use in many academic centers. Room temperature probes with automated matching and tuning at least allow optimal sensitivity across the variable polarity deuterated solvents necessary for NMR spectroscopy on different solvent batches of NPs.

By contrast to the above-mentioned sample requirements for isolated compounds, the rapid ^{1}H NMR spectroscopic screening, or metabolite fingerprinting, of crude and semi-purified extracts requires 1–5 mg of dried residue to be reconstituted in deuterated solvent.[4,10] However, chemical shifts may vary due to solvent effects and efficient solvent suppression is required. Injection of precious NP samples via a flow NMR robotic sampler and flow NMR probe is limited where reliability issues from pumping systems may exist. Tube autosamplers are often considered safer options.

13.3 ADVANCES IN NMR ACQUISITION AND DATA PROCESSING

The hardware development of probe electronics and higher field magnets has greatly decreased the amount of sample needed for 2D NMR spectroscopy to identify and characterize pure NPs from the 20–50 mg range to under 1 mg. Owing to the chemical diversity of NPs, some classes can be readily recognized only by chemical shift regions in the ^{1}H NMR spectrum: aromatic signals for flavonoids, anomeric protons for carbohydrates and 0–3 ppm for terpenes and steroids. Figure 13.1 depicts two classical examples of the challenges for NMR spectroscopy in assigning NPs. The full structure elucidation of a purified, isolated NP often still requires the full armory of classical 1D (^{1}H, ^{13}C-DEPT, and ^{15}N) and 2D (gCOSY, gTOCSY, NOESY, HSQC, HMBC, and HNMBC) datasets.[2] At that, full stereochemical assignment at all carbons may not be achieved. The main improvements in pulse sequences, acquisition and processing methods of most relevance to NP research are amply reviewed elsewhere.[3]

Two notable developments for accessing the second NMR dimension so critical to NP structure elucidation relate to integrating new pulse sequences with hardware advances in COSY and INADEQUATE. In realtime ultrafast 2D COSY NMR, the detection is based on echo-planar imaging so every H–H correlation is observed in one scan. A proof of concept LC-NMR application generated two scan COSY spectra acquired every 12 s as flavonoids eluted from an HPLC column, to identify naringin, epicatechin, and naringenin.[11] For assignment of NP molecules with low observable protonation such as ellagic acid and hypericin (Figure 13.1), the evolution of the classically insensitive INADEQUATE experiment has evolved to

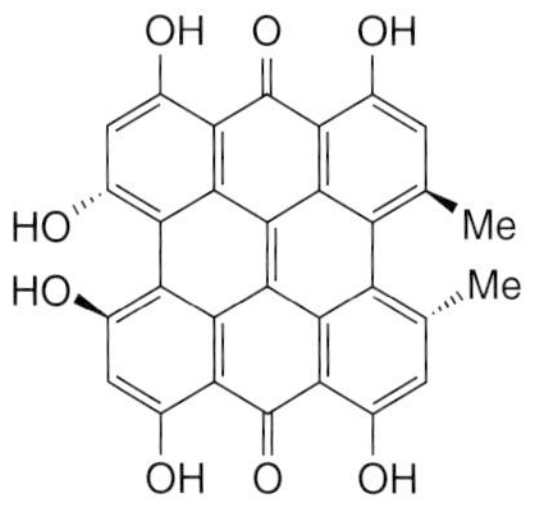

(a)   Hypericin

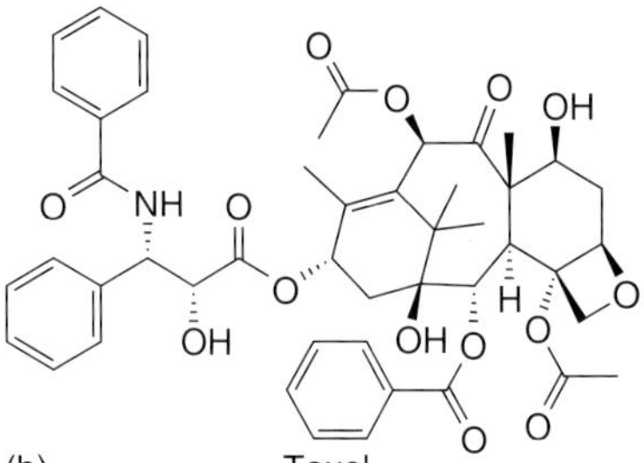

(b)   Taxol

Figure 13.1. Challenging chemical structures of natural product hits for NMR characterization. (a): The natural product hypericin, one of the active ingredients of the ancient anxiety medication St. John's Wort *Hypericum perforatum*. Despite a relatively simple molecular formula of $C_{30}H_{16}O_8$, hypericin gives few observable ^{1}H NMR signals in DMSO-d$_6$ at 2.67 ppm (6H singlet), 6.47 ppm (2H singlet), and 7.34 ppm (2H singlet). (b): Paclitaxol (Taxol) is a structurally complex lipophilic natural product originally extracted from the Pacific yew tree *Taxus brevifolia*. The compound shows good activity against ovarian and breast cancer, and a wide spectrum of carcinomas of the lung, colon, prostate, and brain. It required solid-state NMR spectroscopy and ^{13}C cross-polarization magic angle spinning (CPMAS) to distinguish between the two polymorphic forms that exist

HSQC-ADEQUATE for the accurate determination of adjacent and long-range carbon connectivity. This can be decisive for elucidating demanding NP structures. It is still considered time-consuming despite hardware advances, although the revised structure of the natural antibiotic coniothyrone was solved in approximately 17 h with only 1.2 mg in 30 µl of solvent, using the 1.7 mm MicroCryoProbe™.[12] For researchers interested in learning from the fragment-assembly approach to NP assignment, a helpful illustrative example using a full NMR spectroscopic dataset on Taxol is given elsewhere.[2]

By contrast, the high-throughput screening (HTS) by 1D ^{1}H NMR spectroscopy of crude solvent extracts may be achieved on as little as 1 mg of dried extract in deuterated solvent at 400–600 MHz ^{1}H observation. However, this success depends on the dynamic range of the sample and the efficiency of solvent suppression for extracts containing protonated solvent interferences. In the 1990s, the need for efficient pulse sequences to combat the towering water/D$_2$O and ^{13}C satellites from acetonitrile/methanol signals from HPLC solvent mixtures, that otherwise overwhelm receiver ranges, was a key driver of early requirements to achieve successful LC-NMR results.[13]

13.4 ^{1}H NMR SPECTROSCOPIC SCREENING OF NATURAL PRODUCT EXTRACTS

Recent developments in current analytical platforms and NMR sensitivity for NP mixture analysis permit integration of metabolomics into classical phytochemistry for dereplication of extracts and semi-pure fractions.[10] The term *metabolite fingerprinting* refers to the untargeted rapid classification of samples using multivariate data analysis (MVDA) in a hypothesis-generating approach.[4] On the other hand, metabolite profiling focuses on the targeted analysis of a large group of metabolites either related to a specific metabolic pathway or a class of compounds and follows specific hypotheses. The relatively low sensitivity of ^{1}H NMR spectroscopy has not hampered the development of simple and reproducible methods for routinely profiling crude natural extracts. This makes it a unique and nondestructive tool for efficient dereplication, given the various solvents used for NP extracts. Unlike hyphenated chromatography-MS, prior chromatography is not required. As little as 1 mg of dried extract in deuterated solvent is required for 400–600 MHz ^{1}H NMR data to be generated in about 1 min acquisition. This may provide direct quantitative information owing to the intensity of the proton signal being related to the molar concentration of given signals representing metabolites. Provided standard sample preparation protocols are used with specific solvents and/or buffers, NMR chemical shifts are highly reproducible allowing comparison of data with NP databases. As yet, a few class-specific ^{1}H and ^{13}C NMR spectral databases exist. Databases on flavonoids, taxane diterpenoids, and terpenes have been developed by NP groups with specific expertise.

These and other metabolomic applications of NMR spectroscopic profiling of natural extracts are reviewed elsewhere.[4,10]

NMR spectroscopy can monitor inherent variability of NP sources due to climate, geography and seasons, amongst other ecologies. Thus NMR-principal components analysis (PCA) of 12 *Cannabis sativa* cultivars showed clear discrimination on the basis of important cannabinolic metabolites.[14] ^{1}H NMR spectral profiling of commercial herbal products can also be used to protect formulation IP in the botanicals sector, as reported for western herb formulations of lemonbalm[15] and the traditional Chinese medicine ginseng.[16] NMR spectroscopic data can also inform intellectual property strategies, starting with plant authentication and the correct identification of novel chemical structures. Chemometric analysis of NMR spectral data on seven different solvents was used to optimize bioassay-guided fractionation of 1 g crude extracts from *Scutellaria baicalensis,* a plant formulated to treat cancer patients in China.[17]

13.5 MICRO NMR SPECTROSCOPY AND CRYOGENIC NMR PROBES

Novel probe coil designs have been applied to both static and flow probe configurations resulting in two distinct advances: cryogenic probeheads (where cooling reduces thermal noise on coil electronics) such as the 1.7 mm TXI MicroCryoProbe™ having approximately 10 μl volume) and capillary NMR probes (5 μl sample required for 1.5 μl active volume).[4,9,18] An alternative for indirect coupling of chromatography to NMR spectroscopy is to collect LC fractions by constant time intervals or triggered by UV/MS, for off-line NMR spectroscopy in either probe type. The miniaturization of sample volumes (typically 1.5–5 μl) for NMR spectral acquisition appears to be a burgeoning approach to handling 5–100 μg amounts of precious NP samples. However, limiting factors are solubility and the handling of volumes much below 1 μl. Microcoil NMR spectroscopy, using solenoidal microcoil, microstrip, and microslot probes, is based on the increase of coil sensitivity for smaller coil diameters (approximately 1/d).[18] Microcoil NMR probes for 400–600 ^{1}H MHz deliver mass-based sensitivity increases of 8- to 12-fold when compared with routine 5 mm NMR probes. Moreover, this economy scaling permits fully deuterated solvents systems for microflow NMR applications, thus negating the more

stringent solvent suppression sequences required to avoid ^{13}C satellites of organic solvent signals. Pure compounds may be manually injected but it is preferable to use automated injection methods, which include injection with no column in-line or capillary-LC (cLC), capillary electrophoresis (CE, also known as capillary zone electrophoresis or CZE) and capillary electrochromatography (CEC). These latter 'electrically driven' separations also eliminate the need for solvent to be delivered via a pump and can access limits of detection in the nanogram to low microgram range when using stopped-flow techniques.

In continuing efforts to explore minor components of plants, platforms to miniaturize NP drug discovery have integrated the outputs from high-throughput NP chemistry methods with the capabilities of 1 mm Cap-NMR technology. Hu *et al.*[19] prepared an NP library from the extract of *Penstemon centranthifolius*, and subsequently isolated and identified six known iridoid glycosides using 25–300 μg of material. This required approximately 5 μg to perform COSY and NOESY experiments; HMQC- or HSQC (30 μg)-NMR spectra; HMBC (75–100 μg); and around 200 μg of compound to perform ^{13}C- and DEPT-NMR experiments. This off-line high-throughput preparative approach has since been enhanced using the TCI 1.7 mm Micro-CryoProbe to identify polyoxygenated cyclohexenes as minor constituents from the African shrub *Monanthotaxis congoensis*, that show moderately antiproliferative activity (7–14 μM) against human cancer cell lines NCI-H460 and M14.[20]

13.6 NMR SPECTROSCOPY OF NATURAL PRODUCTS FOLLOWING CHROMATOGRAPHY

As outlined earlier, the more common alternative for precious NP material that constitutes minor plant constituents (<1%) is for off-line NMR where samples are first prepared via classical extraction, SPE, and preparative techniques and then transferred to NMR tubes for analysis by capNMR probe or MicroCryoProbe.[8,12,19,20] This permits extraction and sample preparation to be carried out in different NP laboratories for transport to NMR laboratories that have higher NMR specification systems (such as higher magnet strength for greater spectral resolution and cryoprobes for enhanced sensitivity).

There are, however, significant benefits to NP drug discovery from the advances in interfacing flow NMR probes with directly coupled chromatography including HPLC, capillary LC, UPLC, or SPE. The broad applications in NP chemistry are reviewed elsewhere.[4] While hyphenating LC-UV with NMR spectroscopy can be a powerful structure determination tool, coupling other spectrometers in-line can greatly enhance information recovery from chromatographic separations (albeit requiring higher capital investment and system complexity).[4,13] LC-NMR-MS can provide both fragmentation patterns and, with high-resolution mass spectrometers, data on atomic composition, providing the potential to derive molecular formulas, and uncovering 'NMR invisible' atoms (including halogens, sulfur, and oxygen among others). Many operation modes of directly coupled LC-NMR-MS exist, depending on the available hardware and the configuration of the MS in the system. A wide diversity of NP applications have been reviewed to demonstrate the different operational LC-NMR(-MS) modes including onflow, stopped flow, time-slicing, loop storage, and online SPE trapping.[4]

An early application of stopped-flow LC-NMR for NP chemistry was the on-line separation and structure elucidation of naphthodianthrones, flavonoids, and other constituents of an extract from *Hypericum perforatum* L. using LC-NMR-MS.[21] A conventional reversed-phase HPLC system using ammonium acetate as the eluent buffer was used, and ^{1}H NMR spectra were obtained on a 500 MHz NMR instrument. NMR acquisition was directed by UV and MS chromatographic peaks. MS and MS/MS analyses were performed using negative electrospray ionization and all of the major known constituents in extracts from *Hypericum perforatum* L. were identified, along with two new substances which had not previously been reported as constituents of extracts of this plant. The value of online deuterium exchange experiments using the mass spectrometer was particularly helpful in determining the number of NMR-silent hydroxyl groups on these metabolites.

Parking HPLC peaks in online storage loops for automated presentation to the NMR flow probe was a logical progression but the expense of deuterated solvents was the key driver for incorporating online SPE. This allows greater freedom for the chromatographer in resolving HPLC peaks via necessary buffers for pH control and the use of nondeuterated solvents. Once HPLC peaks have been multiply loaded onto SPE cartridges on multiple successive injections, the SPE cartridges are dried under nitrogen and the analytes eluted

to the NMR flow probe using low volumes of deuterated solvent (150–350 μl per analyte).[4] In an interesting study, LC-NMR spectroscopy with loop storage was compared with LC-SPE-NMR spectroscopy (using a cryo-flow probe) for the identification of isomeric tropane alkaloids extracted from the bark of *Schizanthus grahamii*.[22] Both approaches enabled the alkaloids to be characterized, while demonstrating that increased sensitivity was only seen for the SPE-trapping approach where compounds were well retained on the SPE trapping phase. In this example about 50% of the target analytes were not well retained and were therefore lost.

Related to stopped-flow LC-NMR are the 'time-slicing' modes. The HPLC pump is programmed to start and stop every few intervals (seconds or minutes, depending on required signal to noise ratio) allowing for acquisition of NMR spectra in between. This can be applied where incomplete HPLC resolution is achieved between successively eluting HPLC peaks, it may be possible on the basis of the ^{1}H NMR spectrum to correctly assign the different eluting compounds. Alternatively, for dereplication purposes where little UV or MS data is available on the sample in advance, NMR spectra can be acquired on peaks eluting from the chromatographic column regardless of the presence of a UV chromophore (as in steroids and terpenes) or absence of ionizable groups (as for carbohydrates in MS). A related off-line application of time-based trapping of HPLC-separated compounds onto SPE cartridges and subsequent elution to NMR tubes was carried out to emulate the function of HPLC-NMR for dereplication purposes.[23] Off-line HPLC-SPE prepared samples for a 600 MHz 1.7 mm MicroCryoProbe. The resulting ^{1}H NMR spectra were compared to an in-house-developed database with matching algorithms, based on partitioning of the spectra and allowing for changes in the chemical shifts. This approach was reported for an extract of *Carthamus oxyacantha* (wild safflower), containing an array of spiro compounds, and an extract of the endophytic fungus *Penicillum namyslowski*, containing griseofulvin and analogs. The database matching of the resulting spectra positively identified expected compounds, while the number of false positives was few and easily recognized.

For efficient dereplication of NP mixtures, onflow, or continuous flow chromatography allows for ^{1}H NMR spectral acquisition 'on the fly' as the eluent moves analytes through the active flow probe volume. A typical on-flow run results in a 2D plot that displays NMR frequencies on one axis and chromatographic separation time on the other axis. The best results require a lower flow rate in the range 0.2–0.5 ml min^{-1}, an HPLC column of higher loading capacity (>250 mm length) and the cost of deuterated organic solvent (methanol-d$_4$ or acetonitrile-d$_3$) as well as D$_2$O may be considered prohibitive. The dereplication of antiinflammatory furanocoumarins from the crude ethanolic extract of *Angelica dahurica* Fisch. Ex Hoffm. (Apiaceae) was achieved using on-flow LC-NMR-MS.[24] A cryoflowprobe (60 µl flow cell) was employed and all the data was generated from an injection of 2.5 mg equivalent of an ethanolic extract.

The use of cryo-flow probes and online SPE-trapping for NP characterization has further benefits. A typical example of this type of application in HPLC-DAD-SPE-NMR for the identification of the constituents of plant extracts is provided by studies on the constituents of *Kanahia laniflora*.[25] Here reversed-phase gradient HPLC, with in-line SPE-trapping of each peak of interest, was undertaken on partially purified extracts. The trapped peaks were eluted with CD$_3$CN into a 30 µl inverse ^{1}H–^{13}C-flow probe and 1D- and 2D-NMR spectroscopy were performed (COSY, HSQC, and HMBC) which allowed the identification of seven constituents (four flavonol glycosides and three 5α-cardenolides). The most technologically complete platform reported thus far has been LC-DAD-high resolution MS-SPE-NMR equipped with microvolume, cryogenic, flow probes. The combined advantage of a cryogenically cooled probe, miniaturization, and multiple trapping enabled the first reported application of HPLC-SPE-NMR analysis using direct-detected ^{13}C NMR spectra that allow more routine access to the ^{13}C skeleton.[26] The direct ^{13}C NMR detection of triterpenoids from a *Ganoderma lucidum* extract was achieved in hyphenation mode and HPLC column loading, accumulative SPE trappings, and the effect of different elution solvents were evaluated and optimized. It was found that a column loading of approximately 600 µg of a prefractionated triterpenoid mixture, six trappings, and an acquisition time of 13 h resulted in spectra with adequate signal-to-noise ratios to detect all ^{13}C signals. Dereplication strategies reviewed elsewhere include online LC-NMR-MS dereplication of flavonoids from roots of *Sophora flavescens* and 33 anthocyanins from *Vitis* sp. Grapes.[4]

13.7 DEREPLICATION AND NATURAL PRODUCT NMR DATABASES

Despite a rich history of lead molecules and high potency hits, the HTS approach to NPs in recent years has arguably failed to live up to investor hopes. There are many economic and competitive barriers to obtain high-quality NP screening libraries.[1] These may be libraries of authenticated samples (though not always, depending on licensing conditions) comprising isolated compounds and, often, semi-purified solvent extracts. The diverse sources span plant, microbial, marine, reptile, insect, and animal. Unlike the systems biology NMR spectral approaches that led to open access relational databases such as the human metabolome database (HMDB),[10] the far wider chemodiversity thwarts community efforts to design suitable open access NP databases. Indeed, predictive NMR software such as nmrdb.org is reliable only for the smaller, simpler molecules and not the vast chemical diversity of most NP molecules. A key feature to the success of dereplication strategies is computational support for data handling, processing, and structure elucidation and despite the availability of user-friendly and sophisticated software packages for effective data mining, they are not yet widely used for dereplication purposes in NP.[9] It is argued that this is owing to NPs unique and unexpected spectral patterns and the residual complexity observed.

Dereplication methods aside,[4] the early and correct stereochemical characterization of known NP bioactives may, in turn, govern intellectual property strategies. This requires convergent MS and NMR spectral datasets for structure elucidation. However, this relies on quality spectra, preferably for comparison against fully assigned reference standards where available. The NMR spectra must be of adequate spectral resolution and the structures correctly annotated. The difficulties in building open access NP databases are now obvious: most are by necessity commercial or in-house, being specific to magnet strength, temperature, specific deuterated solvents, and internal reference standards. Included compounds are restricted to available reference standards from Sigma or specialist suppliers, or focused on chemical groups or isolated from specific organisms. ^{1}H NMR spectra at lower NMR magnet strength (400 MHz) typically lack sufficient resolution for assignment of steroids, polypeptides, and terpenes. Generating the ^{13}C NMR spectral data necessary for full assignment of these chemical classes relies, as ever, on sufficient and pure material. The recent addition of hyphenated NMR/MS platforms[13] to the arsenal provides a further layer of complexity where not only ^{1}H NMR chemical shift, integration and coupling constants, but HPLC solvent NMR signals and MS-indexed LC retention times must be cross-referenced in relational databases such as Bruker's AMIX (Analysis of MIXtures) software platform. A welcome addition has been the development of covariance NMR to identify features across 2D NMR experiments for characterizing mixtures. This was extended by Nicholson *et al.* to correlate variable intensities across multiple independent samples, resulting in a covariance matrix of an entire dataset termed statistical total correlation spectroscopy (STOCSY).[27] In the field of NPs, a recent derivation of this covariance was used to identify associations between NMR and MS data in a mapping approach named statistical heterospectroscopy (SHY). This used spectral features to distinguish industrial citrus oils from Italian and Argentinean origins.[28]

The general strategy for data mining across NP databases requires algorithms suitable for structure search, matching, and identification of compounds.[9] The main structural databases contain more than 200 000 natural compounds.[29] Another popular relational NP databases is NAPRALERT® with more than 200 000 literature reports including ethnomedical information, pharmacological/biochemical information of extracts of organisms in vitro, in situ, in vivo, in humans (case reports, nonclinical trials) and clinical studies. Similar information is available for selected secondary metabolites from natural sources. Current NP databases are often restricted by commercial (Wiley's SpecInfo, ACD/Labs and Bruker's NMR database, Chenomx NMR suite) or in-house platforms and vary by the efficacy of the algorithms used. NP groups tend to build local databases around the available instrumentation and compound classes of relevance to the evolution of their own research: MarinLit and AntiBase focus on marine, fungal, and microorganism NPs.[9] These databases allow dereplication using searches based on substructures, NMR spectral features, and calculated ^{13}C and ^{1}H NMR shift data. Another open access online organic spectra database that includes assigned NPs is SDBS. This contains 15 400 ^{1}H NMR and 13 600 ^{13}C NMR spectra. However, two-thirds of the compounds, mostly chemical reagents, are relatively small secondary metabolites containing between 6 and 16 carbon atoms. Pubchem and ChemSpider (32 million structures) are open access repositories that include mostly high abundance NPs. As yet, few class-specific ^{1}H and ^{13}C NMR spectral databases exist. Those databases on flavonoids,[30] terpenes,[31] and taxane diterpenoids[32] were developed by NP groups with specific expertise.

13.8 CONCLUSIONS

Arguably, it is the exacting requirements on NMR spectroscopy posed by the challenges of mining such chemodiverse NPs from complex natural matrices that has driven NMR advances over the last 30 years. These advances span hardware, pulse sequences, and hyphenated systems that continue to accelerate the characterization and identification of NP hits present in complex mixtures. Indeed there is now considerable literature available illustrating applications of the technology across pharmaceutical, biotechnology, and NP research sectors. It is to be hoped that the current gap in NP databases will be ably filled by improved chromatographic separations at the miniaturized scale, such as those involving ultra-high performance LC (UHPLC) combined with appropriate microflow NMR cryoprobes for enhanced sensitivity. Community initiatives toward open access sharing of NP NMR spectral data, not only on pure compounds but also on crude extracts, could ensure continued development in this area. In relation to the latest concepts of translational medicine, the metabolism of inactive NPs to pharmacologically active metabolites after ingestion by mammalian systems may provide a rich vein to tap during preclinical experiments exploring the NP 'hit hyperspace', and is worthy of future development.

RELATED ARTICLES IN EMAGRES

Revisiting Carbon-detected NMR Experiments in Light of Technological Advances in Modern Instrumentation

REFERENCES

1. F. E. Koehn, *Prog. Drug Res.*, 2008, **65**, 175.

2. N. Bross-Walch, T. Kuhn, D. Moskau, and O. Zerbe, *Chem. Biodiv.*, 2005, **2**, 147.

3. R. C. Breton and W. F. Reynolds, *Nat. Prod. Rep.*, 2013, **30**, 501.

4. J. L. Wolfender, G. Marti, A. Thomas, and S. Bertrand, *J. Chromatogr. A.*, 2014. http://dx.doi.org/10.1016/j.chroma.2014.10.091.

5. M. S. Butler, A. A. B. Roberts, and M. A. Cooper, *Nat. Prod. Rep.*, 2014, **31**, 1612.

6. M. Heinrich, in *Comprehensive Natural Products II: Chemistry and Biology*, ed. J. Reedijk, Waltham, MA, US: Elsevier, 2013, 351.

7. A. Adetutu, W. A. Morgan, and O. Corcoran, *J. Ethnopharmacol.*, 2011, **137**, 50.

8. F. Bucar, A. Wube, and M. Schmid, *Nat. Prod. Rep.*, 2013, **30**, 525.

9. M. Halabalaki, K. Vougogiannopoulou, E. Mikros, and A. L. Skaltsounis, *Curr. Opin. Biotechnol.*, 2014, **25**, 1.

10. D. G. Cox, J. Oh, A. Keasling, K. L. Colson, and M. T. Hamann, *Biochim. Biophys. Acta*, 2014, **1840**, 3460.

11. L. H. K. Queiroz Junior, D. P. K. Queiroz, L. Dhooge, A. G. Ferreira, and P. Giraudeau, *Analyst*, 2012, **137**, 2357.

12. G. E. Martin, A. V. Buevich, M. Reibarkh, S. B. Singh, J. G. Ondeyka, and R. T. Williamson, *Magn. Resonan. Chem.*, 2013, **51**, 383.

13. O. Corcoran and M. Spraul, *Drug Discov. Today*, 2003, **8**, 624.

14. Y. H. Choi, H. K. Kim, A. Hazekamp, C. Erkelens, A. W. Lefeber, and R. Verpoorte, *J. Nat. Prod.*, 2004, **67**, 953.

15. A. Sanchez-Medina, C. J. Etheridge, G. E. Hawkes, P. J. Hylands, B. A. Pendry, M. J. Hughes, and O. Corcoran, *J. Pharm. Pharm. Sci.*, 2007, **10**, 455.

16. J. Yuk, K. L. McIntyre, C. Fischer, J. Hicks, K. L. Colson, E. Liu, D. Brown, and J. T. Arnason, *Anal. Bioanal. Chem.*, 2012, **1–11**.

17. J. Gao, H. Zhao, P. J. Hylands, and O. Corcoran, *J. Pharm. Biomed. Anal.*, 2010, **53**, 723.

18. M. Kuhnle, K. Holtin, and K. Albert, *J. Sep. Sci.*, 2009, **32**, 719.

19. J. F. Hu, E. Garo, H. D. Yoo, P. A. Cremin, L. Zeng, M. G. Goering, M. O'Neil-Johnson, and G. R. Eldridge, *Phytochem. Anal.*, 2005, **16**, 127.

20. C. M. Starks, R. B. Williams, S. M. Rice, V. L. Norman, J. A. Lawrence, M. G. Goering, M. O'Neil-Johnson, J.-F. Hu, and G. R. Eldridge, *Phytochemistry*, 2012, **74**, 185.

21. S. H. Hansen, A. G. Jensen, C. Cornett, I. Bjornsdottir, S. Taylor, B. Wright, and I. D. Wilson, *Anal. Chem.*, 1999, **71**, 5235.

22. S. Bieri, E. Varesio, J.-L. Veuthey, O. Munoz, L.-H. Tseng, U. Braumann, M. Spraul, and P. Christen, *Phytochem. Anal.*, 2006, **17**, 78.

23. K. T. Johansen, S. G. Wubshet, and N. T. Nyberg, *Anal. Chem.*, 2013, **85**, 3183.

24. S. W. Kang, C. Y. Kim, D. G. Song, C. H. Pan, K. H. Cha, D. U. Lee, and B. H. Um, *Phytochem. Anal.*, 2010, 322.

25. C. Clarkson, D. Staerk, S. H. Hansen, and J. W. Jaroszewski, *Anal. Chem.*, 2005, **77**, 3547.

26. S. G. Wubshet, K. T. Johansen, N. T. Nyberg, and J. W. Jaroszewski, *J. Nat. Prod.*, 2012, **75**, 867.

27. O. Cloarec, M. E. Dumas, A. Craig, R. H. Barton, J. Trygg, J. Hudson, C. Blancher, D. Gauguier, J. C. Lindon, E. Holmes, and J. Nicholson, *Anal. Chem.*, 2005, 1282.

28. G. Marti, J. Boccard, F. Mehl, B. Debrus, L. Marcourt, P. Merle, E. Delort, L. Baroux, H. Sommer, S. Rudaz, and J. L. Wolfender, *Food Chem.*, 2014, **150**, 235.

29. H. Chapman, Dictionary of Natural Products on DVD (23:1), CRC Press, Taylor & Francis Group, Florida, USA. 2014. http://dnp.chemnetbase.com/.

30. V. V. Mihaleva, T. A.de Beek, F.van Zimmerman, S. Moco, R. Laatikainen, M. Niemitz, S. P. Korhonen, M. A.van Driel, and J. Vervoort, *Anal. Chem.*, 2013, **85**, 8700.

31. F. Qiu, A. Imai, J. B. McAlpine, D. C. Lankin, I. Burton, T. Karach, N. R. Farnsworth, S.-N. Chen, and G. F. Pauli, *J. Nat. Prod.*, 2012, **75**, 432.

32. J. T. Fischedick, S. R. Johnson, R. E. B. Ketchum, R. B. Croteau, and B. M. Lange, *Phytochemistry*, 2014, http://dx.doi.org/10.1016/j.phytochem.2014.11.020.

PART D
Lead to Drug Candidate

Chapter 14

NMR-based Structure Determination of Drug Leads and Candidates

Torren M. Peakman

torren@nmrpeak.co.uk

14.1 INTRODUCTION

Two recent articles in Chemical & Engineering News[1,2] clearly highlight the importance of correct

NMR in Pharmaceutical Sciences. Edited by Jeremy R. Everett, John C. Lindon, Ian D. Wilson, and Robin K. Harris
© 2015 John Wiley & Sons, Ltd. ISBN: 978-1-118-66025-6
Also published in eMagRes (online edition)
DOI: 10.1002/9780470034590.emrstm1406

structure assignment in the pharmaceutical industry. In this case, the structure of a patented anticancer compound known as TIC10 or *ONC201* (**1**), that is about to enter human clinical trials, has been reassigned as the 'angular' isomer (**2**).[3] The implications of this error are considerable[1,2] and there is little doubt that a large number of medicinal chemists have shaken their heads in disbelief at this revelation. Indeed, one posting (Corante Weblogs. In the Pipeline: drug discovery. http://pipeline.corante.com/archives/2014/05/22/a_horrible_expensive_and_completely_avoidable_drug_development_mixup.php) has described it as 'a horrible, expensive, and completely avoidable drug development mix-up'.

1 (Linear isomer)

2 (Angular isomer)

Within the public domain, this error is far from a one off! The Pfizer anticancer compound bosutinib

(**3**), currently undergoing Phase III clinical trials for chronic myeloid leukemia, has been marketed by a number of suppliers with some of these inadvertently selling the isomeric structure (**4**) containing a 3,5-dichloro-4-methoxyanilide rather than a 2,4-dichloro-5-methoxyanilide.[4] This has unfortunately led to a large number of academic results being invalidated without mentioning the time and cost involved.[4] An interesting discussion on how NMR spectroscopy could have been used to prevent this happening has appeared on Ryan's blog (Ryan's Blog on NMR Software. http://acdlabs.typepad.com/my_weblog/2012/06/the-bosutinib-isomer-a-case-for-nmr-asv.html).

3

4

These two striking examples highlight the importance of correct structure assignment of drug leads and candidates and many in the pharmaceutical industry have no doubt lost count of the number of wrong compounds, supplied or even synthesized in house, that they have come across.

Pharmaceutical companies possess large numbers of compounds in their compound collection, typically between tens and hundreds of thousands. It is probably fair to say that in many of these collections, the number of samples with the wrong structure to that documented is 10–20%. Indeed, this scenario occurred in the case of TIC10 with the compound being obtained from the National Cancer Database.[1,2] Over the past two decades, parallel chemistry has no doubt reduced the number of incorrect structures in compound collections. Following the merger of Glaxo Wellcome and SmithKline Beecham in 2000, the new company GlaxoSmithKline embarked on a review of their combined compound collection.[5] Compounds were

initially screened by liquid chromatography–mass spectrometry (LC-MS) with those being pure and having the correct mass to the documented structure being assumed to be correct, while those being pure but having the wrong mass being submitted for NMR structure determination.[5] The reader will no doubt note that isomeric structures such as those highlighted earlier would have slipped through the net! This exercise, however, clearly shows how seriously companies such as GlaxoSmithKline take the issue of correct structural identity[5] (see Chapter 11). Interestingly, there is now a trend emerging in the pharmaceutical industry towards reducing the number of compounds synthesized and thinking more deeply about why such a compound is being prepared in the first place. With this increased emphasis on design, the need for correct structure assignment is clearly ever more crucial. Although not an area covered in this chapter, NMR studies can be fruitful in determining some of the key physical properties of small drug-like molecules that can drive this understanding.

14.2 BACKGROUND TO STRUCTURE DETERMINATION BY NMR IN THE PHARMACEUTICAL INDUSTRY

So when should structure determination by NMR become a key tool in a drug discovery program? Let us assume that a screen against a particular target has been completed and a number of hits from the compound collection have resulted. A medicinal chemist will look at selecting the most interesting of these to progress further, with the ultimate aim of finding several distinct entities as starting points from which to launch synthetic chemistry campaigns to make molecules with improved drug-like properties. Returning to my comment that typically 10–20% of compounds in compound collections have the wrong structure to that documented, structure confirmation would clearly be worthwhile. Clearly, this was not the case with TIC10 with the structure being assumed to be that registered (**1**) and not the revised structure (**2**).[1,2] Medicinal chemists will want to resynthesize and retest the interesting hit compounds to confirm activity and to eliminate the chance of any contaminant in the original material triggering a false positive. It is worthwhile expanding on this point about being sure of the correct structure at the beginning. Starting with an incorrect structure could easily send a project down the wrong path. What about the scenario of a chemist starting with the wrong intermediate as no doubt occurred in the case of the intended commercial synthesis of bosutinib (**3**) that actually transpired to be the isomeric structure (**4**)? There has been a fair amount of discussion of when to acquire more data than just a ¹H NMR and an LC-MS. Some people have advocated doing this on final compounds. This is a fair point and certainly would give a lot of confidence to the project. If the extra data proved that the structure was wrong then there has been a significant waste of manpower and money in finding the error at the end point. Certainly, it is better to have found the error but finding the mistake earlier in the day would have been advantageous. Imagine the scenario where the starting material is wrong and a chemist spent a week or more taking it through to the final product only to find out that it possessed the wrong structure! To avoid just this situation occurring, companies such as Pfizer utilize online liquid chromatography–evaporative light scattering–mass spectrometry (LC-ELSD-MS) systems in order to confirm the identity (MS, neglecting isomers), the purity (LC), and the concentrations of all hits from high-throughput screens at the stage of multiple concentration activity confirmation. NMR and MS software vendors now market packages for automatically assessing the consistency of experimental data with the proposed structures and this can be of great value in organizations dealing with large numbers of compounds or contracting out their synthetic chemistry.

When we talk about structure confirmation by NMR what exactly do we mean? Is this simply a ¹H NMR spectrum that is fully consistent with the proposed structure in terms of the number of hydrogens and multiplicities or something more? The difference in the ¹H NMR spectra of structures (**1**) and (**2**) is subtle[3] and similarly between (**3**) and (**4**) (Ryan's Blog on NMR Software. http://acdlabs.typepad.com/my_weblog/2012/06/the-bosutinib-isomer-a-case-for-nmr-asv.html) and in cases like this something more is required. Experience tells us that there are thousands of examples for which the ¹H NMR spectrum is consistent with more than just a single structure. With today's modern NMR spectrometers, a wide range of experiments are available with just a single button click that can potentially be used to provide more conclusive evidence. Knowing which experiment(s) to pick and how to interpret them is not always quite so straightforward and it is all too easy for inexperienced chemists to acquire a whole plethora of data. I hope to highlight in the following sections examples of how

NMR spectroscopy can be used in the pharmaceutical industry to provide correct structure assignment that will go a long way in preventing errors such as those highlighted earlier from happening.

14.3 INFORMATION FROM NMR EXPERIMENTS FOR STRUCTURE DETERMINATION

Some of the key information furnished by NMR that is commonly used for structure determination of small drug-like molecules is outlined in the following text:

- Chemical shifts (¹H, ¹³C, ¹⁵N, ¹⁹F, ³¹P)
- Integration and absolute quantitation (¹H, ¹⁹F, ³¹P)
- Multiplicities and coupling constants (¹H, ¹³C, ¹⁹F, ³¹P)
- 2D correlations via J-couplings (¹H–¹H, ¹H–¹³C, ¹H–¹⁵N, ¹H–¹⁹F, ¹H–³¹P, ¹⁹F–¹³C, ¹⁹F–¹⁵N, ¹⁹F–¹⁹F)
- 2D correlations via ROE/NOE (¹H–¹H, ¹H–¹⁹F)
- Selective 1D correlations via J-couplings and ROE/NOE (¹H–¹H, ¹H–¹⁹F)

Chemical shift data, integration, multiplicities, and coupling constants are all obtained from simple 1D spectra with ¹H accounting for the majority of cases. Quantitation is an area that has seen a marked rise in usage over the last decade and is frequently termed qNMR, although it does not concern us here[6] (see Chapter 5). Numerous experiments are available to obtain the information listed earlier. Indeed, this approach is what the reader would find in most text books on the use of NMR for structure determination. I am going to take a slightly different approach by considering some of the typical structural problems encountered in the pharmaceutical industry. Where possible these examples are of drug candidates or synthetic intermediates or by-products.

14.4 CONSTITUTION. USE OF ¹³C NMR CHEMICAL SHIFT PREDICTIONS AND ¹H–¹³C CHEMICAL SHIFT CORRELATIONS FROM HSQC AND HMBC SPECTRA

14.4.1 ¹³C NMR Chemical Shift Predictions and HSQC Spectra

Consider (**5**) and (**6**), two typical intermediates isolated from the same reaction that differ in their constitution with the *para*-methoxybenzyl protecting group on either the nitrogen as in (**5**) or the oxygen as in (**6**). The ^{1}H NMR spectra are consistent with either structure and although there are differences in chemical shifts, it would be difficult to distinguish between them with any degree of certainty. Those experienced in ^{13}C NMR spectroscopy would no doubt be aware of potential key differences between the two structures and these can be assessed routinely through the use of ^{13}C chemical shift prediction that is available in many NMR software packages (predicted ^{13}C NMR chemical shifts were obtained using MNova9). A particularly attractive difference is the benzylic carbon, that is predicted at δ ^{13}C 48.4 for the N-substituted and 69.8 ppm for the O-alkylated product. It is not necessary to acquire a ^{13}C NMR spectrum but simply a heteronuclear single quantum coherence (HSQC) experiment that correlates hydrogens with their attached carbons, i.e., a correlation over one bond. From the simple HSQC

experiment, the two compounds can easily be distinguished with experimental values of δ 45.5 for (**5**) and 67.5 ppm for (**6**).

It is worthwhile exploring the HSQC experiment in a little more detail. Typically, the experiment is acquired in phase-sensitive, edited mode with CH and CH$_3$ correlations phased opposite to that of CH$_2$. Note that with modern versions of the experiment using adiabatic pulses, the CH$_2$ groups of cyclopropyl groups phase in the same manner as CH and CH$_3$ groups.[7] It is worthwhile noting the increased use of nonuniform sampling for this type of experiment.[8] This use of the HSQC experiment, together with ^{13}C NMR chemical shift prediction, is particularly useful in distinguishing possible reaction products that differ in constitution such as (**5**) and (**6**) and is a typical problem encountered in medicinal chemistry. Another example in which two products (**7**) and (**8**) were obtained is shown below. Again, the ^{1}H NMR spectra of the two products are consistent with either structure but the ^{13}C NMR chemical shift prediction highlights a major difference in the methyl group. In the OCH$_3$ analog (**6**), the predicted ^{13}C NMR chemical shift is δ 56.3 ppm, while in the NCH$_3$ analog (**7**), it is δ 41.8 ppm. These predictions match well with the experimental data of δ 55.5 and 43.3 ppm, respectively.

NCH$_2$
^{1}H δ 5.42 ppm (observed)
^{13}C δ 45.5 ppm (observed)
^{13}C δ 48.4 ppm (predicted)

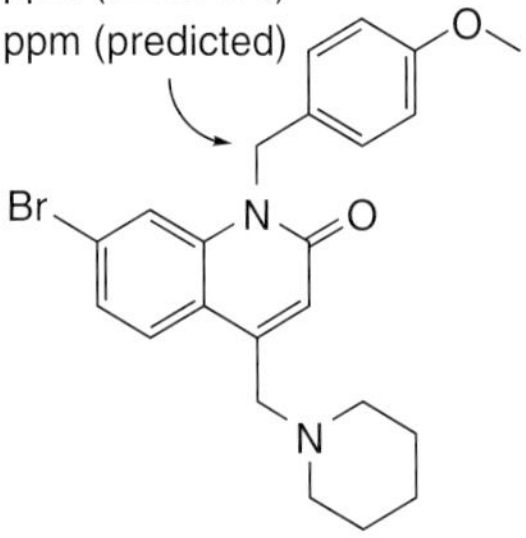

5

OCH$_2$
^{1}H δ 5.44 ppm (observed)
^{13}C δ 67.5 ppm (observed)
^{13}C δ 69.8 ppm (predicted)

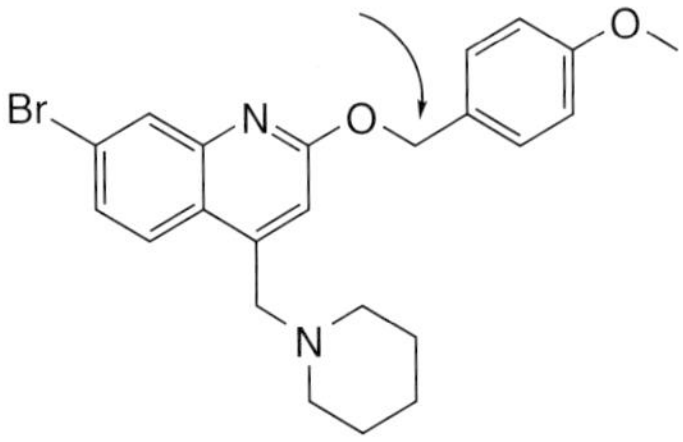

6

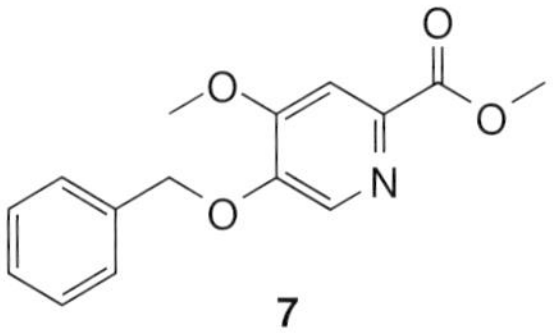

7

OCH$_3$
^{1}H δ 3.98 ppm (observed)
^{13}C δ 55.5 ppm (observed)
^{13}C δ 56.3 ppm (predicted)

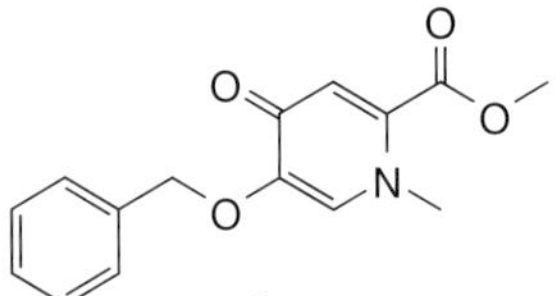

8

NCH$_3$
^{1}H δ 3.93 ppm (observed)
^{13}C δ 43.3 ppm (observed)
^{13}C δ 41.8 ppm (predicted)

14.4.2 ^{13}C NMR Chemical Shift Predictions and HMBC Spectra

Consider the situation where the key differences in carbon chemical shifts are only for carbons with no attached hydrogens. Two compounds (**9**) and (**10**) can be isolated from cyclopropylmethylation of dichloropurine. A ^{13}C NMR chemical shift prediction (predicted ^{13}C NMR chemical shifts were obtained using MNova9) suggests that the ring junction carbons are significantly different between the two products. This is born out experimentally with the ^{13}C NMR chemical shifts identified from long-range ^{1}H–^{13}C correlations (i.e., correlations through 2 and 3 bonds) using the heteronuclear multiple-bond correlation (HMBC) experiment. Note again that it is not necessary to acquire a ^{13}C NMR spectrum. The key correlations in this case are shown below in blue.

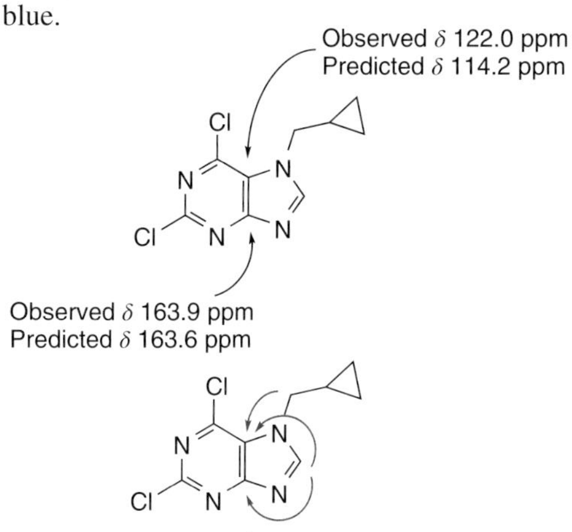

Observed δ 122.0 ppm
Predicted δ 114.2 ppm

Observed δ 163.9 ppm
Predicted δ 163.6 ppm

Observed δ 130.9 ppm
Predicted δ 131.1 ppm

Observed δ 152.2 ppm
Predicted δ 153.4 ppm

9

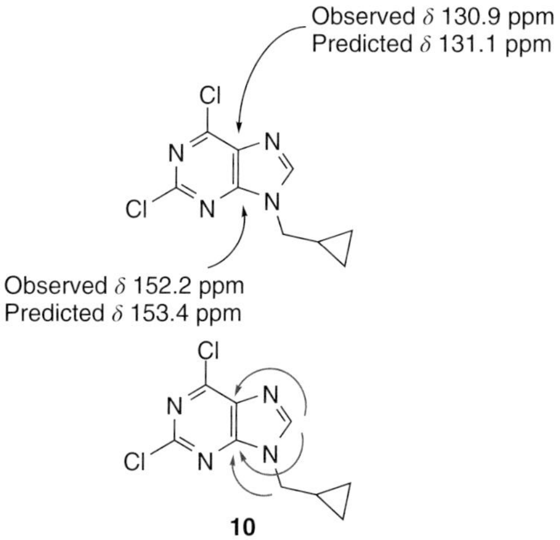

10

14.4.3 Considerations When Using ^{13}C NMR Chemical Shift Predictions

The above-mentioned examples highlight the usefulness of ^{13}C NMR chemical shift prediction and how similar structures can be distinguished using ^{1}H–^{13}C chemical shift correlation experiments such as HSQC and HMBC. No doubt the reader is thinking about the scenarios where the ^{13}C NMR chemical shifts are similar. First off let us consider how much weight we want to give to the ^{13}C NMR chemical shift prediction and then to the match with experimental data. A ball park figure of a predicted difference of at least 10 ppm between two related structures is often a good starting point to decide if the problem can potentially be solved by this approach. Similarly, a difference in experimental data between the two structures of at least 10 ppm would be necessary to give confidence. NMR software packages with chemical shift prediction capabilities usually have an option to highlight the expected accuracy and, furthermore, have database augmentation capabilities so that chemical shift data from newly assigned compounds can be added to improve predictions.

14.4.4 Cases Where ^{13}C NMR Shift Predictions Fail to Distinguish Between Isomers

Consider the pyrazine carboxylic acids (**11**) and (**12**), isolated from a single reaction, in which the carboxylic acid and *para*-fluorphenyl groups are either *para* or *meta*. Carbon-13 NMR chemical shift predictions for the two structures are similar and this is born out experimentally. In such situations, we need to look more closely at the data. In the HMBC spectrum, one of the compounds shows long-range correlations between the two pyrazine CH groups. These correlations are only consistent with the meta isomer (**12**). Other long-range ^{1}H–^{13}C correlations shown below in blue also distinguish between the two positional isomers.

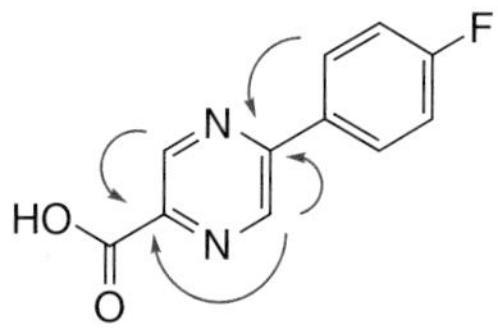

11

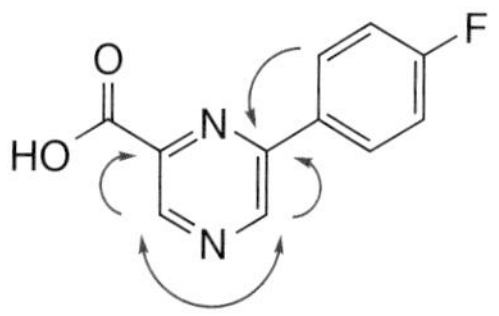

12

14.4.5 Use of ^{13}C NMR Chemical Shift Predictions When Only One Product Is Obtained

In many cases, only a single product is isolated from a reaction but with two or more structural possibilities. The ^{13}C NMR shift prediction can still be useful in such cases, although it is often pertinent to carry out a detailed analysis of the HMBC correlations to confirm the structure. Compounds (**13**) and (**14**) were potential reaction products with only one being isolated. The predicted ^{13}C NMR chemical shift of the methylene carbon is very similar in both compounds but the quaternary carbon is notably different. In the HMBC spectrum, a long-range ^{1}H–^{13}C correlation, shown below in blue, highlights a quaternary carbon at δ 69.6 ppm that is only consistent with the t-alcohol (**13**) with a predicted shift of δ 70.6 ppm as opposed to the predicted shift in the primary alcohol (**14**) of δ 53.3 ppm.

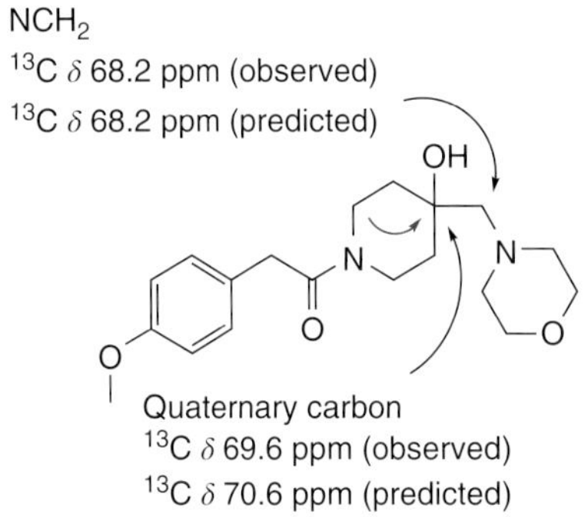

NCH$_2$
^{13}C δ 68.2 ppm (observed)
^{13}C δ 68.2 ppm (predicted)

Quaternary carbon
^{13}C δ 69.6 ppm (observed)
^{13}C δ 70.6 ppm (predicted)

13

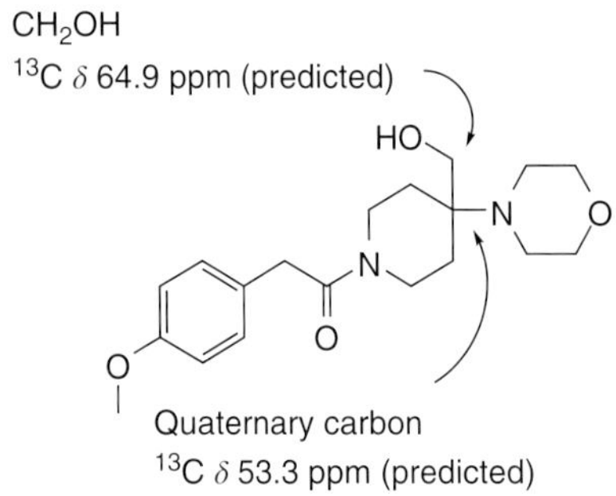

CH$_2$OH
^{13}C δ 64.9 ppm (predicted)

Quaternary carbon
^{13}C δ 53.3 ppm (predicted)

14

14.5 STEREOCHEMISTRY PROBLEMS. PROTON–PROTON THROUGH-SPACE CORRELATIONS

So far we have considered structural problems that can be solved by through-bond correlations (J-couplings). Many stereochemical problems cannot be solved by this approach, such as ring junction stereochemistry and stereochemistry of disubstituted alicyclic rings. (Relative stereochemistry in six-membered alicyclic rings can be made when the ring is in a chair conformation by taking advantage of the large difference in coupling constants between axial–axial hydrogens (about 12 Hz) and axial–equatorial and equatorial–equatorial hydrogens (about 4 Hz). Sometimes this information is not forthcoming from splitting patterns observed in the ^{1}H NMR spectra or by examination of the patterns of cross peaks in phase-sensitive ^{1}H–^{1}H COSY spectra.) In these instances, the required relative stereochemistry information can be obtained by through-space ^{1}H–^{1}H correlations as observed in ROESY spectra with genuine correlations being of opposite phase to the diagonal. Before interpreting ROESY spectra, careful assignments of ^{1}H NMR chemical shifts are required, typically using ^{1}H–^{1}H and ^{1}H–^{13}C through-bond correlations. Steroids represent a class of biologically active compounds that have been of interest to the pharmaceutical industry for decades with one common feature being the nature of the A/B ring junction. When the ring junction is saturated, the stereochemistry can be *cis* or *trans*. An example of determination of ring junction stereochemistry is shown below for a pair of tricyclic steroidal mimics possessing either a *cis* (**15**) or a *trans* (**16**) A/B ring junction. In both

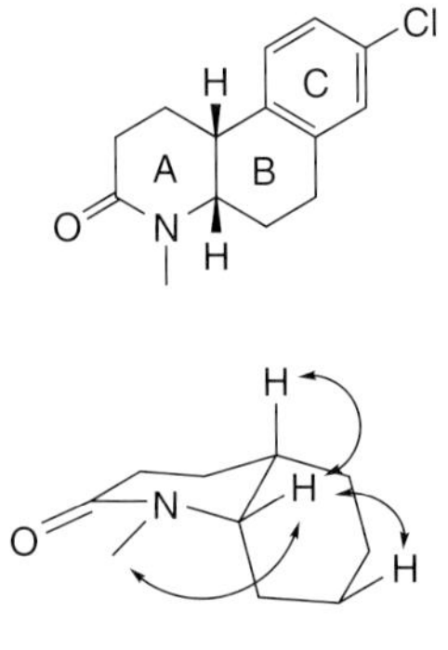

15 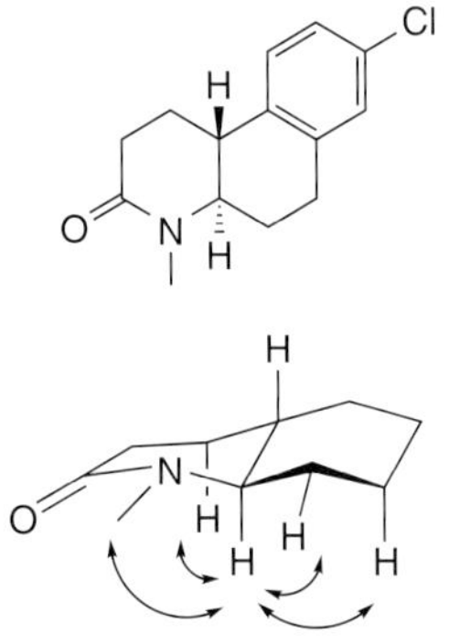   **16**

cases, a careful assignment of ^{1}H NMR chemical shifts was made as noted earlier and assignment of ring junction stereochemistry made from interpretation of through-space ^{1}H–^{1}H correlations observed in the ROESY spectra as illustrated for the AB rings.

14.6 USE OF THE ^{15}N ISOTOPE

Although I have made extensive use of ^{1}H–^{13}C through-bond correlations, it is worthwhile highlighting one example that makes use of ^{15}N rather than ^{13}C. Nitrogen is a common constituent of drug molecules with the ^{15}N isotope ($I = 1/2$) being an attractive option for structural studies using NMR spectroscopy. The structure of the tetra-substituted benzene intermediate (**17**) was determined using a combination of experiments with ^{1}H–^{15}N HMBC being crucial for locating the position of the nitro group as shown in blue and a ^{1}H–^{1}H through-space correlation highlighting the location of the methoxy group in red.

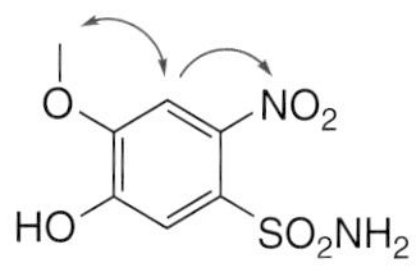

17

14.7 EXAMPLES WHERE ^{1}H NMR SPECTRA ARE BROAD OR CONTAIN MORE THAN ONE SPECIES IN SOLUTION

14.7.1 Tautomerism, Isomerism, and Atropisomerism

Some classes of compounds give ^{1}H NMR spectra that are complicated by either broad resonances or containing more than one species in solution. Many pharmaceutical compounds fall into this category, necessitating investigation by those experienced in structure determination. A large number of these examples (Figure 14.1) arise from either tautomerism or restricted rotation. The latter can arise as a result of the double-bond character of the N–CO bond as in t-amides or from steric hindrance potentially leading to atropisomers. Tautomerism is observed in nitrogen

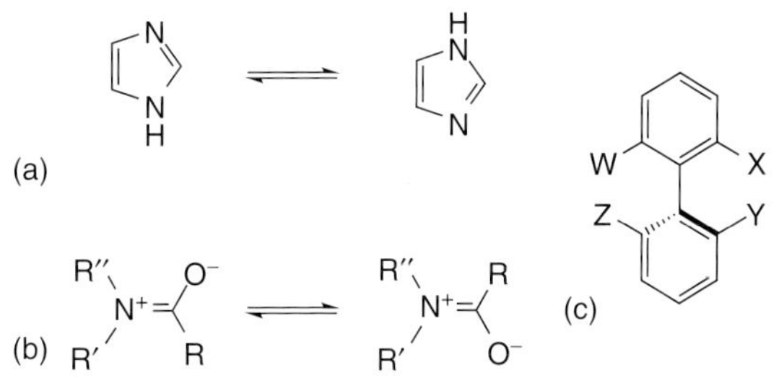

Figure 14.1. Examples of compound classes that exhibit either broad ^{1}H NMR spectra and/or ^{1}H NMR spectra containing more than a single species in solution. (a) Imidazole tautomerism. (b) Isomerism in t-amides. (c) Atropisomerism in biphenyls

heterocycles such as imidazoles and pyrazoles. In such situations, improved ^{1}H NMR spectra can be obtained by addition of a small amount of acid that increases the rate of interconversion of the tautomers on the NMR time scale. Isomerism is observed in t-amides as a result of restricted rotation owing to the partial double-bond character of the N–CO bond. Biphenyls represent the classic case of atropisomerism when at least three of the four *ortho* positions are substituted. The condition for the existence of atropisomerism is fulfilled when the barrier to rotation is sufficiently high enough to allow separation of the stereoisomers and where the half-life is >1000 s.

It is worthwhile mentioning that some pharmaceutical compounds afford unexpectedly broad and complex ^{1}H NMR spectra that are sometimes difficult to rationalize. In such cases, a simple change in solvent can sometimes produce a much clearer ^{1}H NMR spectrum and this should always be a first approach.

14.7.2 t-Amides

The number of t-amides in the collections of pharmaceutical companies has increased markedly due to the relative ease of amide coupling that is amenable to parallel chemistry easily generating thousands of analogs. The ^{1}H NMR spectra of these compounds are frequently complicated by the presence of two species in solution that can lead to confusion as to whether the sample contains a single component or two. Many chemists are taught at university that all one needs to do is simply acquire the NMR data at a higher temperature when a single species will be observed. Although there are many instances where this approach is successful, there are at least as many and possibly more examples, where a sufficiently high temperature is not reached to allow fast interconversion on the NMR scale. In many cases, the ^{1}H NMR spectrum is sufficiently well resolved at room temperature to assign the two contributing species. Frequently cooling the sample if possible rather than warming gives better results. 2D ROESY spectra show exchange peaks, with the same phase as the diagonal, between resonances that are interconverting on the NMR time scale. This can be a useful and simple means of diagnosing such behavior. Consider the following compound (**18**)

that was obtained as the major product from synthesis of a pharmaceutical intermediate. The compound was not actually the desired product but it was important to understand the chemistry of the reaction. The ^{1}H NMR shows the presence of two species in solution (about 3 : 2 ratio) and both species show the same array of ^{1}H–^{1}H and long-range ^{1}H–^{13}C correlations in the COSY and HMBC spectra, respectively. Heating the sample to 383 K afforded a broad ^{1}H NMR spectrum. The room temperature ROESY spectrum shows exchange peaks between the individual resonances of the two species. All of these data are consistent with the substituted formamide whose formation could then easily be rationalized from the chemistry.

The situation with compounds containing two t-amide linkages adds another degree of complexity and in situations like this, cooling the sample is invariably the preferred option instead of warming. An interesting example concerns inhibitors of the

18

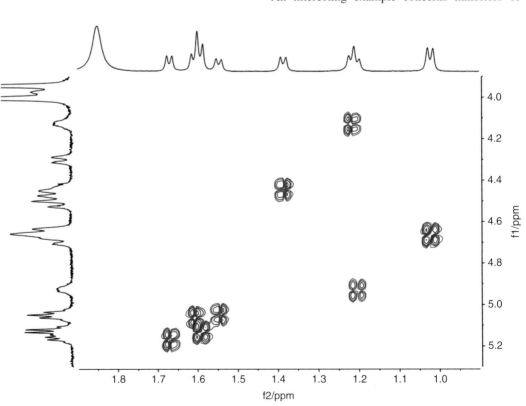

Figure 14.2. Partial 500 MHz ^{1}H–^{1}H COSY spectrum in tetrachloroethane-d$_2$ at 253 K of **19** showing cross peaks arising from coupling with the two methyl groups. Four species are observed in solution as a result of isomerism about the two t-amide linkages. The resonances for the methyl group of the piperazine are observed at 1.0–1.4 ppm

gp120-CD4 interaction of the type (**19**) containing two t-amide linkages and two chiral centers.[9]

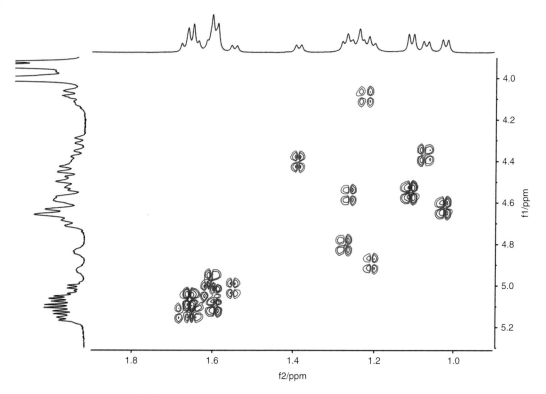

19

The ^{1}H NMR spectra at room temperature for **19** and its analogs are broad and even at 383 K, broad unresolved spectra were obtained. In contrast, ^{1}H NMR spectra acquired at 253 K were much sharper. This observation proved invaluable in discovering the presence of diastereoisomers in **19**, before development of a chiral HPLC method, that could be traced to a batch of chiral methylpiperazine of low enantiomeric excess (ee). Cross peaks are observed in the ^{1}H–^{1}H COSY spectrum acquired at 253 K for four piperazine CH–CH$_3$ groups as a result of isomerism about the

two t-amide linkages for samples of a single (or both) enantiomers (Figure 14.2), but eight in the presence of diastereoisomers (Figure 14.3).

14.7.3 Atropisomerism

The ^{1}H NMR spectrum of the triazolobenzodiazepine vasopressin V1a antagonist **20**[10] displayed broad resonances at room temperature that raised concerns of purity and/or the possibility of atropisomerism. Restricted rotation about the phenyl-triazole bond (cf. biphenyl in Figure 14.1 with substitution on three of the four *ortho* positions) generates a chiral plane rendering the two methylene groups of the seven-membered ring and the four methylene groups of the piperidine prochiral. Using tetrachloroethane-d$_2$ as solvent, a series of ^{1}H NMR spectra was acquired over a temperature range from 248 to 363 K (Figure 14.4).

Figure 14.3. Partial 500 MHz ^{1}H–^{1}H COSY spectrum in tetrachloroethane-d$_2$ at 253 K of **19** of low diastereomeric excess (de) as a result of a starting batch of methylpiperazine of low ee showing cross peaks arising from coupling with the two methyl groups. Eight species are observed in solution as a result of the presence of diastereoisomers and isomerism about the two t-amide linkages. The resonances for the methyl group of the piperazine are observed at 1.0–1.4 ppm

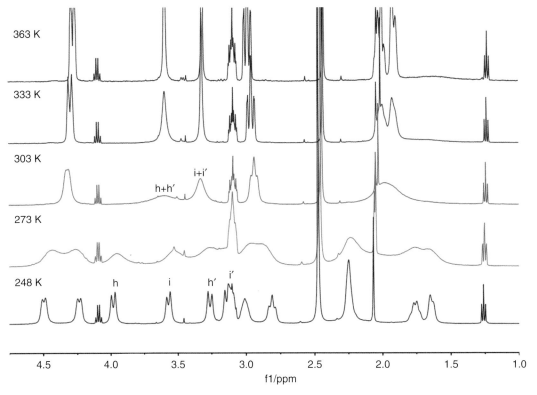

Figure 14.4. Partial variable temperature 500 MHz ^{1}H NMR spectra in tetrachloroethane-d$_2$ for the Vasopressin V1a antagonist (**20**). The methylene hydrogens of the seven-membered ring are labeled h,h′ and i,i′, respectively. At 248 K, hydrogen i′ overlaps with one of the piperidine hydrogens

At 248 K, the hydrogens of the two methylene groups in the seven-membered ring form a pair of AB systems that coalesce to two A$_2$ systems at higher temperatures (283 and 293 K). From the geminal coupling constants and frequency separations measured at 248 K and the observed temperatures of coalescence (T_{coal}), an approximate value of k_{coal} was calculated as 801 s^{-1}. Using this value for k_{coal} in the Eyring equation gives an estimate of the free energy of activation of 55 kJ mol^{-1}. This is lower than that

required to satisfy the condition for the existence of atropisomerism, so (**20**) could be considered as a single compound rather than a pair of stereoisomers. The ^{1}H NMR spectrum at 363 K was sharp allowing a straightforward assignment in agreement with the constitution. This example illustrates the potential usefulness of NMR spectroscopy in assessing the existence of atropisomers.

14.8   CONCLUSIONS

Incorrect structure assignment of drug leads and candidates can potentially affect drug development[1,2] and invalidate research efforts.[4] NMR spectroscopy is a key tool in structure determination and although there are no right or wrong ways in using the results from this technique, the examples highlighted herein demonstrate that a degree of methodology can be applied by considering the problem in question and using the appropriate experiment(s). With many pharmaceutical companies contracting out their synthetic chemistry, there is clearly the risk of a gap arising in the validation of structures. Historically, many companies were notably proactive in proving the structure of compounds passing into development with production of an analytical data package. Clearly, there is a need for a return to such diligence and an increased awareness of how molecular structure determinations have been made.

REFERENCES

1. S. Borman, *Chem. Eng. News*, 2014, **92**, 7.

2. S. Borman, *Chem. Eng. News*, 2014, **92**, 32.

3. N. T. Jabob, J. W. Lockner, V. V. Kravchenko, and K. D. Janda, *Angew. Chem. Int. Ed.*, 2014, **53**, 6628.

4. B. Halford, *Chem. Eng. News*, 2012, **90**, 34.

5. S. J. Lane, D. S. Eggleston, K. A. Brinded, J. C. Hollerton, N. L. Taylor, and S. A. Readshaw, *Drug Discov. Today*, 2006, **11**, 267.

6. G. F. Pauli, S.-N. Chen, C. Simmler, D. C. Lankin, T. Gödecke, B. U. Jaki, J. B. Friesen, J. B. McAlpine, and J. G. Napolitano, *J. Med. Chem.*, 2014, **57**, 9220.

7. R. D. Boyer, R. Johnson, and K. Krishnamurthy, *J. Magn. Reson.*, 2003, **165**, 253.

8. J. C. Hoch, M. W. Maciejewski, M. Mobli, A. D. Schuyler, and A. S. Stern, *Encyclopedia of Magnetic Resonance*, 2012, **1**, 181. DOI: 10.1002/9780470034590.emrstm1239.

9. D. H. Williams, F. Adam, D. R. Fenwick, J. Fok-Seang, I. Gardner, D. Hay, R. Jaiessh, D. S. Middleton, *et al.*, *Bioorg. Med. Chem. Lett.*, 2009, **19**, 5246.

10. D. M. Beal, J. S. Bryans, P. S. Johnson, J. Newman, C. Pasquinet, T. M. Peakman, T. Ryckmans, T. J. Underwood, and S. Wheeler, *Tetrahedron Lett.*, 2011, **52**, 5913.

Chapter 15

Mixture Analysis in Pharmaceutical R&D Using Hyphenated NMR Techniques

Ian D. Wilson and John C. Lindon

Computational and Systems Medicine, Department of Surgery and Cancer, Imperial College London, Sir Alexander Fleming Building, South Kensington, London SW7 2AZ, UK

15.1 INTRODUCTION

High-performance liquid chromatography (HPLC) and its higher resolution variants [in particular ultra (high)-performance liquid chromatography (UPLC)] currently represent the dominant techniques used for the separation and analysis of chemical mixtures. HPLC, and increasingly UPLC, when hyphenated to mass spectrometry (MS) in the guise of liquid

NMR in Pharmaceutical Sciences. Edited by Jeremy R. Everett, John C. Lindon, Ian D. Wilson, and Robin K. Harris
© 2015 John Wiley & Sons, Ltd. ISBN: 978-1-118-66025-6
Also published in eMagRes (online edition)
DOI: 10.1002/9780470034590.emrstm1391

chromatography (LC)–MS provides the first line of attack in the characterization and identification of key pharmaceutical components (active natural products, drug impurities, xenobiotic metabolites, etc.). However, unequivocal identification often requires more structural information than can readily be obtained from MS alone in order to be able to confidently assign the molecular structures of some unknowns (e.g., positional isomers). In such circumstances, structure determination may require both NMR spectroscopic and MS data. One way of obtaining the required NMR data is the isolation and purification of the analyte(s), and this represents a well-established and conventional approach. However, an arguably more efficient, and certainly more elegant, approach involves the hyphenation of a suitable LC separation with an NMR spectrometer for direct on-line structure determination of unknowns (or confirmation of known or expected compounds) present in mixtures of varying degrees of complexity.[1] In particular, by eliminating the need for tedious and often time-consuming isolation steps, the use of on-line LC–NMR spectroscopy can provide significant time savings. In addition, by also including on-line MS (and indeed other spectroscopies) in the LC–NMR system, a so-called hypernated (multiply hyphenated) approach can be assembled with which full structural characterization might be possible in a single chromatographic run.[1] Such a hypernated system has the additional advantage that spectroscopic data on the target analyte(s) can be obtained simultaneously, thereby improving the integrity of the data by removing potentially confounding factors such as those that might result from, e.g., sample instability or small differences in chromatographic retention time owing to the performance of the analysis on separate LC–NMR and LC–MS systems. Much of this potential was amply demonstrated following the introduction of commercial HPLC–NMR probes in the last decade of the twentieth century. These probes, generally, with flow cell volumes in the range 30–120 μl, provided a number of impressive demonstrations of the potential of LC–NMR and LC–NMR–MS in pharmaceutical applications ranging from studies on formulated drugs to drug metabolism studies,[1,2] as well as studies on natural products.[3]

In addition to LC–NMR, other innovations have included the linking of NMR spectrometers with separations such as supercritical fluid chromatography (SFC), where supercritical fluid CO_2 is used as a solvent,[4] capillary electrophoresis (CE),[5,6] and capillary LC.[6] In addition, developments in NMR hardware such as cryo-flow probes, loop collection of analyte peaks, and on-line peak trapping via solid-phase extraction (SPE) have also been adopted.[7]

15.2 PRACTICAL AND TECHNICAL CONSIDERATIONS IN LC–NMR

15.2.1 Solvent Selection for Chromatography

In practice, there are some constraints on the choice of solvents that can be used to make up the mobile phases employed for the chromatographic separations used for LC–NMR. These choices are governed by the type of separation employed and the contribution that these solvents make to the resulting NMR spectra. Thus, complex solvent systems, containing multiple organic solvents and organic additives (such as the organic acids and ion-pair reagents used routinely to control solvent pH and retain polar ionic analytes), can often deny the NMR spectroscopist access to many regions of the resulting spectra. In order to retain the largest possible spectroscopic windows, it may therefore be necessary to modify the separations that have been used for sample analysis to simplify them for LC–NMR. For the bulk of pharmaceutical applications, reversed-phase separations have been used, generally based on deuterium oxide (D_2O) and either

methanol or acetonitrile (as either conventional protonated solvents or their deuterated equivalents). Clearly even these solvents contribute potentially interfering signals to the resulting spectra and solvent suppression, using an appropriate NMR pulse sequence that provides double presaturation to suppress the signals for both the organic solvent and residual water protons must be used. While the use of these solvents may still mean that parts of the spectrum for the analyte are lost to the spectroscopist nevertheless both acetonitrile and methanol represent good choices for ^{1}H LC–NMR. In particular, acetonitrile and methanol represent the most popular choices for the organic modifiers used for reversed-phase chromatography, and their singlet methyl proton resonances can be suppressed relatively easily. If there is a need for their use in their fully deuterated forms to reduce the peak intensity further and make solvent suppression easier, then both deuterated acetonitrile and methanol are relatively inexpensive, justifying their use when high sensitivity is needed.

Where the decision is made to use protonated organic solvents, these should be carefully checked before use to ensure that they do not contain unwanted ^{1}H NMR-detectable impurities that will also contribute to the background. In addition to the main resonances from these solvents, the natural abundance ^{13}C satellite peaks due to the one-bond [1]H–[13]C spin couplings for molecules containing a ^{13}C isotope present in the methyl carbons remain even when the main solvent peak has been suppressed, raising the possibility of further interference with signals from trace analytes. Such considerations provide a further reason for opting for solvent systems produced entirely from fully deuterated solvents. Alternatively, the suppression irradiation frequency can be set over the main solvent peak and the ^{13}C satellite peaks in a cyclical manner or, if an inverse geometry probe equipped with a ^{13}C coil is used, broadband ^{13}C decoupling to collapse the satellite peaks can be employed.

Ideally in LC–NMR applications, isocratic elution, where the solvent composition is kept constant, should be used. However, for complex mixtures, where the components cover a wide range of polarities, it is often the case that gradient elution, where the proportion of the organic solvent in the mobile phase is increased over time, is used (solvent gradients also have the benefit of sharpening up analyte peaks compared to isocratic separations thereby increasing the substance concentration in eluting peaks). However, the use of gradient elution provides a further practical problem for LC–NMR as the position of the ^{1}H NMR frequency of the solvents will be subject to change during chromatography as the solvent composition alters. A solution to this problem is provided by software that enables automatic solvent peak tracking and suppression to be undertaken.

In the case of on-line SPE, the use of D_2O and deuterated organic solvents is less critical as the SPE column can be flushed with a deuterated solvent to remove protonated chromatographic eluents before the elution of the analyte with an appropriate deuterated solvent for spectroscopy. Another method that can be used to reduce (or eliminate) the organic solvent in the LC mobile phase is to perform the separation at elevated temperatures (high temperature or HT-LC).[8]

As indicated earlier, the ionic nature of some analytes may require the pH of the solvent to be adjusted to ensure chromatographic retention and for LC–NMR, this can usually be achieved using nonprotonated inorganic buffers, e.g., phosphate. Alternatively, an organic acid such as trifluoroacetic acid (TFA) can be used, as it will not contribute interfering signals to any ^{1}H NMR spectrum.

15.2.2   Modes of HPLC–NMR

15.2.2.1   On-line LC–NMR Approaches: On-flow and Stopped-flow Techniques

LC–NMR is generally used with detection by ^{1}H or ^{19}F NMR, with ^{1}H spectra providing the most informative data. In practice, the separation can be performed using either isocratic or gradient elution with spectra acquired using continuous on-flow methods, stopped-flow spectroscopy, or analyte collection based on peak detection by other means, e.g., ultraviolet (UV) or MS detection, either into capillary loops for later NMR measurement or onto SPE phase sorbents for concentration and postchromatographic elution into the flow probe.[7]

Clearly, the easiest experimental setup to implement in LC–NMR is the use of isocratic chromatography combined with continuous flow detection. For this strategy to be successful, the compounds of interest in the mixture need to be present in the eluent entering the flow probe at concentrations that are high enough to result in acceptable NMR spectra 'on the fly' from a limited number of FIDs or scans during the short period of time when the peak is passing through the flow probe (although the residence time can obviously

be extended somewhat using low flow rates). Isocratic methods are best suited to mixtures where spectra are required for a number of components of similar physicochemical properties and thus the maximum resolving power of the chromatographic system is to be utilized. If the analytes are present in sufficient concentration, then spectra can be obtained on the fly as indicated earlier.

In practice, using a 600 MHz NMR spectrometer in combination with a 'conventional' 4.6 mm i.d. HPLC column and a flow rate of 1 ml min^{-1}, as well as chromatography providing peak widths of about 10–30 s, a detection limit of 5–10 µg of a typical 400 molecular mass compound can be obtained (using an acquisition time of below 1 s, giving between 8 and 24 transients/spectrum) in continuous flow mode.[1,2] In the case of complex mixtures where there are large physicochemical differences between the compounds of interest, simple isocratic methods are inappropriate and solvent gradients, where the eluotropic strength of the chromatographic mobile phase is increased during the separation to elute more strongly retained analytes, become the method of choice. An obvious problem for continuous flow NMR detection during gradient elution is that shifts in the positions of the resonances of the solvent peaks (residual HOD and the methyl signals for methanol or acetonitrile, etc.) occur as the solvent composition changes through the gradient. To ensure effective solvent suppression, these changes in the position of the signals for the solvents must be compensated for and a variety of strategies have been developed to deal with this.

In both isocratic and gradient elutions where it is not possible to obtain spectra in continuous flow mode because, e.g., not enough material is present in the peak or 2D NMR spectroscopic techniques, including COSY or TOCSY, are required, then stopped-flow NMR represents the only practical option. As the name implies, in stopped-flow LC–NMR, the HPLC pump is stopped at the point that the chromatographic peak of interest (detected by either, e.g., on-flow UV or MS) enters the NMR flow probe. NMR spectroscopy is then performed on the stationary peak until sufficient FIDs have been obtained to provide an adequate signal-to-noise ratio to acquire the required spectroscopic information.

Indeed, there are numerous examples of the successful application of this approach to structure determination. However, the use of stopped flow has been presented as a disadvantage when spectroscopy is to be performed on several components during the same chromatographic run, because this might result in peak broadening owing to on-column diffusion of the analytes.[2] While there may be some merit in this view, in practice, the effects on the resulting spectra are generally minor (especially in gradient elution) and negligible compared to the band broadening caused by the relatively (compared to normal HPLC practice) large dead volumes provided by the tubing used to connect the LC to the flow probe and by the volume of the flow probe itself (typically ranging from 30 to 240 μl for conventional LC–NMR). Indeed, we have found in practice that it is eminently feasible to acquire the NMR spectra for a number of peaks in the same chromatographic separation using sequential stopped-flow experiments with excellent results, and no evidence of unacceptable band broadening. The development of cryogenic probes has resulted in a significant increase in NMR sensitivity, as a result of the reduction in the electronic noise present in the system, and the production of cryogenic flow probes has provided similarly improved sensitivity for LC–NMR.[9] Where chromatographic resolution is less than complete, such that two or more components elute only partially separated, a technique called *timeslicing* can be applied whereby having taken spectra from the leading edge of the peak in stopped-flow mode, flow is resumed for a few seconds and fresh spectra are obtained. If there is partial separation between the peak components, this will be readily apparent in differences in the resulting ^{1}H NMR spectra, and structure determination may be possible as a result.[10]

15.2.2.2 At-line LC–NMR Approaches: Loop Collection and In-line SPE

There are a range of techniques where the chromatographic separation can be considered not as an integral part of the analysis but as a means of sample collection/preparation for conventional 'static' NMR spectroscopy. In these systems, the peaks of interest for spectroscopy are captured either in simple loops or on in-line SPE cartridges as described in the following section. Although these methodologies are described as LC–NMR, neither the loop storage nor the SPE devices need actually be connected to the NMR flow probe during chromatography, and indeed they need not even be in the same laboratory. To the purist, they are therefore best described as 'at-line' or 'in-line' techniques.

The simplest form of automated at-line collection system is using loop collection. In this mode, all that is done is that peaks of interest eluting from the column, and having been observed via, e.g., UV, MS, or radioactivity monitors, are diverted into capillary loops where they are retained until such time as they are moved into the flow probe of the NMR spectrometer. The number of peaks that can be stored in this way is limited only by the number of storage loops available on the interface, and the subsequent order of analysis is at the discretion of the investigator. Having been moved from the loop into the flow probe, they can then be subjected to all of the usual 1D and 2D NMR experiments used for structure determination.

An alternative to loop collection is to trap eluting peaks onto a suitable chromatographic sorbent in a form of on-line SPE.[7,9,11] This methodology has some advantages over loop collection in that it allows for the preconcentration of analytes on the trapping column (increases in sensitivity when using LC–SPE–NMR of about four- to fivefold have been reported[11]). Indeed, where the concentration of the compound(s) of interest is insufficient, several chromatographic runs can be performed and the same peak repeatedly trapped on the SPE column in order to obtain sufficient material. In order to ensure that the peak of interest is retained, it is usual to mix the eluent containing the peak of interest with a less eluotropic mobile phase before entry onto the SPE column. Once the operator has collected enough of the required analyte(s) they can be recovered from the cartridges using a highly eluotropic, deuterated, solvent into the flow probe for spectroscopy. The elution step offers a number of advantages compared to loop storage as it enables chromatography to be undertaken using inexpensive protonated solvents, including those with more than one organic modifier, and also means that all the collected peaks can be subjected to NMR spectroscopy in the same deuterated solvent.

As with the loop collection system described earlier, the at-line nature of peak collection means that there is no need for the SPE device to be physically connected to the NMR spectrometer while peaks are being collected. There are now, following the commercialization of this type of system, a large number of examples of the use of this SPE-trapping methodology in the literature.[9,12–14] There has also been application of the SPE collection methodology combined with the use of cryo-flow probe technology for samples with limited availability.[9,13] Thus, the combination of LC–SPE–cryo-NMR has been shown for the in vitro metabolism of an experimental MAP-kinase inhibitor (using hepatic microsomes), where only 40 μg of the drug was available for biotransformation in the system. This methodology enabled the authors to identify several of the metabolites considered to be present in quantities of 1–10 μg.[13]

In an interesting study, LC–NMR with loop storage was compared with LC–SPE–NMR (using a cryo-flow probe) for the identification of isomeric tropane alkaloids extracted from the bark of *Schizanthus grahamii*.[15] Both approaches enabled the alkaloids to be characterized, while demonstrating that increased sensitivity was only seen for the SPE-trapping approach for those compounds that could be well retained on the phase used for SPE trapping. In this example, about 50% of the target analytes were not well retained and were therefore lost.

15.2.3 Microprobes and Capillary LC Methods

A desire to increase sensitivity in LC–NMR has resulted in a number of developments in miniaturization resulting in the appearance of microcoil NMR probes, having active probe volumes as small as 1.5–5 μl, linked to separations performed using CE and capillary liquid chromatography (cap-LC or cLC).[5,6,16,17] A major advantage of these miniaturized techniques is that such small quantities of solvent are required that running costs with respect to mobile phases become trivial and thus this allows fully deuterated solvents to be used as the default option. In addition to cap-LC, the coupling of NMR spectrometers equipped with microcoil flow probes to both CE (also known as capillary zone electrophoresis or CZE) and capillary electrochromatography (CEC) has also been successfully demonstrated. These 'electrically driven' separations also have the advantage that the need for solvent to be delivered via a pump is eliminated. Experimental data have demonstrated that all three types of miniaturized separation (cap-LC–, CE–, and CEC–NMR) are very sensitive (in NMR terms) with limits of detection in the nanogram to low microgram range when using stopped-flow techniques.[16,17] The major limitation of capillary methods is that only small sample volumes can be applied (a few nanoliters or less) with the result that either the initial sample concentration has to be high or methods of on-column/capillary preconcentration are required. While by no means numerous, examples of the use of cap-LC–NMR, CE–NMR, and CEC–NMR are available in the literature and include

the analysis of natural products and metabolites of acetaminophen (paracetamol).[18]

15.3 HIGH-PERFORMANCE LIQUID CHROMATOGRAPHY–NUCLEAR MAGNETIC RESONANCE–MASS SPECTROMETRY (HPLC–NMR–MS)

While the hyphenation of LC with NMR can be very powerful in its own right as a means of structure determination, there is no doubt that by also coupling other spectrometers in-line, further advantages can be obtained in terms of information recovery from chromatographic separations (albeit at the expense of higher capital investment and system complexity). The simplest of these combinations with NMR is arguably the in-line addition of a UV/Vis diode array detector (DAD) that, as well as providing UV spectra that can be of some assistance in structure determination, provides a very convenient means of monitoring the separation. On the whole, the solvent systems employed for LC–NMR are well suited to the requirements of UV/Vis spectroscopy, and as such UV detectors are routinely included as part of LC–NMR systems. That said, while the UV/Vis DAD is simple to use, and relatively inexpensive, the resulting UV spectra do only contain a limited amount of structural information. As a result, there is value in also trying to include more information-rich spectrometric techniques, in particular MS.[1,2,19] The addition of mass spectral data on eluting peaks in an LC–NMR–MS setup confers considerable benefits for structure determination via, e.g., both fragmentation patterns and, with accurate mass instruments, data on atomic composition, providing the potential to derive molecular formulae, and showing the presence of 'NMR invisible' atoms (e.g., halogens, sulfur, and oxygen).[1,2] The multiple hyphenation (or in this case 'hypernation') of LC–NMR and MS (and often also DAD) is relatively straightforward, and indeed there are numerous examples in the literature demonstrating this. The largest practical difficulty is to select chromatographic solvent systems that provide both the required separations and that are also compatible with both NMR and mass spectrometers. In the case of MS, the requirement is that any buffers employed for controlling the solvent pH should be volatile; otherwise, the ionization source of the MS will rapidly become contaminated. In the case of NMR, as alluded to earlier, buffer components should ideally not contribute interfering signals in the resulting spectra that would blind the spectroscopist to resonances present in the compound of interest. For example, the use of inorganic buffers (e.g., sodium phosphate/phosphoric acid mixtures), while ideal for NMR as they contribute no signals at all to the ^{1}H NMR spectrum, would rapidly degrade the performance of the mass spectrometer. A useful compromise that can be used to provide acidic mobile phases is formic acid where the singlet in the NMR spectrum (observed at about δ 8.5) causes minimal interference. In general, most LC–NMR–MS systems have the two spectrometers linked in parallel to the LC separation. Because of differences in spectrometer sensitivity, it is usual to direct 95% of the LC eluent to the NMR flow cell with the remainder sent to the ion source of the MS. In this arrangement, the flow of sample to spectrometer can be arranged so that both streams of eluent arrive simultaneously in the spectrometers. Alternatively, the system can be configured such that peaks arrive at the mass spectrometer first, allowing the operator to select samples for NMR spectroscopy based on mass spectral data. There are also examples where an in-line configuration is employed whereby the MS is placed after the NMR flow probe. This type of layout simplifies the 'plumbing' aspects of the system but clearly precludes the use of MS data to select peaks for NMR spectroscopy.[1,2] Even more complex hypernated systems have been described that have linked UV, IR, NMR, and MSs in a single instrumental setup. Such LC–UV–IR–NMR–MS demonstration systems have been assembled and applied to mixture analysis on analytes such as nonsteroidal anti-inflammatory drugs (NSAIDs), polymer additives, mixtures of pharmaceuticals, and extracts of plants containing bioactive steroids (ecdysteroids) with the chromatographic separation performed using either conventional reversed-phase eluents or superheated D_2O.[1,2] While the hypernation of LC–NMR and MS provides a clear advantage over LC–NMR and LC–MS conducted separately such that the structural information is acquired on the same sample and the same separation simultaneously, this is achieved at the cost of somewhat increased system complexity.

15.4 APPLICATIONS OF LC–NMR

Following the widespread availability of reliable commercial systems for LC–NMR, there have been numerous pharmaceutical applications of the technique. These include the identification of impurities and degradation/reaction products,[10,12] as well as analysis of samples from in vivo and in vitro studies of drug metabolism.[9,13,14]. In the case of drug metabolism, the use of LC–NMR for the determination of the identity of the metabolites of the drug ibuprofen contained in off-line solid-phase extracts of human urine represents one of the first applications of the technique.[20] SPE was used here as a means of preconcentrating, and partially purifying the metabolites from the urine before LC. Later studies, using LC–NMR–MS on the urinary metabolites of ibuprofen resulted in an even more comprehensive description of the human metabolism of the drug.[21] While ^{1}H NMR spectroscopy is probably the most useful general approach, the presence of fluorine (or fluorines) enables a combination of ^{19}F and ^{1}H NMR spectroscopy to be used as illustrated in the study of the metabolites of the drug flurbiprofen present in human urine following oral dosing.[22] Similarly, limited studies have also been reported for other NMR active nuclei such as ^{31}P and ^{2}H NMR where LC–NMR spectroscopy has been used to examine the metabolism of phosphorus- and deuterium-containing compounds.[23] The first application of a cryo-flow NMR probe to study in vivo drug metabolism used reversed-phase gradient chromatography combined with a 500 MHz NMR instrument for the metabolic profiling of samples of human urine collected after oral administration of 500 mg of acetaminophen (paracetamol). This study used both on-flow and stopped-flow NMR spectroscopy to identify the drug itself and five metabolites, together with several endogenous compounds.[24] In the case of in vitro metabolism, successful applications of LC–NMR have been demonstrated for tissue slices, cell suspensions,[23] and subcellular fractions such as microsomes.[13] LC–NMR has also been used extensively in the past for the structural determination of the positional isomers resulting from the transacylation of ester glucuronides. Ester glucuronides (in the form of β-1-O-acylglucuronides) are common metabolites of drugs bearing a carboxylic acid as a substituent and are considered to be potentially toxic as a result of these transacylation reactions, which occur under weakly alkaline conditions, because covalent adduct formation with proteins can result. The result of such transacylations is a complex mixture formed of the 2-, 3-, and 4-O-acyl glucuronides and their corresponding α- and β-anomers, plus free glucuronic acid as a result of spontaneous hydrolysis of the glucuronide isomers. Using LC–NMR, it has been possible to unambiguously assign the structures obtained from these acyl migration reactions. Given that the system comprises a mixture of equilibrating species, all with the same molecular mass, prior purification of individual substances or mass spectrometry would not have proved successful, and the on-line LC–NMR approach was the only way to characterize these substances. An example is shown in Figure 15.1, which depicts the NMR chromatogram obtained by continuous-flow 750 MHz HPLC–NMR of an O-(4-fluorobenzoyl)-glucopyranuronic acid mixture. In this case, the equilibrium mixture was subjected to on-flow HPLC–NMR spectroscopy, and the NMR spectra of the various isomers can be observed as they elute. The vertical axis corresponds to retention time and the horizontal axis is the normal ^{1}H chemical shift (the chemical shift region δ 3.3–6.2 ppm is displayed corresponding to the protons on the glucuronide ring). The band observed at δ 4.83 arises from residual HOD in the solvent and has been suppressed by saturation. The nomenclature for the assignments in the NMR chromatogram is as follows: 4,1′,α means the 1′-proton on the glucuronide ring of the α-4-O-acyl isomer, similarly 3,3′,β means the 3′-proton on the glucuronide ring of the β-3-O-acyl isomer, and so on.[25] Subsequently, each individual glucuronide isomer can be held in the flow probe in stopped-flow mode and its reactivity can be monitored over time using ^{1}H NMR spectroscopy as it re-requilibrates to the other isomers, thus simplifying the determination of the complex kinetic parameters.[26]

As discussed earlier, the double hyphenation of both NMR and MS to HPLC has resulted in a number of publications in the drug metabolism and natural product fields. As an example of the results that can be obtained using such a combined system, Figure 15.2 shows one of the first applications, studying the human metabolism of the drug acetaminophen (paracetamol). This shows on the left in (a) to (e) the 500 MHz ^{1}H NMR spectra obtained by extracting individual rows from the continuous-flow chromatogram of a urine extract, indicating the NMR spectra of acetaminophen glucuronide, acetaminophen sulfate, hippurate, and phenylacetylglutamine. On the right in (f) and (g) are shown the positive-ion mass spectra of acetaminophen glucuronide and the MS–MS spectrum of the m/z 334 ion used to confirm the identity of the metabolite.[27]

However, as described earlier, many of the more recent applications in LC–NMR have not used direct on-flow or stopped-flow techniques but have instead employed SPE systems designed for the in-line collection of analytes (often driven by MS or UV detection) with subsequent elution into the NMR flow probe for spectroscopy. However, while drug metabolism studies have often been investigated using LC–SPE–NMR (often in combination with MS), the use of cryo-flow probes for natural product characterization has also benefitted from this technology. A typical example of this type of application of HPLC–DAD–SPE–NMR for the identification of the constituents of plant extracts is provided by studies on the constituents of *Kanahia laniflora*.[28] Here reversed-phase gradient HPLC, with in-line SPE trapping of each peak of interest, was undertaken on partially purified extracts. The trapped peaks were eluted with CD_3CN into a 30 µl inverse ^{1}H–^{13}C-flow probe and 1D and 2D NMR spectroscopies were performed [COSY, heteronuclear single quantum coherence (HSQC), and heteronuclear multiple bond coherence (HMBC)], which allowed the identification of seven constituents (four flavonol glycosides and three 5α-cardenolides).

15.5 CONCLUSIONS

HPLC–NMR, in various forms and formats, including the on-line, at-line, and off-line techniques that have been developed, has enabled the rapid and efficient characterization and identification of analytes present in complex mixtures. Indeed, there is now a considerable body of literature available illustrating applications of the technology in pharmaceutical, biomedical, and natural product research. The combination of LC–NMR with other spectroscopies in multiply hyphenated systems (hypernation) such as, e.g., HPLC–NMR–MS, while less common, increases the power of the technique for compound identification. It is to be hoped that improved chromatographic separations, such as those involving ultra-high-performance liquid chromatography (UHPLC), combined with appropriate flow probes with enhanced sensitivity (for use, e.g., with capillary separations) either alone or in combination will ensure continued development in this area.

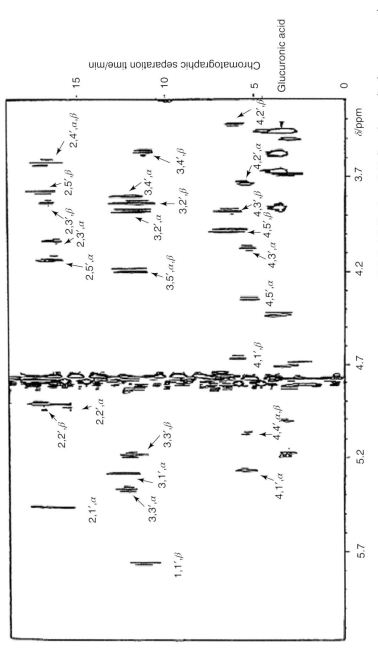

Figure 15.1.   NMR chromatogram obtained by continuous-flow 750 MHz HPLC–NMR of an *O*-(4-fluorobenzoyl)-glucopyranuronic acid mixture. The vertical axis is retention time. The displayed chemical shift region δ 3.3–δ 6.2 corresponds to the protons on the glucuronide ring. The band observed at δ 4.83 arises from residual HDO in the solvent and has been suppressed by saturation. The nomenclature for the assignments in the NMR chromatogram is as follows: 4,1′,α means the 1′-proton on the glucuronide ring of the α-4-*O*-acyl isomer; similarly 3,3′,β means the 3′-proton on the glucuronide ring of the β-3-*O*-acyl isomer, etc. (Reprinted with permission from U.G. Sidelmann, C. Gavaghan, H.A.J. Carless, M. Spraul, M. Hofmann, J.C. Lindon, I.D. Wilson, and J.K. Nicholson. 750-MHz directly coupled HPLC–NMR: Application for the sequential characterization of the positional isomers and anomers of 2-, 3-, and 4-fluorobenzoic acid glucuronides in equilibrium mixtures. *Anal. Chem.*, 1995, 67, 4441. Copyright 1995 American Chemical Society)

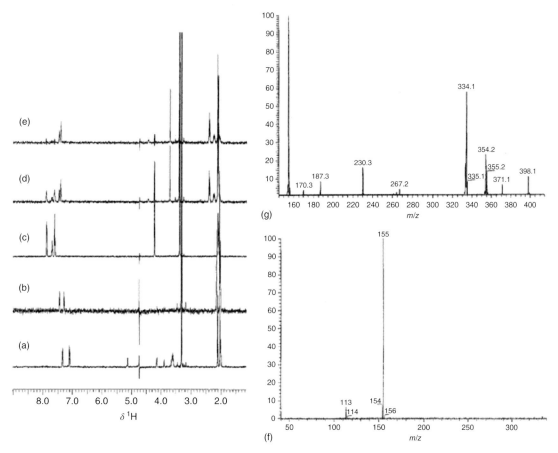

Figure 15.2. 500 MHz ^{1}H NMR spectra extracted from a continuous-flow chromatogram of human urine after administration of acetaminophen (paracetamol). (a) 7.6 min, acetaminophen glucuronide, (b) 8.6 min, acetaminophen sulfate, (c) 13.2 min, hippurate; (d) 13.6 min, phenylacetylglutamine and hippurate; and (e) 13.6 min with subtraction of hippurate resonances. (f) Positive-ion mass spectrum of acetaminophen glucuronide obtained during continuous-flow HPLC–NMR–MS detection and (g) the MS–MS spectrum of the m/z 334 ion. Adducts $(M + ND_4)^+$ and $(M + ND_4 + ACN\text{-}d_3)^+$ are seen at m/z 354 and 398, respectively. There is evidence from the multiplet nature of some of the peaks of incomplete deuteration of exchangeable hydrogens. (Adapted with permission from J.P. Shockcor, S.E. Unger, I.D. Wilson, P.J.D. Foxall, J.K. Nicholson, and J.C. Lindon. Combined HPLC, NMR spectroscopy, and ion-trap mass spectrometry with application to the detection and characterization of xenobiotic and endogenous metabolites in human urine. *Anal. Chem.* 1996, 68, 4435. Copyright 1996 American Chemical Society)

RELATED ARTICLES IN EMAGRES

Flow NMR

Electrophoretic NMR

Natural Products

Supercritical Fluids

NMR Probes for Small Sample Volumes

Flow Probes for NMR Spectroscopy

HPLC-NMR Spectroscopy

REFERENCES

1. I. D. Wilson, in NMR Spectroscopy in Pharmaceutical Analysis, eds U. Holzgrabe, I. Wawer, and B. Diehl, Elsevier: Amsterdam, The Netherlands, 2008.

2. J. C. Lindon, J. K. Nicholson, and I. D. Wilson, *J. Chromatogr. B*, 2000, **748**, 233.

3. J. L. Wolfender, E. F. Queiroz, and K. Hostettmann, in Bioactive Natural Products, ed S. Colgate, CRC Press: Boca Raton, FL, 2008.

4. K. Albert, *J. Chromatogr. A*, 1997, **785**, 65.

5. A. M. Wolters, D. A. Jayawickrama, A. G. Webb, and J. V. Sweedler, *Anal. Chem.*, 2002, **74**, 5550.

6. M. Kuhnle, K. Holtin, and K. Albert, *J. Sep. Sci.*, 2009, **32**, 719.

7. L. Griffiths and R. Horton, *Magn. Reson. Chem.*, 1998, **36**, 104.

8. T. Teutenberg, High Temperature Liquid Chromatography: A Users Guide for Method Development, Royal Society of Chemistry: Cambridge, 2010.

9. M. Godejohann, L.-H. Tseng, U. Braumann, J. Fuchser, and M. Spraul, *J. Chromatogr. A*, 2004, **1058**, 191.

10. N. Mistry, I. M. Ismail, M. S. Smith, J. K. Nicholson, and J. C. Lindon, *J. Pharm. Biomed. Anal.*, 1997, **16**, 697.

11. M. Sandvoss, B. Bardsley, T. L. Beck, E. Lee-Smith, S. E. North, P. J. Moore, A. J. Edwards, and R. J. Smith, *Magn. Reson. Chem.*, 2005, **48**, 762.

12. J. Larsen, D. Staerk, C. Cornett, S. H. Hansen, and J. W. Jaroszewski, *J. Pharm. Biomed. Anal.*, 2009, **49**, 839.

13. B. Kammerer, H. Scheible, G. Zurek, M. Godejohann, K.-P. Zeller, C. H. Gleiter, W. Albrecht, and S. Laufer, *Xenobiotica*, 2007, **37**, 280.

14. A. W. Nicholls, I. D. Wilson, M. Godejohann, J. K. Nicholson, and J. P. Shockcor, *Xenobiotica*, 2006, **36**, 615.

15. S. Bieri, E. Varesio, J.-L. Veuthey, O. Munoz, L.-H. Tseng, U. Braumann, M. Spraul, and P. Christen, *Phytochem. Anal.*, 2006, **17**, 78.

16. J. Schewitz, K. Pusecker, P. Gfrorer, U. Gotz, L.-H. Tseng, K. Albert, and E. Bayer, *Chromatographia*, 1999, **50**, 333.

17. A. D. Jayawickrama and J. V. Sweedler, *J. Chromatogr. A*, 2003, **1000**, 819.

18. J. Schewitz, P. Gfrorer, K. Pusecker, L.-H. Tseng, K. Albert, E. Bayer, I. D. Wilson, N. Bailey, G. B. Scarfe, J. K. Nicholson, and J. C. Lindon, *Analyst*, 1998, **123**, 2835.

19. G. Schlotterbeck and S. M. Ceccarelli, *Bioanalysis*, 2009, **1**, 549.

20. M. Spraul, M. Hofmann, P. Dvortsack, J. K. Nicholson, and I. D. Wilson, *Anal. Chem.*, 1993, **65**, 327.

21. E. Clayton, S. Taylor, B. Wright, and I. D. Wilson, *Chromatographia*, 1997, **47**, 264.

22. M. Spraul, M. Hofmann, I. D. Wilson, E. Lenz, J. K. Nicholson, and J. C. Lindon, *J. Pharm. Biomed. Anal.*, 1993, **11**, 1009.

23. I. D. Wilson, J. C. Lindon, and J. K. Nicholson, in Handbook of Drug Metabolism, 2nd edn, eds P. G. Pearson and L. C. Wienkers, Marcel Dekker: New York, Basel, 2009, 373.

24. M. Spraul, A. S. Freund, R. E. Nast, R. S. Withers, W. E. Mast, and O. Corcoran, *Anal. Chem.*, 2003, **75**, 1536.

25. U. G. Sidelmann, C. Gavaghan, H. A. J. Carless, M. Spraul, M. Hofmann, J. C. Lindon, I. D. Wilson, and J. K. Nicholson, *Anal. Chem.*, 1995, **67**, 4441.

26. U. G. Sidelmann, S. H. Hansen, C. Cavaghan, H. A. J. Carless, J. C. Lindon, R. D. Farrant, I. D. Wilson, and J. K. Nicholson, *Anal. Chem.*, 1996, **68**, 2564.

27. J. P. Shockcor, S. E. Unger, I. D. Wilson, P. J. D. Foxall, J. K. Nicholson, and J. C. Lindon, *Anal. Chem.*, 1996, **68**, 4435.

28. C. Clarckson, D. Staerk, S. H. Hansen, and J. W. Jaroszewski, *Anal. Chem.*, 2005, **77**, 3547.

Chapter 16

Conformation and Stereochemical Analysis of Drug Molecules

Gary J. Sharman

Eli Lilly & Company, Erl Wood Manor, Sunninghill Road, Windlesham, Surrey GU20 6PH, UK

16.1 INTRODUCTION

It is perhaps not an exaggeration to say that there is a tendency among some chemists to regard stereochemistry and conformation as the poor relations of molecular structure, somehow less important than the regiochemistry and connectivity of molecules. Of course, there is really no basis for this view, other than perhaps the difficulty of controlling stereochemistry, separatings stereoisomers or the fact that they are more familiar with techniques for determining

---

Written by Gary Sharman in his capacity as an employee of Eli Lilly and Co.

NMR in Pharmaceutical Sciences. Edited by Jeremy R. Everett, John C. Lindon, Ian D. Wilson, and Robin K. Harris
© 2015 John Wiley & Sons, Ltd. ISBN: 978-1-118-66025-6
Also published in eMagRes (online edition)
DOI: 10.1002/9780470034590.emrstm1392

regiochemistry! Stereochemistry and conformation can and do influence the physicochemical properties of a molecule, such as reactivity or solubility, and these effects can be just as dramatic as, for example, moving a substituent around an aromatic ring or changing a methyl for an ethyl ester. Within the realm of drug-like molecules and pharmaceuticals, stereochemistry and conformation have a particular influence on a molecule's ability to interact with a receptor. Thus, a medicinal chemist could in principle use conformational information to help them make better molecules. Further, correct assignment of stereochemistry is essential for a complete characterization of drug substances and can form an important part of characterization data, route development, and regulatory documentation to prove the compound is what it is said to be.

In discussions of stereochemistry, it is perhaps useful to distinguish absolute stereochemistry from relative stereochemistry. Generally, NMR is silent on the question of absolute stereochemistry: the NMR spectra of enantiomers are identical under normal analysis conditions. That is not to say however that NMR has no role to play in determining absolute stereochemistry. In order to use NMR to determine absolute stereochemistry, it is necessary to 'make' diastereomers, which can then be distinguished. This can be done by forming noncovalent complexes with chiral shift reagents or by forming covalent derivatives, such as the well-known Mosher's esters. Generally speaking, such approaches require a known standard to determine which resonance belongs to which enantiomer. However, certain derivatives can be used to determine absolute stereochemistry if some assumptions about the conformation of the derivatives are made. A full discussion of methods for determination of absolute stereochemistry is beyond the scope of this work, and interested readers can consult a related article within this book. From this point on discussion will be confined to relative stereochemistry.

This chapter is about stereochemistry and conformation. We often use these terms together, but of course they are not the same thing. Stereochemistry is a primary property of a molecule, due to the three-dimensional arrangement of its atoms relative to one another. Conformation is the shape that molecule actually adopts in a particular environment. Conformation is a secondary property – it is entirely determined by primary structure, along with the environment a molecule finds itself in (e.g., what solvent it is dissolved in). That said, the two are often interrelated. We use the same techniques to study both things, and equally it is often not possible to decouple the two: it may be necessary to determine both in order to solve a problem. For example, even if our only aim is to determine stereochemistry, interpretation of the various NMR parameters that define it is only possible with a model of the conformation. The distinction between stereochemistry and conformation may seem like an obvious one, but it has some important consequences in the design of drug-like molecules, where often what we would like to do is be able to manipulate the conformation of a molecule directly, but we cannot do this. The only way to influence the conformation is via changing the molecular structure (which of course includes stereochemistry). This however is bound to change other interactions too.

16.2   EXPERIMENTAL TECHNIQUES

NMR is such an important tool in conformational and stereochemical studies because it provides a number of readily accessed parameters that provide information about the shape of the molecule. The most well known of these are NOEs and scalar couplings, but a wealth of other techniques such as residual dipolar couplings (RDCs) or even simple chemical shifts can also be brought to bear.

16.2.1   Nuclear Overhauser Enhancements (NOEs)

Every undergraduate chemist will know about the qualitative use of NOEs to obtain information about relative distances in molecules. Protein NMR experts will also be very familiar with their quantitative use, but quantitative NOEs can also be used very effectively for small molecule conformational studies. A number of methods can be used to relate NOEs to distances in a molecule and vice versa. The simplest involves simply comparing the measured NOE for an unknown distance to that of a known one, for example, the diastereotopic protons of a methylene group (equation 16.1).

$$r_{ij} = r_{\text{ref}} \cdot \sqrt[6]{N_{\text{ref}}/N_{ij}} \qquad (16.1)$$

The distance between two protons i and j (r_{ij}) can be calculated from the measured NOE between i and j (N_{ij}), the known reference distance between two other protons (r_{ref}) and the NOE corresponding to that reference distance (N_{ref}).

Deriving distances in this way can be surprisingly accurate.[1] A problem with fixed distances is that protons are often also coupled and zero quantum-like artifacts can be a problem for accurate integration, but methods exist to deal with this.[2] Other methods that measure the initial rate of NOE buildup (to get over spin diffusion effects) can also play a role and perhaps the most powerful method is to use the so-called full relaxation matrix approach.[3] This does not make the two-spin approximation inherent in the other methods but instead fully treats the interrelated cross relaxation between all protons in the molecule. Further details are beyond the scope of this work and may be found in references herein.[4]

16.2.2   Scalar (J) Couplings

Scalar couplings are of course familiar to any chemist with a rudimentary knowledge of NMR, and it is well known that they can provide important information about dihedral angles within a molecule. Various versions of the famous Karplus relationship[5] are generally the basis for a quantitative analysis of couplings constants. Typically, such equations are parameterized for a particular type of torsion (e.g., Hα-NH in peptides), but general equations such as the Altona equation[6] that take account of electronegativity of substituents and their effect on coupling are more generally applicable. Many chemists will be familiar with the use of proton–proton couplings in this regard, but proton–carbon couplings are also a ready source of information. They can be particularly useful where there are insufficient proton–proton couplings to define torsions. Again, Karplus-like equations that are suited to particular functional arrangements can be used to improve accuracy,[7] and recently an 'Altona-like' general CH equation has been published.[8] Perhaps one of the reasons that CH couplings are less used than their H–H counterparts is that measuring them is seen as more difficult. While proton–proton couplings can generally be assigned and measured from a simple proton spectrum, the same is not true for proton–carbon couplings. One might immediately think to measure them from a 'coupled' (i.e., not decoupled) carbon spectrum, but sensitivity is very low and assigning which coupling is which can be very hard. A number of two-dimensional methods have been proposed over the years such as heteronuclear long-range coupling (HETLOC) and J-HMBC. Recently, some very general methods such as heteronuclear single quantum multiple bond correlation (HSQMBC) have been developed, which make measurement much more straightforward. An overview of all the techniques available is beyond the scope of this work and interested readers can consult references herein.[9] The important message here is that it is now relatively easy to obtain CH couplings with modern probe technology and experiments.

In addition to these empirical relationships, recent advances in ab initio calculations of coupling constants have been made.[10,11] This combined with the ever-increasing power of computers means that such approaches for small molecules are eminently feasible. The big advantage of such an approach is that in principal it does not rely on empirical data in the same way the coefficients in Karplus equations are derived and the accuracy can therefore be very good.

16.2.3 Residual Dipolar Couplings (RDCs)

RDCs are a very powerful method for determining the shape of molecules.[12] Dipolar couplings are not normally observed in solution NMR as isotropic tumbling averages them to zero. However, careful choice of an alignment medium can break the symmetry of this tumbling and lead to NMR spectra that exhibit some remnant of the dipolar coupling, which can be measured. RDCs are probably less generally used than NOEs or scalar couplings. There are perhaps two reasons for this. Sample preparation is not as straightforward as for regular NMR samples. In addition, RDCs generally do not have the same intuitive interpretation as NOEs and couplings do, meaning a 'manual' type analysis of them is not really possible. For example, while we can infer from a large scalar coupling that two protons are *anti*, there is no such general and straightforward interpretation of a large RDC, which depends on all the distances within a molecule and its overall alignment with the field (although some reports suggest that such relationships may exist[13]). The flip side of this is that RDCs contain information that is not local and can therefore relate distant parts of the molecule in a way that couplings or NOEs cannot. One example of the importance of this is Archazolide A, a large macrocycle with seven stereogenic centers located at three distant positions. The combination of NOEs and *J* values solved the relative configurations of adjacent centers, but failed to distinguish between remote centers. The ambiguity was solved by RDCs, comparing the values with those calculated for the conformation deduced from J-couplings and NOE restraints for each configuration.[14] RDCs is a fast developing field and they have shown themselves to be a very powerful tool in stereochemical analysis.[15] Recent advances in dealing with molecular flexibility, a potential problem for RDCs also bode well for their increased use in the future.

16.2.4 Chemical Shifts

Chemical shifts are the simplest NMR parameter to measure, although relating them to stereochemistry and conformation in a quantitative way is perhaps not quite as straightforward as coupling constants and NOEs. That said, every undergraduate is taught of ring current effects and how they can affect chemical shifts, and examples of unusual shifts such as those for methyl groups shifted upfield when they are in the face of aromatic rings are well known. To make the analysis more quantitative, some kind of calculation is required. A number of methods exist using some form of parameterized model (e.g., CHARGE),[16] but perhaps the most powerful methodology is to use ab initio calculations of chemical shift. A nice example of such an approach is that of Barone *et al.*,[17] which use calculated shifts weighted by conformer energies for stereo isomers of a flexible molecule. Comparison of overall error across the matrix of possible theoretical/experimental combinations shows the error for the correct pairings to be the best (Figure 16.1).

When comparing calculated to experimental shifts, it is important to recognize that there will always be one answer that is 'better' than the others. However, the real question is not which is best, but whether the degree to which it betters the others is significant or not. A particularly nice way of dealing with this is the so-called DP4 approach of Goodman *et al.*[18] In this method, the residuals (i.e., difference between observed and calculated shifts) are calculated, and their distribution examined to say how likely such a structure is, using a Bayesian probability model. The approach can therefore provide not only a parameter from which we can choose 'the best' fit but also an actual probability that the experimental data is able to distinguish one stereo isomer from another.

16.2.5 Detection of Hydrogen Bonds

Knowledge of hydrogen bonding can be a key piece of information to help define a conformation – much can be inferred if we know a hydrogen bond to be or not be present. Effectively, it gives another restraint on the possible conformations. Two methods in particular

for determining the presence of hydrogen bonds are the temperature dependence of exchangeable proton residues and protection from exchange (Figure 16.2). In the former of these methods, we make use of the fact that hydrogen-bonded protons have a chemical shift that is only weakly dependent on temperature (typically $1-2$ ppb K^{-1}), whereas non hydrogen-bonded ones move much more as the temperature is changed ($6-10$ ppb K^{-1}).[19] To make firm conclusions, it is often useful to have an internal reference for these values – for example, two OHs in a molecule, where one cannot be hydrogen bonded would provide a control against which to measure the H-bonded one. Another way of getting the same information is protection from exchange. For small molecules, the protection is usually not so pronounced that we can use direct kinetic analysis as a resonance exchanges out in real time. However, we can measure things indirectly by presaturating a water peak and noting if any protons that are hydrogen-bonding candidates are reduced in intensity. Again a suitable control is often needed here as the intrinsic rate of exchange will depend on local electronic effects, but given the right circumstances such hydrogen bonding information can be very useful in defining a conformation. A further method is the so-called SIMPLE approach, which uses isotope shifts

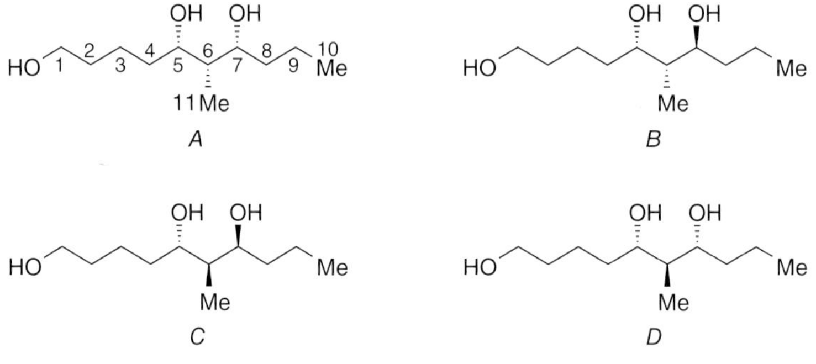

	A_{calcd}	B_{calcd}	C_{calcd}	D_{calcd}
A_{exp}	6.77	17.95	17.29	19.98
B_{exp}	26.68	10.91	15.89	16.08
C_{exp}	26.50	18.47	9.93	15.10
D_{exp}	23.42	12.42	12.68	7.96

Figure 16.1. Sum of absolute errors for four stereoisomers *A–D*, for all possible experimental – calculated combinations. The correct assignment fits significantly better in all cases. (Reproduced with permission from Ref. 17. © WILEY-VCH Verlag GmbH, Weinheim, Fed. Rep. of Germany, 2002)

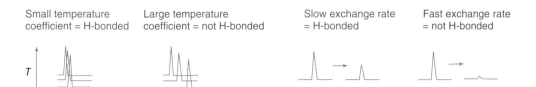

Figure 16.2. Differences in temperature coefficients or exchange rates can be indicators of the presence of hydrogen bonds

on partial deuteration of exchangeable protons to detect hydrogen bonding.[20,21]

It is also possible that scalar couplings may be observed between atoms (i.e., H and N) involved in a hydrogen bond, offering very direct and specific evidence of an interaction. Although such couplings are well known in the field of biomolecules such as DNA, reports of them in small molecules are not so common, perhaps because of unfavorable exchange kinetics typical of small molecules. However, there have been some reports of such interactions[22] and as such, an unexpected coupling could be a useful piece of information to define a conformation.

16.3   ANALYSIS OF NMR RESULTS

As has been illustrated previously, NMR can provide a wealth of experimental parameters that can define the stereochemistry and conformations of a molecule. What is now required is some kind of structural model against which they can be compared or against which they can be used as constraints.

16.3.1   Configuration Analysis

The most intuitive way of interpreting NMR data concerning stereochemistry and conformation is the configuration analysis method.[23] In this approach, the various staggered conformers around a rotatable bond are drawn as Newman projections, for all possible stereochemistries. Expected couplings for ^3J interactions are assigned as large or small based on simple geometric arguments – *anti* conformations are labeled as 'large', whereas *gauche* conformations are labeled as 'small'. The analysis can be extended to two-bond couplings by noting that an electronegative atom in an *anti* position will also give a small coupling. Both HH and HC couplings can be considered; experimental values are likewise assigned as large (>7 Hz) or small (<7 Hz) and compared, hopefully resulting in a clear fit to

one model but not the other. Even if one conformer does not fit, it may be possible to rule out one isomer based on inconsistency. The procedure is outlined in Figure 16.3. Although relatively 'simple' and qualitative, this can nevertheless be an excellent approach to solve problems of stereochemistry.

16.3.2   Molecular Modeling

In order to put the analysis of NMR data on a more quantitative footing, it is necessary to generate three-dimensional structures computationally. As far as making a comparison of experimental to theoretical values, there are two ways we may proceed. The first might be termed the 'protein NMR' approach

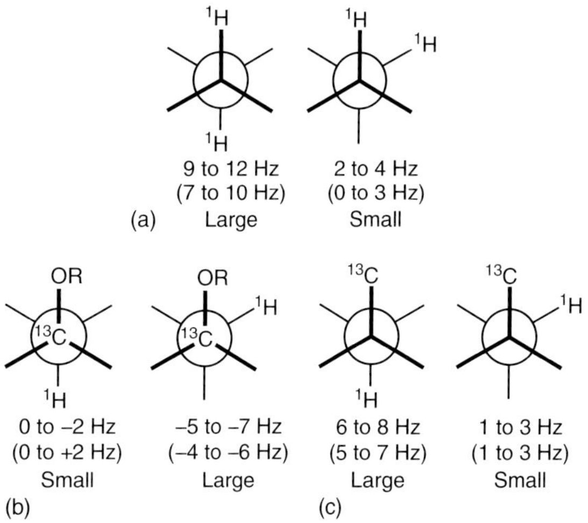

Figure 16.3. Dependence of $^3J_{H,H}$, $^2J_{C,H}$, and $^3J_{C,H}$ coupling constants on dihedral angle. (a) Vicinal $^1H-^1H$ coupling, (b) Geminal $^{13}C-^1H$ coupling constants, $^2J_{C,H}$, and (c) Vicinal $^{13}C-^1H$ coupling constants. Figures in parentheses represent values of 1,2-dioxygenated systems. (Reprinted with permission from Ref. 23. Copyright 1999 American Chemical Society)

of using the experimental data to drive a calculation to a best fit solution (or solutions). The second is the more traditional 'small molecule' NMR approach of determining which of a set of pregenerated structural models fits the data best. Both approaches are useful. For the latter approach, typically structure generation is done with a software package that uses a molecular mechanics force field to generate three-dimensional models. Even for relatively simple molecules, minimization of the force field energy is rarely enough to generate the lowest energy conformers, as it is very easy to fall into local energy minima. Usually, a Monte Carlo type conformational search is used to find the global energy minimum and so generate the lowest energy conformation. In most cases, there will also be a number of other conformers within an energy window which are likely to contribute to the experimental NMR spectrum.

16.3.3   The Problem of Multiple Conformations

Drug molecules often have a number of rotatable bonds and may well have a large number of conformations present in solution. Most conformational processes are fast on the NMR timescale, so we do not observe separate subspectra for conformers, but a weighted average dependent on the mole fraction of each contributing species. A notable exception to this rule is tertiary amides where separate NMR signals for cis and trans forms are almost always observed at room temperature owing to slow rotation about the

partial double bond between N and C. This can be a very common linkage in drug-like molecules, as amide bond formation is a favorite way of exploring an SAR and introducing diversity in medicinal chemistry.

It is thus very possible that no single structure will satisfy a set of restraints. This makes it very hard to use NMR data as restraints in the way that they are often used in protein NMR, because those restraints will drive toward an impossible structure (Figure 16.4). A number of ways of dealing with this have been developed and are outlined in Table 16.1.

The simplest method is to hope that one conformer is dominant and that the experimental spectrum will thus approximate that expected for this structure. In many cases, this simple analysis is enough and 'the' conformation can be assigned in this way. Small contributions from other species are subsumed in to the overall error. This approach can however fall over rather badly when a small contribution from a conformer with a very short H–H distance is present. Owing to the $1/r^6$ averaging for NOEs, this will have a disproportionate effect on the observed NOE and may lead to substantial errors.[24]

The next method in terms of complexity is to use Boltzmann averaging. The energies of individual conformers are taken from the force field and are used to calculate the expected mole fraction according to Boltzmann's formula (equation 16.2):

$$v_i = \frac{e^{-E_i/RT}}{\sum_{j=1}^{j=n} e^{-E_j/RT}} \tag{16.2}$$

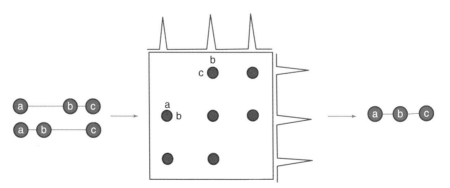

Figure 16.4.   For a 1 : 1 mixture of conformers where distances between atoms a, b, and c differ as shown, the NOESY spectrum would apparently show equal NOEs between resonances, all corresponding to short distances. Deriving a single structure from this data would give a 'virtual' conformer that differs from the two that are actually present. In real situations, this data may even be contradictory and impossible to reconcile to a single structure

Table 16.1. Methods of handling flexibility in the study of stereochemistry and conformation of small molecules

Method	Comments	Advantages	Disadvantages
Fit to one conformer	This simple approach assumes one conformer dominates and can approximate the ensemble	Easy to do. Readily interpretable by nonexpert	Will fail if its central assumption (that one conformer can approximate the ensemble) is not valid
Use Boltzmann average	Next step in complexity – we calculate average values for an ensemble and compare to experimental	Relatively straightforward to apply	Relies on force field energies
Derive conformer populations	Force field generates possible conformers, but experimental data used to derive their populations	Does not (overly) rely on force field energies	Can be underdetermined (a large range of feasible solutions may remain). Force field must generate all possible conformers
Averaged restraints	NMR data used as restraints that must be satisfied for ensemble rather than individual structures	No reliance on force field	Needs a lot of data to get unique solutions

The mole fraction of a particular conformer (v_i) can be calculated from the energy of that conformer (E_i) and the energies of all conformers (E_j).

The expected value of an NMR parameter is then calculated as a weighted average across the set of conformers (equation 16.3):

$$J_{\text{calc}} = \sum_{i=1}^{i=n} v_i J_i \qquad (16.3)$$

The overall calculated coupling (J_{calc}) is calculated from the calculated coupling of each conformer (J_i) multiplied by its mole fraction (v_i).

This calculation is fairly straightforward to perform and will of course take adequate account of the kind of disproportionate contributions mentioned earlier. The big issue with this approach is that it relies entirely on the energies of the force field. Owing to the exponential relationship between mole fraction and energy, a fairly small energy error can lead to a large error in predicted mole fraction. For example, if the relative energies of two conformers are the same, their populations will be 1 : 1. However, if that energy was miscalculated by just 5.7 kJ mol^{-1}, at room temperature, we would calculate the populations as 9 : 1 – which could have a massive influence on the weighted averages. Force fields have improved in recent years, but such an error might certainly be reasonably common in a conformational search.[25] One might mitigate against this risk by examining the solution obtained with different force fields,

or even taking energies from ab initio geometry optimizations of each conformer.

Another possibility is to use time averaged restraints.[26] That is, we do use the NMR data to drive the solution to the problem, but we do not insist that any single structure satisfies all restraints. Instead, the restraints are compared to the calculated properties for the ensemble. Such an approach is implemented in many protein modeling packages and has been used extensively for biomolecules such as DNA.[27] Applications to small molecules have been more limited, but a recent example of a similar approach was described by Blundell *et al.*[28] who used a combination of NOEs, couplings, and RDCs to determine solution conformations of streptomycin.

Another approach that can be used is to make the following postulate: although the force field used to generate conformers may not get their relative energies absolutely right, it will nonetheless identify all conformers that *might* contribute to the ensemble. Thus, we can generate a set of conformers and then ignore their exact energies, and find the set of mole fractions that best reproduce the experimental data. Such an approach was first outlined in the program NAMFIS.[29] Essentially, it is a linear algebra problem to find the set of coefficients, for which a number of well-optimized routines can be used.

The latter two approaches both suffer from potential ambiguity in the fitting problem that falls under the heading 'under determination'. That is, it will not necessarily be clear whether any flexibility is due to real flexibility or lack of information. Likewise, the results often present the best fit solution, but it is rather less easy to convey if other solutions, although inferior in terms of overall error, are nevertheless entirely consistent with the data. Thus, it is sometimes more useful to think about 'feasible ranges' (or to put it the other way around, solutions that are not possible) rather than what *the* best solution is.

If the problem at hand is an assignment of stereochemistry rather than a conformational determination, flexibility can provide a particular problem. We may proceed by assuming one stereochemistry and performing our analysis. We then make the opposite assumption. Hopefully, one assumption leads to a feasible solution, while the other does not adequately reproduce the data.[30] A useful guide here is to ask is there a measureable difference between the NMR spectra of the diastereomers. If there is not, proceeding further is clearly futile!

16.3.4 The Problem of Assignment

Generally, it is good to establish an assignment of all resonances before going on to any conformational analysis. Using through-bond experiments such as heteronuclear multiple-bond correlation (HMBC)/heteronuclear single quantum correlation (HSQC) and COSY along with the usual interpretation approaches, this can often be achieved. We can then go on to analyze any NOEs or couplings safe in the knowledge that our assignments are correct. Sometimes, however, the assignment and stereochemical/conformational problem cannot be decoupled in this way. That is the assignment and conformational questions become convoluted. One common example of this is diastereotopic methylene protons. Where conformational flexibility is present, it can often be challenging to be certain which proton is the pro-S and the pro-R, i.e., to assign which is which. Where an assignment cannot easily be made, we can take a similar approach to that described earlier, and assume a certain assignment and determine if any viable solution are possible to the conformational problem, and then to assume the opposite and likewise look for viable solutions. In an ideal world, one will give sensible answers and the other not, but of course this is not guaranteed to be the case.

16.3.5 Relevant Solvents

A question that often comes up in the discussion of conformation in relation to biological activity is 'what solvent did you use?' A moment's reflection would suggest that a biologically relevant buffer such as phosphate buffered saline should be used in order to ensure the conformation is relevant to what will be present in the body. The problem here is that many drug-like molecules, particularly in early discovery phases, do not have high enough solubility in such media and so acquisition of the necessary data is either slowed or impossible owing to poor signal to noise. In such circumstances, one possible approach is to obtain a proton spectrum in the relevant medium and then to titrate in an organic cosolvent to improve solubility. If any coupling constants or chemical shifts relevant to the conformation do not change, this is good evidence that the conformation is not affected by the solvent – the medium in which solubility is higher can then be used to acquire any experiments requiring greater sensitivity. Another thing that should be considered is the question of whether an aqueous solvent really is the relevant medium. For example, many drug targets are membrane bound receptors, and some ligands bind within the membrane. For such ligands, one might argue that the conformation in a solvent such as chloroform is actually more relevant.

16.3.6 Conformation of the Bound State

All of the discussion around conformation up until this point has been about that of a molecule in free solution. The relationship between free and bound states is discussed in detail later; however, if direct knowledge of the bound state conformation is what is required, NMR can do this using the so-called transferred NOE experiment.[31,32] In this method, a NOESY is performed in the presence of a protein. The ligand is generally present in large excess and the experiment requires that it be exchanging between free and bound forms at a rate that is fast compared to the buildup of NOEs. Under these conditions, the large negative NOE enhancements from the bound state come to dominate over the much weaker positive NOEs of the free molecule, and the pattern of NOEs for the bound state are observed on the free ligand (Figure 16.5).

For projects that are structurally enabled, that is, those that have a structure of the desired ligand binding site, typically from X-ray crystallography, a common

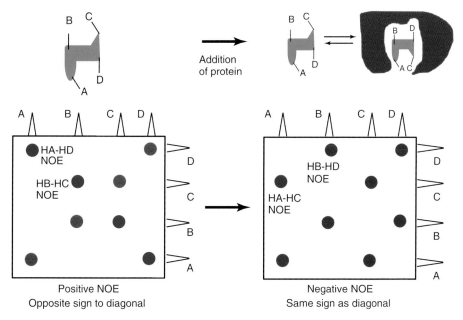

Figure 16.5. Schematic showing how the transferred NOE experiment can give information about conformation of the bound state. In the presence of protein, the rapid equilibrium of free and bound states means the NOEs of the free ligand are reflective of the bound state

design strategy is to dock potential ligands into the target site and carry out a virtual screen. Alternatively, we may have carried out an actual screen and have a number of known hits, but not know how they bind. For conformationally flexible molecules, docking can be a very hard problem as there are many ways a ligand could fit into a receptor. NMR can again help us in this situation, as we may calculate NMR parameters for the bound state and either compare them to experimental ones or use them to drive the solution. The saturation transfer difference (STD) effect is one such piece of information.[33] The STD effect arises when resonances on a protein are saturated, and that saturation is transferred to a ligand binding with that protein. The STD effect is widely used for screening – that is, the detection of binding, but it potentially contains much more information. So-called Epitope maps can be created, which show how the STD effect varies across the atoms of a molecule, from which qualitative information on binding mode can be obtained. However, it is also possible via a theoretical treatment of the STD effect to calculate the exact values and to use this to drive bound structures.[34] This gives information not just on the gross orientation of a ligand in a binding site, but also the conformation of the ligand itself.

Another possibility for studying the bound state might be to look at chemical shift perturbations of ligand resonances on binding to rank binding poses and conformations. Bound ligand shifts can be obtained by titration for ligands in fast exchange. We have had some success using this approach with a model system,[35,36] but its general applicability remains to be proven.

16.4 APPLICATIONS OF STEREOCHEMICAL AND CONFORMATIONAL STUDIES IN THE PHARMACEUTICAL INDUSTRY

16.4.1 Proof of Structure and Reactivity

One might argue that the 'core' use of NMR in the study of stereochemistry of drug molecules is proof of structure. NMR is probably the most used method of establishing relative stereochemistry from early discovery through to regulatory submissions, for final compounds and for intermediates.[25] Many of the most useful experiments can easily be carried out in an open access environment without the need for a specialist. Its use in this context is widespread and in many cases routine. Another area of widespread usage is to help better understand chemical reactions. Sometimes, reactions to synthesize drug compounds do not go as planned. During the first synthesis of molecules for primary biological screening, or particularly during scale up for toxicological studies, it is not unusual to find a reaction that gives poor yields or an unexpected product, and often a chemist will want to know why it went wrong or what the product is in order to improve their reaction. Sometimes, differences in reactivity between diastereomers can be observed, and an analysis of their conformation and stereochemistry can provide insights into what is going wrong. An example of such an application is provided by the following tale of a 'routine' deprotection that went wrong.[37] While one isomer gave the correct product, the other gave a structure, which although initially consistent in terms of its NMR spectrum, gave some unexpected correlations in COSY and HMBC, which did not fit the expected product. Stereochemical and conformational analyses of the two species by NMR, facilitated by the open source program Janocchio,[37] showed that in one isomer the carboxylic acid group was ideally placed to attack the double bond in the molecule and so affect a rearrangement to give the observed product. Such a mechanism was not available to the other diastereomer, which therefore gave the expected product (Figure 16.6).

16.4.2 Molecular Design

Perhaps the single most important reason medicinal chemists want to know the stereochemistry and conformation of a molecule is to use the information to better understand the structure–activity relationships (SAR) of their molecules, and thus design better ones. Figure 16.7 illustrates the thermodynamics of the binding process and how conformation affects it, and forms a useful frame of reference for the following discussion. It illustrates that whilst there is no *a priori* reason why a ligand must be in the bound conformation before binding, all things equal, such preorganized molecules must bind more tightly as a natural consequence of the first law of thermodynamics. The way we might use these ideas in molecular design depends crucially on whether we have direct information on the bound conformation. If we do, perhaps from an X-ray structure, we may want to compare free and bound conformations to assess the degree to which a molecule is preorganized. This in turn could allow hypotheses to be made about new molecules that might be more like the bound conformation, and therefore have higher binding affinity. The issue here is that the only way to influence the conformation is via changing the molecular structure (which might include the stereochemistry). This is bound to change other interactions too, affecting binding affinity, and it may thus be difficult to apportion any changes to improved conformation or other factors. The second approach, which we might

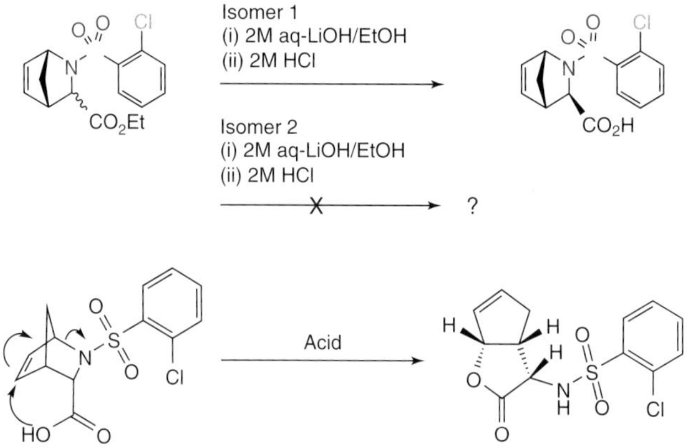

Figure 16.6. Two diastereomers show different reactivity: one gives the expected product, the other an unknown. NMR analysis of the conformation and stereochemistry provides an explanation as to why one isomer gives an alternative product. (Reproduced with permission from Ref. 37. © John Wiley & Sons, Ltd., 2007)

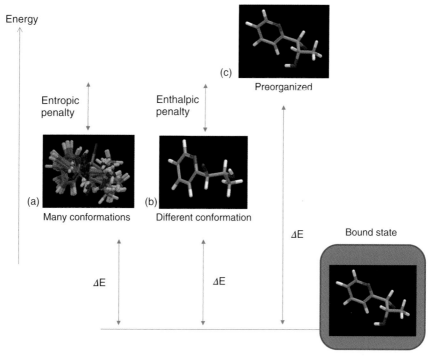

Figure 16.7. Where a molecule adopts (a) many conformations or (b) is in the wrong conformation for binding there is a cost (entropic or enthalpic, respectively) to get the molecule into the bound conformation (c). This cost is repaid on binding, but all things equal this cost will lead to a less favorable overall binding energy (ΔE) than if the molecule were preorganized

use in the absence of a knowledge of the bound conformation, is to assume that the solution conformation is a model for the bound one (i.e., that the molecule is preorganized). The solution structure therefore affords a three-dimensional model from which we may infer binding site interactions, and so design new structures. The validity of this assumption is discussed in some of the examples discussed in the following text.

An example of how NMR studies of free ligand conformation can be used to aid molecular design is provided by beta-secretase (BACE), an important target in Alzheimer's disease research. In a study carried out by Espinosa *et al.*,[38] coupling constants and NOEs were used in a configuration analysis type approach to examine conformational preferences of a number of ligands. They were able to show that the apparently flexible hydroxyethylene isostere was in fact strongly preorganized in solution, with a number of extreme coupling constants indicative of an extended conformation. Refinement of the conformation by molecular dynamics simulation then showed that the solution conformation was essentially the same as that of

the bound ligand (Figure 16.8). This work was important from a design perspective as medicinal chemists may have been tempted to try some of their favorite strategies to improve binding affinity: for example, by making cyclic derivatives designed to restrict conformational entropy and give more of the molecule in the bound conformation. This work suggested such an approach was unnecessary as the apparently flexible ligand was already strongly disposed to be in the correct conformation.

A similar example involving HIV protease inhibitors[39] shows that throughout the drug discovery and development cycle from hit to candidate to clinical compound, the core of the molecules studied retained a favored conformation. This favored conformation was the same as that of the bound state. Further, an analysis of the SAR reveals that substitutions that favor a solution conformation similar to that of the bound conformation impart extra binding affinity.

NMR can also play an important role in supporting or ruling out direct conformational hypotheses. The so-called magic methyl effect is a well-known effect

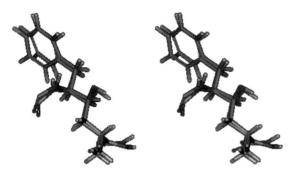

Figure 16.8. Stereo view of the NMR-derived conformation in solution for the hydroxyethylene core of a BACE inhibitor (green) superimposed on the X-ray structure of the same ligand from a BACE complex (blue). (Reprinted with permission from Ref. 38. Copyright (2005) American Chemical Society)

in medicinal chemistry[40] where the single-point replacement of a proton with a methyl group leads to an unexpectedly large (>10-fold) increase in affinity. It is termed *magic* as the increased binding affinity from Van der Waals interactions and the hydrophobic effect cannot explain more than an order of magnitude increase in affinity. An often-touted hypothesis to explain such an effect is that the extra affinity is due to a change in conformation, such that the methyl-containing species is predisposed to the bound conformation. We have seen examples (Figure 16.9) where such a hypothesis has been supported or refuted.

Of course most examples of the use of NMR to study conformation and stereochemistry for drug discovery remain proprietary, so the number of articles representing it in the literature vastly underestimates its actual use.

16.4.3 Possible Limitations of Design Approaches

As it is hoped has been illustrated, NMR provides a number of powerful ways to study the stereochemistry

and conformations of drug-like molecules. However, the real question a medicinal chemist often wants to answer is not 'what is the conformation of this molecule?' but 'how do I use the information to design a better molecule?' The link between the conformation of a molecule in solution and a better molecule is not a straightforward one, and assumes that the solution conformation is predictive of the bound conformation or preorganization is important. Although a number of examples of such a situation were discussed in the previous section, one does equally not have to look too hard into the literature to find counter-examples. For example, an often cited computational chemistry paper[41] states that some degree of reorganization is easily tolerated and very common in flexible ligands. In fact, the authors go as far as to say that not only is it uncommon for compounds to bind in the global minimum conformation, but also that bound conformations often do not even correspond to local minima in free solution. More recent work suggests that these energy penalties may be somewhat overestimated,[42,43] but nevertheless are still important. Other work[44] supports the idea that solution conformations are rarely the same as bound, suggesting though that common 'anchor points' mean that they may still be useful as pharmacophore models, but also concluding that predicted conformations are perfectly adequate in this regard. The true situation probably lies between these various viewpoints and is likely to be example dependent, but what is clear is that NMR can probably provide the most accurate picture of conformation, and all things being equal, designing molecules that mimic the bound state in free solution cannot be a bad thing. The approach could be seen as stacking the odds in your favor whether such a conformation is strictly required or not. Furthermore, one has to remember that X-ray 'structures' themselves are really X-ray *models* – bound ligands are not determined at atomic resolution and fine details of conformation may well be a result of fitting the X-ray-derived electron density maps.

The devil's advocate may also say that a conformational analysis of a series of molecules may help us

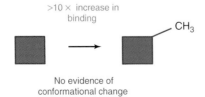

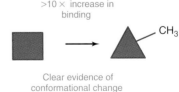

Figure 16.9. NMR can provide evidence for and against conformational hypotheses

understand the shape required to interact with the receptor, but a model that requires an experimental conformation in order to predict binding is not very useful, as we are required to synthesize the molecule in order to determine its conformation. Having synthesized, one might argue that we simply test it in the relevant biological assays, a process that is considerably faster and requires less material than doing conformational studies. What we really require is a good way of *predicting* conformation accurately so that we can design molecules and triage them against such models before they are synthesized. NMR of course can play a role here by helping to understand which theoretical models give the best prediction of conformation or to optimize their parameters to improve predictions in the future.

Finally, it should be noted that obtaining potency is rarely an issue for medicinal chemists. Designing a drug is about much more than potency, and indeed most projects fail because of the difficulties of optimization against multiple objectives: toxicological issues, metabolic problems, lack of selectivity against counter targets, or problems with bioavailability and distribution are often far more challenging than obtaining potency against a particular target. An understanding of conformation may not be helpful to make progress against these other objectives. The counter argument to this is that although chemists are very good at improving potency, they may well achieve this in the 'wrong' way. That is, rather than adding quality interactions that an understanding of conformation may bring, they increase molecular weight or logP to achieve better potency – changes that decrease ligand efficiency and that may well be associated with some of the other issues such as toxicity or increased metabolism mentioned earlier.

The aforementioned discussion highlights the roles NMR-based conformational studies currently play in drug discovery: they are excellent at rationalizing SAR and they can help with the design of molecules, but they are not a panacea to solve all problems.

16.5 CONCLUSIONS

Questions concerning stereochemistry and conformation are important things to be addressed in the design and development of drug molecules. From correctly defining the molecular structure to helping design the next generation of molecules against a target, NMR measurements can provide key characterization information or can support or refute hypotheses. NMR is most powerful when allied with molecular modeling, and when applied thoughtfully can help advance pharmaceutical research.

ACKNOWLEDGMENTS

Thanks to Juan Espinosa and David Evans for their helpful comments during the preparation of this manuscript.

RELATED ARTICLES IN EMAGRES

Residual Dipolar Couplings

REFERENCES

1. C. P. Butts, C. R. Jones, E. C. Towers, J. L. Flynn, L. Appleby, and N. J. Barron, *Org. Biomol. Chem.*, 2011, **9**, 177.

2. M. J. Thrippleton and J. Keeler, *Angew. Chem. Int. Ed.*, 2003, **42**, 3938.

3. M. Forster, *J. Comput. Chem.*, 1990, **12**, 292.

4. D. Neuhaus and M. P. Williamson, *The Nuclear Overhauser Effect in Structural and Conformational Analysis*, Wiley-VCH, New york, 2000, vol. 2.

5. M. Karplus and D. H. Anderson, *J. Chem. Phys.*, 1959, **30**, 6.

6. C. A. G. Hasnoot, F. A. A. M. De Leew, and C. Altona, *Tetrahedron*, 1980, **36**, 2783.

7. R. Wasylishen and T. Schaefer, *Can. J. Chem.*, 1973, **51**, 961.

8. G. Palermo, R. Riccio, and G. Bifulco, *J. Org. Chem.*, 2010, **75**, 1982.

9. T. Parella and J. F. Espinosa, *Prog. Nucl. Magn. Reson. Spectrosc.*, 2013, **73**, 17.

10. G. Bifulco, C. Bassarello, R. Riccio, and L. Gomez-Paloma, *Org. Lett.*, 2004, **6**, 1025.

11. W. Deng, J. R. Cheeseman, and M. J. Frisch, *J. Chem. Theory Comput.*, 2006, **2**, 1028.

12. R. M. Gschwind, *Angew. Chem. Int. Ed.*, 2005, **44**, 4666.

13. J. Yan, A. D. Kline, H. Mo, M. J. Shapiro, and E. R. Zartler, *J. Org. Chem.*, 2003, **68**, 1786.

14. C. Fares, J. Hassfeld, D. Menche, and T. Carlomagno, *Angew. Chem. Int. Ed.*, 2008, **47**, 3722.

15. B. Boettcher and C. M. Thiele, *Encycl. NMR*, 2012, **8**, 4736.

16. R. J. Abraham, M. Mobli, and R. J. Smith, *Magn. Reson. Chem.*, 2003, **41**, 26.

17. G. Barone, D. Duca, A. Silvestri, L. Gomez-Paloma, R. Riccio, and G. Bifulco, *Chemistry*, 2002, **8**, 3240.

18. S. G. Smith and J. M. Goodman, *J. Am. Chem. Soc.*, 2010, **132**, 12946.

19. S. H. Gellman, G. P. Dado, G. B. Liang, and B. R. Adams, *J. Am. Chem. Soc.*, 1991, **113**, 1164.

20. J. C. Christofides and D. B. Davies, *J. Am. Chem. Soc.*, 1983, **105**, 5099.

21. J. R. Everett, *J. Chem. Soc., Chem. Commun.*, 1987, 1878.

22. X. S. Huang, X. Liu, K. L. Constantine, J. E. Leet, and V. Roongta, *Magn. Reson. Chem.*, 2007, **45**, 447.

23. N. Matsumori, D. Kaneno, M. Murata, H. Nakamura, and K. Tachibana , *J. Org. Chem.*, 1999, **64**, 866.

24. C. P. Butts, C. R. Jones, and J. N. Harvey, *Chem. Commun. (Cambridge, UK)*, 2011, **47**, 1193.

25. G. J. Sharman and I. C. Jones, *Magn. Reson. Chem.*, 2001, **39**, 549.

26. A. E. Torda, R. M. Scheek, and W. F. Van Gunsteren, *Chem. Phys. Lett.*, 1989, **157**, 289.

27. U. Schmitz, N. B. Ulyanov, A. Kumar, and T. L. James, *J. Mol. Biol.*, 1993, **234**, 373.

28. C. D. Blundell, M. J. Packer, and A. Almond, *Bioorg. Med. Chem.*, 2013, **21**, 4976.

29. D. O. Cicero, G. Barbato, and R. Bazzo, *J. Am. Chem. Soc.*, 1995, **117**, 1027.

30. G. J. Sharman, *Magn. Reson. Chem.*, 2007, **45**, 317.

31. F. Ni, *Prog. Nucl. Magn. Reson. Spectrosc.*, 1994, **26**, 517.

32. F. Ni and H. A. Scheraga, *Acc. Chem. Res.*, 1994, **27**, 257.

33. M. Mayer and B. Meyer, *Angew. Chem. Int. Ed.*, 1999, **38**, 1784.

34. J. Angulo and P. M. Nieto, *Eur. Biophys. J.*, 2011, **40**, 1357.

35. G. J. Sharman and J. Morgan, Ab initio chemical shift calculations as a tool within pharmaceutical research, SMASH NMR conference, 2012.

36. J. Morgan, *Investigation into the Application of Fluorine NMR Spectroscopy in Medicinal Chemistry.* [MSc]. University of York, 2012.

37. D. A. Evans, M. J. Bodkin, S. R. Baker, and G. J. Sharman, *Magn. Reson. Chem.*, 2007, **45**, 595.

38. P. Vidal, D. Timm, H. Broughton, S.-H. Chen, J. Martin, A. Rivera-Sagredo, J. R. McCarthy, J. M. Shapiro, and F. J. Espinosa, *J. Med. Chem.*, 2005, **48**, 7623.

39. S. R. LaPlante, H. Nar, C. T. Lemke, A. Jakalian, N. Aubry, and S. H. Kawai, *J. Med. Chem.*, 2014, **57**, 1777.

40. H. Schonherr and T. Cernak, *Angew. Chem. Int. Ed. Engl.*, 2013, **52**, 12256.

41. E. Perola and P. S. Charifson, *J. Med. Chem.*, 2004, **47**, 2499.

42. K. T. Butler, F. J. Luque, and X. Barril, *J. Comput. Chem.*, 2009, **30**, 601.

43. J. Liebeschuetz, J. Hennemann, T. Olsson, and C. R. Groom, *J. Comput. Aided Mol. Des.*, 2012, **26**, 169.

44. M. Vieth, J. D. Hirst, and C. L. Brooks III, *J. Comput. Aided Mol. Des.*, 1998, **12**, 563.

Chapter 17

NMR Methods for the Assignment of Absolute Stereochemistry of Bioactive Compounds

Jose M. Seco and Ricardo Riguera

Department of Organic Chemistry and Center for Research in Biological Chemistry and Molecular Materials (CIQUS), University of Santiago de Compostela, E-15782 Santiago de Compostela, Spain

17.1 INTRODUCTION

The importance of the absolute stereochemistry of organic compounds is particularly relevant for their biological and pharmaceutical properties. Although X-ray diffraction is the most valuable method,[1] it

NMR in Pharmaceutical Sciences. Edited by Jeremy R. Everett, John C. Lindon, Ian D. Wilson, and Robin K. Harris
© 2015 John Wiley & Sons, Ltd. ISBN: 978-1-118-66025-6
Also published in eMagRes (online edition)
DOI: 10.1002/9780470034590.emrstm1398

requires monocrystals, sometimes not available, and provides stereochemical information in the solid state. Other spectroscopic techniques have been introduced that provide reliable assignment in solution. One of those is circular dichroism (CD),[2] a technique available in many laboratories, and particularly useful in the Harada-Nakanishi approach; another very related one is vibrational circular dichroism (VCD)[3] that unfortunately is not so widely present in laboratories and the third one – the object of this article – is NMR.[4-7]

None of those three techniques can be claimed to be as general as X-ray diffraction but their use is much more immediate in the day-to-day work of a laboratory. This is particularly so for NMR owing to the extensive use and presence of these instruments that allows a chemist to check the absolute stereochemistry of a compound within a couple of hours or even less after having been isolated or synthesized.

In this article, we will present the basic information that will allow one to decide if NMR is the method of choice to assign the absolute configuration of their compounds and how to carry out the assignment using arylmethoxyacetic acids (AMAAs) as auxiliary reagents.

To this end, we will present the different classes of substrates whose absolute configuration can be analyzed by NMR, classified by functional group, chiral auxiliary reagents required in every case, and the procedure for the correct assignment.

A literature search of papers published within the past 15 years indicates that more than 200 bioactive natural compounds have been assigned using this NMR methodology.

In this article, we present a selection of those structures to illustrate the scope of this methodology for assignment of absolute configuration.

Apart from those natural products, many synthetic compounds have been studied too. Their structures, NMR spectral data, and so on can be found in the literature.[4–6]

17.2 THE ABSOLUTE CONFIGURATION AND NMR SPECTRA

The NMR spectra of two enantiomers are identical; thus, the first step to distinguish enantiomers by NMR is to introduce some chiral element that converts the enantiomeric relationship into diastereomeric.

There are two ways to address this problem (Figure 17.1). One is to add to the NMR tube a chiral solvent or a chiral reagent, which associates differently to each enantiomer of the substrate producing diastereomeric complexes that lead to different spectra (Figure 17.1a). This is the so-called chiral

solvating agent (CSA) approach,[7] amply used for enantiomeric excess determination and much less for stereochemical assignment because the high specificity of every substrate–CSA interaction prevents the generalization of a CSA to other substrates.[7] The use of CSAs has been the subject of many reviews[7] and will not be discussed here. The second approach is the object of this article and differs in that instead of using weak bonds to associate the chiral substrate to the chiral auxiliary reagent as with CSA, covalent bonds are employed, and the auxiliary reagent is now named as a chiral derivatizing agent (CDA),[4–6] (Figure 17.1b).

It might be thought that this involves more bench work and complications than just mixing CSA and substrate; however, in fact, the use of CDA constitutes a much more general method for absolute configuration assignment, particularly if the CDAs are of the AMAA class of reagents, that are the most widely studied and tested.

The procedure for assignment using the NMR spectra of the CDA derivatives consists of the separate derivatization of the substrate (unknown enantiomer) with the two enantiomers of the CDA (step 1, Figure 17.1b), and the comparison of the NMR spectra of the two resulting derivatives[4–6] (step 2, Figure 17.2). These, being diastereomers, have different spectra and, most importantly, present chemical shifts that correlate with their absolute stereochemistry.

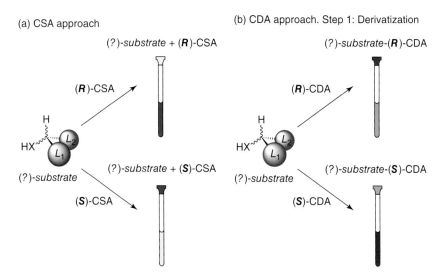

Figure 17.1. Step 1 of the general procedure for the assignment of the absolute configuration by NMR: derivatization

Step 2: Comparison of NMR spectra

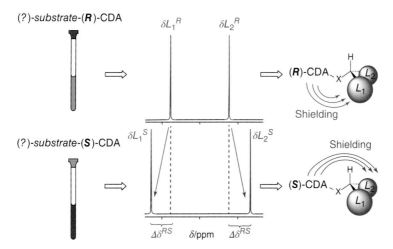

Figure 17.2. Step 2 of the general procedure for the assignment of the absolute configuration by NMR: comparison of the NMR spectra

In general, the NMR spectrum of one of the AMAA derivatives presents the signals for L_1 and L_2, at closer chemical shifts than in the other derivative. This is schematically illustrated in Figure 17.2, for a substrate with substituents L_1 and L_2 that, for simplicity, are represented as giving singlets in the NMR spectra.

These different shifts for L_1 and L_2 in the two CDA derivatives are originated by the aromatic shielding effect produced by the aryl ring of the AMAA part, which selectively affects only one of the substituents in each derivative. Therefore, those shieldings inform about the spatial position of L_1 and L_2 with respect to the aryl ring and allow to determine the absolute configuration at the substrate part of the derivative.

The theoretical and conformational bases and implications of this phenomena have been amply discussed elsewhere[4-6] and will not be commented on here.

In practice, identification of the spatial location of L_1 and L_2 can be easily carried out by subtraction (step 3, Figure 17.3a) of the chemical shifts for L_1 and L_2 in the (R)-CDA derivative minus the values in the (S)-CDA derivative (Figure 17.3a). This difference $\Delta\delta^{RS}$ ($\Delta\delta^{RS} = \delta R - \delta S$) is negative for one substituent and positive for the other (positive for L_2 and negative for L_1 in Figure 17.3b), and that sign is used to locate L_1/L_2 on its place around the asymmetric carbon (step 4, Figure 17.3b). In the example of Figure 17.3(b), this correlation $\Delta\delta^{RS}$ sign/spatial location is presented in the form of a tetrahedral model where the substituent with the positive difference $\Delta\delta^{RS}$ (step 4, Figure 17.3b) occupies the back position (L_2) and the one with the negative $\Delta\delta^{RS}$ the front one (L_1).

(a) Step 3: Calculation of $\Delta\delta^{RS}$ (b) Step 4: Assignment of configuration

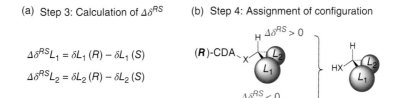

$$\Delta\delta^{RS}L_1 = \delta L_1(R) - \delta L_1(S)$$

$$\Delta\delta^{RS}L_2 = \delta L_2(R) - \delta L_2(S)$$

Figure 17.3. Steps 3 and 4 of the general procedure for the assignment of the absolute configuration by NMR

Thus, the general procedure implies

1. Step 1: separate derivatization with the two enantiomers of a chiral auxiliary reagent (R)- and (S)-CDA (Figure 17.1b);
2. Step 2: comparison of the NMR chemical shifts of the L_1 and L_2 substituents, linked to the asymmetric carbon of the substrate in the two resulting derivatives (Figure 17.2);
3. Step 3: calculation of the chemical shift differences $(\Delta\delta^{RS})$ for L_1 and L_2 (Figure 17.3a);
4. Step 4: placement of the substituent with the positive and the one with the negative $\Delta\delta^{RS}$ on their spatial locations according to the tetrahedral correlation model (Figure 17.3b).

It is quite important to point out that each couple substrate/CDA corresponds to a particular correlation between chemical shifts and spatial location, and that if a substrate is derivatized with different CDAs, their NMR spectra may respond to different correlations.

In the following sections, we will show what are the compound classes that can be assigned by this methodology, what are the most adequate CDAs for each class, and the corresponding correlation between linking the signs of $\Delta\delta^{RS}$ of L_1 and L_2 and their spatial position.

17.3 CDAS AND SUBSTRATES

In the literature, there are many publications describing CDAs and their applications but we will limit our presentation to the AMAAs[4–6] and compounds with quite similar structure presented in Figure 17.4, because they constitute the most widely studied group, have been validated with numerous substrates of known absolute configuration, and present the wider scope of application.[4–6]

Among the reagents of Figure 17.4, some (MPA, MTPA, BPG, and *trans*-2-PCH) are commercially available in chiral form, whereas others (9-AMA, 1-NMA, 2-NMA, 2-NTBA, and 9-AHA) can be easily prepared.[6] All those auxiliary reagents present either a carboxylic or a hydroxyl group in their structure. These functions are used for the covalent bonding of the CDA with the substrate whose absolute configuration is to be determined.

Figure 17.5 lists the structures of the chiral substrates whose absolute configuration can be determined using

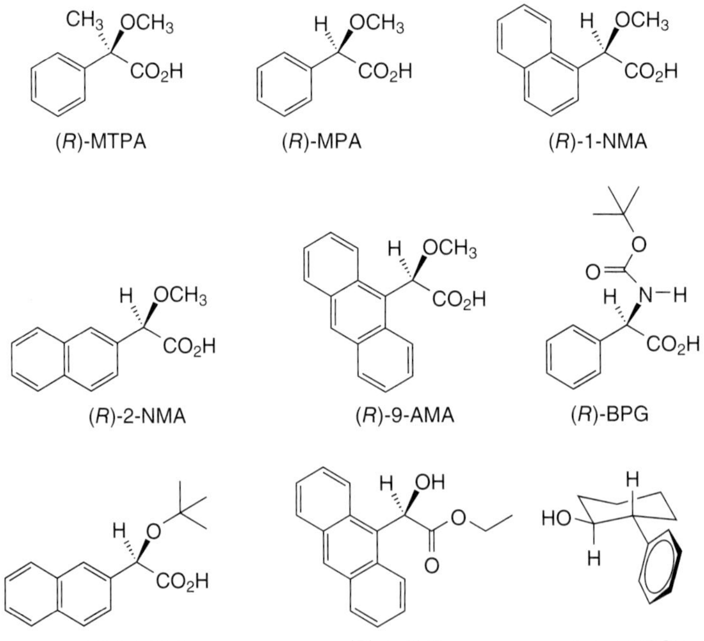

Figure 17.4. CDA for the NMR assignment of absolute configuration by NMR

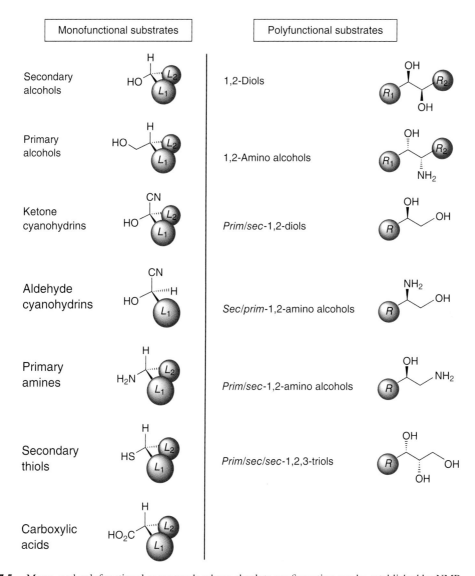

Figure 17.5. Mono- and polyfunctional compounds whose absolute configuration can be established by NMR

the CDAs shown in Figure 17.4. This is constituted by alcohols, amines, thiols, cyanohydrins, carboxylic acids, diols, aminoalcohols, and triols, and therefore their assignment is based on their derivatization with the R- and the S-enantiomers of the appropriate CDA (Figure 17.4) and comparison of the NMR spectra of the resulting esters, amides, or thioesters.

The derivatization follows standard reaction procedures, and experimental details can be found in the literature.[6] In addition, resin-bound CDA reagents can be easily prepared,[8] are stable for months, and allow the assignment to be carried out in very short time – from minutes to a couple of hours – without any bench work being necessary. The use of the resin-bound CDA (mix and shake method) is very convenient for laboratories wanting to assign a large series of compounds in a short time or to carry out a systematic study.

In Figure 17.5, there are a number of monofunctional compounds (in this context, monofunctional means that only one group is derivatizable with the CDA and, therefore, the CDA derivative includes only one CDA unit in its structure) and some di- and trifunctional ones with one or two asymmetric carbons that are all assigned at the same time.

While for the monofunctional substrates,[5,9–17] the NMR signals relevant for assignment are always those of L_1/L_2, for the di- and trifunctional substrates,[4]

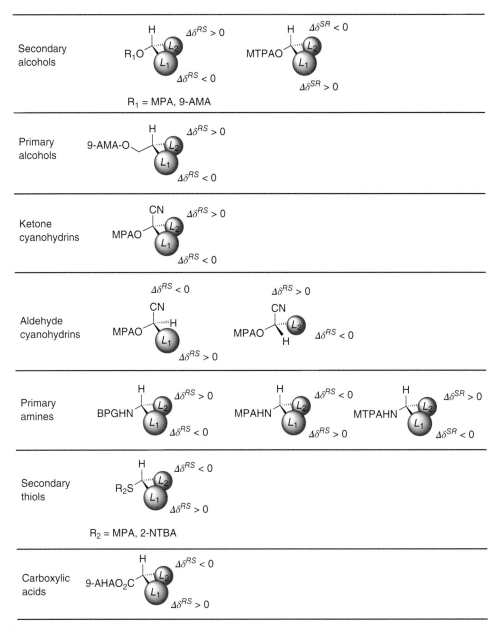

Figure 17.6. Recommended CDAs and correlation models for monofunctional substrates

other signals should be used. We will describe first the methodology for assignment of monofunctional compounds and later on that for polyfunctional compounds.

Unfortunately, and in spite of the close chemical structure, the CDAs listed in Figure 17.4 have no general applicability to other classes of substrates such as tertiary alcohols and secondary amines, although some particular cases can be solved.

A few key points should be borne in mind regarding the correct use of this methodology. One is that not all those CDAs can be used for all the substrates in Figure 17.5. For the assignment of a particular class of substrates, only certain CDAs have been proven to produce reliable results (i.e., for sec alcohols, MPA, MTPA, and 9-AMA can be used,[17,18] but BPG has been proven only for amines[16]). The correlation between the NMR spectra (sign of $\Delta\delta^{RS}$ for L_1/L_2) and the spatial location of the L_1/L_2 substituents is specific for every couple substrate class/CDA. By historical reasons, when MTPA is used as CDA, the chemical shift differences are measured subtracting the chemical shifts in the (S)-MTPA derivative minus those in the (R)-MTPA derivative – the opposite to that with other CDAs – leading to $\Delta\delta^{SR}$ instead of $\Delta\delta^{RS}$.

Figure 17.6 shows the tetrahedral model expressing the correlation between the sign of $\Delta\delta^{RS}$ and the spatial location of L_1/L_2 for the monofunctional substrates of Figure 17.5 (secondary alcohols,[5,9,10] primary alcohols,[5,11] aldehydes[12] and ketone cyanohydrins,[13] thiols,[5,14] primary amines[5,15,16], and carboxylic acids[5,17]), derivatized with the adequate CDAs of Figure 17.4.

Those correlations result from experimental and theoretical studies (semiempirical and ab initio calculations, shielding effect calculations, dynamic NMR, etc.), and their validity for the assignment of absolute configuration has been validated with a number of substrates of known absolute stereochemistry.[5,9–17]

17.4 THE USE OF ^{13}C NMR FOR ASSIGNMENT

As mentioned at the beginning, the differences in chemical shifts observed for the (R)- and the (S)-CDA derivatives of a substrate have their origin in the aromatic shielding effect produced by the aryl ring of the CDA on the substrate substituents.[4,5] Therefore it is not surprising that similar shieldings could be observed in the ^{13}C NMR spectra. This means that the

CDA derivatives of a certain compound can be examined by ^{1}H and/or ^{13}C NMR and that the ^{1}H $\Delta\delta^{RS}$ and ^{13}C $\Delta\delta^{RS}$ sign distribution would be the same for a given stereochemistry.[18] Thus, the graphical models expressing the correlation between the signs of $\Delta\delta^{RS}$ and the spatial position of the substituents shown in Figure 17.6 work for ^{13}C as well as for ^{1}H chemical shifts, although in same cases, the small ^{13}C $\Delta\delta^{RS}$ values limit its usefulness.[18] Apart from being a complement to the ^{1}H data, the ^{13}C $\Delta\delta^{RS}$ has a specific advantage in that it allows the assignment of compounds with very little or no protons in L_1/L_2 as in perdeuterated compounds.[18]

17.5 THE ASSIGNMENT OF ABSOLUTE CONFIGURATION OF MONOFUNCTIONAL COMPOUNDS BY DOUBLE DERIVATIZATION

The procedure for assignment follows the lines stated at the beginning of this article (Figures 17.1b, 17.2, and 17.3). Thus, in a real case, the practitioner should first produce the derivatives (step 1, Figure 17.1b) of the compound with the CDA that corresponds to that particular substrate (Figure 17.6), then take the ^{1}H NMR spectra (or the ^{13}C) of the (R)- and the (S)-derivatives (step 2, Figure 17.2). The comparison of the chemical shifts of the signals for L_1 and L_2 in both derivatives leads to the differences measured as $\Delta\delta^{RS}$, and finally, L_1 and L_2 are placed in their place according to their $\Delta\delta^{RS}$ signs in the graphical correlation model (steps 3 and 4, Figure 17.3).

Figure 17.6 shows the recommended CDAs for the assignment of the monofunctional compounds (secondary alcohols,[5,9,10] primary alcohols,[5,11] aldehydes[12] and ketone cyanohydrins,[13] thiols,[5–14] primary amines[5,15,16], and carboxylic acids[5,17]) and the corresponding graphical correlation models.

As an example to illustrate this methodology, we show next the procedure and spectra corresponding to the assignment of (−)-borneol using MPA as CDA (Figure 17.7).

Following the general steps described earlier, a sample of one enantiomer of borneol is separately derivatized with (R)- and (S)-MPA (step 1, Figure 17.7) and the ^{1}H NMR spectra of the MPA esters recorded (step 2, Figure 17.7). The signals are identified, paying special attention to the protons located at the two substituents of the chiral carbon carrying the hydroxyl

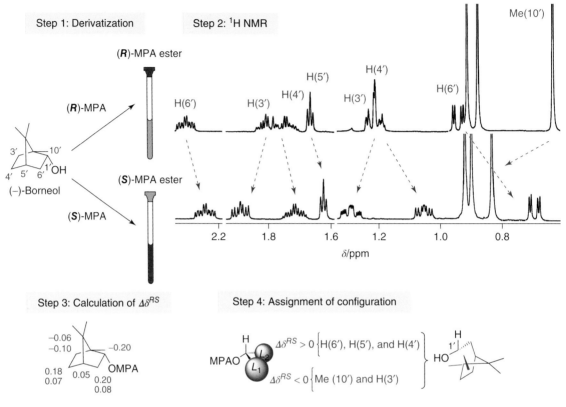

Figure 17.7. Main steps in the configurational assignment of (−)-borneol using MPA

group [C(1′)], which constitute the L_1 and L_2 substituents.

Thus, in borneol, the signals for assignment are those from Me (10′) and H(3′) at one side and H(4′), H(5′), and H(6′) at the other (Figure 17.7).

Subtraction of the chemical shifts of each of those protons in the (*R*)-MPA ester minus the chemical shift in the (*S*)-MPA ester leads to the $\Delta\delta^{RS}$ for each proton (step 3, Figure 17.7), which in this case happens to be positive for H(4′), H(5′), and H(6′) and negative for Me (10′) and H(3′).

The graphical model expressing the correlation between the $\Delta\delta^{RS}$ signs of L_1/L_2 and their spatial location in MPA esters of secondary alcohols[5,9,10] is shown in Figure 17.6 and indicates that the protons with the negative $\Delta\delta^{RS}$ sign should be located on the substituent on the front side of the tetrahedral model, whereas those with positive $\Delta\delta^{RS}$ sign should be placed on the back side of the 'tetrahedron'.

In this way, Me (10′) and H(3′) are on the front substituent (L_1) of the 'tetrahedron' (step 4, Figure 17.7), and H(4′), H(5′), and H(6′) on the back place (L_2) (step 4, Figure 17.7). That spatial location corresponds in the case to the R-enantiomer [(−)-borneol].

The same methodology is followed for the assignment of any one of the other substrates included in Figure 17.6. As mentioned earlier, ^{13}C NMR chemical shifts can also be used and the [13]C $\Delta\delta^{RS}$ signs compared with the same tetrahedral model (Figure 17.6) used for ^{1}H NMR.

The validity of this method for the assignment of absolute configuration has been demonstrated with a series of substrates of known stereochemistry that follow the correlations described in Figure 17.6 and whose structures can be examined in the literature.[5,9–17]

The simplicity of this NMR procedure, the small amount of sample needed, and the possibility of recovering the sample after the assignment by hydrolysis of the derivatives make this procedure attractive especially in the search for bioactive natural products. In Figures 17.8–17.13, we show a selection of the bioactive compounds isolated from different marine and terrestrial sources whose absolute configuration has been assigned by comparison of the NMR of their (*R*)- and (*S*)-CDA derivatives following the lines of Figure 17.7 and the correlations expressed in Figure 17.6. The compounds in Figures 17.8–17.11 are secondary alcohols in all cases, those in Figure 17.12 are α-chiral primary amines, and in Figure 17.13, a selection of bioactive α-chiral primary alcohols and carboxylic acids are shown.

17.6 SINGLE DERIVATIZATION METHODS FOR ALCOHOLS AND AMINES

The preceding procedure requires the preparation of two derivatives of the chiral substrate. In the particular case of secondary alcohols and primary amines with an α-chiral carbon, it is possible to simplify the procedure and obtain the absolute configuration using only one derivative, either the (*R*)- or the (*S*)-CDA.[5,19–23] Naturally, in addition to the simplification in bench work, this may represent a great advantage when the amount of substrate is very small (for procedures requiring just mixing and shaking the substrate with resin-bound CDA in the NMR tube[8]).

There are two different ways to carry out the assignment using only one derivative of the substrate. The first one to be described is useful for secondary alcohols and requires the use of 9-AMA as CDA.[20] It is based on the analysis of the chemical shifts experienced by the substrate after esterification with either (*R*)- or (*S*)-9-AMA, and the only disadvantage is that 9-AMA is not commercially available and has to be prepared.

The second approach can be applied to the assignment of both secondary alcohols and α-chiral primary amines. It requires either (*R*)- or (*S*)-MPA as auxiliary reagent and is based on the controlled displacement of the conformational equilibrium produced by changing the NMR probe temperature[21] (for secondary alcohols) or by complexation with barium salts[22,23] (useful for both secondary alcohols and α-chiral primary amines).

17.6.1 Esterification Shifts: Assignment of Secondary Alcohols as 9-AMA Esters

This single derivative procedure[19,20] consists of the comparison of the chemical shifts observed for L_1/L_2 in the starting secondary alcohol and in its 9-AMA ester derivative [either the (*R*)- or the (*S*)-9-AMA]. The *L* substituent of the secondary alcohol located at the same side as the anthryl group of 9-AMA in the NMR representative conformer shows strong shielding, whereas the other *L* substituent suffers clearly smaller shielding. These differences are expressed in the (*R*)-9-AMA ester derivatives as $\Delta\delta^{AR}$ [$\Delta\delta L^{AR} = \delta L$ in the secondary alcohol minus δL in the (*R*)-9-AMA ester] or $\Delta\delta^{AS}$ if we use (*S*)-9-AMA [$\Delta\delta L^{AS} = \delta L$ in the alcohol minus δL in the (*S*)-9-AMA ester].

Figure 17.14 shows schematically the steps for assignment following this procedure and the correlation between the spatial arrangement of L_1/L_2 and the corresponding $\Delta\delta L^{AR}$ in the (*R*)-9-AMA ester derivatives [$\Delta\delta L^{AS}$ in the (*S*)-9-AMA derivatives].

An example illustrating the application of this procedure for the determination of the absolute configuration of (*R*)-3,3-dimethylbutan-2-ol is presented in Figure 17.15.

The ^{1}H NMR spectra of (*R*)-3,3-dimethylbutan-2-ol and of its (*R*)-9-AMA ester derivative are shown (steps 1 and 3, Figures 17.14 and 17.15). The signals important for the assignment (L_1/L_2) are those of the *t*-butyl group at one side of the asymmetric carbon and the Me(1′) at the other. Comparison of the two ^{1}H NMR spectra (step 4) shows that Me(1′) resonates virtually at the same position in the free alcohol and in the (*R*)-9-AMA ester ($\Delta\delta^{AR}$ 0.00 ppm), whereas the signal for the *t*-Bu protons shows a large difference (0.71 ppm). As $\Delta\delta^{AR}L_2 \ll \Delta\delta^{AR}L_1$ (step 5, Figures 17.14 and 17.15), the protons with the larger difference (*t*-Bu) should be placed at the front side of the 'tetrahedron' (L_1) and those with smaller difference [Me(1′)] at the back (L_2), leading to the (3*R*) configuration shown (Figure 17.15).

17.6.2 Low-temperature NMR for the Assignment of Secondary Alcohols as MPA Esters

This approach[19,21] is based on the comparison of the ^{1}H NMR spectra of one MPA ester derivative, either the (*R*)- or the (*S*)-MPA, of the *sec*-alcohol at two different temperatures.

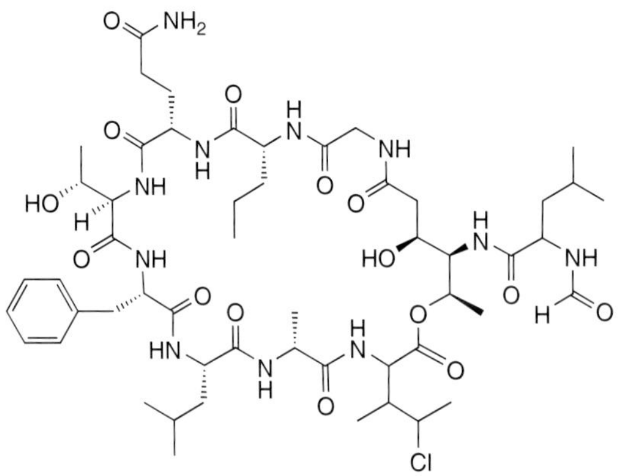

Cyclolithistide A, an antifungal cyclic depsipeptide from the marine sponge *Discodermia japonica* *J. Nat. Prod.*, 2014, **77**, 154

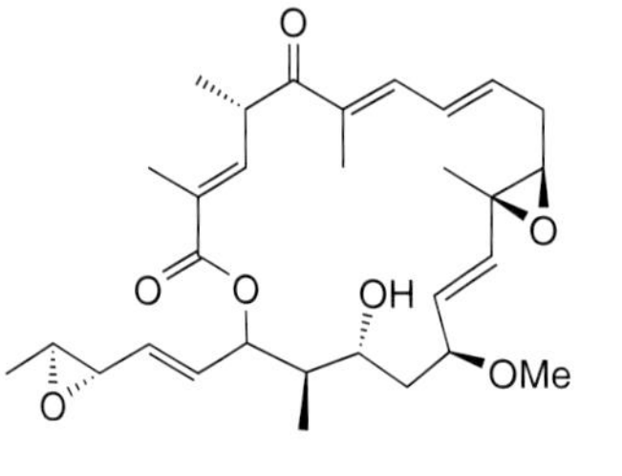

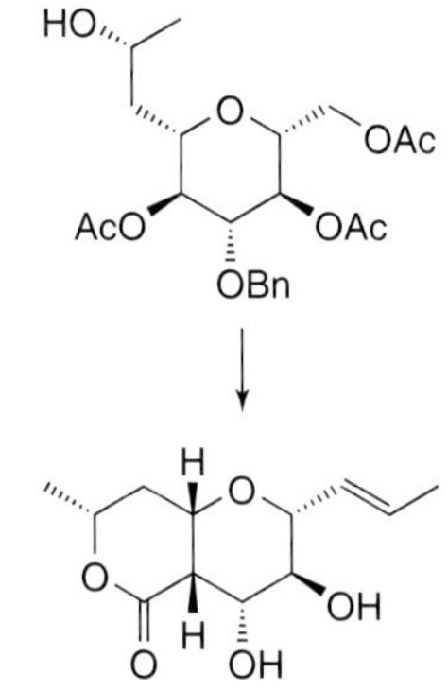

Dinemasone B, naturally occurring antibiotic with biological activities against *Bacillus megaterium* (Gram-positive), *Microbotryum violaceum* (fungus), and *Chlorella fusca* (alga). Isolated from the endophytic fungus *Dinemasporium strigosum. J. Org. Chem.* 2013, **78**, 9354

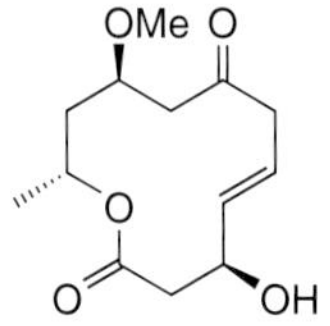

Maltepolides, with cytostatic activity on L929 mouse fibroblast cell lines. Isolated from myxobacterium *Sorangium cellulosum* Soce 1485. *Angew. Chem., Int. Ed.* 2013, **52** 5402

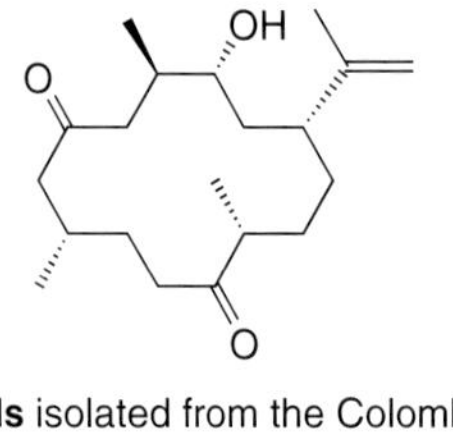

Dendrodolide B isolated from *Dendrodochium* sp., a fungus associated with the sea cucumber *Holothuria nobilis Selenka*. Growth inhibitory activity against SMMC-7721 and HCT116 cells. *J. Org. Chem.* 2013, **78**, 7030

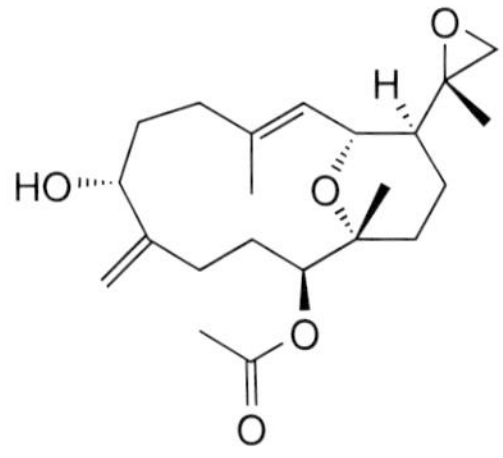

Cembranoids isolated from the Colombian Caribbean octocoral *Pseudoplexaura flagellosa*. Antifouling compounds *Tetrahedron*, 2011, **67**, 9112

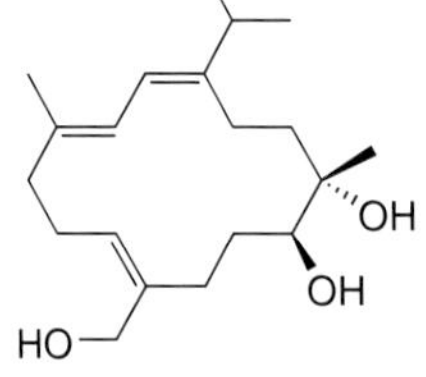

Cembradiene diterpenoid from the sea whip *Eunicea* sp. Antibacterial activity. *Tetrahedron Lett.*, 2011, **52**, 2515

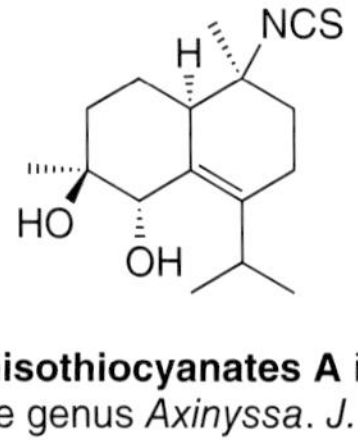

Sinulariols A, **D**, and **L**, cembrane-type diterpenoids from the chinese soft coral *Sinularia rigida*. Antifouling activity. *Tetrahedron*, 2011, **67**, 6018

Axinisothiocyanates A isolated from a sponge of the genus *Axinyssa. J. Nat. Prod.*, 2008, 608

Figure 17.8. Selected examples of bioactive natural products (chiral secondary alcohols) whose absolute configuration has been assigned by comparison of the NMR spectra of their (*R*)- and (*S*)-CDA ester derivatives

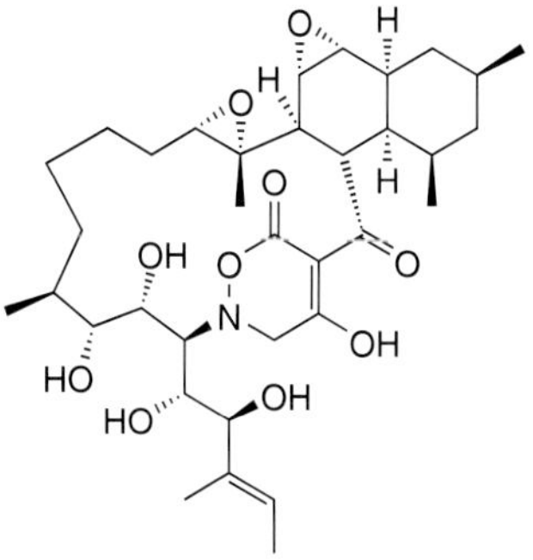

Alchivemycin A, polycyclic polyketide from the culture extract of a plant-derived actinomycete *Streptomyces* sp. Antimicrobial activity against *Micrococcus luteus*. *Org. Lett.*, 2010, **12**, 3402

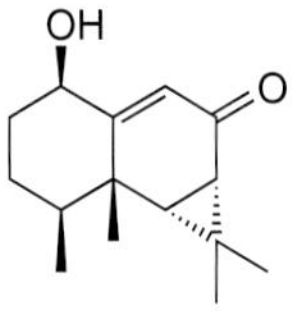

Axinysones A and **B**, sesquiterpenes isolated from the sponge *Axinyssa isabela*. *J. Nat. Prod.*, 2008, **71**, 2004

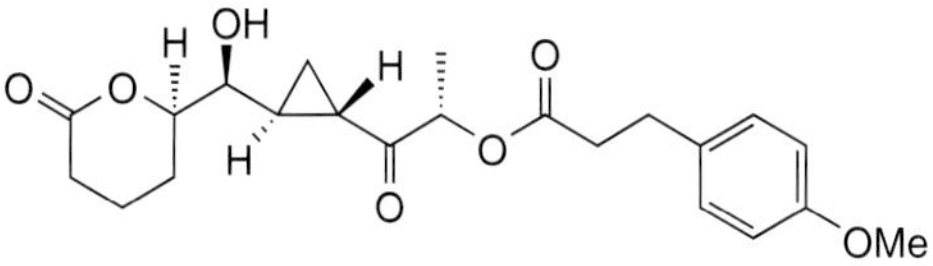

Bioactive **brevipolides A-J** from *Hyptis brevipes*. *J. Nat. Prod.*, 2013, **76**, 72

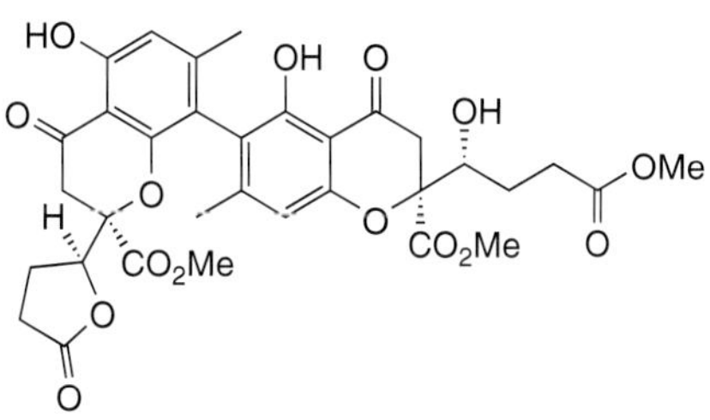

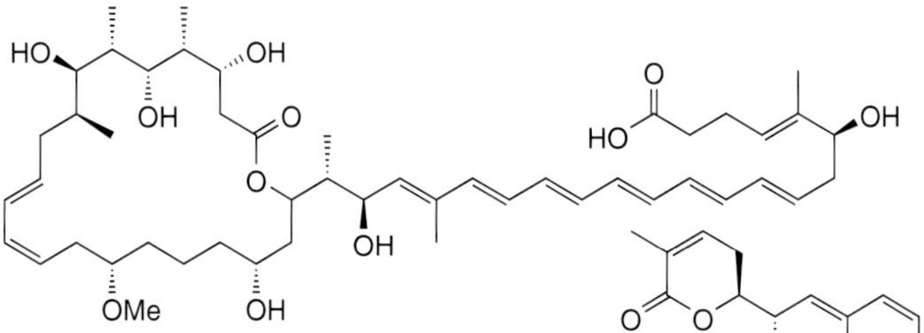

Etnangien, a macrolide antibiotic. *Org. Biomol. Chem.*, 2013, **11**, 2116

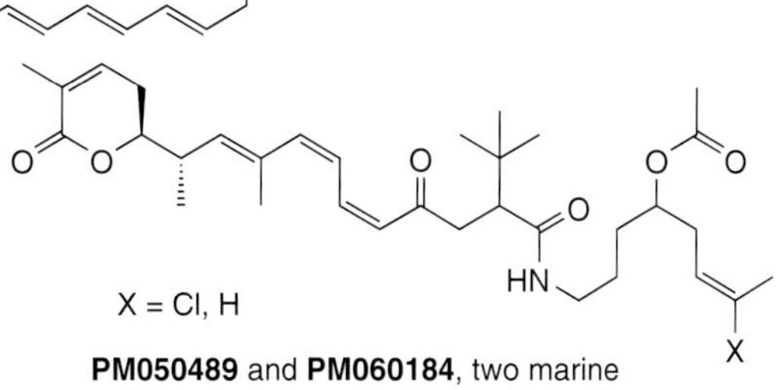

Noduliprevenone, heterodimeric chromanone with cancer chemopreventive potential. Isolated from the fungus *Nodulisporium* sp., an endophyte of a Mediterranean alga. *Chem. Eur. J.*, 2008, **14**, 9860

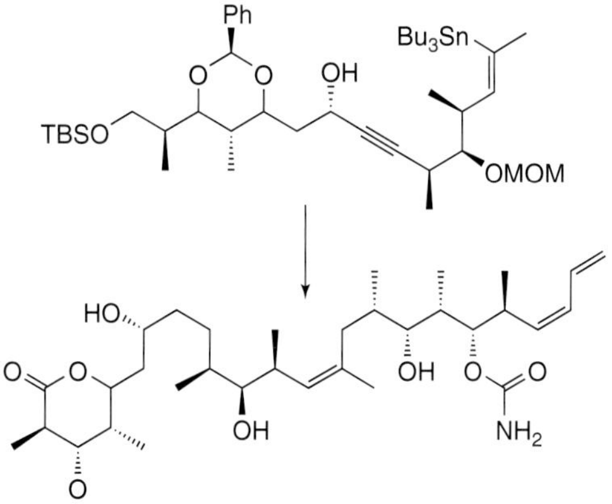

Discodermolide (DDM), marine anticancer polyketide from the Caribbean deepwater sponge *Discodermia dissoluta*. *Chem. Eur. J.*, 2008, **14**, 11092

X = Cl, H

PM050489 and **PM060184**, two marine anticancer compounds isolated from the Madagascan sponge *Lithoplocamia lithistoides*. *J. Am. Chem. Soc.* 2013, **135**, 10164

Figure 17.9. Selected bioactive chiral secondary alcohols isolated from natural sources whose absolute configuration has been assigned by comparison of the NMR of their (*R*)- and (*S*)-CDA derivatives

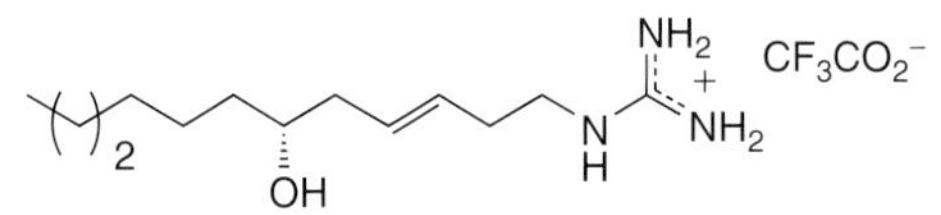

Pulchranin A, from the marine sponge *Monanchora pulchra*. Inhibitor of TRPV-1 channels. *Tetrahedron Lett.*, 2013, **54**, 1247

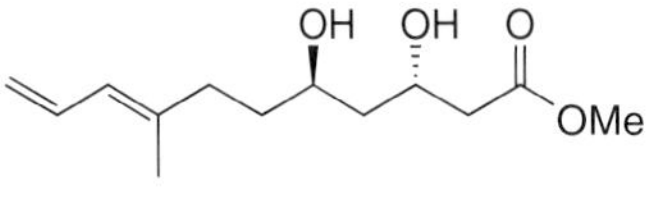

Ieodomycins A-D, antimicrobial marine metabolites isolated from *Bacillus* sp. *J. Org. Chem.*, 2013, **78**, 7274

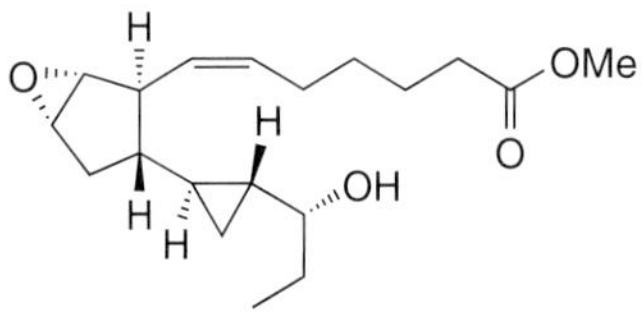

Cymatherol A, isolated from marine brown alga *Cymathere triplicata. Phytochemistry*, 2012, **73**, 134

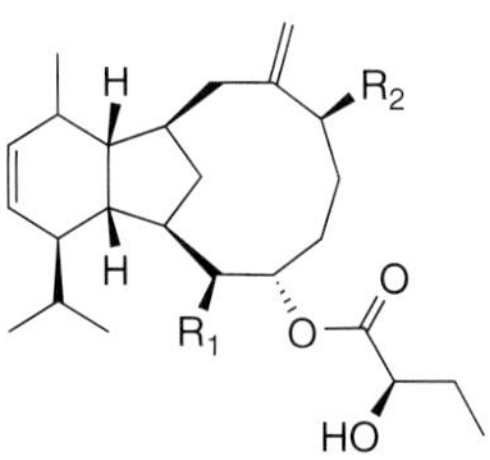

Cladielloides A (R_1 = OAc, R_2 = OH) and **B** (R_1 = OH, R_2 = OAc) with moderate cytotoxicity toward CCRF-CEM tumor cells. Isolated from the Indonesian octocoral, *Cladiella* sp. *Mar. Drugs*, 2010, **8**, 2936

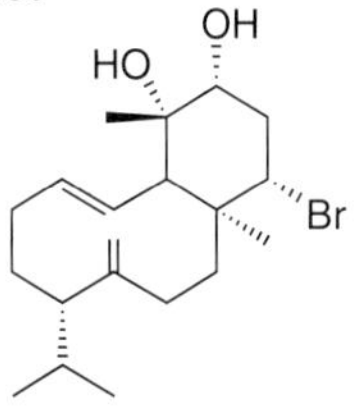

Bromocorodienol-antitumoral activity against four human apoptosis-resistant (U373, A549, SKMEL-28, OE21) and two human apoptosis-sensitive (PC-3, LoVo) cancer cell lines. Isolated from the organic extract of *Sphaerococcus coronopifol*ius. *Bioorg. Med. Chem.* 2010, **18**, 1321

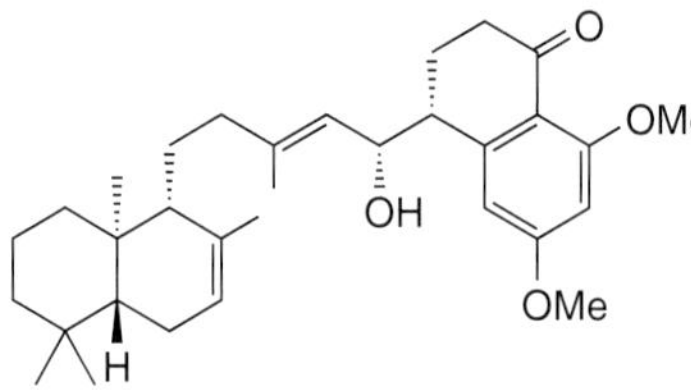

Actinoranone, a cytotoxic meroterpenoid from a marine adapted *Streptomyces* sp. *Org. Lett.*, 2013, **15**, 5400

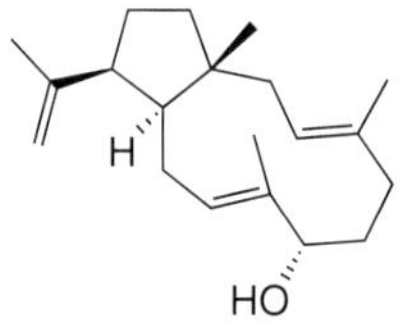

Antibacterial **Dolabellane** isolated from the brown alga *Dilophus spiralis. Eur. J. Org. Chem.*, 2012, **27**, 5177

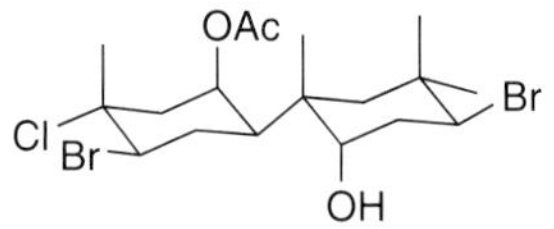

Halogenated metabolites isolated from the Brazilian red alga *Laurencia catarinensis*. Cytotoxicity against HT29, MCF7, and A431 cell lines. *J. Nat. Prod.*, 2010, **73**, 27

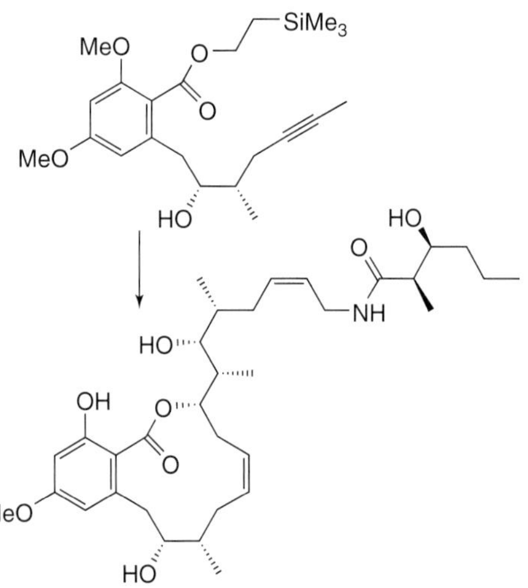

Cruentaren A, a highly cytotoxic F-ATPase inhibitor. *Chem. Eur. J.*, 2009, **15**, 12310

Figure 17.10. Selected examples of bioactive natural products (chiral secondary alcohols) whose absolute configuration has been assigned by comparison of the NMR spectra of their (*R*)- and (*S*)-CDA ester derivatives

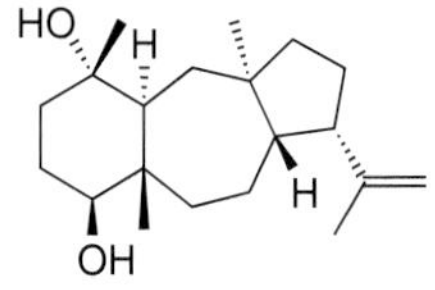

Dolastane from the brown alga *Dilophus spiralis*: Cytotoxic against L16 and A549. *Tetrahedron*, 2008, **64**, 3975

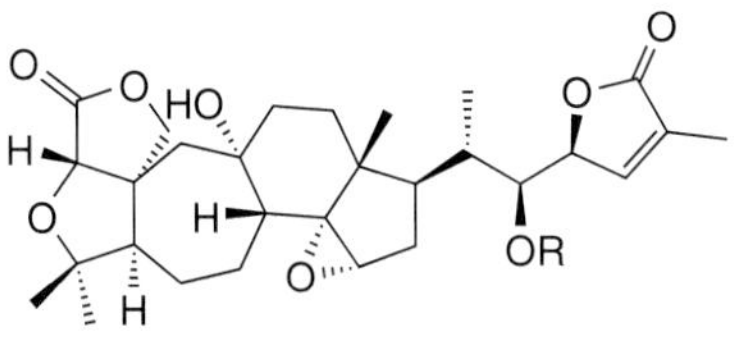

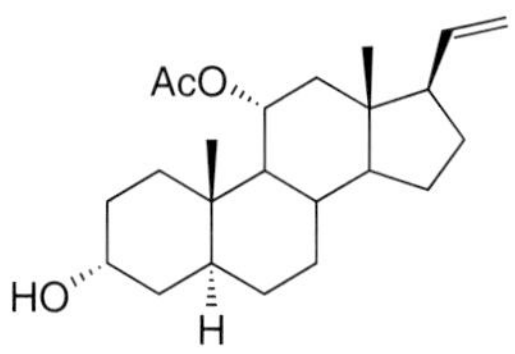

Pregnane steroid isolated from the gorgonian *Eunicella cavolini*. Partial growth inhibitory effects toward MCF-7 human breast cancer cells. *Tetrahedron*, 2008, **64**, 11797

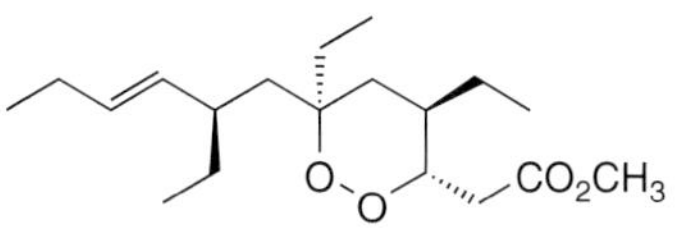

Micrandilactone B isolated from *Schisandra chinensis*. Weak anti-HIV-1 activity. *Org. Lett.* 2007, **9**, 2079

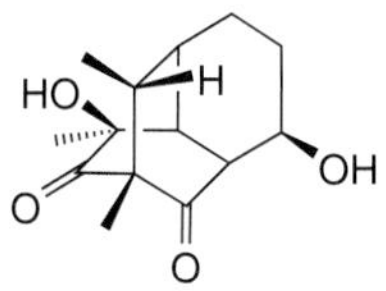

Plakortide Q, polyketide cycloperoxide from the Marine Sponge*Plakortis simplex*, with antimalarial activity. *Eur. J. Org. Chem*. 2005, 5077

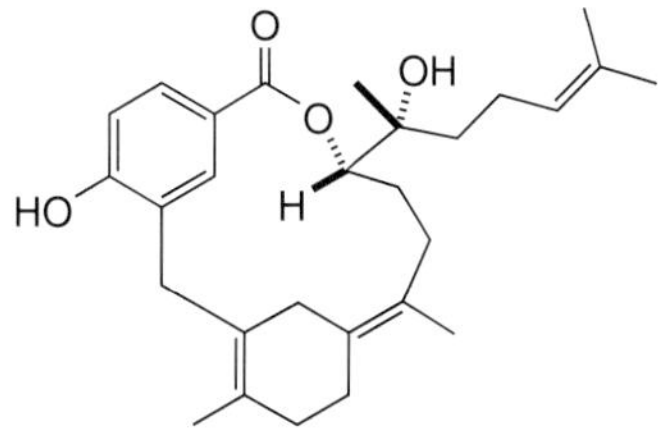

Penicillone B isolated from *Penicillium terrestre*. Cytotoxic activity toward P338 and A-549 cell lines. *Tetrahedron Lett*. 2005, **46**, 4993

Callophycolide A, an antimalarial meroditerpene isolated from the tropical red macroalgae *Callophycus serratus*. *Bioorg. Med. Chem. Lett*. 2010, **20**, 5662

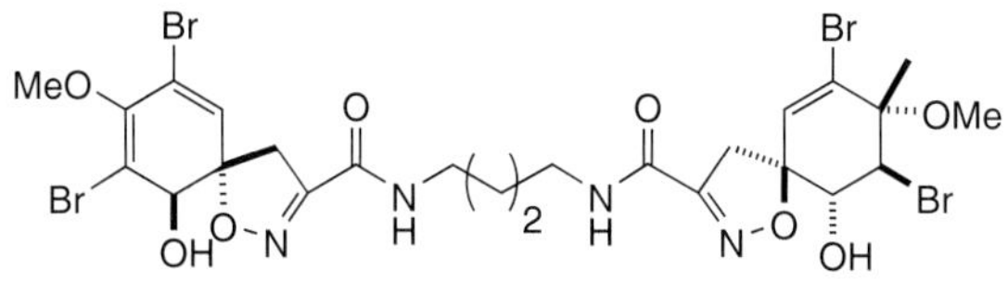

Aplysinone B, dibromotyrosine-derived metabolite, isolated from the sponge Aplysina gerardogreeni with cytotoxic activity against human tumor cell lines. *Bioorg. Med. Chem*. 2007, **15**, 5275

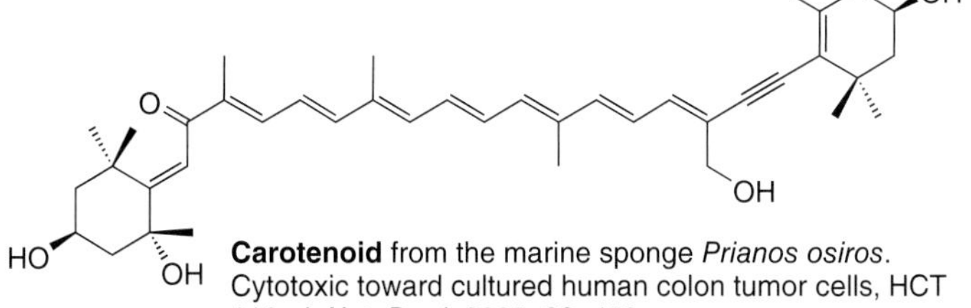

Carotenoid from the marine sponge *Prianos osiros*. Cytotoxic toward cultured human colon tumor cells, HCT 116. *J. Nat. Prod*. 2005, **68**, 450

Figure 17.11. Selected examples of bioactive natural compounds (chiral secondary alcohols) isolated whose absolute configuration has been assigned by comparison of the NMR of their (*R*)- and (*S*)-CDA derivatives

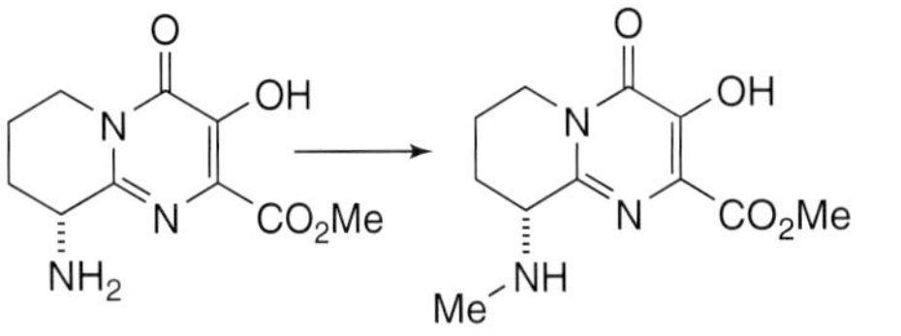

Tetrahydropyridopyrimidone, scaffold for HIV-1 integrase inhibitors. *Tetrahedron Lett.*, 2007, **48**, 6552

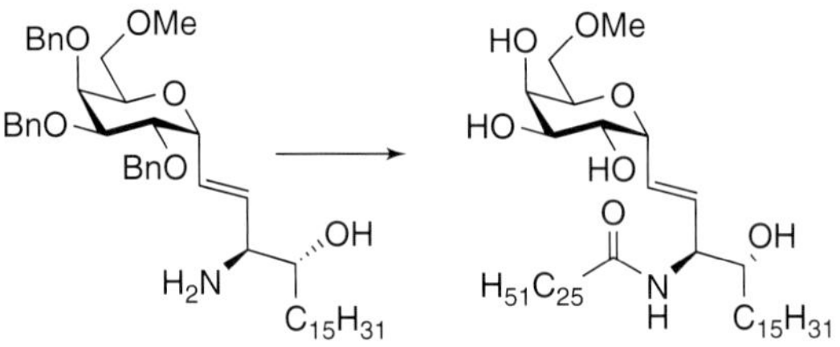

C-glycoside analog of *R*-galactosylceramide (R-GalCer, KRN7000). Glycolipid antigen for activating invariant natural killer T (iNKT) cells. *Org. Lett.*, 2012, **14**, 620

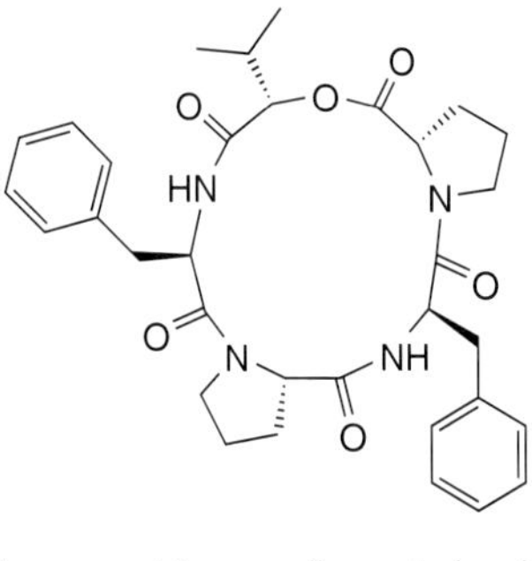

Alternaramide, a cyclic pentadepsipeptide from the marine-derived fungus *Alternaria* sp. SF-5016. Antibiotic activity against *Bacillus subtilis* and *Staphylococcus aureus*. *J. Nat. Prod.*, 2009, **72**, 2065

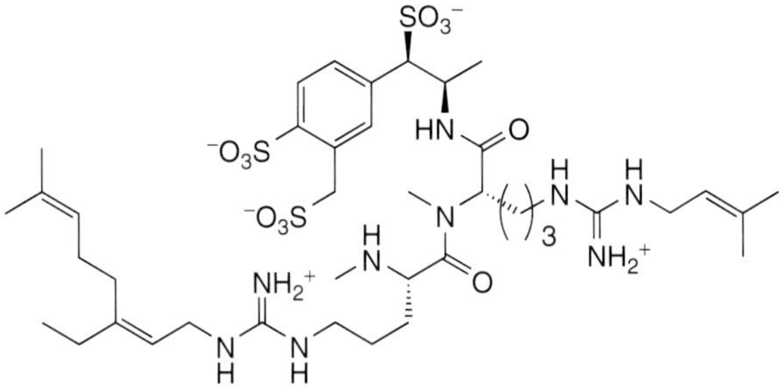

Aeruginoguanidine 98-A, from the cyanobacterium *Microcystis aeruginosa*. Cytotoxic against the P388 murine leukemia cells. *Tetrahedron*, 2002, **58**, 7645

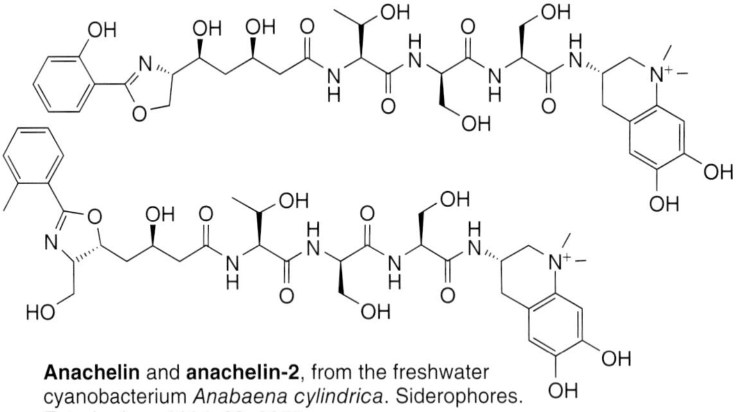

Anachelin and **anachelin-2**, from the freshwater cyanobacterium *Anabaena cylindrica*. Siderophores. *Tetrahedron*, 2004, **60**, 9075

Figure 17.12. Selected examples of bioactive natural compounds (α-chiral primary amines) whose absolute configuration has been assigned by comparison of the NMR spectra of their (*R*)- and (*S*)-CDA derivatives

α-Chiral primary alcohols

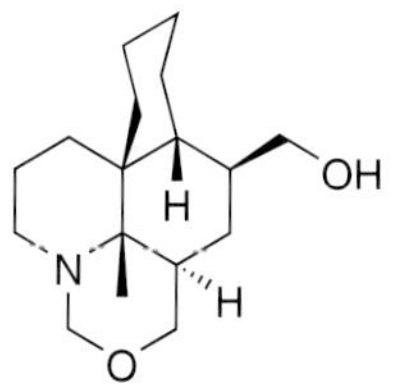

Myrioneurinol, from the leaves of *Myrioneuron nutans* with antimalarial activity against *Plasmodium falciparum. Tetrahedron: Asymmetry*, 2003, **14**, 503

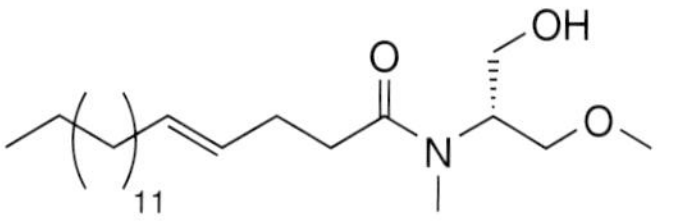

Serinolamide A, agonist of CB1 cannabinoid receptor. Isolated from the marine cyanobacteria *Lyngbya majuscula. J. Nat. Prod.*, 2011, **74**, 2313

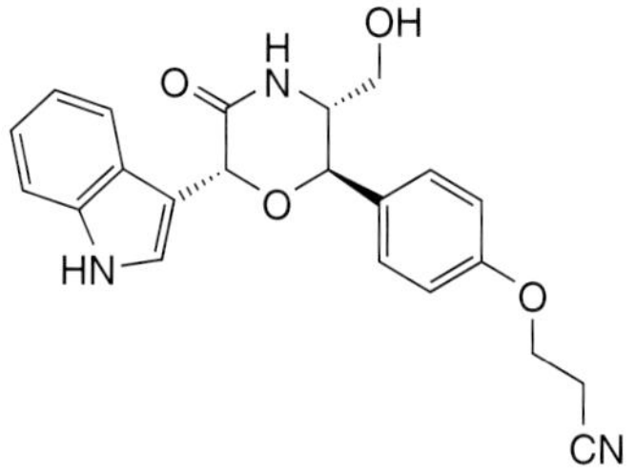

Oxazinin 1, from toxic mussels (*Mytilus galloprovincialis*). Cytotoxic. *Tetrahedron*, 2006, **62**, 7738; *Tetrahedron*, 2001, **57**, 8189

α-Chiral carboxylic acids

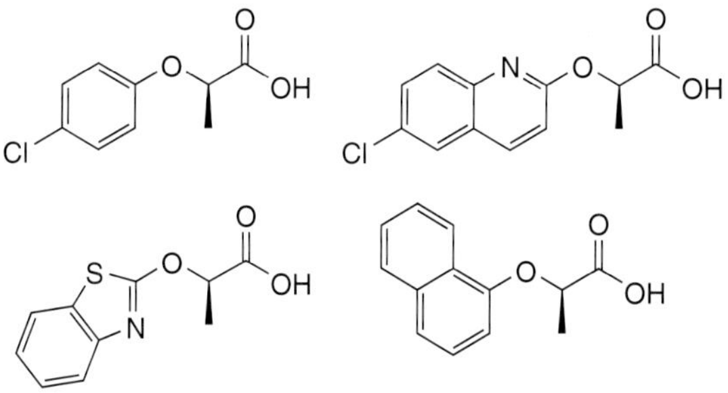

Chiral analogues of clofibric acid, the active metabolite of clofibrate. *Tetrahedron: Asymmetry*, 2008, **19**, 989

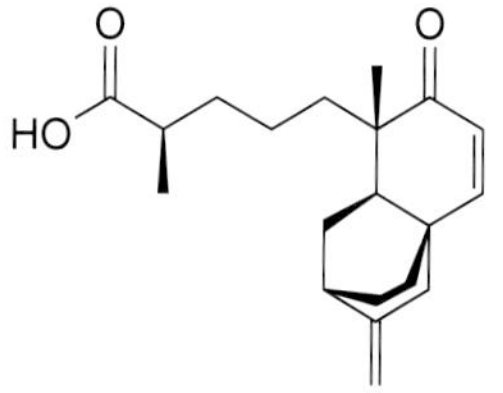

Platencin SL6, inhibitor of bacterial and mammalian fatty acid synthases. *J. Nat. Prod.*, 2012, **75**, 2158.

Figure 17.13. Selected examples of bioactive natural compounds (α-chiral primary alcohols and carboxylic acids) whose absolute configuration has been assigned by comparison of the NMR spectra of their (*R*)- and (*S*)-CDA derivatives

Figure 17.16 shows the idealized spectrum of the (*R*)-MPA ester derivative of a secondary alcohol taken at room (T_1) and at low temperature (T_2). A decrease in the temperature of the NMR probe leads to an increase of the population of the more stable conformer (*sp*) and therefore a greater shielding of the *L* group located in front of the phenyl ring (L_1, Figure 17.16a). On the other hand, the substituent located at the opposite site to the phenyl ring (L_2, Figure 17.16b) is deshielded at low temperature. As a result, L_1 and L_2 change their chemical shifts with the temperature and those differences ($\Delta\delta L^{T_1 T_2}$) reflect their spatial position around the asymmetric carbon. The $\Delta\delta L^{T_1 T_2}$ for L_1/L_2 are defined as $\Delta\delta L^{T_1 T_2} = \delta L^{T_1} - \delta L^{T_2}$ ($T_1 \gg T_2$). Figure 17.16(d) and (e) shows the correlation between the signs

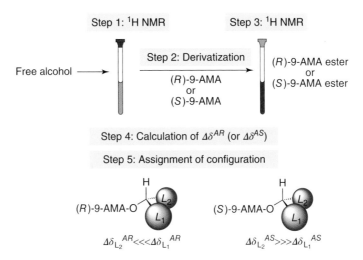

Figure 17.14. Main steps in the configurational assignment of secondary alcohols using the esterification shifts with 9-AMA

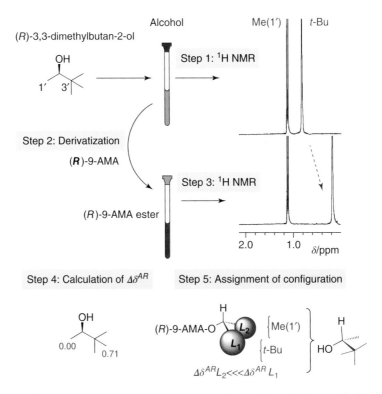

Figure 17.15. Assignment of the absolute configuration of (*R*)-3,3-dimethylbutan-2-ol as (*R*)-9-AMA ester by the esterification shifts NMR procedure

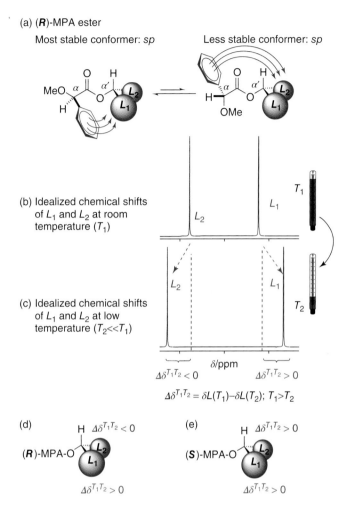

Figure 17.16. (a) Main conformers (*sp* and *ap*) in the equilibrium of the (*R*)-MPA ester of a secondary alcohol. (b, c) Simulated NMR spectrum of the (*R*)-MPA ester derivative of a secondary alcohol at room (T_1) and at low temperature (T_2). (d, e) Graphical correlation models

of $\Delta\delta L^{T_1 T_2}$ and the spatial position of L_1/L_2 in a secondary alcohol derivatized as (*R*)-MPA ester (Figure 17.16d) or as (*S*)-MPA ester (Figure 17.16e). In this way, the $\Delta\delta L^{T_1 T_2}$ signs are used to correlate the absolute configuration in basically the same way as $\Delta\delta^{RS}$ in the double derivatization procedures.

The step-by-step procedure is summarized in Figure 17.17. The starting point is the preparation of one MPA derivative of the secondary alcohol (step 1, Figure 17.17). A first NMR spectrum of this derivative should be taken at room temperature (step 2,

Figure 17.17) in CS_2/Cl_2CD_2 (4:1) as solvent, followed by a second spectrum of the same sample at lower temperature (usually 220–200 K is enough, step 3, Figure 17.17). The $\Delta\delta^{T_1 T_2}$ parameters are obtained (step 4, Figure 17.17) and their signs used to place L_1/L_2 around the asymmetric carbon according to the graphical correlation models shown in Figure 17.16(d) and (e) (step 5, Figure 17.17).

An example illustrating the application of this procedure for the determination of absolute configuration of (*R*)-butan-2-ol derivatized with (R)-MPA is presented in Figure 17.18.

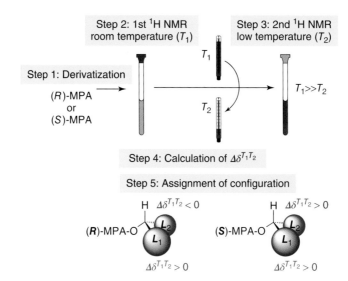

Figure 17.17. Main steps in the configurational assignment of secondary alcohols using the single derivatization procedure based on modification of the temperature

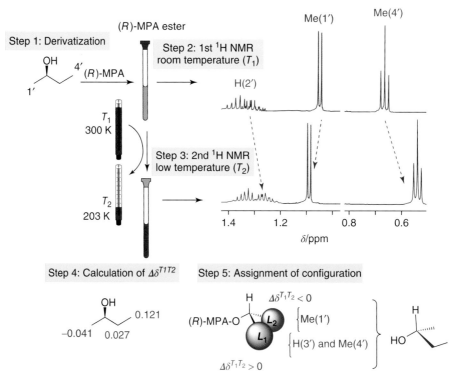

Figure 17.18. Main steps in the configurational assignment of (*R*)-butan-2-ol using (*R*)-MPA by the low-temperature NMR procedure

The (R)-MPA ester of a pure enantiomer of butan-2-ol is prepared (step 1, Figure 17.18) and the ^{1}H NMR spectra of the ester recorded at two different temperatures (300 and 203 K, steps 2 and 3, Figure 17.18). Assignment of the signals and comparison of the two ^{1}H NMR spectra (step 4, Figure 17.18) indicates that Me(1′) resonates at higher field at lower temperature and presents a negative $\Delta\delta^{T_1 T_2}$ value (−0.041 ppm), whereas H(3′)/Me(4′) shift to lower field and present positive $\Delta\delta^{T_1 T_2}$ values (0.027 and 0.121 ppm, respectively). Comparison with the graphical model for (R)-MPA esters (step 5, Figure 17.18) allows one to place H(3′)/Me(4′) instead of L_2 and Me(1′) instead of L_1, leading to the (R) absolute configuration shown in Figure 17.18.

17.6.3 Complexation Shifts: Assignment of Secondary Alcohols and Amines with MPA

The absolute configuration of secondary alcohols[19,22] and α-chiral primary amines[19,23] can also be assigned without changing the probe temperature, by addition of barium perchlorate to the NMR tube containing one MPA derivative in MeCN-d_3 as solvent. The basis for the method lies on the displacement of the conformational equilibrium by preferential complexation leading to changes in the chemical shifts of L_1/L_2 that correlate with their spatial location[19,22,23] (Figure 17.19).

This correlation is expressed through the difference between the chemical shifts of L_1/L_2 in the alcohol or amine, in the absence of barium salt minus the chemical shifts resulting by addition of the barium salt ($\Delta\delta^{Ba}$) to the NMR tube. Each substituent presents a sign of $\Delta\delta^{Ba}$ depending on its spatial location and the enantiomer of the MPA used.

The essentials of the procedure and the graphical models indicating the correlation between the $\Delta\delta^{Ba}$ signs and the spatial position of L_1/L_2 are summarised in Figure 17.19.

In the case of an (R)-MPA ester, the addition of Ba(ClO$_4$)$_2$ to the NMR tube shifts the conformational equilibrium in favor of the sp conformer and in consequence, the L substituent located in front of the phenyl ring of the (R)-MPA in the sp conformer is more shielded by addition of Ba^{2+} (L_1, positive $\Delta\delta^{Ba}$, Figure 17.19a), whereas the protons located in the other side will show a deshielding (L_2, negative $\Delta\delta^{Ba}$, Figure 17.19a).

The addition of the barium salt can be used for the assignment of absolute configuration of α-chiral primary amines too. In the (R)-MPA amides, the main conformer is the ap and the addition of Ba(ClO$_4$)$_2$ shifts the conformational equilibrium in favor of the sp conformer (Figure 17.19c). In this way, the shielding effects act exactly in the same way as in alcohols: a positive $\Delta\delta^{Ba}$ means that the L group concerned is on the same side as the phenyl ring in the sp conformer (L_1, Figure 17.19c), whereas the protons located in the other side will show a negative $\Delta\delta^{Ba}$ (L_2, Figure 17.19c).

Naturally, when (S)-MPA is used, the opposite set of signs result for the same amine (negative $\Delta\delta^{Ba}$ for L_1 and positive for L_2, Figure 17.19(b) and (d)).

In practice, the procedure for the assignment requires (Figure 17.20) preparation of one MPA derivative, the (R)- or the (S)- (step 1, Figure 17.20), and recording of its NMR spectrum in MeCN-d_3 (step 2, Figure 17.20), then solid Ba(ClO$_4$)$_2$ is added to the NMR tube and a new spectrum is taken (step 3, Figure 17.20). Comparison of the two ^{1}H NMR spectra leads to the $\Delta\delta^{Ba}$ signs for L_1/L_2 (step 4, Figure 17.20) that can be placed in their spatial location around the asymmetric carbon accordingly to the correlation models shown in Figure 17.19.

A scheme with the main steps for the configurational assignment of secondary alcohols and α-chiral primary amines using the complexation shifts is presented in Figure 17.20.

An example illustrating the application of this procedure to the assignment of an α-chiral primary amine, (−)-isopinocampheylamine, derivatized with (S)-MPA, is shown next.

Figure 17.21 presents the partial NMR spectra of the (S)-MPA amide of (−)-isopinocampheylamine in deuterated acetonitrile, before and after addition of barium perchlorate. The signals for protons located at the two sides of the asymmetric carbon – that serve as diagnostic signals for assignment – are clearly identified in both spectra [H(6′) and Me (8′)]. Comparison of the chemical shifts indicates that complexation moves Me(8′) to lower field (negative $\Delta\delta^{Ba}$) while H(6′) moves to higher field (positive $\Delta\delta^{Ba}$). Application of the correlation model (Figure 17.21d) places Me(8′) at the front side of the 'tetrahedron' (i.e., L_1) and protons H(6′) at the back side (i.e., L_2), leading to the stereochemistry shown in Figure 17.21.

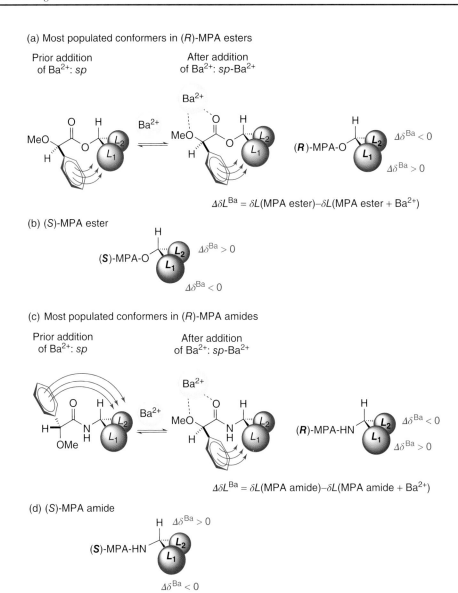

Figure 17.19. Conformational equilibria and $\Delta\delta^{Ba}$ signs for the MPA derivatives of secondary alcohols (a, b) and α-chiral primary amines (c, d)

In Figures 17.22 and 17.23, we present a selection of bioactive natural products (*sec*-alcohols and α-chiral *prim*-amines) whose absolute configuration has been determined using these single derivatization procedures. The structures of selected synthetic compounds used to validate the procedure are included in the original papers.[5,19–23]

17.7 THE ASSIGNMENT OF POLYFUNCTIONAL COMPOUNDS

Application of the CDA-based NMR-based methodology to the assignment of absolute configuration of compounds carrying more than one functional group

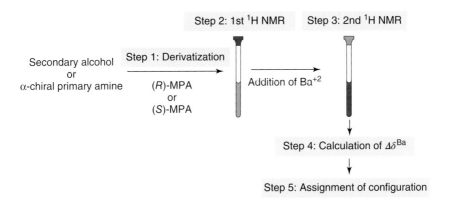

Figure 17.20.   Main steps in the configurational assignment of secondary alcohols and α-chiral primary amines using the complexation shifts

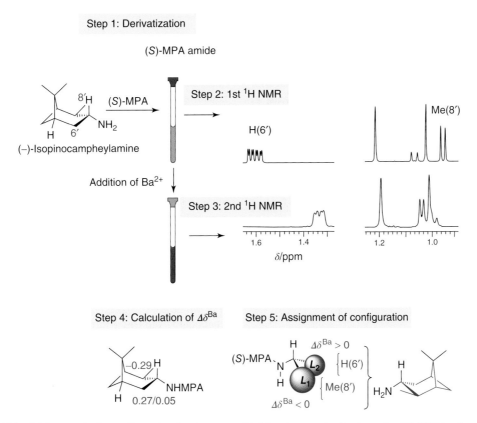

Figure 17.21.   Main steps in the configurational assignment of (−)-isopinocampheylamine using (*S*)-MPA by the complexation with Ba²⁺ NMR procedure

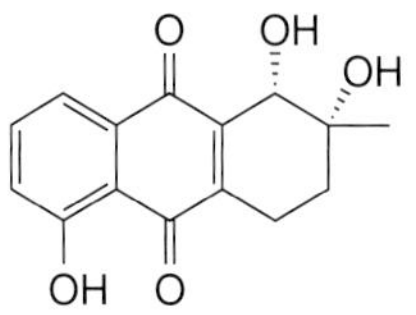

Altersolanol P, Gram-positive antibacterial activity against Gram-positive and growth inhibition of Gram-negative *Haemophilus influenzae*. *J. Nat. Prod.*, 2014, **77**, 497

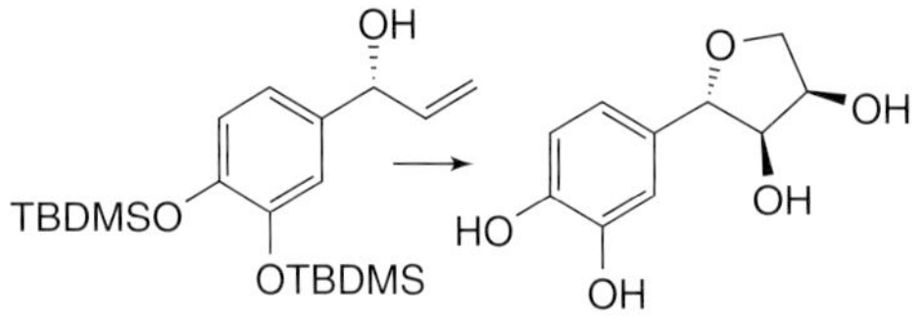

(−)-Gloeosporiol, natural antioxidant from the fungus *Colletotrichum gloeosporioides*. *Tetrahedron*, 2010, **66**, 8068

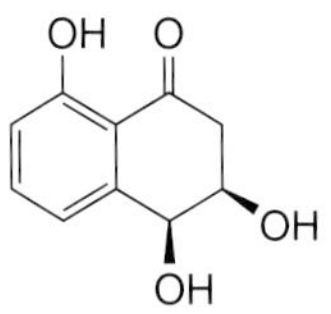

(+)-(3R,4S)-cis-4-hydroxy-6-deoxyscytalone, a phytotoxic metabolite isolated from *Colletotrichum acutatum*. Phytotoxic. *Tetrahedron*, 2009, **65**, 3392

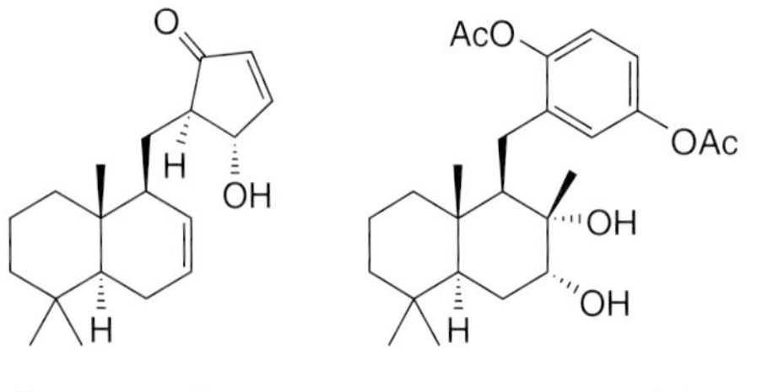

Merosesquiterpenes from two sponges of the genus *Dysidea*. *J. Nat. Prod.*, 2005, **68**, 653

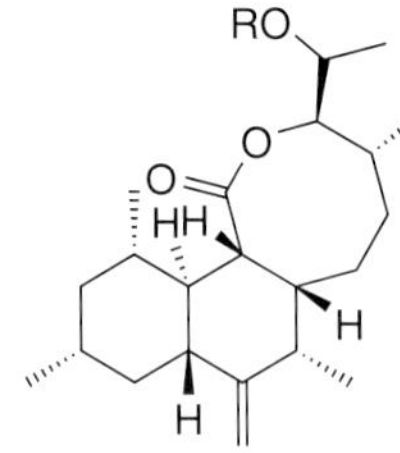

Satosporin A, R = β-D-Glu(1→3)-β-D-Glu
Satosporin B, R = β-D-Glu
Satosporin C, R = H

Glucosylated polyketides from the actinomycete *Kitasatospora griseola* MF730-N6. *Org. Lett.*, 2013, **15**, 3864

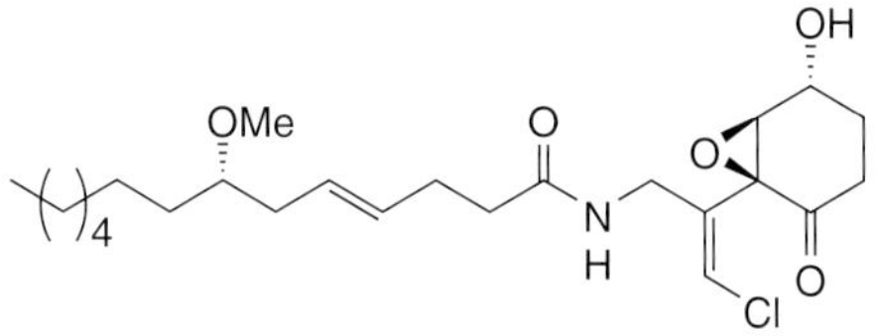

8-Epi-Malyngamide C, from the marine cyanobacterium *Lyngbya majuscula*, with moderate cytotoxicity to NCI-H460 human lung tumor and neuro-2a cancer cell lines. *Phytochemistry*, 2010, **71**, 1729

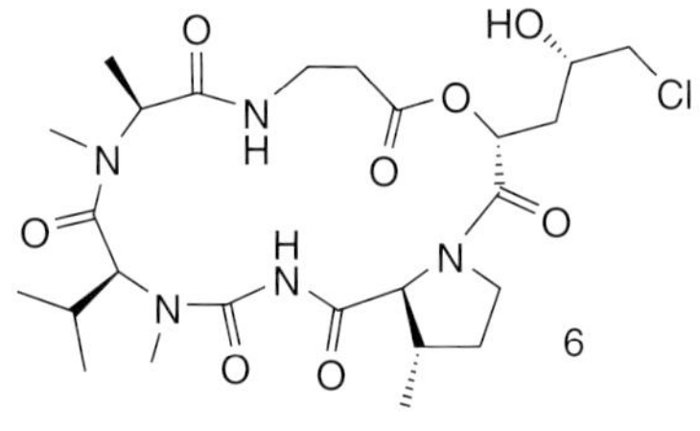

Destruxin E Chlorohydrin, from the marine fungus *Beauveria felina*. *J. Antibiot.*, 2006, **59**, 553

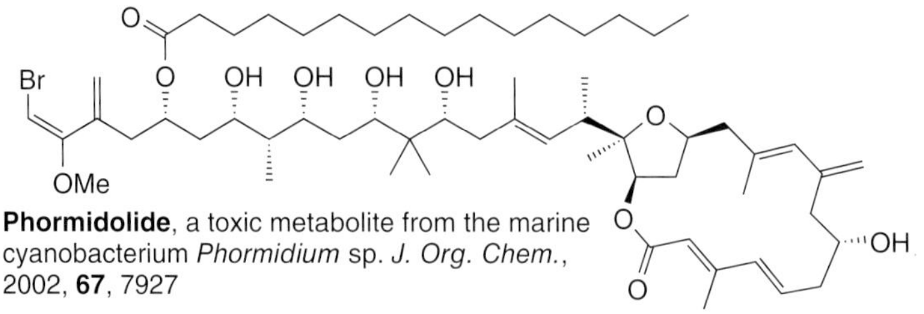

Phormidolide, a toxic metabolite from the marine cyanobacterium *Phormidium* sp. *J. Org. Chem.*, 2002, **67**, 7927

Figure 17.22. Selected examples of bioactive natural compounds (chiral secondary alcohols) whose absolute configuration has been assigned by low-temperature NMR of the MPA ester derivative

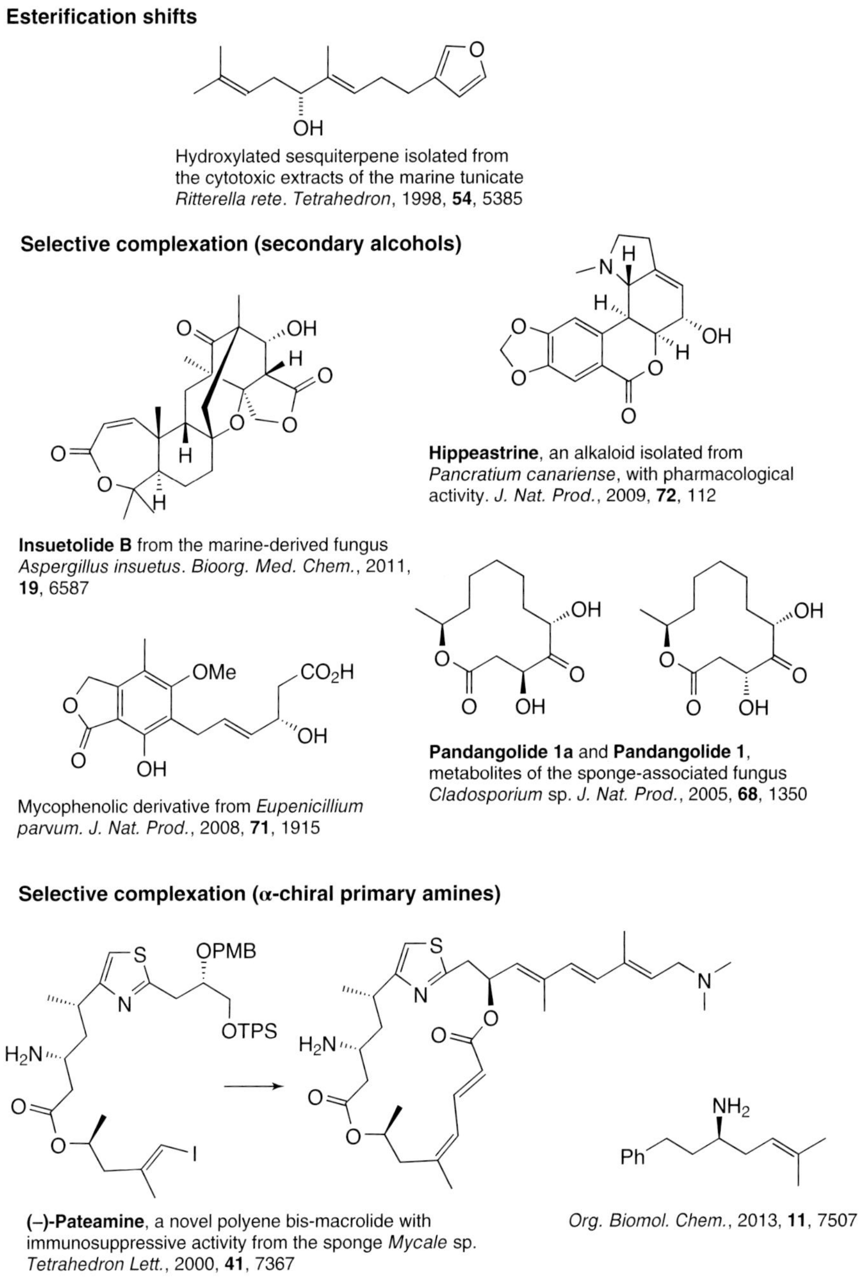

Esterification shifts

Hydroxylated sesquiterpene isolated from the cytotoxic extracts of the marine tunicate *Ritterella rete*. *Tetrahedron*, 1998, **54**, 5385

Selective complexation (secondary alcohols)

Hippeastrine, an alkaloid isolated from *Pancratium canariense*, with pharmacological activity. *J. Nat. Prod.*, 2009, **72**, 112

Insuetolide B from the marine-derived fungus *Aspergillus insuetus*. *Bioorg. Med. Chem.*, 2011, **19**, 6587

Pandangolide 1a and **Pandangolide 1**, metabolites of the sponge-associated fungus *Cladosporium* sp. *J. Nat. Prod.*, 2005, **68**, 1350

Mycophenolic derivative from *Eupenicillium parvum*. *J. Nat. Prod.*, 2008, **71**, 1915

Selective complexation (α-chiral primary amines)

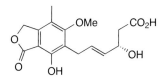

(–)-**Pateamine**, a novel polyene bis-macrolide with immunosuppressive activity from the sponge *Mycale* sp. *Tetrahedron Lett.*, 2000, **41**, 7367

Org. Biomol. Chem., 2013, **11**, 7507

Figure 17.23. Selected examples of bioactive natural compounds (chiral secondary alcohols and α-chiral primary amines) whose absolute configuration has been assigned by analysis of the esterification shift of the 9-AMA ester derivatives or by complexation of the MPA ester or amide derivative

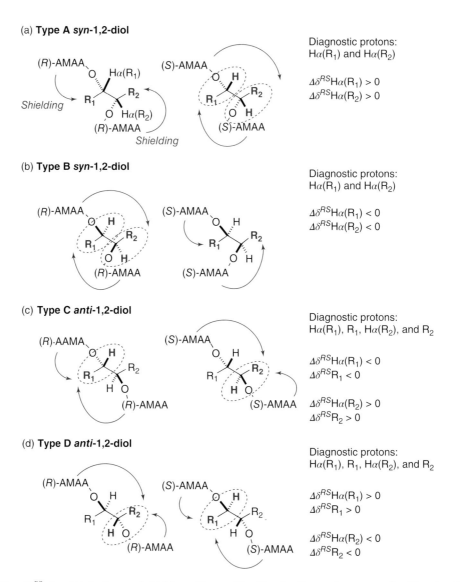

Figure 17.24. $\Delta\delta^{RS}$ sign distributions and shielding effects suffered by the diagnostic protons of a difunctional compound (*sec/sec*-1,2-diol) when it is derivatized with two auxiliary units

derivatizable with the CDA (i.e., diols, aminoalcohols, etc.) deserves certain considerations. The first is that the models expressing the correlation between the chemical shifts and the spatial location of the substituents in monofunctional compounds cannot be applied when more than one CDA unit is present in the derivative.[24]

This is easily understood when we examine a *sec/sec*-diol with two asymmetric carbons as an example. The simultaneous derivatization of the two hydroxyl groups with MPA as auxiliary produces the corresponding bis-(*R*)- and bis-(*S*)-MPA esters (Figure 17.24). These two derivatives have two auxiliary units (MPA) in their structures; therefore, there are two phenyl rings producing independent shieldings on the rest of the molecule. As a result, the shifts observed for L_1/L_2 are now the result of the combination of those two shieldings, as opposed to that produced by a single aryl ring as in the monofunctional compounds discussed earlier (Figure 17.24). In addition, the signals important for the assignment are not in all cases the same as in monofunctional substrates.

The second aspect to be noted is that while in the monofunctionalized derivatives of Figure 17.5, there are two stereochemical possibilities (two enantiomers) that correlate with two different $\Delta\delta^{RS}$ sign distributions, in a *sec/sec*-diol, with two asymmetric carbons, there are four possible stereochemistries; therefore, the NMR method should provide four different and specific $\Delta\delta^{RS}$ sign distributions.

In Figure 17.24, we represent how the shieldings produced by the two auxiliary units affect the rest of the molecule depending on the stereochemistry. Thus, in a type A 1,2-diol (Figure 17.24a), $\underline{H}\alpha(R_1)$ and $\underline{H}\alpha(R_2)$ are shielded in the bis-(*S*)-AMAA ester but

not in the bis-(*R*)-AMAA ester; therefore, their $\Delta\delta^{RS}$ is positive. In a type B 1,2-diol (Figure 17.24b), $\underline{H}\alpha(R_1)$ and $\underline{H}\alpha(R_2)$ are shielded in the bis-(*R*)-AMAA ester but not in the bis-(*S*)-AMAA ester, and present a negative $\Delta\delta^{RS}$.

For type C diols (Figure 17.24c), R_1 and $\underline{H}\alpha(R_1)$ are more shielded in the bis-(*R*)-AMAA ester, whereas R_2 and $\underline{H}\alpha(R_2)$ are more shielded in the bis-(*S*)-AMAA ester. Therefore, the $\Delta\delta^{RS}$ for R_1 and $\underline{H}\alpha(R_1)$ are negative, whereas those for R_2 and $\underline{H}\alpha(R_2)$ are positive.

Finally, for type D diols (Figure 17.24d), $\Delta\delta^{RS}$ for R_2 and $\underline{H}\alpha(R_2)$ are negative, whereas those due to R_1 and $\underline{H}\alpha(R_1)$ are positive.

In addition to *sec/sec*-diols, several other polyfunctional compounds follow a correlation between the NMR spectra of their CDA derivatives and the absolute configuration. These correlations have been validated with series of substrates of known

$\Delta\delta^{RS}$ ($\Delta\delta^{SR}$) **Sign distribution**

(a) *sec/sec*-1,2- and 1,*n*-diols

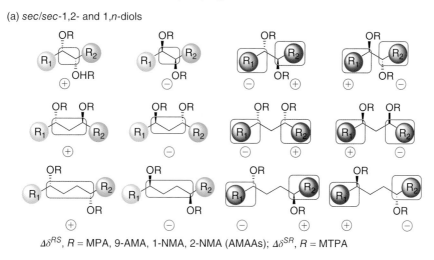

$\Delta\delta^{RS}$, R = MPA, 9-AMA, 1-NMA, 2-NMA (AMAAs); $\Delta\delta^{SR}$, R = MTPA

(b) *sec/sec*-1,2-amino alcohols

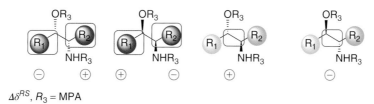

$\Delta\delta^{RS}$, R_3 = MPA

Figure 17.25. $\Delta\delta^{RS}$ ($\Delta\delta^{SR}$) sign distributions in every isomer of (a) acyclic *sec/sec*-1,2-diols and *sec/sec*-1,*n*-diols and (b) acyclic *sec/sec*-1,2-amino alcohols. The signals for stereochemical diagnosis are those of $\underline{H}\alpha(R_1)$ and $\underline{H}\alpha(R_2)$, R_1, and R_2

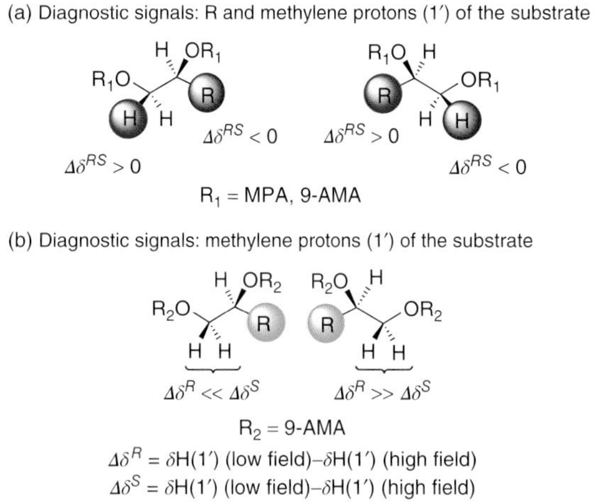

(a) Diagnostic signals: R and methylene protons (1′) of the substrate

$\Delta\delta^{RS} < 0$ $\Delta\delta^{RS} > 0$

$\Delta\delta^{RS} > 0$ $\Delta\delta^{RS} < 0$

R_1 = MPA, 9-AMA

(b) Diagnostic signals: methylene protons (1′) of the substrate

$\Delta\delta^R \ll \Delta\delta^S$ $\Delta\delta^R \gg \Delta\delta^S$

R_2 = 9-AMA

$\Delta\delta^R = \delta H(1')$ (low field)$-\delta H(1')$ (high field)
$\Delta\delta^S = \delta H(1')$ (low field)$-\delta H(1')$ (high field)

Figure 17.26. $\Delta\delta^{RS}$ signs, $\Delta\delta^R$, and $\Delta\delta^S$ values specific for each isomer of *prim/sec*-1,2-diols. The assignment is carried out by double derivatization using as signal for diagnosis (a) R and the methylene protons of the substrate and (b) the separation of the methylene protons of the substrate

stereochemistry and by theoretical and experimental studies. Overall, the absolute configuration of *sec/sec*-1,2-diols,[4,25] *sec/sec*-1,2-amino alcohols[4,26]

with two asymmetric carbons; *prim/sec*-1,2-diols,[4,27] *prim/sec*-1,2-aminoalcohols and *sec/prim*-1,2-aminoalcohols[4,28,29] with one asymmetric carbon but two derivatizable functional groups; and *prim/sec/sec*-1,2,3-triols[4,30] with three derivatizable groups and two asymmetric carbons can be determined by NMR of adequate CDA derivatives.

Relevant information for the NMR assignment by double derivatization of those classes of compounds is enclosed in Figures 17.25–17.28. The general structures as well as the sign distribution of the diagnostic signals, in the CDA derivatives of acyclic *sec/sec*-1,2-diols and *sec/sec*-1,n-diols, are shown in Figure 17.25(a); those for acyclic *sec/sec*-1,2-amino alcohols are shown in Figure 17.25(b); for *prim/sec*-1,2-diols in Figure 17.26; for *prim/sec*-1,2-aminoalcohols and *sec/prim*-1,2-aminoalcohols in Figure 17.27; and for the assignment of *prim/sec/sec*-1,2,3-triols the information is shown in Figure 17.28.

From a practical viewpoint, the assignment of the absolute configuration of polyfunctional compounds comprises (i) the preparation of the two bis-CDA derivatives (tris-CDA esters in the case of triols) by simultaneous derivatization with the CDA of all the functional groups of the substrate,

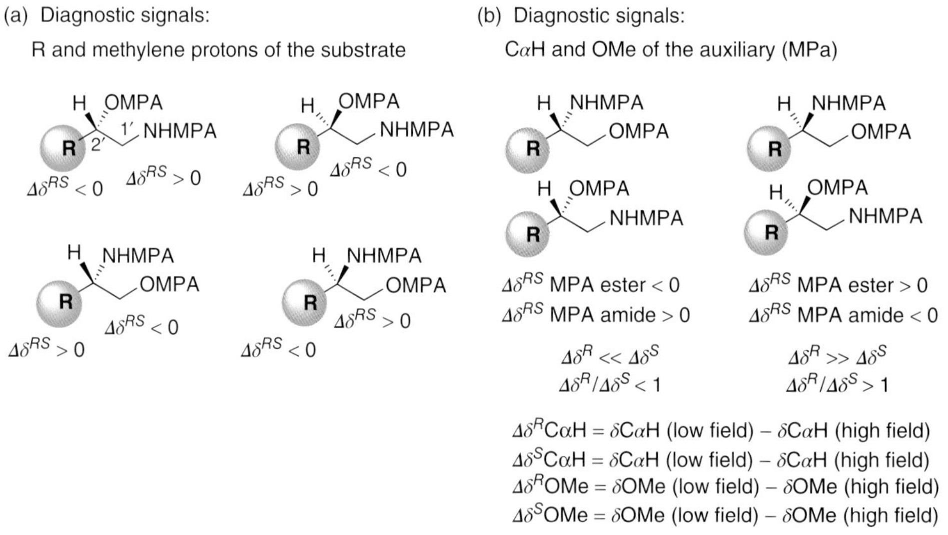

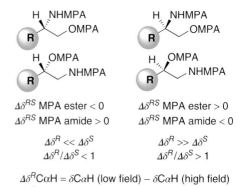

(a) Diagnostic signals:

R and methylene protons of the substrate

$\Delta\delta^{RS} < 0$ $\Delta\delta^{RS} > 0$ $\Delta\delta^{RS} > 0$ $\Delta\delta^{RS} < 0$

$\Delta\delta^{RS} > 0$ $\Delta\delta^{RS} < 0$ $\Delta\delta^{RS} < 0$

(b) Diagnostic signals:

CαH and OMe of the auxiliary (MPa)

$\Delta\delta^{RS}$ MPA ester < 0 $\Delta\delta^{RS}$ MPA ester > 0
$\Delta\delta^{RS}$ MPA amide > 0 $\Delta\delta^{RS}$ MPA amide < 0

$\Delta\delta^R \ll \Delta\delta^S$ $\Delta\delta^R \gg \Delta\delta^S$
$\Delta\delta^R/\Delta\delta^S < 1$ $\Delta\delta^R/\Delta\delta^S > 1$

$\Delta\delta^R$CαH = δCαH (low field) $-$ δCαH (high field)
$\Delta\delta^S$CαH = δCαH (low field) $-$ δCαH (high field)
$\Delta\delta^R$OMe = δOMe (low field) $-$ δOMe (high field)
$\Delta\delta^S$OMe = δOMe (low field) $-$ δOMe (high field)

Figure 17.27. $\Delta\delta^{RS}$ signs, $\Delta\delta^R$, and $\Delta\delta^S$ values, specific for each stereoisomer of *prim/sec*- and *sec/prim*-1,2-amino alcohols. The assignment is carried out by double derivatization using (a) the protons of the substrate and (b) the protons of the auxiliary reagent, as diagnostic signals

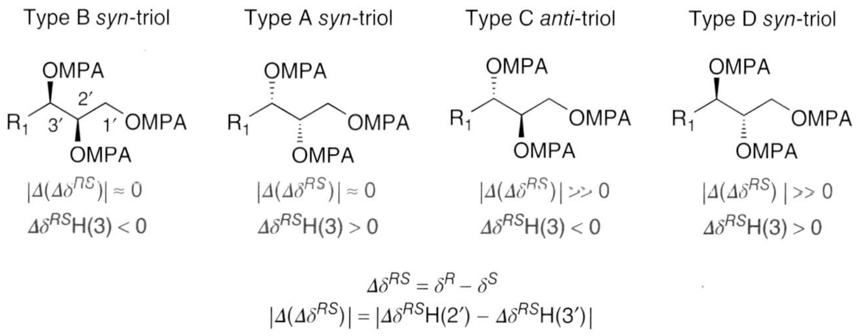

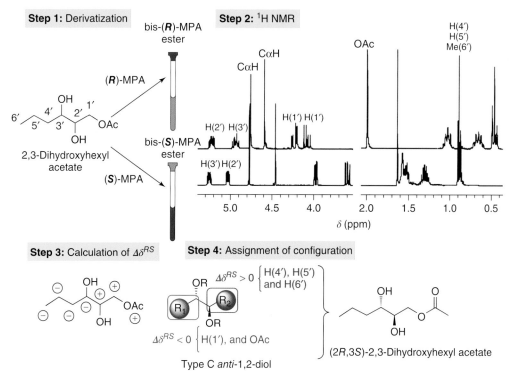

Figure 17.28. $\Delta\delta^{RS}$ sign of H(3′) and $|\Delta(\Delta\delta^{RS})|$ values for the assignment of *prim/sec/sec*-1,2,3-triols by double derivatization

Figure 17.29. Main steps in the absolute configuration assignment of 2,3-dihydroxyhexyl acetate using double derivatization with MPA

(ii) comparison of the corresponding NMR spectra and calculation of the chemical shift differences ($\Delta\delta^{RS}$, $\Delta\delta^{SR}$, $\Delta\delta^{R}$, $\Delta\delta^{S}$, and $|\Delta(\Delta\delta^{RS})|$), and (iii) comparison of the sign distributions with those shown in Figures 17.25–17.28 that represent the correlation

NMR/stereochemistry for every case and leads to the absolute configuration.

An example illustrating the procedure is shown next with the assignment of the absolute configuration of the *sec/sec*-1,2-diol 2,3-dihydroxyhexyl acetate.

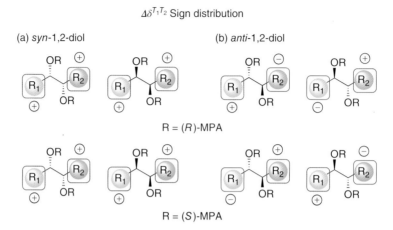

Figure 17.30. $\Delta\delta^{T_1 T_2}$ sign distributions for the assignment of *syn* (a) and *anti* (b) acyclic *sec/sec*-1,2-diols by single derivatization. Diagnostic signals correspond to R$_1$ and R$_2$

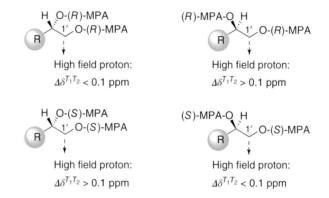

Figure 17.31. $\Delta\delta^{T_1 T_2}$ sign distributions for the assignment of *prim/sec*-1,2-diols by single derivatization. Diagnostic signals correspond to the methylene higher field proton, H(1')

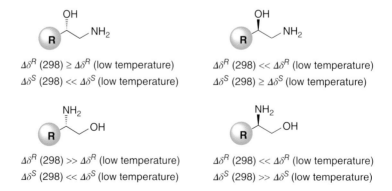

Figure 17.32. $\Delta\delta^R$ and $\Delta\delta^S$ values for the assignment of *prim/sec*- and *sec/prim*-1,2-amino alcohols by single derivatization. The diagnostic signals are those of the CαH of the MPA units

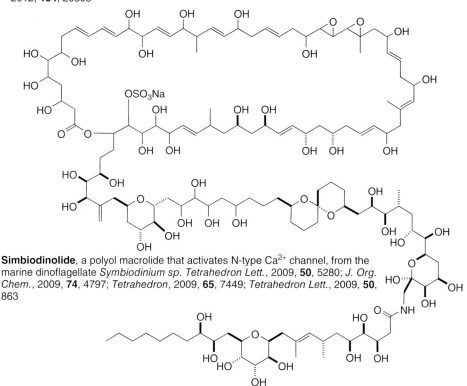

Piperidine alkaloid, from the whole plants of *Lobelia chinensis. Fitoterapia,* 2014, **93**,168

Falcarindiol, with antimycobacterial activity from the roots of *Heracleum maximum. Journal of Ethnopharmacology,* 2013, **147**, 232

Thermolide D, nematocidal PKS-NRPS metabolite from the thermophilic fungus *Talaromyces thermophilus. J. Am. Chem. Soc.,* 2012, **134**, 20306

Mycangimycin, with antifungal actifungal activity from *Streptomyces* sp. *Org. Lett.,* 2009, **11**, 633

Simbiodinolide, a polyol macrolide that activates N-type Ca²⁺ channel, from the marine dinoflagellate *Symbiodinium sp. Tetrahedron Lett.*, 2009, **50**, 5280; *J. Org. Chem.*, 2009, **74**, 4797; *Tetrahedron*, 2009, **65**, 7449; *Tetrahedron Lett.*, 2009, **50**, 863

Figure 17.33. Selection of *sec/sec*-1,2- and *sec/sec*-1,*n*-diols diols of natural origin, whose absolute configuration has been determined by NMR of their CDA derivatives

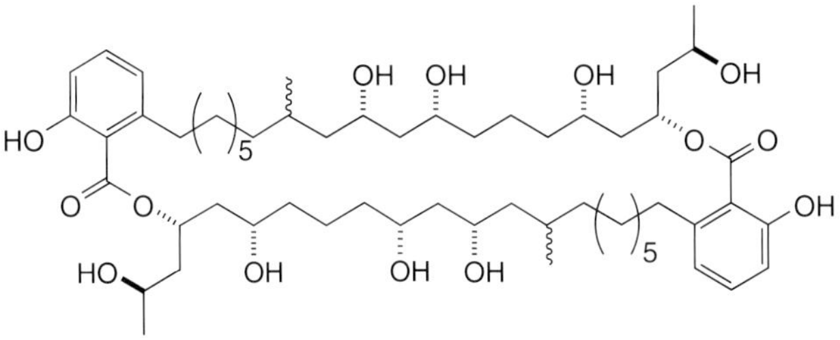

Marinomycins A, antitumor-antibiotics from a marine actinomycete of the genus 'Marinispora'. *J. Am. Chem. Soc.*, 2006, **128**, 1622

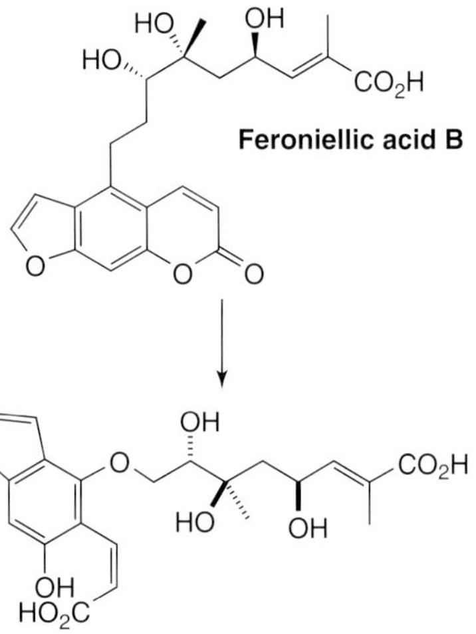

Achaetolide

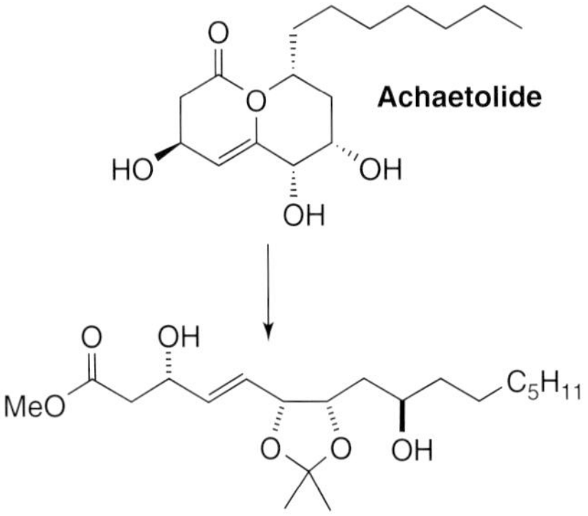

Achaetolide, isolated from a fermentation broth of *Ophiobolus* sp. *Tetrahedron,* 2009, **65**, 7464

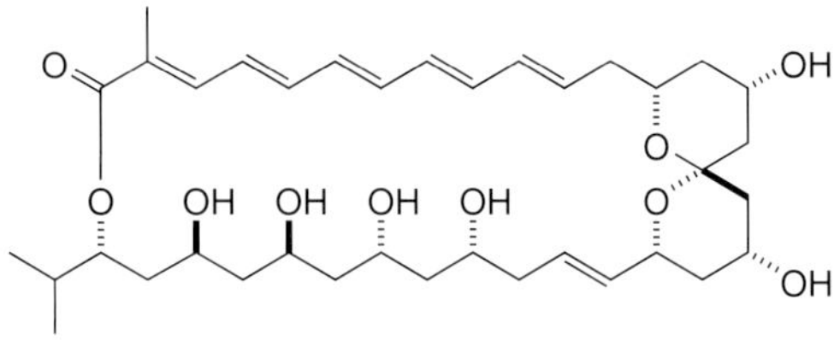

Feroniellic acid B

Feroniellic acid B from *Feroniella lucida*. *Tetrahedron Lett.*, 2008, **49**, 3133

Marinispolide A from a marine actinomycete of the genus *Marinispora*. *J. Org. chem.*, 2009 **74**, 675

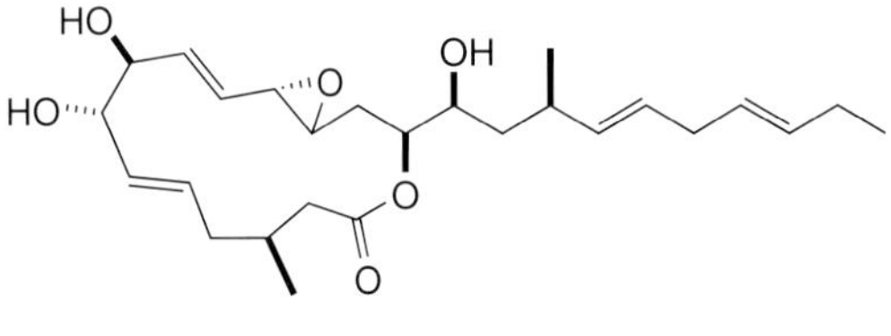

Iriomoteolide-3a isolated from a marine benthic dinoflagellate *Amphidinium* sp. Cytotoxic activity against tumor cells. *J. Org. Chem.*, 2008, **73**, 1567

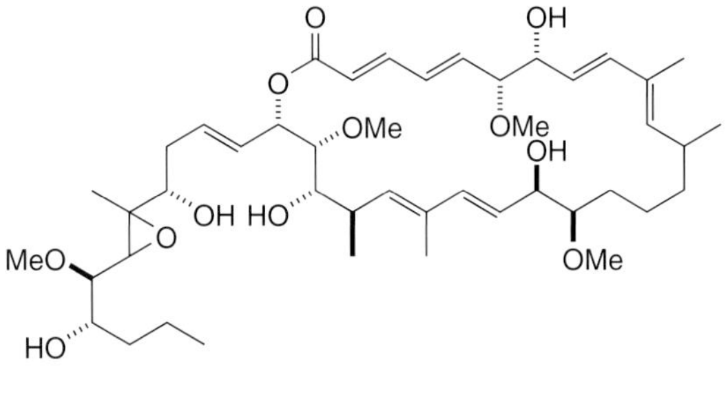

Arenicolide A, from the marine actinomycete *Salinispora arenicola. J. Org. Chem.*, 2007, **72**, 5025

Figure 17.34. Selection of *sec/sec*-1,2- and *sec/sec*-1,*n*-diols isolated from natural sources and whose absolute configuration has been determined by NMR of their CDA derivatives

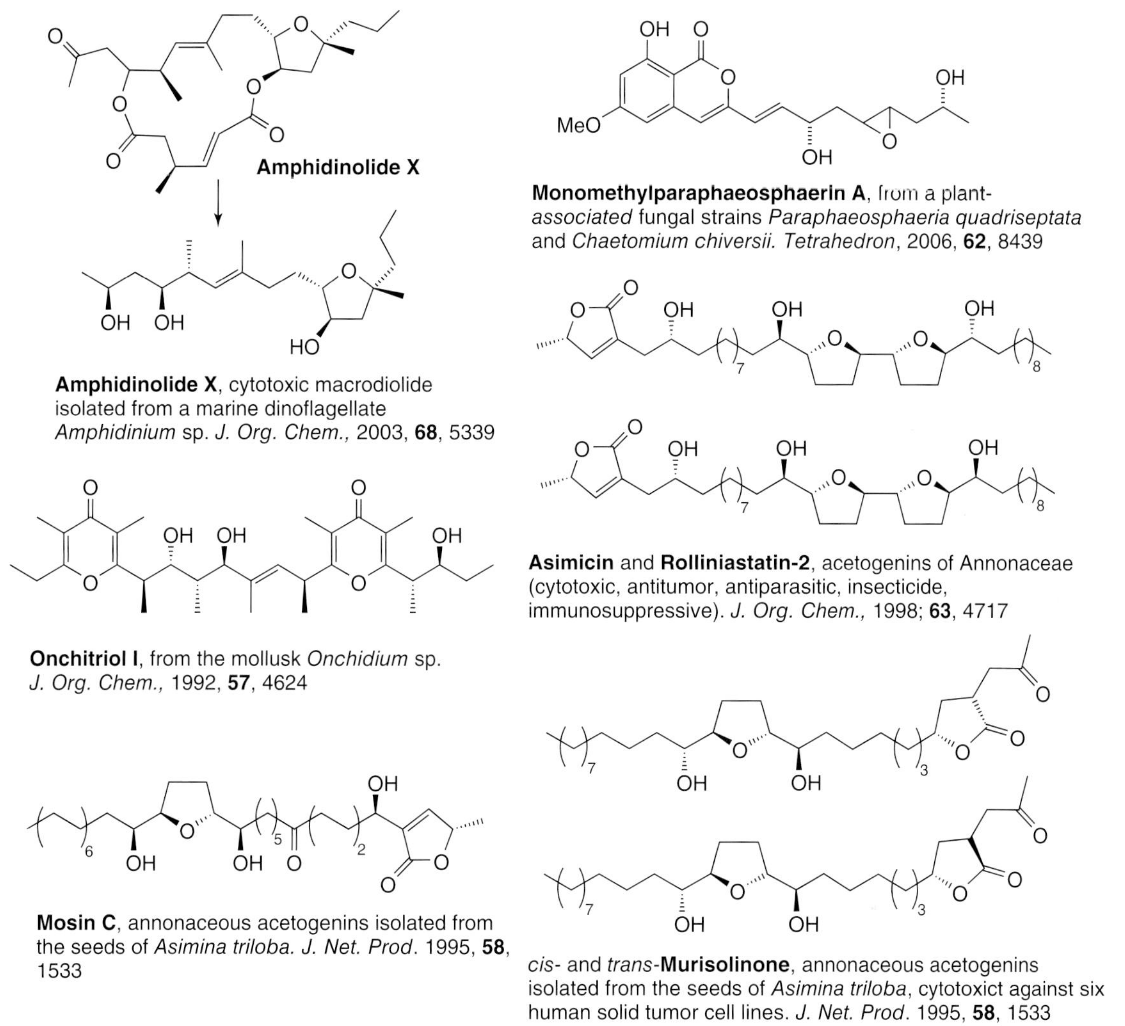

Amphidinolide X

Amphidinolide X, cytotoxic macrodiolide isolated from a marine dinoflagellate *Amphidinium* sp. *J. Org. Chem.*, 2003, **68**, 5339

Monomethylparaphaeosphaerin A, from a plant-associated fungal strains *Paraphaeosphaeria quadriseptata* and *Chaetomium chiversii. Tetrahedron*, 2006, **62**, 8439

Asimicin and **Rolliniastatin-2**, acetogenins of Annonaceae (cytotoxic, antitumor, antiparasitic, insecticide, immunosuppressive). *J. Org. Chem.*, 1998; **63**, 4717

Onchitriol I, from the mollusk *Onchidium* sp. *J. Org. Chem.*, 1992, **57**, 4624

Mosin C, annonaceous acetogenins isolated from the seeds of *Asimina triloba. J. Net. Prod.* 1995, **58**, 1533

cis- and *trans-***Murisolinone**, annonaceous acetogenins isolated from the seeds of *Asimina triloba*, cytotoxict against six human solid tumor cell lines. *J. Net. Prod.* 1995, **58**, 1533

Figure 17.35. Selection of *sec/sec*-1,2- and *sec/sec*-1,*n*-diols isolated from natural sources whose absolute configuration has been assigned by NMR of their CDA derivatives

A sample of 2,3-dihydroxyhexyl acetate of unknown absolute configuration is separately derivatized with two equivalents of (*R*)-MPA and two equivalents of (*S*)-MPA and the ^{1}H NMR spectra of the resulting bis-MPA esters registered (steps 1 and 2, Figure 17.29a and b).

The next step consists of the calculation of the chemical shift differences (step 3, Figure 17.29c), leading to a positive $\Delta\delta^{RS}$ for H(2′), H(1′), and the AcO and a $\Delta\delta^{RS}$ negative for H(3′), H(4′), H(5′), and H(6′).

Finally, comparison of distribution of the experimentally obtained $\Delta\delta^{RS}$ signs with those shown in Figure 17.25 leads to the absolute configuration (2*R*, 3*S*) shown in Figure 17.29(d).

Single derivatization procedures based on low-temperature NMR of MPA ester derivatives[4,19] can also be used for the assignment of some of those bifunctional compounds (*sec/sec*-1,2-diols,[4,19,25] *prim/sec*-1,2-diols,[4,19,27] and *prim/sec*- and

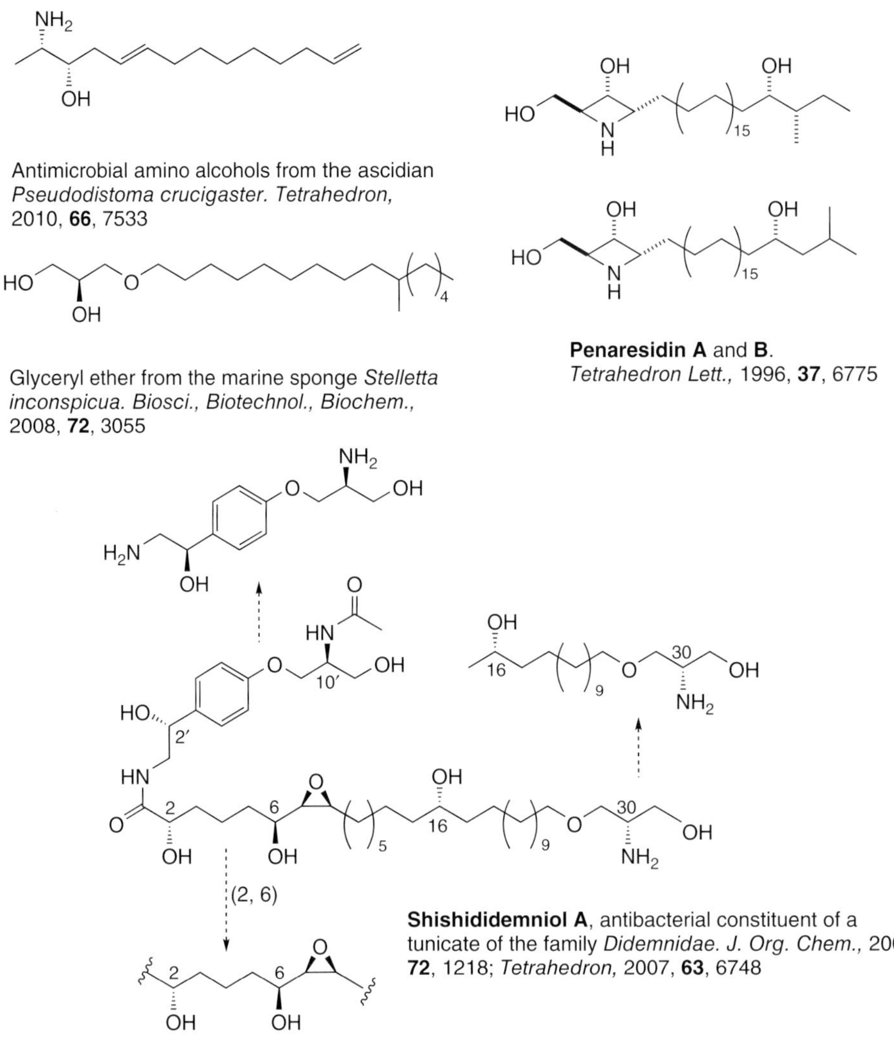

Antimicrobial amino alcohols from the ascidian *Pseudodistoma crucigaster*. Tetrahedron, 2010, **66**, 7533

Glyceryl ether from the marine sponge *Stelletta inconspicua*. Biosci., Biotechnol., Biochem., 2008, **72**, 3055

Penaresidin A and **B**. Tetrahedron Lett., 1996, **37**, 6775

Shishididemniol A, antibacterial constituent of a tunicate of the family *Didemnidae*. J. Org. Chem., 2007, **72**, 1218; Tetrahedron, 2007, **63**, 6748

Figure 17.36. Selection of bioactive natural *sec/sec*-1,2-amino alcohols, *prim/sec*-1,2-diols, and 1,2- and *sec/prim*-1,2-aminoalcohols whose absolute configuration has been determined by NMR of their CDA derivatives

sec/prim-1,2-amino alcohols[4,19,29]), using MPA as CDA. Figures 17.30, 17.31, and 17.32 show the $\Delta\delta^{T_1 T_2}$ sign distributions for every isomer of acyclic *sec/sec*-1,2-diols (Figure 17.30), *prim/sec*-1,2-diols (Figure 17.31), *prim/sec*-1,2-aminoalcohols, and *sec/prim*-1,2-aminoalcohols (Figure 17.32).

The low-temperature procedure for the assignment of *sec/sec*-1,*n*-diols,[4,19] is based on the $\Delta\delta^{T_1 T_2}$ sign distribution of R_1 and R_2 as diagnostic signals (Figure 17.30). Unfortunately, this procedure does not allow one to distinguish between the two enantiomers forming the *syn*-pair.

For *prim/sec*-1,2-diols[4,19,27] (Figure 17.31), the diagnostic signals are those of the methylene protons and the procedure is based on the $\Delta\delta^{T_1 T_2}$ of the methylene proton that resonates at higher field.

Finally, the absolute configuration of *prim/sec*- and *sec/prim*-1,2-amino alcohols[4,19,29] (Figure 17.32) can be assigned by low-temperature NMR of a single derivative, analyzing the evolution with temperature of the signals due to the CαH protons of the two MPA units. In these substrates, the NMR parameter expressing the correlation with the stereochemistry is the separation between those two singlets at room and at a lower temperature.

From a practical viewpoint, as in the case of monofunctional substrates, the procedure requires the preparation of only one MPA derivative (step 1) and comparison of its spectrum taken at room temperature (step 2) in CS_2/Cl_2CD_2 (4 : 1) as solvent and at lower temperature (usually 220–200 K is enough, step 3). The absolute configuration is obtained by comparison of the signs of $\Delta\delta^{T_1 T_2}$, $\Delta\delta^R$, or $\Delta\delta^S$ (step 4) with those shown in Figures 17.30, 17.31, and 17.32 for all the isomers.

In Figures 17.33–17.36, a selection of polyfunctional biologically active substances, whose absolute configuration has been assigned by NMR, is presented.

As with monofunctional compounds, ^{13}C NMR chemical shifts can also be used to determine the absolute configuration of polyfunctional compounds. Unfortunately, the small $\Delta\delta$ values obtained in many cases restrict the usefulness of this procedure to *sec/sec*-1,2-diols and *sec/sec*-1,2-amino alcohols.[18]

In all those cases, the absolute configuration can be derived by comparison of the experimental ^{13}C $\Delta\delta^{RS}$ signs with the tetrahedral models used for proton NMR methods (Figure 17.25).

RELATED ARTICLES IN EMAGRES

Analysis of High-Resolution Solution State Spectra

Chiral Discrimination Using Chiral Ordering Agents

Lanthanide-Induced Shifts (LIS) in Structural and Conformational Analysis

Natural Products

Shielding Calculations

Shielding in Small Molecules

REFERENCES

1. H. D. Flack and G. Bernardinelli, *Chirality*, 2008, **20**, 681.

2. N. Berova, K. Nakanishi, and R. W. Woody, Circular dichroism: Principles and Applications, 2nd edn, Wiley-VCH: New York, 2000.

3. T. B. Freedman, X. Cao, R. K. Dukor, and L. A. Nafie, *Chirality*, 2003, **15**, 743.

4. J. M. Seco, E. Quiñoá, and R. Riguera, *Chem. Rev.*, 2012, **112**, 4603.

5. J. M. Seco, E. Quiñoá, and R. Riguera, *Chem. Rev.*, 2004, **140**, 17.

6. J. M. Seco, E. Quiñoá, and R. Riguera, *Tetrahedron: Asymmetry*, 2001, **12**, 2915.

7. T. J. Wenzel, Discrimination of Chiral Compounds Using NMR Spectroscopy, Wiley: Hoboken, NJ, 2007.

8. S. Porto, J. M. Seco, J. F. Espinosa, E. Quiñoá, and R. Riguera, *J. Org. Chem.*, 2008, **73**, 5714.

9. S. Latypov, J. M. Seco, E. Quiñoá, and R. Riguera, *J. Org. Chem.*, 1995, **60**, 504.

10. S. Latypov, J. M. Seco, E. Quiñoá, and R. Riguera, *J. Org. Chem.*, 1996, **61**, 8569.

11. S. K. Latypov, M. J. Ferreiro, E. Quiñoá, and R. Riguera, *J. Am. Chem. Soc.*, 1998, **120**, 4741.

12. I. Louzao, J. M. Seco, E. Quiñoá, and R. Riguera, *Chem. Commun.*, 2006, (13), 1422.

13. I. Louzao, R. García, J. M. Seco, E. Quiñoá, and R. Riguera, *Org. Lett.*, 2009, **11**, 53.

14. S. Porto, E. Quiñoá, and R. Riguera, *Tetrahedron*, 2014, **70**, 3276.

15. J. M. Seco, S. Latypov, E. Quiñoá, and R. Riguera, *J. Org. Chem.*, 1997, **62**, 7569.

16. J. M. Seco, E. Quiñoá, and R. Riguera, *J. Org. Chem.*, 1999, **64**, 4669.

17. M. J. Ferreiro, S. K. Latypov, E. Quiñoá, and R. Riguera, *J. Org. Chem.*, 2000, **65**, 2658.

18. I. Louzao, J. M. Seco, E. Quiñoá, and R. Riguera, *Chem. Commun.*, 2010, **46**, 7903.

19. J. M. Seco, E. Quiñoá, and R. Riguera, in Structure Elucidation in Organic Chemistry: On Search for the Right Tools, eds M. M. Cid and J. Bravo, Wiley-VCH Verlag GmbH & Co. KGaA: 2014, Chap. 7, pp. 241–277. Simplified NMR procedures for the assignment of the absolute configuration, ISBN: 978-3-527-33336-3.

20. J. M. Seco, E. Quiñoá, and R. Riguera, *Tetrahedron*, 1999, **55**, 569.

21. S. K. Latypov, J. M. Seco, E. Quiñoá, and R. Riguera, *J. Am. Chem. Soc.*, 1998, **120**, 877.

22. R. García, J. M. Seco, S. A. Vázquez, E. Quiñoá, and R. Riguera, *J. Org. Chem.*, 2002, **67**, 4579.

23. R. García, J. M. Seco, S. A. Vázquez, E. Quiñoá, and R. Riguera, *J. Org. Chem.*, 2006, **71**, 1119.

24. J. M. Seco, E. Quiñoá, and R. Riguera, *Tetrahedron: Asymmetry*, 2000, **11**, 2781.

25. F. Freire, J. M. Seco, E. Quiñoá, and R. Riguera, *J. Org. Chem.*, 2005, **70**, 3778.

26. V. Leiro, F. Freire, E. Quiñoá, and R. Riguera, *Chem. Commun.*, 2005, (44), 5554.

27. F. Freire, J. M. Seco, E. Quiñoá, and R. Riguera, *Org. Lett.*, 2010, **12**, 208.

28. V. Leiro, J. M. Seco, E. Quiñoá, and R. Riguera, *Org. Lett.*, 2008, **10**, 2729.

29. V. Leiro, J. M. Seco, E. Quiñoá, and R. Riguera, *Org. Lett.*, 2008, **10**, 2733.

30. F. Freire, E. Lallana, E. Quiñoá, and R. Riguera, *Chem. Eur. J.*, 2009, **15**, 11963.

Chapter 18

Applications of Preclinical MRI/MRS in the Evaluation of Drug Efficacy and Safety

Thomas M. Bocan[1,2], Lauren Keith[3], and David M. Thomasson[3,4]

[1]US Army Medical Research Institute of Infectious Diseases, 1425 Porter Street, Frederick, MD 21702, USA
[2]The Geneva Foundation, 917 Pacific Ave, Suite 600, Tacoma, WA 98402, USA
[3]National Institute of Allergy and Infectious Disease, 8200 Research Plaza, Frederick, MD 21702, USA
[4]Tunnel Government Services, 6701 Democracy Boulevard, Suite 515, Bethesda, MD 20817, USA

18.1 INTRODUCTION

The drug discovery process while appearing linear is a series of interrelated, bidirectional steps culminating in the declaration of an investigational new drug. Upon selection of a disease target and initiation of drug screening and chemistry efforts, secondary and tertiary assays are developed in animals to assess the efficacy and safety of potential lead candidates. Efficacy endpoints are often functional, behavioral, biochemical, or histological and are typically disease or safety endpoint specific. Biomarkers within animals are evaluated as part of the efficacy and safety evaluation and to build confidence that the drug or target will be successful in the clinic. Taken collectively, biomarkers that indicate drug exposure at the target site of action, target occupancy, and functional modulation of the target have been shown to increase the probability of success of the clinical candidate.[1] Imaging approaches are uniquely positioned to provide these three types of biomarkers, which are often referred to as the *three pillars of drug survival*. Positron emission tomography (PET) can directly demonstrate drug exposure and target occupancy while MRI and MRS can provide a sensitive measure of target modulation. In this article, we will (i) briefly review the MRI/MRS approaches and methods of quantification that could be applied to drug discovery; and (ii) summarize how MRI/MRS have been applied to drug discovery across such areas as neuroscience, cardiovascular, oncology, inflammation, pulmonary, and infectious diseases.

NMR in Pharmaceutical Sciences. Edited by Jeremy R. Everett, John C. Lindon, Ian D. Wilson, and Robin K. Harris
© 2015 John Wiley & Sons, Ltd. ISBN: 978-1-118-66025-6
Also published in eMagRes (online edition)
DOI: 10.1002/9780470034590.emrstm1422

18.2 REVIEW OF MRI/MRS METHODS

Numerous MRI and MRS tools exist that can be used for evaluation of tissue and organ structure, composition and function within the animal during disease progression and/or drug intervention. In the following sections, we will review the key methods available for assessing anatomy, function, and biochemical changes.

18.2.1 Morphological Imaging, Relaxation Weighted T1, T2, T2*

Anatomical imaging in which tissue contrast is weighted by certain tissue relaxation parameters is the benchmark application for MRI. These methods can provide detailed images with exquisite soft tissue contrast and the interpretation of these images can be highly sensitive to disease processes. However, additional measurements can provide quantitative information regarding tissue characteristics. To accomplish this, data acquisition must shift from relaxation weighted images to the direct mapping of specific contrast generating parameters such as T1, T2, or T2*. As these parameters are physical characteristics of tissues, quantitative relaxometry measurements are independent of imaging equipment vendor and acquisition technique. Standardizing imaging parameters is beneficial because, while there are a variety of techniques available for relaxometry, specific applications may require choosing a technique that provides optimal information density per unit time, and this can be organ specific.

18.2.2 Magnetization Preparation Techniques

Several strategies can be employed to improve contrast to noise, sensitivity, and specificity for biomarkers of interest. Generally, these strategies are referred to as *magnetization preparation techniques*. For example, inversion recovery sequences are used when the longitudinal relaxation (*T*1) parameters of the tissue within a volume element (voxel) or between nearby voxels are nonhomogeneous. FLuid Attenuated Inversion Recovery (FLAIR) uses the known, long *T*1 of cerebrospinal fluid (CSF) to acquire an image of the central nervous system such that the CSF signal is nulled.

Black-Blood (BB) imaging uses a similar approach to null the otherwise bright blood signal when interrogating the myocardium or atherosclerotic plaques.

Another class of magnetization preparation techniques is based on presaturation of protons in molecules other than water. On the basis of the ability to exchange saturated protons with the local water molecules, these presaturated protons affect water proton relaxation and can be used to detect the presence and relative concentration of larger molecules. These techniques are broadly classified as chemical exchange saturation transfer (CEST).[2]

18.2.3 Blood Volume, Flow, and Perfusion

Blood volume (BV), blood flow (BF), and perfusion are biomarkers for disease processes such as inflammation, hemorrhage, infarct, or angiogenesis. There are numerous approaches to measure these parameters. The most appropriate technique depends on the organ system being studied and the physiological and pathological conditions of that organ. For example, dynamic susceptibility contrast (DSC) techniques can assess BV or BF, but require very homogeneous magnetic fields and a closed one compartment system without leaky vessels such as the brain. For quantitative flow measurements in vasculature with high blood velocity, phase-based techniques are most appropriate. For disease processes with high-permeability surface products (extraction fraction = 1), dynamic contrast-enhanced T1 perfusion is most appropriate and as such continue to be used by the imaging community.[3] If the use of exogenous, gadolinium-based contrast material is contraindicated, spin labeling techniques such as arterial spin labeling (ASL) or pseudo-continuous arterial spin labeling (pCASL),[4] can be beneficial despite signal to noise ratios lower than contrast material-enhanced techniques.

18.2.4 Diffusion

Diffusion imaging describes the freedom with which water molecules move through their local environment. This can be done qualitatively, with diffusion weighted imaging, or quantitatively with parametric mapping of diffusion amplitude and direction. These measurements have potential as a biomarker of therapeutic response in many organ systems.[5] Techniques evolve from the signal changes in response to magnetic field gradients applied along multiple axes in a preparation phase of the imaging sequence, where subsequent spatial encoding can be achieved by gradient echo, spin echo; or echo planar data acquisition. In order to minimize the effects of confounding, physiological processes such as perfusion recommended sequence parameters are published by various consensus panels. With the application of diffusion sensitizing gradients in a sufficient number of directions, a tensor may be constructed to map structures with anisotropic diffusion such as white matter tracts in the brain that may correlate to brain function.[6]

18.2.5 Magnetic Resonance Spectroscopy (MRS)

In vivo spectroscopy can be used to measure metabolite concentration, either relatively or explicitly. The technique is based on the NMR signal generated by protons in metabolite molecular structures that have distinct resonance frequencies. Techniques range from sampling a single volume element (voxel) or simultaneously sampling one, two, or three-dimensional arrays referred to as *chemical shift imaging (CSI)*. Quantification is typically based on relative peak heights or area under peaks; but signal processing techniques can affect the statistical significance of the outcomes.[7] Signal-to-noise improvements are necessary for in vivo studies and are based on field cycling (averaging), spectral editing, and spin decoupling. In addition to spatial separation in CSI, one can also employ multidimensional spectral separations using traditional multidimensional spectroscopy, e.g., correlation spectroscopy (COSY) in vitro or in vivo.[8]

18.2.6 Functional MRI

Functional MRI (fMRI) employs the blood oxygenated level dependent (BOLD) MRI technique which measures BF and BV through detection of deoxyhemoglobin. When a drug is utilized in the fMRI protocol the procedure is often termed *pharmacological MRI (phMRI)*.[9] fMRI and phMRI are used to assess neurological activation due to task performance or drug administration. The image

contrast is a representation of either the transition from oxy- to deoxyhemoglobin (BOLD) or capillary perfusion (ASL). Processing of fMRI data is based on the statistical significance of correlations between stimuli and the acquired MRI images. Unlike BOLD MRI which has the capacity to monitor global changes in BF and volume, ASL MRI measures flow in specific user defined brain slices and the signal is not dependent on the blood pO_2 or pCO_2. There are several acquisition and postprocessing considerations that can improve image and data quality[10] including limiting motion and cardiac or respiratory effects.

18.2.7 Non-hydrogen Imaging Applications

Hydrogen MRI/MRS techniques are the most prevalent owing to the biological abundance of hydrogen atoms ensuring adequate SNR for most applications. However, other nuclei with spin $\neq 0$ can be used for both MRI and MRS. MRI or MRS of these nuclei requires broadband amplifiers and specially tuned RF coils; however, newer direct to digital technology shows significant improvement over conventional designs.[11] Both sodium and fluorine have been imaged,[12] while phosphorus and carbon have been studied with MRS. In both cases, most applications are being performed at higher magnetic fields to compensate for the low SNR. Enrichment and use of exogenous contrast agents or hyperpolarization are techniques to improve SNR. However, hyperpolarized imaging requires specialized pulse sequences due to the loss of polarization with repeated imaging excitations that decay the overall spin population.

As noted earlier, numerous MRI and MRS tools exist which can be used alone or in combination for the evaluation of disease pathology and drug intervention. Such tools can aid in establishing a compound's mechanism of action, efficacy, and safety and providing evidence of a physiological response to drug intervention. For greatest value in drug discovery, MRI/MRS must be quantitative in order to discern drug dose effects and stages of disease progression. In the next sections, examples with some pictorial representations (Figures 18.1 and 18.2) of how the various imaging methods have been applied to drug discovery and potentially drug safety are summarized.

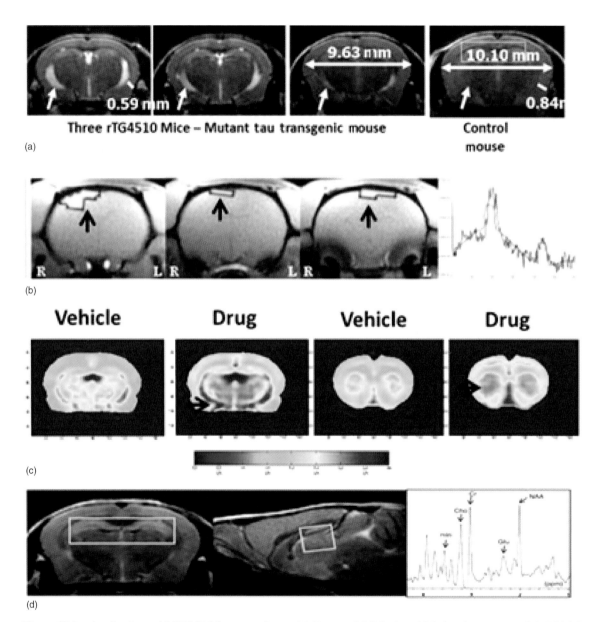

Figure 18.1. Applications of MRI/MRS in neuroscience. (a) Structural MRI of an Alzheimer's mouse model, TG4510. Arrows denote lateral ventricles; bars represent cortical thickness; double-pointed arrows show brain cross sectional thickness. (b) BOLD MRI signal associated with right forepaw electrical stimulation. Arrows denote areas of activation. Graph equals stimulus intensity and timing. (c) ASL MRI of drug-induced increases (red regions) in brain blood flow; blue (low) to red (high) CBF. (d) MRS of glutamine/glutamate spectra. Box represents regions of interest from which the spectra were collected.

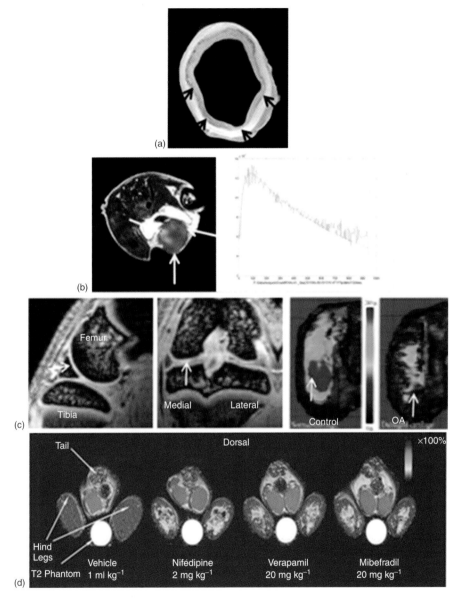

Figure 18.2. Applications of MRI in atherosclerosis, oncology, inflammation, and drug safety. (a) Ex vivo MRI of a formalin-fixed aorta from a cholesterol-fed rabbit. Arrows denote the intima-media boundary below the atherosclerotic lesion. Resolution is 25 μm. (b) dceMRI of a tumor implanted into a SCID mouse. Left pane is a computed Ktrans map of the tumor (denoted by arrows) overlaid onto a T2 weighed image. The rim of the tumor appears bright and has higher Ktrans values. Right pane is the intensity change of the tumor associated with influx of contrast agent. (c) Left two panels are representative in vivo MR images of a rat knee joint. Arrows denote cartilage. Right panel is a three-dimensional representation of the tibial medial cartilage thickness in normal and osteroarthritis rat model. Blue represents thicker cartilage (arrow) while red is thinned cartilage (arrow). (d) T2 MRI image of rat muscle following infusion of vehicle and three different calcium channel blockers. The images are pseudocolored on top of the grayscale anatomical image to represent changes in edema. Varying degrees of edema as depicted by changes in color (red-high, blue-low) are noted across the four images.

18.3 APPLICATIONS OF MRI/MRS IN DRUG DISCOVERY

18.3.1 Neuroscience

Imaging is widely used in the discovery and development of drugs for the treatment of neurological disorders ranging from degenerative diseases such as Alzheimer's disease (AD) and Parkinson's disease to psychiatric disorders such as schizophrenia, depression, and anxiety. Preclinical evaluations of therapeutic agents have helped to better define mechanisms of action, demonstrated drug efficacy, identified specific brain regions where target modulation is effective, and defined brain regions for future clinical evaluation.

Anatomical ^{1}H MRI is a primary use of MRI given the high spatial resolution and sensitivity to anatomic changes. Structural MRI has been used numerous times to characterize brain morphology associated with AD. In a mouse model of AD, MRI identified reductions in both whole brain and hippocampal volume and increases in ventricular volume by 5 months of age (Figure 18.1a). With shrinkage of the brain, there is an increase in neurodegeneration which can be detected by diffusion tensor MRI (dtiMRI) and other MRI tools such as DWI and measurement of ADC. With axonal damage and demyelination, dtiMRI can detect the changes in flow of water down the axon and increased flow across the axon in accordance with the loss of the myelin sheath. Combination of anatomical MRI and MRI sequences to assess axonal water flow can highlight structural connections between neuronal centers, a measure of neuronal damage and link structural to functional changes.

The application of preclinical fMRI or phMRI is challenging. Some investigators have used anesthetics only briefly while setting up the procedure and then transitioned to paralytics during the fMRI procedure to immobilize the animal. Such an approach has been successful for the evaluation of sensory stimuli such as pain (Figure 18.1b), auditory and olfactory; however, it is questionable whether other forms of brain activation can be discerned above the stress response associated with the procedure. A group of investigators have employed fMRI and phMRI to evaluate drug affects in conscious, restrained animals. The animals are trained in simulated environments to become accustomed to the restraint and the sounds of the MRI procedure. In performing such experiments, image analysis and regions of interest determinations must be controlled to account for movement and measures of

stress hormones and physiological parameters need to be monitored to control for changes that can impact the interpretation of the data.

Recent work has been published utilizing fMRI to characterize intrinsic brain networks across species that are both in common and not susceptible to anesthesia. Identification of the default mode network (DMN) where the fMRI BOLD signal is depressed during cognitive tasks and activated during rest has been identified and the DMN architecture, i.e., connectivity between various brain regions, is similar between humans, nonhuman primates and rats and not affected by anesthesia.[13] Other intrinsic neural networks such as the cerebellar, somatosensory, motor, auditory, and visual also exist. Treatment of rats with D-amphetamine, fluoxetine, or nicotine modified the fMRI response in various intrinsic networks and appears predictive of their beneficial effects in the treatment of schizophrenia. It is also interesting to note that dtiMRI has been used to demonstrate the structural connections associated with the intrinsic neural networks based on establishing tracts of water flow along the axes of the interconnecting neurons.

fMRI or phMRI have been used in the discovery of new antipsychotics and defining the primary mechanism of action of CNS-acting drugs designed to improve cognition.[14] The selective M_4 muscarinic acetylcholine receptor positive allosteric modulator, VU0152100, reduced the amphetamine-induced activation in the nucleus accumbens, caudate-putamen, hippocampus and medial thalamaus, and identified brain networks among the various brain regions that were M_4 mediated.[14] Similar types of evidence were obtained using phMRI following administration of indanone-A, a selective potentiator of the metabotropic glutamate receptor subtype 2, following phenylcyclidine administration. Thus, the fMRI and phMRI signal are highly dependent on vasomotor tone and agents which are vasoconstrictors or vasodilators can contribute to the BOLD fMRI signal.

As BOLD fMRI is highly dependent on the ratio of oxy- to deoxyhemoglobin, the main caveat in performance of these types of experiments is that arterial pO_2 and pCO_2 must be monitored and maintained within physiological ranges. An alternative approach to measuring neuroactivation through assessment of cerebral BF without dependence on the blood oxygen state is the use of ASL MRI under controlled conditions to avoid physiological responses, e.g., hypercapnia increases BF and ASL MRI signal (Figure 18.1c).

Comparison of BOLD fMRI and ASL MRI has been performed and shown to be in good agreement when evaluating the sensory cortex following electric forepaw stimulation in paralyzed animals to simulate a pain response. Thus, application of fMRI or phMRI has proven useful in the translation of preclinical observations to the clinic through defining intrinsic neural networks that can be modulated even under anesthesia, and through assessment of CBF/CBV following neuronal activation by drug or exogenous stimulus.

An additional measure of brain function is the application of MRS. These mechanism-based imaging biomarkers have been used to demonstrate biochemical changes associated with neurodegeneration and drug action at the target through measurement of specific biological pathways. With AD progression in TG4510 mice, MRS measures of the ratio of myoinositol/creatine increased as the size of the brain decreased and such biochemical changes were a reflection of an increase in gliosis as confirmed by histology. In the development of agents for treatment of psychotherapeutic disorders, measurement of global brain pools of glutamine and glutamate by MRS has been used to link the effect of a Kv7 potassium channel opener with alterations in the glutaminergic pathway (Figure 18.1d). Retigabine, a Kv7 potassium channel opener, hyperpolarizes the neuron, reduces action potential and, acting through the glutaminergic pathway decreases hippocampus glutamate concentrations by 20% which is detectable by MRS at 7 T. Application of MRS to assessment of changes in the glutaminergic pathway is a sensitive measure to link drug action with a biological pathway; however, the local synaptic change must be associated with a global brain change in the specific neurotransmitter due to contrast resolution.

Finally, research in the area of discovering and refining MRI contrast agents to detect molecular events is ongoing. While additional work is required to improve potency and bioavailability, investigators have attempted to engineer the heme domain of the bacterial cytochrome P450-BM3 to recognize dopamine and therefore act as a sensor of dopamine release from the neuron upon depolarization and neurotransmitter release.[15] In vivo studies have been performed to demonstrate proof of concept of such an approach but it required intracerebral injection of the MR contrast agent to show an effect. Peripheral targets not requiring the MRI biosensor to cross the blood brain barrier, and significantly more potent biosensors, may be needed for MRI to be considered capable of molecular imaging.

18.3.2 Cardiovascular and Metabolic Disorders

MRI and MRS have been applied preclinically and clinically to assess the progression of disease and therapeutic intervention in the areas of atherosclerosis and diabetes. Atherosclerotic plaque anatomy and composition have been interrogated both ex vivo and in vivo by combining a series of relaxation weighted approaches such as T1W, T2W, and PDW. These relaxation weighted approaches characterize the composition of plaques based on their chemical composition, water content, physical state, i.e., fibrotic or lipid-enrichment, molecular motion, and water diffusion. Utilizing the above-mentioned techniques with BF suppression, referred to as the *black blood technique*, provides improved contrast such that the lumen is black and the plaque components are bright. Selective enhancement of plaque macrophages can be achieved by injection of ultrasmall superparamagnetic particles of iron oxide (USPIO) which become phagocytized by macrophages and cause areas of hypointensity on a T2W image.[16] Targeted approaches for assessment of endothelial dysfunction, adhesion molecules, and vascular lipoprotein uptake have been developed. Endothelial dysfunction and the resulting hyperpermeability can be visualized using MRI contrast agents such as standard gadolinium chelates and the more novel, Gadofosveset,[17] a gadolinium agent which binds to serum albumin, that tracks the uptake of albumin at sites of injury or dysfunction that are typically predisposed to atherosclerosis development. Microparticles of iron oxide (MPIO) conjugated to P-selectin or vascular cell adhesion molecule-1 (VCAM-1) can further elucidate the early events of atherosclerosis by defining areas of increased inflammatory cell adhesion.[18] While the coronary arteries are the most clinically relevant vessels, most MR studies have focused on the carotid, aorta, and peripheral vessels due to the motion artifact associated with the beating heart and the reduced signal to noise.

Application of MRI to the evaluation of therapeutic intervention in animal models of atherosclerosis progression/regression has provided insight into plaque changes rather than being limited to angiographically determined changes in luminal stenosis which may not be a true reflection of disease severity and/or presence of unstable lesions (Figure 18.2a). In one study, rabbits were cholesterol-fed for 4 months before switching to a low cholesterol diet for the remaining 16 weeks. T2W MRI was performed at 4, 12, and 20 months and a 29% reduction in plaque thickness was detected at 20 months relative to 4 months.[19] Treatment of preestablished lesions in cholesterol-fed rabbits with the thromboxane A2 antagonist, S18886, for 5 months, reduced plaque burden 18%, lowered plaque lipid content, i.e., reduced signal intensity in T2W and increased T1W, and increased plaque fibrosis, i.e., high signal intensity in T1W and T2W.[20] Selective reduction in plaque macrophage content in the absence of changes in the lesion area was demonstrated in apoE knockout cross human CCR knockin mice (huCCR2kil/ApoE$^{-/-}$) using USPIO following treatment with the CCR2 antagonist, GSK1344386B.[21] The in vivo MRI findings were confirmed by ex vivo MRI and histological evaluation. Thus, these intervention studies have demonstrated the sensitivity of MRI in detecting changes in lesion size and composition.

The application of MRS while shown to be a sensitive marker for measuring glutamate changes in the brain can be expanded to measurement of other nuclei/molecules such as ^{1}H, ^{31}P, ^{13}C, or ^{13}C-glucose to broaden the target species and/or disease area, e.g., liver fat content, pH or ^{13}C-glucose and liver metabolism associated with diabetes. While MRI is sensitive to detecting structural changes associated with diabetes, i.e., kidney damage, and changes in peripheral tissue perfusion, MRS has broader application to studying the metabolic changes associated with diabetes and metabolic syndrome. Liver fat content can be quantified by the difference between T1W and T2W fat suppressed ^{1}H images. Intramyocellular lipid (IMCL) levels measured by ^{1}H-MRS, while challenging due to the fact that the geometric orientation of the muscle fibers must parallel the axis of the magnet, is highly correlated with insulin sensitivity. A negative correlation between IMCL and insulin sensitivity has been documented in patients with diabetes.[22] Measures of muscle ATP utilization, pH, and energetics under resting and exercised state in normal and pathologic states can also be evaluated using ^{31}P-MRS. Intravenous infusion of ^{13}C-glucose in normal and diabetic patients and evaluation by MRS has demonstrated that suppression of lipolysis with acipimox increases whole body glucose uptake with no change in glucose to glycogen synthesis.[23] These data taken together demonstrate that with

MRS one can noninvasively and serially evaluate changes in liver and muscle lipid content, energetics and glucose–glycogen metabolism under normal and diabetic states, and upon therapeutic intervention has the potential to better understand the mechanism of action of the new drug.

18.3.3 Oncology

Structural and fMRI measures of angiogenesis and tumor necrosis have been utilized both preclinically and clinically to evaluate tumor progression and therapeutic efficacy of pharmacologic agents. Utilizing a T2-weighted spin echo sequence, the cyclin-dependent kinase inhibitor, PHA-848125, was shown to reduce tumor size in a mouse model, K-Ras^{G12D}LA2, of naturally occurring tumors which mimic human lung adenocarcinoma.[24] Dynamic contrast-enhanced MRI (DCE-MRI) which is a measure of blood vessel perfusion and permeability using a gadolinium-based contrast agent has been used extensively to evaluate antiangiogenic therapies[25] (Figure 18.2b). Antibodies to growth factors, tyrosine kinase inhibitors, and antiangiogenic agents targeting endothelial cells and the immune system have all been shown in animal models by DCE-MRI to alter tumor vascularization.[25] Vascular disrupting agents such as tubulin binding and tumor necrosis factor-α (TNF-α) inducing agents and radiation, photosensitizers or targeted chemotherapeutics have been shown by DCE-MRI to reduce blood vessel growth.[25] More target specific agents such as PTK787/ZK22584, an inhibitor of all three vascular endothelial growth factor receptors (VEGFR),[26] and PF-03084014, a γ-secretase inhibitor,[27] were all shown to reduce the vascular DCE-MRI determined perfusion/permeability-related parameter, K^{trans}, which is indicative of a reduction in tumor angiogenesis. It is also interesting to note that the clinically efficacious dose of PTK787/ZK222584 was predicted by the preclinical DCE-MRI observations.[26] While changes in tumor perfusion tend to precede gross changes in tumor size an intermediate phase related to cell death can be monitored by MRI. Diffusion weighted MRI (DWI-MRI) which is sensitive to the movement of water can be used to determine the apparent diffusion coefficient (ADC) within tumors. With cell death caused by cytotoxic agents, the ADC measure increases because water movement is less constrained and the increased ADC value is representative of a reduced tumor cell density. The true power of MRI imaging can be seen when applying both DCE-and DWI-MRI to identify the appropriate timing of administration of antiangiogenic and cytotoxic agents for maximal tumor size reduction.

Metabolic tumor characterization by MRS is an evolving approach to better understanding signal transduction pathways, more target directed approaches, and developing a more personalized medicine approach to oncology treatment.[28] The most common MRS approaches focus on measuring choline–phosphocholine concentration using ^{1}H and ^{31}P MRS, glucose concentration employing intravenous infusion of ^{13}C-glucose or hyperpolarized ^{13}C precursors and lactate levels by ^{1}H-MRS. Evaluation of intracellular and extracellular tumor pH using ^{31}P MRS has proven successful in segregating tumors that may be responsive to therapy. Application of MRS to the evaluation of pharmacologic treatment is best performed at 7+T due to greater resolving power of the metabolites. Therefore, while PET/CT has been routinely used in the diagnosis of tumors, MRI/MRS has the potential to better define the mechanisms of drug action, to allow for more repeat measurements to gauge tumor progression/regression and to map changes in metabolic pathways.

18.3.4 Inflammation

Inflammation underlies several diseases such as osteoarthritis (OA), rheumatoid arthritis (RA), pulmonary diseases, atherosclerosis, and infectious diseases. While PET approaches focus on monitoring the inflammatory cell infiltration, MRI assesses the consequences of inflammation, e.g., cartilage degradation, edema, and alveolar enlargement. In OA, many studies have employed MRI to assess the structural integrity and composition of the articular cartilage[29] (Figure 18.2c). T2 mapping and DWI provide information about the collagen fibers and water content while the glycosaminoglycan (GAG) content is measured indirectly by delayed gadolinium enhanced MRI of cartilage (dGEMRIC), directly using T1ρ mapping, GAG chemical exchange saturation transfer (gagCEST), or [23]sodium MRS. Cartilage MR imaging studies are typically performed in humans or dogs due to their larger cartilage thickness compared to rodents. Resolution of the cartilage boundaries given similar contrast of cartilage and synovial fluid is challenging but achievable by combining several MRI approaches, e.g., dGEMRIC and T2 mapping. While measures

of joint space narrowing have been used to track OA progression, three-dimensional measures of cartilage volume and erosions have proven to be more sensitive to subtle changes. Significant debate on the appropriate method for performing the three-dimensional measurements has occurred over the decade, e.g., fully automated segmentation, manual segmentation, automated with manual quality control. Consistency in image segmentation technique is crucial since minor changes in the cartilage-synovial fluid boundary can mask or overestimate a drug effect. Disease modifying osteoarthritis drugs (DMOADS) targeting the preservation of cartilage and subchondral bone such as inhibitors of matrix metalloproteinase, growth factors, inflammatory pathways, and cytokines; antibodies against TNF and steroids have all been evaluated. While OA is characterized by MR detectable changes in cartilage volume and quality as assessed by dGEMRIC, RA is a disease associated with joint edema and bone inflammation. T2 weighted and short tau inversion recovery (STIR) sequences can detect joint synovitis, tenosynovitis, and synovial effusions as well as bone inflammation presenting as bone marrow edema/osteitis. With addition of gadolinium contrast agents and employing T1 weighted with fat saturation sequences, areas of increased vascularity can be defined. It is also possible to employ the dGEMRIC protocol to identify early cartilage changes in the hand and wrist of RA patients. MR has been shown to be a very sensitive technique to identify early RA changes and when the measures of edema and bone inflammation are taken together, MR is a good predictor of both lesion progression and clinical benefit of therapeutic intervention.

Structural imaging of the lung by MRI is much more challenging owing to the abundance of air in the lung, the weak signal generated by the low tissue density and the motion artifacts associated with the cardiac cycle and breathing. Gating can be used to overcome the motion artifacts to allow for measurements of edema and fibrosis which are a common result of pulmonary inflammation. Utilization of hyperpolarized helium ([3]He) which is costly and rare and xenon ([129]Xe) which is cheaper and more abundant has enabled the structural and functional evaluation of the lung in normal, chronic obstructive pulmonary disease (COPD) and asthma patients and preclinical animal models. By measuring the ADC following hyperpolarized [3]He ventilation, alveolar size can be determined. With COPD, alveolar size increases as noted by an increase in ADC. Using hyperpolarized ^{129}Xe, alveolar size and functional measures of gas diffusion and lung perfusion can be measured because hyperpolarized ^{129}Xe is soluble, can cross the alveolar membrane into the circulation, and the MR relaxation properties of gaseous and dissolved ^{129}Xe are distinct. A new evolving contrast agent, hyperpolarized krypton (^{83}Kr), has proven useful in generating information about the surface characteristics of the lung alveoli because ^{83}Kr T1 relaxation is affected by the surface chemistry, hydration, and temperature. Inhalation of fluorinated gases such as sulfur hexafluoride, hexafluoroethane, or perfluoropropane and MR imaging may prove to be an alternative to hyperpolarized ^{3}He and ^{129}Xe in the future. While the MR tools such as the hyperpolarized gases exist for assessing lung function, drug intervention has focused on the assessment of lung edema caused by inflammation utilizing budesonide, mometasone, or a selective PDE4 inhibitor.[30] The reason for the limited use of hyperpolarized gas MR imaging in drug discovery may be owing to the need for colocalizing the gas hyperpolarizer with the MRI, the cost and presence of patents governing the use of the technique.

While no work has been published to date, it is possible that MRI and MRS can be applied to the study of infectious diseases and especially BSL3/4 agents for which there are limited therapies, and clinical development is complex because clinical efficacy studies in humans may not be feasible or ethical. As inflammation and respiratory distress tend to predominate in the sequella of the body's response to infectious agents, MRI tools, noted earlier and throughout the article, can be adapted and applied to the study of countermeasures to BSL3/4 pathogens.

18.3.5   MRI/MRS and Drug Safety

Following identification of a viable lead drug candidate and before performance of clinical studies, a series of drug safety studies are performed to determine structural pathologies that are typically assessed by histology. The therapeutic index, i.e., difference between the efficacious dose/plasma drug level and the toxic dose/plasma drug level, is determined to support further development. Surprisingly, there are a few publications on the use of MRI or MRS in the assessment of drug toxicity. Despite the high spatial resolution of MRI and the fact that imaging can dynamically follow the progression of the pathology, more classical

histological and biochemical approaches have been used to support the investigational new drug filings. It is possible that the high cost of establishing the MR facilities, the classical approaches being viewed as the gold standard for regulatory filings, the need to perform an histologic evaluation first to define the pathology before initiating MRI, the cost of qualifying an imaging approach relative to the gold standard, and the need to perform toxicology studies under Good Laboratory Practices (GLP) may account for the limited use of MRI in drug safety studies.

Upon identification of the drug-induced toxicology by classical approaches, structural changes in organ size could be monitored dynamically by serial ^{1}H MRI and the mechanisms associated with the pathology can be evaluated using defined MRI techniques. DWI MRI and calculation of ADC can monitor areas of cell death and necrosis. T2 weighted and STIR sequences can discern areas of edema. For instance, MRI was utilized to measure edema associated with acute exposure to calcium channel blockers (CCBs) and demonstrated that T-type CCDs may reduce vasodilatory edema related to L-type CCBs (Figure 18.2d). Hemorrhage and areas of microthrombi can be assessed through changes in contrast related to the presence of the iron moiety of hemoglobin. High signal intensity in T1W and T2W may be associated with areas of fibrosis. Skeletal muscle damage can be monitored through assessment of the energetic state, that is, ADP/ATP ratio, using ^{31}P-MRS. While MRI/MRS can be applied to the evaluation of drug toxicity, it is important to stress that the organ and type of pathology must be known a priori so that the MR sequences can be selected correctly to detect the change of interest rather than detect an anomaly in the tissue or MR signal. MRI/MRS should not be used in a manner comparable to '-omics' to identify signatures of presumed drug-related toxicology because the relevance of the MRI signatures may or may not be related to the drug or drug toxicology and could inadvertently compromise the development of the drug.

18.4   CONCLUSIONS

MRI and MRS are powerful techniques for use in the discovery and development of new pharmaceutical agents. At a basic level, ^{1}H-MRI can provide dynamic, three-dimensional measures of tissue structure at near histologic resolution. While ^{1}H-MRI has been used to evaluate disease pathology and drug efficacy in such areas as Alzheimer's, OA and atherosclerosis, an unrealized opportunity exists to use MRI to track the progression toward safety endpoints and establish whether the toxicology is reversible. fMRI which includes measures of perfusion, diffusion, necrosis and edema, and MRS measures of biochemical endpoints provide one an opportunity to noninvasively and nondestructively assess a compound's mechanism of action, demonstrate an in vivo pharmacologic response to a new drug and to better define the dose–response relationship associated with efficacy or safety. In summary, MRI and MRS have specific advantages; in that, high-resolution anatomic changes can be monitored over the course of a disease and/or drug intervention and functional measures of a pharmacodynamic effect can be established to demonstrate that target engagement results in a pharmacologic response associated with drug efficacy.

18.5 DISCLAIMER

Opinions, interpretations, conclusions, and recommendations stated within the chapter are those of the authors and are not necessarily endorsed by the US Army. The content of this publication does not necessarily reflect the views or policies of the US Department of Health and Human Services or of the institutions and companies affiliated with the authors. LK performed this work as an independent contractor for Advanced Health Education Center and DMT performed this work as an employee of Tunnell Government Services, Inc. Both companies are subcontractors to Battelle Memorial Institute under its prime contract with NIAID, under Contract No. HHSN272200200016I.

18.6 CONFLICT OF INTERESTS

The authors have no conflicts of interest.

ACKNOWLEDGMENT

Funding for the preparation of the article was provided by Department of Defense Chemical Biological Defense Program through the Defense Threat Reduction Agency (DTRA) under USAMRIID project number 1323839 and Battelle Memorial Institute's prime contract with NIAID (HHSN2722002000161).

RELATED ARTICLES IN EMAGRES

Inversion-Recovery Pulse Sequence in MRI

Relaxation Measurements in Imaging Studies

Cerebral Perfusion Imaging by Exogenous Contrast Agents

Blood Flow: Quantitative Measurement by MRI

Assessment of Regional Blood Flow and Volume by Kinetic Analysis of Contrast-Dilution Curves

Methods and Applications of Diffusion MRI

Brain Infection and Degenerative Disease Studied by Proton MRS

Spatial Localization Techniques for Human MRS

Nuclear Overhauser Effect

Brain: Sensory Activation Monitored by Induced Hemodynamic Changes with Echo Planar MRI

Functional MRI at High Fields: Practice and Utility

Image Processing of Functional MRI Data

Radiofrequency Systems and Coils for MRI and MRS

Human Muscle Studies by Magnetic Resonance Spectroscopy

Hyperpolarized Gas Imaging

REFERENCES

1. P. Morgan, P. H. Van Der Graaf, D. E. Feltner, K. S. Drummond, C. D. Wagner, and S. D. A. Street, *Drug Discov. Today*, 2012, **17**, 419.

2. M. Zaiss and P. Bachert, *Phys. Med. Biol.*, 2013, **58**, R221.

3. J. R. B. O'Connor, A. Jackson, G. J. M. Parker, C. Roberts, and G. C. Jayson, *Nat. Rev. Clin. Oncol.*, 2012, **9**, 167.

4. D. C. Alsop, J. A. Detre, X. Golay, M. Gunther, J. Hendrikse, L. Hernandez-Garcia, H. Lu, B. J. MacIntosh, L. M. Parkes, M. Smits, M. J. P. van Osch, D. J. J. Wang, E. C. Wong, and G. Zaharchuk, *Magn. Reson. Med.*, 2015, **73**, 102.

5. C. Messiou and N. M. deSouza, *Cancer Biomark.*, 2010, **6**, 21.

6. D. S. Margulies, J. Bottger, A. Watanabe, and K. J. Gorgolewski, *Neuroimage*, 2013, **80**, 445.

7. E. Mosconi, D. M. Sima, M. I. O. Garcia, M. Fontanella, S. Fiorini, S. van Huffel, and P. Marzola, *NMR BioMed.*, 2014, **27**, 431.

8. L. Macintyre, T. Zhang, C. Viegelmann, I. J. Martines, C. Cheng, C. Dowdells, U. R. Abdelmohsen, C. Gernert, U. Hentschel, and R. Edrada-Ebel, *Mar. Drugs*, 2014, **12**, 3416.

9. B. G. Jenkins, *Neuroimage*, 2012, **62**, 1072.

10. S. Haller and A. J. Bartsch, *Eur. Radiol.*, 2009, **19**, 2689.

11. F. van Liere Patent No. US 8,049,505 B2 (2011).

12. E. J. Ribot, J. M. Gaudet, Y. Chen, K. M. Gilbert, and P. J. Foster, *Intl. J. Nanomedicine*, 2014, **9**, 1731.

13. J. Smucny, K. P. Wylie, and J. R. Tregellas, *Trends Pharmacol. Sci.*, 2014, **35**, 397.

14. N. D. Kelm, S. Damon, T. M. Bridges, B. J. Melancon, J. C. Tarr, J. T. Brogan, M. J. Avison, A. Y. Deutch, J. Wess, M. R. Wood, C. W. Lindsley, J. C. Gore, P. J. Conn, and C. K. Jones, *Neuropsychopharmacology*, 2014, **39**, 1578.

15. M. G. Shapiro, G. G. Westmeyer, P. A. Romero, J. O. Szablowski, B. Kuster, A. Shah, C. R. Otey, R. Langer, F. H. Arnold, and A. Jasnoff, *Nat. Biotechnol.*, 2010, **28**, 264.

16. L. H. Eraso, M. P. Reilly, C. Sehgal, and E. R. Mohler III, *Vasc. Med.*, 2011, **16**, 145.

17. A. Phinikaridou, M. E. Andia, A. Protti, A. Indermuehle, A. Shah, A. Smith, A. Warley, and R. M. Botnar, *Circulation*, 2012, **125**, 707.

18. M. A. McAteer, J. E. Shneider, Z. A. Ali, N. Warrick, C. A. Bursill, C. von zur Muhlen, D. R. Greaves, S. Neubauer, K. M. Channon, and R. P. Choudhury, *Arterioscler. Thromb. Vasc. Biol.*, 2008, **28**, 77.

19. M. V. McConnell, M. Aikawa, S. E. Maier, P. Ganz, P. Libby, and R. T. Lee, *Arteriloscler. Thromb. Vasc. Biol.*, 1999, **19**, 1956.

20. R. Corti, J. I. Osende, J. T. Fallon, V. Fuster, G. Mizsei, H. Jneid, S. D. Wright, W. F. Chaplin, and J. J. Badimon, *J. Am. Coll. Cardiol.*, 2004, **43**, 464.

21. A. R. Olzinski, G. H. Turner, R. E. Bernard, H. Karr, C. A. Cornejo, K. Aravindhan, B. Hoanh, M. A. Ringenberg, P. Qin, K. B. Goodman, R. N. Willette, C. J. Mecphee, B. M. Jucker, C. A. Sehon, and P. J. Gough, *Arterioscler. Thromb. Vasc. Biol.*, 2010, **30**, 253.

22. M. Krssak, K. Falk Peteresen, A. Dresner, L. DiPietro, S. M. Vogel, D. L. Rothman, M. Roden, and G. I. Shulman, *Diabetologia*, 1999, **42**, 113.

23. E. L. Lim, K. G. Hollingsworth, F. E. Smith, P. E. Thelwall, and R. Taylor, *Clin. Sci.*, 2011, **121**, 169.

24. A. Degrassi, M. Russo, and C. Nanni, *Mol. Cancer Ther.*, 2010, **9**, 673.

25. T. Nielsen, T. Wttenborn, and M. R. Horsman, *Pharmaceutics*, 2012, **4**, 563.

26. L. Lee, S. Sharma, B. Morgan, P. Allegrini, C. Schnell, J. Brueggen, R. Cozens, M. Horsfield, C. Guenther, W. P. Steward, J. Drevs, D. Lebwohl, J. Wood, and P. M. J. McSheehy, *Cancer Chemother. Pharmacol.*, 2006, **57**, 761.

27. C. C. Zhang, Z. Yau, A. Giddabasappa, P. B. Lappin, C. L. Painter, Q. Zhang, G. Li, J. Goodman, B. Simmons, B. Pascual, J. Lee, T. Levkoff, T. Nichols, and Z. Xie, *Cancer Med.*, 2014, **3**, 462.

28. S. A. Moestue, O. Engebraaten, and I. S. Gribbestad, *Mol. Oncol.*, 2011, **5**, 224.

29. S. J. Matzar, J.van Tiel, G. E. Gold, and E. H. G. Oei, *Quant. Imaging Med. Surg.*, 2013, **3**, 162.

30. Z. Liu, T. Araki, Y. Okajima, M. Albert, and H. Hatabu, *Eur. J. Radiol.*, 2014, **83**, 1282.

Chapter 19

Practical Applications of NMR Spectroscopy in Preclinical Drug Metabolism Studies

Raman Sharma and Gregory S. Walker

Pfizer-PDM (Pharmacokinetics Dynamics and Metabolism), MS 8220-4374, Eastern Point Road, Groton, CT 06340, USA

19.1 INTRODUCTION

Over the past two decades, the pharmaceutical industry has seen a significant increase in the success rates of drug candidates owing, largely, to an improved ability to predict the absorption, distribution, metabolism, and excretion (ADME) properties of drug molecules. Before 1990, it was estimated that over 40% of drug failures were caused by poor prediction and understanding of drug ADME properties. By the year 2000, this number had dropped significantly to 10%, and since then, numbers as low as 1–3% have been cited.[1] These advances have been made possible, largely, by

NMR in Pharmaceutical Sciences. Edited by Jeremy R. Everett, John C. Lindon, Ian D. Wilson, and Robin K. Harris
© 2015 John Wiley & Sons, Ltd. ISBN: 978-1-118-66025-6
Also published in eMagRes (online edition)
DOI: 10.1002/9780470034590.emrstm1412

increased investment and attention earlier in the drug discovery phase resulting in higher quality compounds entering into development. This early investment also results in a decrease in the overall costs associated with drug discovery and development.

Critical to advances in the understanding of the ADME properties of drugs have been parallel advances in the analytical tools used for data generation. Among these tools, the two most commonly used are mass spectrometry and NMR spectroscopy. Technological enhancements in both of these analytical techniques over the past few decades have provided scientists with the ability to obtain critical quantitative and qualitative information about drug molecules. This discussion will focus mainly on the application of NMR in preclinical ADME drug discovery. Some context will be provided in discussing the different types of experiments/studies usually conducted during this period and the role NMR plays in these studies. Figure 19.1 depicts a high level overview of the typical ADME-related studies that are conducted during the path to drug approval.

19.2 NMR IN THE REALM OF ADME DISCOVERY

During the early and middle stages of drug discovery (also called lead identification and optimization), a number of in vitro experiments are usually conducted, often in a high throughput manner, with the goal

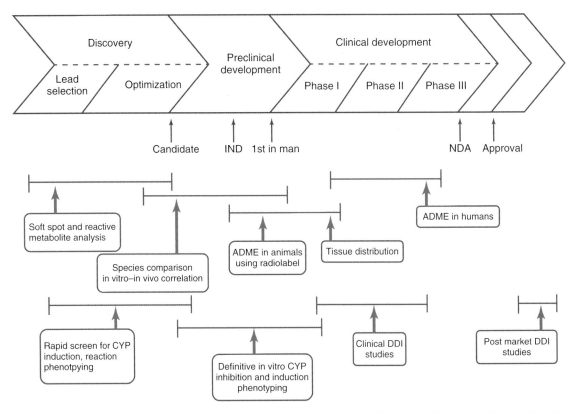

Figure 19.1. High level overview of drug discovery and development process. (Reproduced from John Wiley & Sons, Ltd. © 2011)

of identifying potential ADME liabilities of a compound. On the basis of the data derived from these experiments, molecules can then be redesigned to eliminate or minimize these liabilities. If redesign proves to be unfeasible, discontinuing the molecule earlier in discovery enables resources to be applied to compounds with a greater potential for success in the development phase.

A brief summary of some of the typical types of ADME experiments conducted in the early stages of drug discovery and the role NMR spectroscopy plays in these studies is as follows.

19.2.1 Soft Spot Analysis

There are a number of organs in the body, which contain drug-metabolizing enzymes that convert or 'biotransform' drugs into chemically distinct molecules called *metabolites*. Among these organs, the liver has by far the highest concentration of these enzymes

and is responsible for biotransformation of most drug molecules. In preclinical drug discovery, a drug candidate is often incubated in vitro with liver cells (hepatocytes) or subcellular fractions obtained from the liver (microsomes and cytosol), to enable prediction of how this candidate would be metabolized in vivo. A typical microsomal incubation would consist of incubating a compound at 10 μM in phosphate buffer (pH 7.4), NADPH, and human liver microsomes with appropriate cofactors for 60 min. The reaction would then be terminated with the addition of an organic solvent (usually acetonitrile or methanol). The sample would then be centrifuged and an aliquot of the resulting supernatant would be analyzed by liquid chromatography mass spectrometry (LC/MS) to look for metabolites. There are a number of metabolic modifications that drugs generally undergo, which can be broadly classified into two categories: Phase I (oxidative and reductive) type reactions and Phase II (conjugative) type reactions (Table 19.1). Usually, but not always,

Table 19.1. Common types of biotransformations

Chemical modification	Description	Isotopic mass change
$R-CH_2C_6H_5 \Rightarrow R-H$	Debenzylation	-90.0468
$R-C_2H_5R-H$	Deethylation	-28.0312
$R-CH_3 \Rightarrow R-H$	Demethylation	-14.0157
$R-CH_2-CH_2-R' \Rightarrow R-CH=CH-R'$	Desaturation	-2.0157
$R-CH_2-NH_2 \Rightarrow R-CH_2-OH$	Oxidative deamination to alcohol	0.984
$R-CO-R' \Rightarrow R-CHOH-R'$	Ketone to alcohol	2.0157
$R-X \Rightarrow R-X-CH_3$	(O, N, S) methylation	14.0157
$R-H \Rightarrow R-OH$	Hydroxylation	15.9949
$R-CH=CH-R \Rightarrow R-CH_2-CHOH-R'$	Hydration, hydrolysis (internal)	18.0106
$R-NH_2 \Rightarrow R-NHCOCH_3$	Acetylation	42.0106
$R-COOH \Rightarrow R-CONHCH_2COOH$	Glycine conjugation	57.0215
$R-OH \Rightarrow R-OSO_3H$	Sulfate conjugation	79.9568
$R-COOH \Rightarrow R-CONH-CH_2CH_2SO_3H$	Taurine conjugation	107.0041
$RR'-CH_2 \Rightarrow RR'-CH-SCH_2CHNH_2-COOH$	Cystein conjugation	119.0041
$RR'-CH_2 \Rightarrow RR'-CH-SCH_2CHNCOCH_3-COOH$	*N*-Acetylcysteine conjugation	161.0147
$R-OH \Rightarrow R-O-C_6H_8O_6$	Glucuronide conjugation	176.0321
$+ C_{10}H_{15}N_3O_6S$	Glutathione conjugation	305.0682

Phase I metabolism
Phase II metabolism

these metabolic modifications result in an increase in the polarity of a molecule that facilitates its removal (or 'clearance') from the blood and elimination via excretion into the urine or feces. Obviously, in order for a drug to have its desired effect, it must stay in the body, thus one method of reducing the clearance of a drug is to structurally modify or block the sites of the molecule that are vulnerable to metabolism in order to prevent its biotransformation. Thus, a 'soft spot' analysis is a common experiment in the drug discovery phase, which consists of incubating a compound in the presence of liver fractions (microsomes, hepatocytes, S9, and cytosol) and looking for formation of metabolites. The resulting metabolites are usually identified by LC/MS, and if a definitive assignment cannot be made by this approach (which is often the case), the metabolites of interest can be further isolated from the incubation media using preparative LC and the metabolite structures elucidated through a combination of MS and NMR data. Once the structure of the metabolites is known, chemical design strategies can then be implemented to block the sites of metabolism. Usually, this will result in a reduction of drug clearance but occasionally may have unintended consequences. The changes in molecular structure may make other areas of the molecule more susceptible to drug-metabolizing enzymes and shift the metabolism from one portion of the molecular motif to another.

Obviously, altering a molecule will change its physicochemical properties, which may affect its ability to be absorbed and alter its pharmacological activity. Thus, modification or strategies must consider all of these factors in order to be successful.

The metabolism of compound 1 is an example of where structural elucidation via NMR spectroscopy played a critical role in understanding metabolic spots. Compound 1 was a small molecule GPR 119 agonist candidate for treatment of type II diabetes.[2] Compound 1 exhibited a rapid unexplained half-life in rodent species (rat, mouse, and guinea pig). As rodents are generally used in drug pharmacology and toxicity models, instability in these species is generally deemed to be unacceptable. Metabolic profiling of rat plasma by LC/MS revealed the reason behind the instability to be the conversion of parent drug into a single major mono-hydroxylated metabolite (Figure 19.2). Incubation of compound 1 in various liver fractions (cytosol, microsomes, and hepatocytes) in the presence of specific enzyme inhibitors implicated an enzyme known as *xanthine oxidase* in the formation of the metabolite. Xanthine oxidase is a molybdenum hydroxylase enzyme known for oxidations of nitrogen heterocycles to the corresponding lactams. On the basis of the understanding of the enzymology alone, the location of oxidation was narrowed down to either the cyanopyrimidine

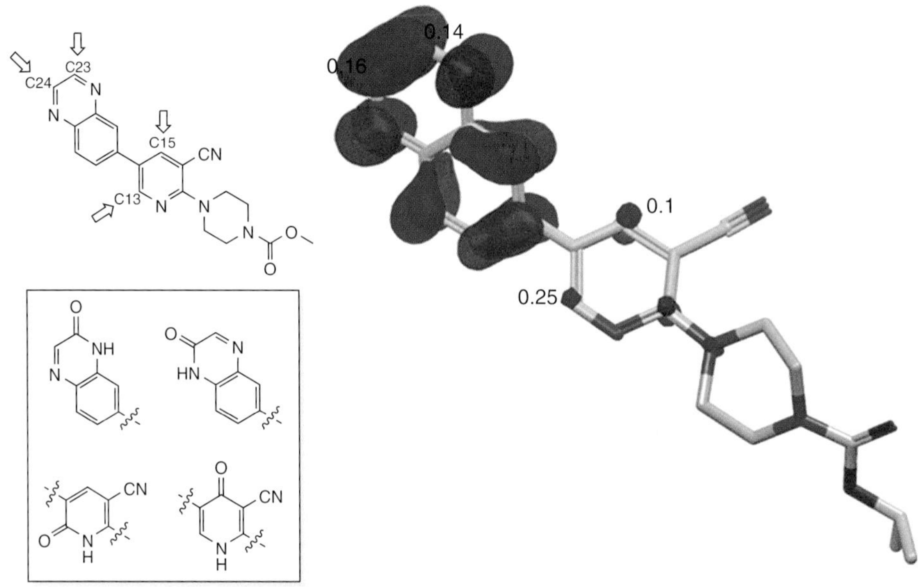

Figure 19.2.   Putative sites of oxidation on Compound 1

or quinoxaline rings of compound 1 (Figure 19.2). The collision-induced spectra (CID) of the metabolite obtained by MS further supported this conclusion; however, MS was unable to discriminate between the quinoxaline and cyanopyrimidine rings (Figure 19.3) as the site of oxidation. Subsequently, a sample of the metabolite of compound 1 was isolated from rat plasma and the structure elucidated by NMR spectroscopy. The cyanopyrimidine ring was quickly ruled out as the site of metabolism. Further analysis of 1D and 2D NMR spectral data sets revealed that the C20, C27, and C28 positions of the quinoxaline ring were also unmodified. The challenging part of the analysis involved differentiating the site of oxidation between the two methine carbons adjacent to nitrogen on the quinoxaline ring (positions C23 and C24), which possessed similar multiplicity and proton and carbon chemical shift values. As 1D NMR spectroscopic data was insufficient for delineating between these two positions, a heteronuclear multiple-bond correlation (HMBC) experiment was conducted, which showed correlation between C24 and C26, implicating C23 as the site of oxidation (Figure 19.4). Subsequent in silico molecular docking of compound 1 into the active site of xanthine oxidase, using a published crystal structure, further corroborated C23 as the site of hydroxylation.[2] As a result of these studies, an analog of compound 1

was synthesized, which lacked the quinoxaline motif (5-(4-cyano-3-methylphenyl)-2-(4-(3-isopropyl-1,2,4-oxadiazol-5-yl)piperidin-1-yl)nicotinitrile (2)) and proved to be stable in rodent plasma while possessing improved GPR119 agonist properties. Such structural information would have been difficult to ascertain in the absence of NMR spectroscopic data.

19.2.2   Reactive Metabolite Screening

Of significant importance in the drug industry is the ability to predict the propensity of a molecule toward forming reactive metabolites. Reactive metabolites have been implicated in the phenomenon known as *idiosyncratic toxicity* (cited as the fourth leading cause of death in the United States).[3] As a result of idiosyncratic toxicity, a number of drugs have either been given black box warnings by the FDA or been withdrawn from the market altogether. Reactive metabolites have been widely implicated as one of the main causes of idiosyncratic toxicity. A number of mechanisms regarding how reactive metabolites cause toxicity have been proposed but it is generally believed that covalent modification (i.e., binding of the reactive portion of the drug to macromolecules, DNA, etc.) elicits a toxic/immune response. Thus, knowing the intrinsic reactivity of a drug, or the potential for it

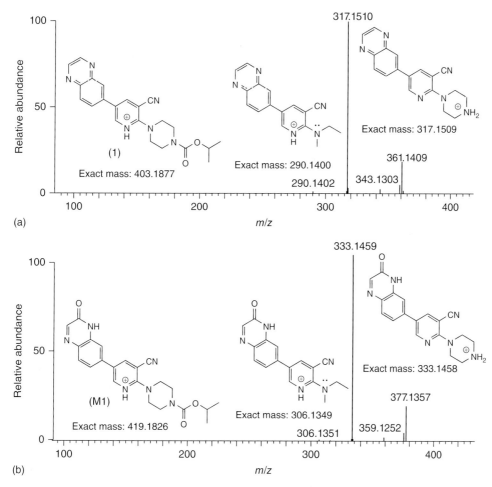

Figure 19.3. Collision-induced dissociation (CID) spectra of compound 1 (a) and its metabolite (b)

to become reactive through enzymatic modification, is important to understand early in the life of a molecule. Reactive metabolite screens are widely employed in the pharmaceutical industry to detect reactive metabolites using endogenous nucleophilic trapping reagents such as glutathione (GSH) or cysteine. Depending on the expected dose of the drug and significance of the reactive pathway with respect to other metabolic pathways, the reactive motif is often modified or eliminated from the drug substance, and in some cases, the compound is terminated from further development. Screens are usually conducted in liver microsomal fractions supplemented with the trapping reagent (GSH or cysteine) and incubated for a sufficient time (30 min–1 h) for generation and trapping of the reactive species. As with soft spot analysis, CID spectra

alone are often incapable of precisely pinpointing the sites of modification. In these situations, the metabolite may be isolated from the appropriate matrix and analyzed by NMR spectroscopy. The utility of NMR spectroscopy for determining the site of conjugation to the trapping reagent was demonstrated with compound 2, a small molecule glucokinase activator candidate, for the treatment of type II diabetes. Compound 2 was observed to be cleared rapidly from circulation in rats. Metabolic profiling of compound 2 in human hepatocytes revealed the major metabolic pathway to involve conjugation of GSH to the parent drug, concomitant with a loss of the methyl sulfone motif. Interestingly, the GSH conjugate was also observed in control incubations that lacked microsomes suggesting that the compound was intrinsically reactive. In

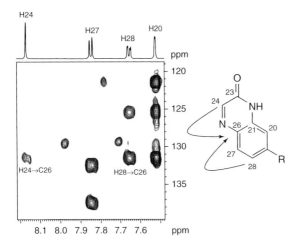

Figure 19.4. ^1H–^{13}C HMBC spectrum compound 1 metabolite

or altering the reactive site, it was necessary to isolate the GSH conjugate from the buffer incubations for analysis by NMR spectroscopy. TOCSY and HMBC experiments (Figure 19.5) were able to clearly demonstrate connectivity between position 1 (the methylene protons adjacent to sulfur) and position B on the pyridine-one ring. This structure suggested a mechanism in which GSH had attacked directly at the carbon attached to the methyl sulfone motif likely through and SN2 type addition–elimination reaction with addition of GSH and elimination of the methyl sulfone (Figure 19.6). As a result of this analysis, subsequent analogs lacking methyl sulfone leaving group were pursued, which were devoid of the reactive metabolite liability while still retaining satisfactory glucokinase activator properties.[4]

19.2.3 Kinetic Measurements of Acyl Glucuronides

Another potential reactive metabolic pathway often implicated in idiosyncratic toxic events is glucuronidation of carboxylic acids. These reactive intermediates, referred to as *acyl glucuronides*, have been linked to toxicity in a certain classes of nonsteroidal

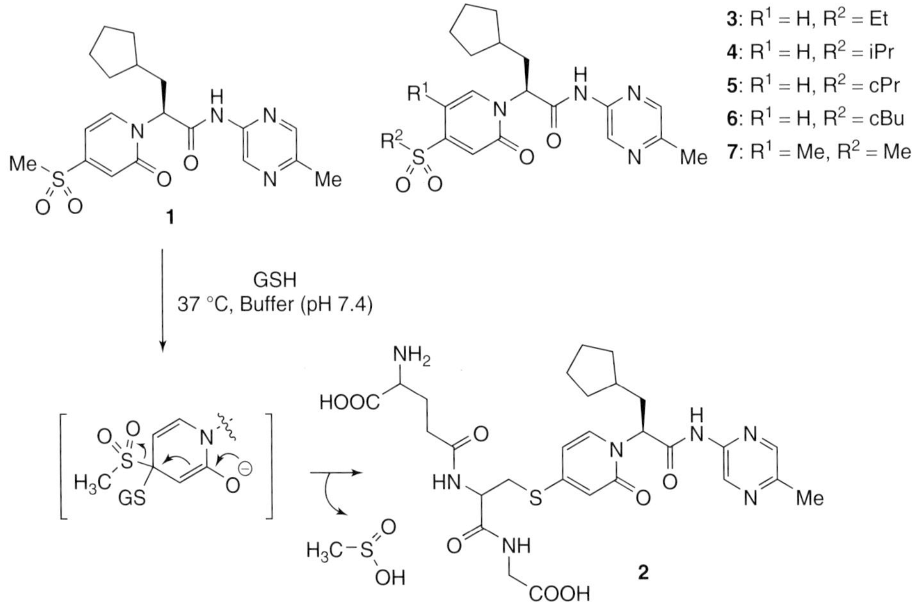

fact, incubation of compound 2 in phosphate buffer (pH 7.4) with 10 mM GSH alone led to complete conversion of compound 2 into the GSH conjugate.

As with the previous example, a precise structural assignment could not be achieved by MS fragmentation alone. In order to determine the site of conjugation and guide chemistry design strategies for blocking

3: R^1 = H, R^2 = Et
4: R^1 = H, R^2 = iPr
5: R^1 = H, R^2 = cPr
6: R^1 = H, R^2 = cBu
7: R^1 = Me, R^2 = Me

Figure 19.5. HMBC and TOCSY correlations for compound 2 glutathione conjugate

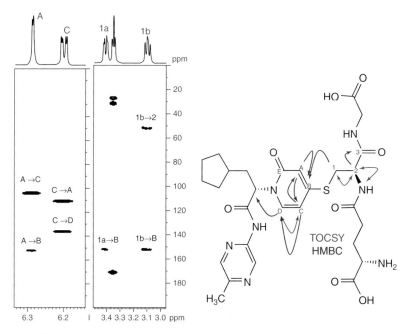

Figure 19.6. Structure of compound 2 and proposed mechanism for formation of the glutathione conjugate

anti-inflammatory drugs that contain carboxylic functional groups.[5] The toxicity is thought to be related to a phenomenon known as *acyl migration.* The Structure–Activity Relationship (SAR) around acyl migration suggests substitution on the carbon alpha to the carboxylic acid reduces the rate of migration. Take, for example, the common over pain medication Ibuprofen contrasted with its analog Ibufenac that has a known history of serious adverse toxic reactions.[6] The only structural difference between the two molecules resides in a methyl substitution at the α position adjacent to the carboxylic acid in Ibuprofen, which is absent in Ibufenac (Figure 19.7). The pronounced difference in toxicity profiles between Ibuprofen and Ibufenac have been attributed to the differences in the migration rate of the acyl glucuronide metabolite formed in both compounds. Drugs that have migration rates with a $T_{1/2}$ of less than 1.5 h have either been withdrawn from the market or received black box warnings. A proposed pathway leading to toxicity of acyl glucuronides is outlined in Figure 19.8, specifically pathway A, where rearrangement of the glucuronide through an intramolecular attack of the carbonyl group by the adjacent alcohol of the sugar causes the glucuronide to 'migrate'. The hemiacetal that results from this process can

then hydrolyze to the corresponding aldehyde, which in turn reacts with amino residues of proteins in the body to illicit a toxic/immune response. Thus, any substitutions at the α position of the acyl group (such as methyl group on Ibuprofen) can create steric hindrance, which reduces the rate of migration. The inherent reactivity of glucuronides follows the general rate order of acetic acid > proprionic acid > benzoic acid.[7] The ability to obtain kinetic rates for migration of acyl glucuronides is critical early in drug discovery and can serve as a benchmark for toxicity risk. An example of this is found with Compound A, a CRTh2 antagonist for treatment of asthma, which possessed a carboxylic acid unsubstituted at the α position that formed an acyl glucuronide. The glucuronide was found to migrate very rapidly as judged by its quick degradation in vitro. As a consequence of the rapid rate of acyl migration, a half-life could not be accurately measured by LC/MS owing to the length of time required to run the sample. Through a simple 1D proton measurement of the anomeric resonance of the glucuronide, a half-life of 0.2 h was easily calculated (Figure 19.9). As the projected human dose of compound A was rather high >100 mg coupled with the fact that glucuronidation was the

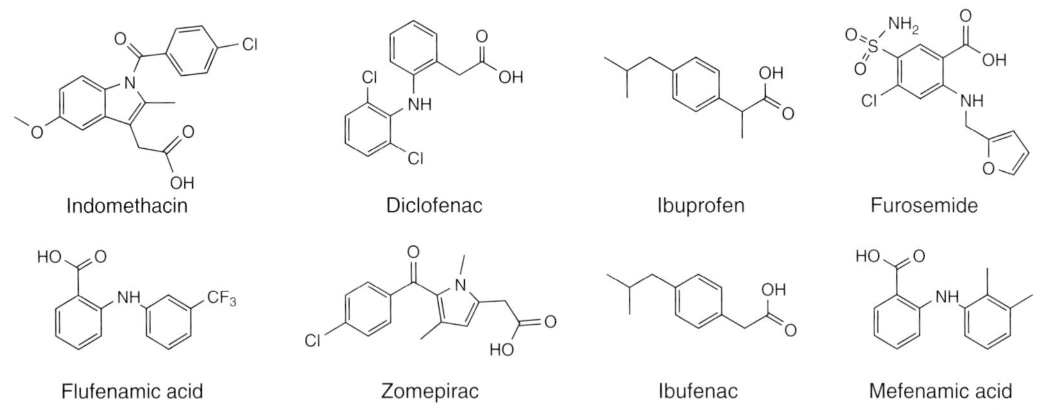

Indomethacin   Diclofenac   Ibuprofen   Furosemide

Flufenamic acid   Zomepirac   Ibufenac   Mefenamic acid

	Literature half-life (h)	NMR determined half-life (h)	References
Diclofenac	0.5	0.6	(23)
Zomepirac	0.6	0.5	(4)
Ibufenac	1.1	1.1	(18)
Indomethacin	1.5	1.4	(25)
Ibuprofen	3.3	4.6	(18)
Furosemide	3.6	5.3	(38)
Flufenamic acid	7.0	10	(25)
Mefenamic acid	22	17	(23)

Compounds with $t_{1/2}$ of less than 1.5 h have either been withdrawn from the market or have been issued black box warnings

Figure 19.7. List of nonsteroidal inflammatory drugs (NSAID'S) and rates of acyl migration calculated by measuring NMR half-life

major metabolic pathway, it was decided to terminate this compound, and other compounds in the series that shared the α unsubstituted carboxylic acid structural motif, and move on to different chemical classes.

19.3 METABOLITE CONCENTRATIONS VERSUS INSTRUMENT SENSITIVITY

As highlighted in the previous examples, an important point to consider is that contrary to other areas of preclinical drug development, where quantities of drug substances are usually available in the milligram scale, drug metabolites generated in in vitro/in vivo systems are significantly lower (microgram per milliliter or less) and exist in highly complex matrices such as liver fractions, blood, urine/feces, or tissue. As can be seen with the previous examples, MS fragmentation, while a very helpful tool in assigning metabolite structure,

is often incapable of definitively elucidating structures based on fragmentation pattern alone. It is in these cases where NMR spectroscopy has proved to be an invaluable tool.

Since its inception in the late 1940s NMR spectroscopy has grown to be arguably the most valuable analytical tool for unambiguous structural characterization. NMR data provides information on the number of certain specific nuclei (see below) within a compound, as well as the micro-chemical environment of each of these through chemical shift and connectivity information through coupling constants. This diversity of information makes NMR spectroscopic data, when used in tandem with MS data, a very useful structural elucidation tool.

When used in preclinical ADME drug discovery, NMR does have its limitations, mostly in the area of sensitivity. Only specific nuclei are NMR active, those with a quantum spin of $1/2$. Hydrogen (^{1}H), carbon (^{13}C), nitrogen (^{15}N), and fluorine (^{19}F) are

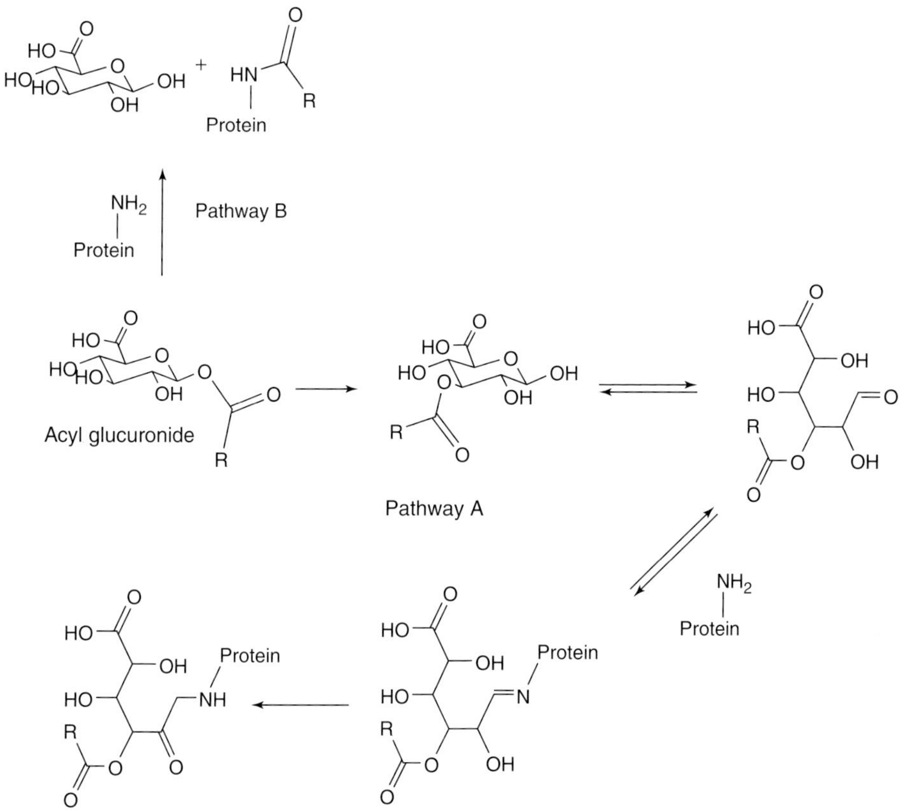

Figure 19.8. Proposed mechanism for acyl glucuronide reactivity

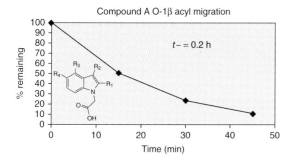

Figure 19.9. Kinetic measurement of compound A acyl glucuronide half-life by NMR spectroscopy

perhaps the most useful NMR active nuclei to a drug metabolism scientist. The sensitivity of experiments targeting these nuclei, in some cases, is further limited by the lack of an abundant active isotope. As an

example, the NMR active nucleus of carbon is the ^{13}C isotopic form, which is only 1.1% of total carbon pool. Obviously this limits the sensitivity of experiments utilizing this nucleus.

In the last 15 years, many advances have been made to overcome this relative lack of sensitivity. New hardware incorporating cryo-probes, NMR probes with small cell volumes and higher field instruments, have increased NMR spectroscopic sensitivity dramatically. Additionally, new more sensitive pulse sequences have enhanced the ability to establish structural connectivity through long range heteronuclear (^{1}H–^{13}C and ^{1}H–^{15}N) correlations.

Another limitation of NMR spectroscopy, when performing experiments for metabolite structural elucidation, is the need for relatively pure sample. The need for an isolated sample has long been a stumbling block for drug metabolism scientists. Although there has been significant improvement in preparative chromatography columns and HPLC systems, a considerable amount of effort is required to isolate a pure metabolite from biological matrices, where it is present only in small quantities. An alternative to preparative isolation is LC–NMR spectroscopy, where in much the same manner as LC–MS, an NMR spectrometer is interfaced with a chromatography system (see Chapter 15). While useful in certain applications, sensitivity, solvent suppression, column loading, and computer automation can all be problematic. However, when a metabolite is unstable, or recalcitrant to isolation, HPLC–NMR spectroscopy may be the best option for acquisition of meaningful NMR spectral data.

19.4 SAMPLE ISOLATION

The isolation of a metabolite for spectroscopic characterization starts by identifying an appropriate source material. The source material may be excreta, serum, plasma, or blood from an in vitro study or more frequently incubates from isolated cell or enzyme system incubations. In all of these cases, screening several systems, usually by HPLC–MS, to identify the material that produces the most metabolite of interest, is necessary.

Once the source material for metabolite isolation has been identified, the sample is usually deproteinated through precipitation with an organic solvent. The insoluble proteins are then removed by centrifugation, and the supernatant decanted. The supernatant contains a mixture of drug-related material (metabolites and parent compound) and endogenous small molecules. This complex mixture is then separated via HPLC. Depending on the volume, the supernatant can either be directly loaded on to an HPLC column using a separate HPLC pump or via multiple small volume injections. The mixture is then separated using gradient elution, and the effluent is collected in fractions of 0.25–1 min. Each fraction is then reanalyzed by HPLC–MS to identify which contains the metabolites of interest. The fractions containing drug-related material are then dried and redissolved in deuterated solvents for NMR spectroscopic analysis.

While sample isolation is critical to the success of any drug metabolism NMR study, a complete discussion of this is beyond the scope of this chapter. The description above provides only a broad overview of the process.

19.5 QUANTITATIVE NMR (QNMR) SPECTROSCOPY

While the discussion above demonstrates the utility of NMR as a qualitative technique in the preclinical drug discovery setting, NMR is also useful as a quantitative technique in this environment. What makes NMR spectroscopy a unique quantitative analytical technique is the uniform response of the hydrogen atoms. Simply, if the 1D 1H acquisition is properly calibrated and executed the response from a single hydrogen for 1 nmol of a steroid molecule will equal the response from a single hydrogen from 1 nmol of a sugar molecule. This enables concentration response standard curves to be constructed from unrelated chemotypes. An example of this is contained in Figure 19.10. Six compounds were prepared in six separate weighings and dissolved in DMSO-d6, resulting in six samples of varying concentrations from approximately 0.1 to 30 mM. 1D 1H data was acquired on each sample using a standard Bruker pulse sequence. From these acquired spectra, specific resonances from each molecule were integrated and normalized to the appropriate number of hydrogens. These compounds and the resonances selected for integration were chosen to maximize the diversity chemical structure, chemical shift, and multiplicity of the data set. The concentration versus response curve from this data is linear over approximately two orders of magnitude.

This concept is important to the drug metabolism scientist because it allows the quantification of a structural unknown, such as a previously uncharacterized metabolite, without a fully qualified reference standard. Within the NMR community the quantitative aspects of NMR spectroscopy are well known. This understanding is much less common within the drug metabolism discipline. The advantages of quantifying an isolated metabolite after its structural characterization has been completed have only been routinely capitalized on over the past 5 years.

qNMR spectroscopy is typically approached in two different ways: either as an assessment of purity or as a determination of concentration. When qNMR spectroscopy is used as a determination of purity, the procedure is gravimetrically based. Typically, milligram quantities of material (a small subset of the total material available) are accurately weighed and

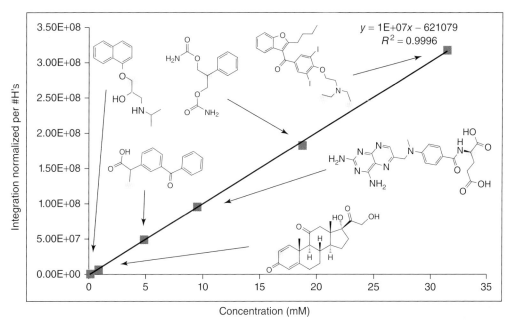

Figure 19.10. 1H NMR Concentration versus response curve of six chemically distinct molecules at various concentrations integrated and normalized to the appropriate number of hydrogens. Individual samples were prepared in DMSO-d6. Data were acquired on a Bruker 600 MHz NMR system with 1.7 mm TCI cryoprobe. The system was controlled, and the data processed using Topspin 3.2. Forty transients were acquired for each spectrum with a total recycle time of 10 s between the transients

volumetrically diluted resulting in a sample with a known mass per volume ratio. Because the initial mass and molecular weight are known, the theoretical concentration can be easily calculated. The nominal concentration can then be determined by various NMR spectroscopic techniques. The percent purity can then be calculated based on the difference between the theoretical concentration of the gravimetrically prepared sample and the concentration determined by NMR spectroscopic analysis. While the percent purity approach is most familiar to medicinal chemist and very useful in situations where there is an abundance of source material, it has limited utility in drug metabolism studies where the isolated materials are in the low microgram range.

In the case of drug metabolism studies, because of the small amounts of material isolated (10's of μg), weighing the sample is not feasible. Instead the isolated material is dissolved in a deuterated solvent (typically DMSO because of its low volatility) and the concentration of the sample is determined directly. This entire sample is then used as a stock for the preparation of analytical standards that can be used in a variety of drug metabolism studies.

Within these two approaches described above there are multiple ways to achieve quantification via NMR spectroscopy. The classic method is the addition of a known amount of an internal standard to the sample.[8] As an alternative to adding an internal standard to the sample, the residual solvent resonance can be calibrated and used as an internal standard. The ERITIC method, where a second RF signal is calibrated and overlaid on to the FID during the acquisition, has also been widely used.[9] The most straightforward methods are based on PULCON (including aSICCO, QUANTUS, and ERETIC2).[10–12]

Compound C is an example of the successful role qNMR spectroscopy may play in drug metabolism studies. LC–MS analysis indicated as major metabolite was a single site oxidation. The major metabolite of compound C was isolated from microsomal incubations and the chromatographic solvents removed. The sample was then prepared for NMR spectroscopic analysis in a dry glove box under argon. 1D and 2D homo- and heteronuclear analysis demonstrated that the single site oxidation was on the *tert*-butyl group designated as C24, 28, and 29. The 1D NMR system had been previously calibrated using 10 mM maleic

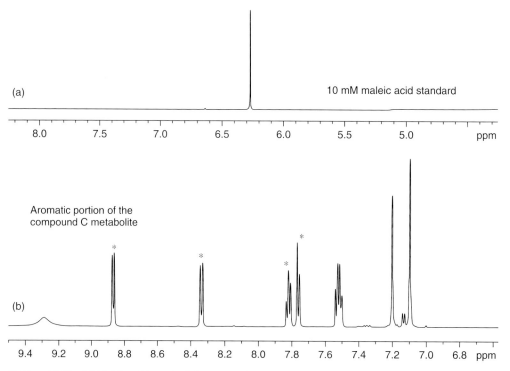

Figure 19.11. (a) 10 mM Maleic acid standard. (b) Aromatic portion of the Compound C metabolite 1 isolate. Data for both spectra were acquired on a Bruker 600 MHz NMR system with 1.7 mm TCI cryoprobe. The system was controlled and the data processed using Topspin 3.2. Data was acquired on a Bruker 600 MHz NMR system with 1.7 mm TCI cryoprobe. The system was controlled and the data processed using Topspin 3.2. Concentrations were determined using Bruker ERETIC2 software contained in Topspin 3.2

acid, thus enabling the calculation of the concentration of the isolate (Figure 19.11). The sample, now with a known structure and known concentration, was then used a substrate to determine the number of ADME-related parameters, including blood-to-plasma ratios, human plasma protein binding, and in vitro efficacy determinations. These data provided a better understanding of the contributions of metabolite C1 to preclinical in vivo efficacy models of compound C.[13]

19.6 DISCUSSION

The role of NMR spectroscopy in drug discovery has evolved considerably over the past few decades. The ability to determine qualitative, kinetic, and quantitative data for drug molecules at low levels has enabled critical information to be gathered early in the drug discovery process providing major influence on decisions to pursue drug candidates. These advances have been relatively recent in the preclinical ADME drug discovery space, where the availability qualified metabolite reference standards, previously not available, have enabled project teams to gain early knowledge about drug/metabolite pharmacology relationships and obtain quantitative levels of metabolites in biological matrices. Newer, more sensitive, NMR probes with smaller cell volumes and higher field instruments have allowed 1D and 2D structural characterization of unweighable amounts of isolated metabolites. An emerging application of NMR spectroscopy in drug discovery has been structural characterization of metabolites generated from various enzyme systems (animal, microbial, and bacterial) for use as chemical matter in pursuing new lead compounds. Undoubtedly, the role of NMR spectroscopy in preclinical ADME discovery continues to evolve as new ways to exploit its capabilities are uncovered.

REFERENCES

1. I. Kola and J. Landis, *Nat. Rev. Drug Discov.*, 2004, **3**, 711.

2. R. Sharma, H. Eng, G. S. Walker, G. Barreiro, A. F. Stepan, K. F. McClure, A. Wolford, P. D. Bonin, P. Cornelius, and A. S. Kalgutkar, *Chem. Res. Toxicol.*, 2011, **24**, 2207.

3. J. Lazarou, B. H. Pomeranz, and P. N. Corey, *J. Am. Med. Assoc.*, 1998, **279**, 1200.

4. J. Litchfield, R. Sharma, K. Atkinson, K. J. Filipski, S. W. Wright, J. A. Pfefferkorn, B. Tan, R. E. Kosa, B. Stevens, M. Tu, and A. S. Kalgutkar, *Bioorg. Med. Chem. Lett.*, 2010, **20**, 6262.

5. M. J. Bailey and R. G. Dickinson, *Chem. Biol. Interact.*, 2003, **145**, 117.

6. M. Castillo and P. C. Smith, *Drug Metab. Dispos.*, 1995, **23**, 566.

7. S. L. Regan, J. L. Maggs, T. G. Hammond, C. Lambert, D. P. Williams, and B. K. Park, *Biopharm. Drug Dispos.*, 2010, **31**, 367.

8. F. Malz and H. Jancke, *J. Pharm. Biomed. Anal.*, 2005, **38**, 813.

9. S. Akoka, L. Barantin, and M. Trierweiler, *Anal. Chem.*, 1999, **71**, 2554.

10. R. D. Farrant, J. C. Hollerton, S. M. Lynn, S. Provera, P. J. Sidebottom, and R. J. Upton, *Magn. Reson. Chem.*, 2010, **48**, 753.

11. R. Espina, L. Yu, J. Wang, Z. Tong, S. Vashishtha, R. Talaat, J. Scatina, and A. E. Mutlib, *Chem. Res. Tox.*, 2009, **22**, 299.

12. G. S. Walker, R. Ryder, R. Sharma, E. B. Smith, and A. Freund, *Drug Metab. Dispos.*, 2011, **39**, 433.

13. G. S. Walker, J. N. Bauman, T. F. Ryder, E. B. Smith, D. K. Spracklin, and R. S. Obach, *Drug Metab. Dispos.*, 2014, **42**, 1627.

Chapter 20

Preclinical Drug Efficacy and Safety Using NMR Spectroscopy

Muireann Coen and Ian D. Wilson

Computational and Systems Medicine, Department of Surgery and Cancer, Imperial College London, Sir Alexander Fleming Building, South Kensington, London SW7 2AZ, UK

20.1 INTRODUCTION

Metabonomics, or the 'the quantitative measurement of the multiparametric metabolic response of living systems to pathophysiological stimuli or genetic modification',[1,2] has found widespread application in the field of toxicology and is increasingly finding applications in drug discovery and development with respect to biomarker discovery.[3] In metabonomics (and the related technique of metabolomics or metabolic profiling), samples of biofluids or tissues are analyzed to produce comprehensive, untargeted, 'global' metabolite profiles. The methods used to obtain these profiles are ideally based on analytical techniques such as NMR spectroscopy or mass spectrometry (MS) that have intrinsically high information content and are therefore capable of providing a comprehensive and detailed biochemical 'fingerprint' of the sample. While 'omics' studies are generally considered to be hypothesis free, there is a clear expectation that these metabolic profiles will reflect the state of the organism under study as the concentrations of metabolites alter in response to changes induced by toxicity, pharmacology, or disease on cellular and organism homeostasis. Understanding these changes will thus, potentially, lead to new hypotheses and insights into the underlying biochemical mechanisms. Metabonomics depends on the generation of 'global' metabolic profiles that reflect the time-related metabolic fluctuations arising as organisms respond to toxic insults or disease progression and as such generates complex data sets that require advanced statistical methods to analyze them. The spectral data are used to build models that can be interrogated to discover biomarkers that may then provide novel mechanistic information and may hold predictive prognostic potential. Here, the NMR-based techniques used for global metabolic profiling are detailed, together with a brief discussion of statistical methods of data treatment and illustrative examples

NMR in Pharmaceutical Sciences. Edited by Jeremy R. Everett, John C. Lindon, Ian D. Wilson, and Robin K. Harris
© 2015 John Wiley & Sons, Ltd. ISBN: 978-1-118-66025-6
Also published in eMagRes (online edition)
DOI: 10.1002/9780470034590.emrstm1408

of the application of these techniques in preclinical toxicology.

20.2 SAMPLE TYPES

The key to any successful metabolic profiling investigation is determined by careful study design, sample collection, and sample treatment, which are essential factors for obtaining relevant and reliable results. The available NMR-based methods can cover many of the sample options available for biomedical or toxicological studies including common biofluids such as urine, bile, blood plasma or serum, saliva, intact tissues, and cell/tissue extracts. Urine is an ideal sample as it can be obtained noninvasively throughout an investigation, thereby providing a time course to enable the onset and severity of toxicity to be determined together with adaptive or regenerative processes (thus pinpointing the optimum time for more detailed and potentially more invasive investigations). Blood plasma/serum provides a more direct 'window' on the organism under study, but is obviously more invasive and, in the case of small animals such as mice or rats, there are practical and ethical limits on the frequency and volume of sample that can be taken. Metabolite profiles can also be obtained from tissue samples, at necropsy, which can prove very informative, in particular, with respect to target organ effects.

20.2.1 Applications of NMR Spectroscopy to Safety Studies

20.2.1.1 Liquid Samples

NMR spectroscopy has particular advantages over many other analytical platforms for metabolite profiling analysis in that, for liquid samples, little sample preparation is required before analysis, the analysis itself is rapid, and also, it is inherently quantitative. In addition, the availability of in-house and online spectral databases of standard metabolites (such as BMRB and HMDB) and the use of 2D NMR spectroscopic experiments make structural identification relatively straightforward. Detailed experimental protocols for NMR spectroscopic analysis of biofluids and tissues are now available.[4] Beginning with studies dating back some 30 years or so, numerous panels of metabolic markers of organ-specific toxicity have been detected using ^{1}H NMR spectroscopy. These

early studies unequivocally demonstrated that it was possible, using NMR spectroscopic-based approaches, to obtain organ- (and site-) specific information on the effects of toxins, including measures of both severity and mechanistic insights, using non-targeted metabolic profiling of biofluids such as urine, as shown in Figure 20.1.[5] Clearly, the technological advances in NMR spectroscopy, that have resulted in high-field strength magnets becoming available (up to 1 GHz), have resulted in much greater sensitivity and signal dispersion, offering even greater opportunities for metabolite profiling of a range of biofluids and cellular extracts. In addition, cryogenically cooled probes (RF coils and preamplifier are cooled to approximately 20 K with liquid helium) have resulted in greatly reduced electronic/thermal noise (ca. four-fold reduction), thereby providing an improved signal to noise (S/N) ratio or equivalently a reduction in the acquisition time (ca. 16-fold for the same S/N). As a result of their increased sensitivity, these cryoprobes can, in addition to ^{1}H NMR spectra, also provide ^{13}C NMR profiles from samples such as urine despite the low natural abundance of ^{13}C nuclei.[6] NMR spectra can also be acquired via automated systems combining 96-well plates, robotic sample handling, and sample transfer via NMR flow probes without the need for operator intervention and are capable of determining ^{1}H NMR spectra for up to about 150 ^{1}H NMR urine samples/day. In addition to ^{1}H and ^{13}C, ^{31}P NMR is also applied in metabolic profiling studies,[7] and if xenobiotic metabolism is also of interest, ^{19}F NMR spectroscopy is applied.[8] Enrichment of nuclei such as ^{15}N is also utilized in NMR spectroscopic-based studies of drug metabolism. For example, study of the urinary hydrazine metabolite profile through ^{15}N NMR revealed numerous metabolites that included acetylhydrazine, diacetylhydrazine, carbazic acid, a pyruvate hydrazone, and a cyclic product formed from 2-oxoglutarate hydrazone. In addition, this approach enabled detection of ^{15}N-enriched urea, which provided evidence for cleavage of the hydrazine N–N bond in vivo and incorporation of the ammonia generated in this process into urea.[9]

20.2.1.2 Solid Samples

High-resolution magic-angle-spinning (HR-MAS) NMR spectroscopy is a technique that enables metabolic profiles to be obtained from intact tissue samples.[10] HR-MAS is based on spinning samples

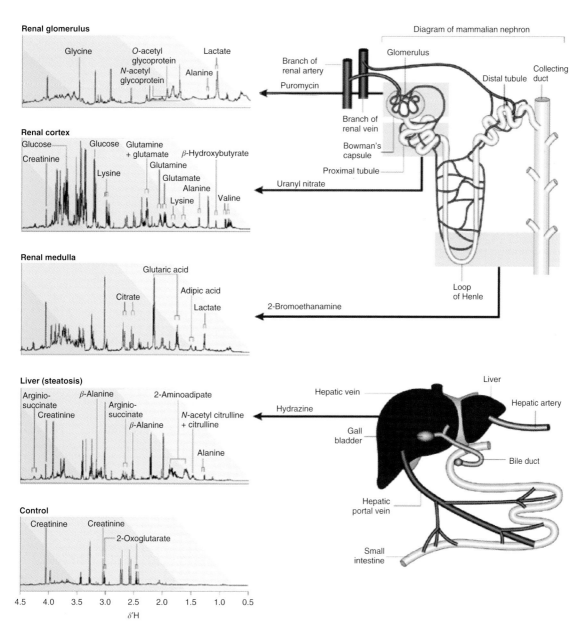

Figure 20.1. Metabonomic detection of toxicity of the liver and the kidney. Stack plot of NMR spectra showing characteristic metabolic fingerprints of tissue-specific toxicity produced by different site-selective xenobiotics given in single doses to rats in relation to an untreated control. Each 600 MHz spectrum represents one time point after dosing for each toxic compound. The xenobiotics affect specific regions within the organs (depicted in the figure by shaded boxes): puromycin affects the renal glomeruli; uranyl nitrate affects the lower regions of the proximal tubules; 2-bromoethanamine affects the renal medulla, including the loop of Henle and the collecting ducts; and hydrazine affects the hepatic parenchymal cells. (Reprinted with permission from Nicholson, J. K.; Connelly, J.; Lindon, J. C.; Holmes, E., Metabonomics: a platform for studying drug toxicity and gene function. Nature Reviews Drug Discovery 2002, 1 (2), 153–161)

at the so-called magic angle of 54.7° to the applied magnetic field. This reduces the anisotropic line-broadening effects that would otherwise be seen, thereby providing 'solution-state-like' spectra. The relatively small samples required (as little as about 10 mg of tissue) make it especially useful for studies in rodents where samples may be limited in size. In addition to ^{1}H, it is also possible to access other nuclei, e.g., heteronuclear ^{1}H–^{31}P statistical total correlation NMR spectroscopy has been performed using HR-MAS to metabolically profile liver samples in studies of galactosamine (galN)-induced hepatotoxicity.[7] This technique has also been applied in the clinical setting as it is readily applicable to the analysis of biopsy samples and has provided novel metabolite-based models that aid diagnosis and enhance stratification of patients.[11]

20.3 STATISTICAL ANALYSIS OF METABONOMIC DATA

The very large amounts of data generated in typical metabonomic studies result in complex spectral data sets that are impractical to analyze without recourse to chemometric methods based on multivariate statistics such as principal components analysis (PCA) for pattern recognition.[12] Such studies have shown that organ-specific toxicities to, e.g., the liver and the kidney could be recognized by statistical methods with the different target organ toxins mapping to different regions of the PCA scores plots. Indeed, both site-specific and severity information were available directly from the urinary metabolite profile. In addition, *via* repeated sampling from the same animals over the time course of studies, the dynamic, temporal, response to the administered toxin could be obtained. These multivariate statistical methods could be employed to classify the large amounts of spectroscopic data generated in a typical study into a representative, low-dimensional space, thereby revealing patterns in the metabolite profiles that can be used to determine the underlying biochemical effects of toxicity or disease. Statistical pattern recognition for metabolic profiling datasets can take several forms.[13] Firstly, there are the so-called unsupervised methods such as PCA, which enable the data for samples to be interrogated with no *a priori* knowledge of sample class membership (e.g., test or control). The alternative to unsupervised approaches are the supervised methods such as partial least squares or projection to latent structures

(PLS), or projection to latent structures discriminant analysis (PLS-DA), that relate a data matrix containing independent variables from, e.g., spectral intensity values (an X matrix) to 'dependent' variables for the samples (e.g., a Y matrix) such as toxicity scores or class membership. These approaches are briefly outlined below.

20.3.1 Unsupervised Statistical Data Analysis

20.3.1.1 Principal Components Analysis

As indicated above, PCA is used extensively in metabonomic analyses. As an unsupervised approach, PCA reduces the dimensionality of a data set, allowing the multidimensional data to be projected onto a hyperplane of lower dimensions (typically 2 or 3) that describes the largest amount of variance contained within the original data set. A typical example of PCA is shown in Figure 20.2 for the modeling of ^{1}H NMR rat urinary spectroscopic data following the administration of isoniazid (INH). This study revealed inter-individual differential response following administration of INH with respect to neurotoxic outcome. This variability in response led to the identification of subgroups termed *CNS responders* and *CNS non-responders* that were further explored through metabolic profiling.[14] The scores plot (Figure 20.2a) shows the distribution of the samples in two principal components, PC1 and PC2, where they are colored by class, i.e., CNS responders (blue) and CNS non-responders (red and yellow). The first principal component (PC1) is simply a linear combination of the original input variables, describing the largest variance in the data set (46%). PC2 is orthogonal to PC1 and gives the next highest degree of variance in the data set (19%). Together, PC1 and PC2 define a plane, and the projection of the individual samples onto this plane gives a convenient two-dimensional visualization of this multidimensional data revealing the inherent clustering of groups of data. The loadings coefficient plot shown in Figure 20.2(b) (individual parts of the full spectral region that contain resonances that discriminate classes are shown) provides a means of extracting information from the scores plot as it highlights the variables that contribute to the clustering of the samples seen in the associated scores plot. In this case, a number of isoniazid metabolites were determined that differentiated the classes based on inter-individual

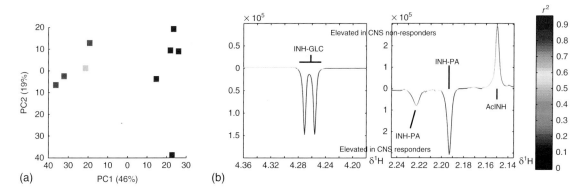

Figure 20.2. PCA model derived from the xenobiotic spectral data following administration of INH (high dose, 400 mg kg^{-1}) at 0–7 h, representing CNS responders and CNS nonresponders. (a) PC1 versus PC2 scores plot. (b) Loadings plot corresponding to PC1. The color scale corresponds to the correlation of determination (r^2) of the variables. Key: (blue) CNS nonresponders; (red) CNS responders; (yellow) CNS nonresponder B (this individual exhibited no overt clinical signs of an adverse CNS effect but shared some of the metabolic features of the CNS responders). (Reprinted with permission from Cunningham, K.; Claus, S. P.; Lindon, J. C.; Holmes, E.; Everett, J. R.; Nicholson, J. K.; Coen, M., Pharmacometabonomic characterization of xenobiotic and endogenous metabolic phenotypes that account for inter-individual variation in isoniazid-induced toxicological response. Journal of Proteome Research 2012, 11 (9), 4630–42. Copyright 2012 American Chemical Society)

variability in toxic response (CNS responders and CNS non-responders).

20.3.2 Supervised Statistical Data Analysis

20.3.2.1 Partial Least Squares

In supervised data analysis, the investigators make use of information that is available to them, e.g., the fact that samples are from either test or control groups, to construct models that maximize the difference between them in order to extract relevant biomarker information. As indicated earlier, PLS or PLS-DA is widely used for the supervised analysis of metabonomic data sets (reviewed in Ref. 13). In this method, a data matrix containing independent variables derived from the analysis of the samples provides an 'X matrix' that is then related to one for a set of dependent variables for the same samples that gives a Y matrix. The Y matrix can consist of measurements of response to a toxin (e.g., a continuous variable such as alanine aminotransferase (ALT) activity reflecting hepatic function) or a healthy vs diseased class (discrete score of 0 or 1). Extensions of PLS include techniques such as OPLS-DA (orthogonal partial least squares discriminant analysis)[15] that build on PLS *via* the pre-filtering or removal of orthogonal (classification-irrelevant) variation present

in the data. By pre-filtering the 'between-class' variation of interest, it enables it to be modeled independently of the 'within-class' variation, thereby improving the interpretability of the resultant models. A recent advance is the application of O-PLS-DA to full-resolution NMR spectral data (i.e., modeling all spectral data points generated) to give loadings plots that reveal both the covariance and correlation weights across the spectral resonances. This provides loadings coefficient plots that are readily interpretable by an NMR spectroscopist and enable discriminatory metabolites to be rapidly identified. Examples of this type of modeling for toxicological applications have been shown for the important hepatotoxins such as galN[16] and methotrexate.[17]

20.3.3 Statistical Correlation Spectroscopy

20.3.3.1 Statistical Correlation Spectroscopy and Statistical Heterospectroscopy

Other ways of examining NMR spectroscopic data include techniques such as statistical total correlation spectroscopy or STOCSY, recently reviewed by Robinette *et al.*[18] STOCSY involves the computation of correlation statistics between the intensity of a given driver peak and all data points in a one-dimensional spectrum and is applied across multiple 1D datasets. This reveals the connectivities between resonances that vary in concentration between samples, thereby providing structural information on molecules of interest from complex and highly overlapped spectra. The STOCSY process is rapid and is not determined or restricted by spin systems or the presence of hetero-nuclei as for traditional 2D NMR experiments.

An illustrative application of STOCSY in biomarker detection for nephrotoxicity is given in studies on the effects of mercuric chloride exposure[19] in the rat. The technique has also been used for the detection and identification of acetaminophen and ibuprofen drug metabolites in human urine samples.[20] The STOCSY approach is demonstrated in Figure 20.3 where the correlation matrix for a one-dimensional dataset is generated from a single data point, in this case, the *N*-acetyl resonance of acetaminophen–glucuronide. The highest correlations are identified between resonances from the same molecule, and lesser correlations are observed to additional acetaminophen metabolites and endogenous metabolites in what could be termed *pathway connectivities*. In a further refinement, a technique for correlating ^{1}H NMR data with that from a heteronucleus (in this case, ^{31}P NMR), known as *HET-STOCSY* (heteronuclear statistical total correlation spectroscopy), was used in the study of galN-induced hepatotoxicity for ^{1}H–^{31}P MAS NMR spectra of intact liver[7] (see Chapter 7).

20.4 ORGAN-SPECIFIC TOXICITIES BY NMR SPECTROSCOPY-BASED METABONOMICS

NMR has been widely used to study the effects of toxicity, and while toxicity is not always limited to a single organ, and indeed the focus of the observed effects of a toxin can evolve over time (with the site changing from one organ to another), for simplicity, it is useful to look at examples where metabolic profiling has been used to discover organ-specific markers of toxicity, which have provided mechanistic insight. Here, coverage is limited to the kidney and the liver, but toxicity to other organs such as pancreas, testes, and brain can also be studied using NMR spectroscopy-based metabonomic methods and approaches,[21] including in combination with, e.g., liquid chromatography–mass spectrometry (LC-MS).

20.4.1 Renal Toxicity

The use of NMR-based metabolite profiling for the detailed investigation of the time course of nephrotoxicity provides many examples of the early application of the use of metabonomics. Analysis of metabolites present in urine to investigate renal toxicity is clearly logical as the composition of this biofluid will be directly related to kidney function. Early examples include studies in the rat on the renal cortical toxin mercuric chloride,[22] the papillary toxin 2-bromoethanamine,[22] and the proximal tubule toxin cephaloridine.[23] In these studies, rat urine was repeatedly collected across the study period to enable the onset, progression, and subsequent recovery of a nephrotoxic lesion to be followed, with complementary histopathology and clinical chemistry. The utility of this approach for sighting studies is well illustrated by the metabonomic studies on the urine of cephaloridine-dosed rats, where by the end of the 14-day dosing period, the kidneys were essentially normal as assessed by histopathology. ^{1}H NMR spectroscopic analysis of the urine, however, revealed very marked changes in metabolic profiles, including a marked glucosuria by day 3 of administration, followed rapidly by recovery of the animals despite continuous administration of the drug. The pattern of metabolites observed was typical of that produced by proximal tubule nephrotoxins and was thus able to provide information on both the time course of the toxic lesion and its site in a relatively noninvasive way. Studies on the nephrotoxic antibiotic imipenem showed interesting effects on the metabolite profiles of cynomologous monkeys administered 180 mg kg^{-1} per day for 7 days, with the excretion of large quantities of β-hydroxybutyrate, acetate, acetone, and acetoacetate.[24] These changes in urinary profiles were accompanied by gross histopathological changes, with tubular necrosis and the presence of intracytoplasmic fat droplets, consistent with a disruption of energy metabolism.

The majority of the metabonomic studies performed in toxicology to date have been undertaken using NMR spectroscopy, and indeed, NMR provided the analytical platform for the study of nephrotoxicity (and hepatotoxicity) in the COMET (Consortium for Metabonomic Toxicology) consortium (described in the following section). However, increasingly high-performance liquid chromatography–mass spectrometry (HPLC-MS)-based analysis, either alone or together with NMR spectroscopy, has also

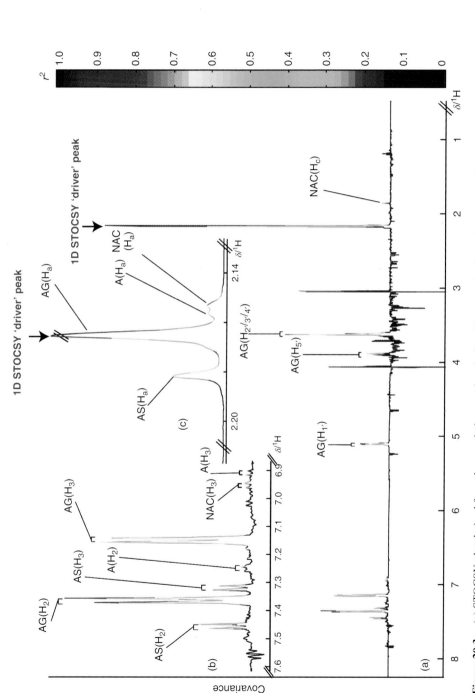

Figure 20.3. (a) STOCSY plot derived from the correlation matrix calculated between the data point at the peak maximums of the *N*-acetyl proton signal of AG (acetaminophen glucuronide) resonance (δ 2.17) and all other data points, as indicated by the arrow, showing strong correlation (red/orange data points) with resonances at δ 3.62, 3.89, 5.1, 7.13, and 7.36. Slightly weaker correlations with A (acetaminophen) and AS (acetaminophen sulphate). (b) Expansion for the aromatic region showing signals for acetaminophen and its related metabolites. (c) Expansions for the δ 2.17 *N*-acetyl resonance of acetaminophen metabolites. (Reprinted with permission from Holmes, E.; Loo, R. L.; Cloarec, O.; Coen, M.; Tang, H.; Maibaum, E.; Bruce, S.; Chan, Q.; Elliott, P.; Stamler, J.; Wilson, I. D.; Lindon, J. C.; Nicholson, J. K. Detection of urinary drug metabolite (xenometabolome) signatures in molecular epidemiology studies via statistical total correlation (NMR) spectroscopy. Analytical Chemistry 2007, 79 (7), 2629–40. Copyright 2007 American Chemical Society)

been applied to this type of work. The different sensitivities and specificities of the two analytical methodologies result in a particularly powerful combination as, used together, NMR and HPLC-MS provide a more complete coverage of the metabolome. This has been illustrated in a series of studies on model nephrotoxins such as gentamicin,[25] mercuric chloride,[26] and cyclosporine.[27] Thus, when gentamicin was administered daily to male rats for 7 days, conventional clinical chemistry indicated increased N-acetyl-β-D-glucosaminidase activity and, at 9 days postdosing, there was clear histopathology indicating damage to the kidneys of all dosed animals.[25] Analysis by ^{1}H NMR spectroscopy and HPLC-ToF-MS (high-performance liquid chromatography–time-of-flight mass spectrometry) revealed time-dependent changes in the urinary metabolic profiles from the dosed animals from day 7 onwards. ^{1}H NMR spectroscopic analysis detected raised glucose concentrations in urine coupled with reductions in the levels of trimethylamine-N-oxide (TMAO) and hippurate detected. HPLC-MS, on the other hand, detected depletion of xanthurenic acid and kynurenic acid and an altered pattern of excretion of sulfate-conjugated metabolites. One cautionary note that should be mentioned in this case, however, is that, in all likelihood, not all of these changes reflected nephrotoxicity, as some undoubtedly were the result of the antibiotic effects of gentamicin on the gut microflora of the dosed animals (e.g., the changes in hippurate, TMAO, and sulfate conjugates). Similarly, the nephrotoxicity resulting from a single dose of mercuric II chloride in the rat revealed marked changes in the pattern of endogenous metabolites detected using both ^{1}H NMR spectroscopy and HPLC-ToF-MS for urine.[26] Three days after dosing, major changes in urinary metabolite profiles were seen, which included increases in the relative concentrations of lactate, alanine, acetate, succinate, trimethylamine (TMA), and glucose together with reductions in the amounts of citrate and 2-oxoglutarate (detected by ^{1}H NMR spectroscopy) and reduced kynurenic acid, xanthurenic acid, pantothenic acid, and 7-methylguanine (identified using LC-MS). However, following early marked changes, this panel of perturbed metabolites gradually became less noticeable as the kidneys recovered from the insult and the metabolite profile gradually returned to a control position as the study progressed. Similarly, a combined HPLC-MS and NMR spectroscopic investigation of cyclosporin A nephrotoxicity also showed excellent agreement and complementary information

for the observed time course of toxicity using both techniques.[27]

20.4.2 Liver Toxicity

Another organ that has enjoyed considerable interest for metabonomic studies has been the liver, in no small measure reflecting the importance of drug-induced liver injury (DILI) in man, and there are a large number of published studies documenting the effects of a range of toxins on liver metabolite profiles. Indeed, the use of ^{1}H NMR spectroscopy for metabolic profiling to study hepatic toxicity was one of the earliest applications of the technique, for example, in demonstrating that urinary hypertaurinuria could be used as a marker for carbon tetrachloride-induced hepatotoxicity.[28] Other changes that can occur in the urinary ^{1}H NMR spectra in response to hepatotoxins are decreases in Krebs cycle intermediates such as citrate, 2-oxoglutarate, and succinate and increases in creatinine, acetate, glucose, and bile acids (with, e.g., galN and α-naphthyl isothiocyanate (ANIT)[29]). Some of these increases (e.g., changes in Krebs cycle intermediates) are also associated with toxicity in other organs and are perhaps not so much liver-specific signals as indicators of perturbed energy metabolism. With ANIT careful interrogation of the data showed that numerous hepatic metabolite changes could be correlated with histopathological changes, for example, elevated levels of hepatic tyrosine, phenylalanine, histidine, and bile acids (C18 methyl resonance) coincided with the pathological observation of bile duct hyperplasia.[30] Subsequently, more detailed, studies, where both urine and plasma samples were analyzed, in addition to the liver, have confirmed, and extended, these findings into a much more integrated picture of the systems level toxicity of ANIT.[31] A number of studies, on a range of toxins, have now also confirmed good concordance between liver toxicity as measured by conventional clinical chemistry, histology, and changes seen using NMR-based metabonomics in a dose-dependent manner for many compounds. A number of studies have now employed MAS NMR to study liver toxicants, and a common pattern is beginning to emerge of reductions in glycogen and glucose coupled with increases in lipids and triglycerides for compounds such as allyl formate, acetaminophen, methapyrilene, ethionine, and thioacetamide, albeit with numerous unique metabolic changes identified for each compound.

The recognition that NMR spectroscopy could provide biomarkers that were organ specific and that different mechanisms of action led to similar patterns of biomarkers for sites and mechanisms of toxicity led to the idea that, with sufficient information, an expert system could be constructed that would allow novel compounds to be assessed via ^{1}H NMR-based metabonomics. This concept resulted in the formation of the COMET project as described below.[32]

20.5 THE COMET CONSORTIUM PROJECT

20.5.1 COMET 1

The application of metabolic profiling for the study of preclinical xenobiotic toxicity was first tackled on a large scale by the first COMET program. The COMET 1 studies were initiated in 2000 and involved collaboration between five major pharmaceutical companies and Imperial College London.[33] This initiative led to the generation of a biobank of biofluids and tissues representing preclinical studies of about 150 compounds and treatments in rodents (rat and mouse). The database of compounds covered a wide chemical space with a focus on hepatic and renal toxins; however, the project also incorporated toxins that targeted multiple organs and diverse treatments that included, for example, hepatectomy, nephrectomy, and food and water restriction. In-life studies were conducted in pharmaceutical companies and involved the assignment of 10 rats or 8 mice to control, low-dose, and high-dose treatment groups. On the basis of in-house data or published literature, the low-dose was chosen to represent a minimal or threshold response, whereas the high-dose generated clear toxicity following a single dose. All studies were supported with clinical chemistry and histopathological analysis. The COMET studies typically involved 7-day studies of acute toxicity (single dose), although a panel of chronic studies (repeat dosing) and 14-day studies were also carried out. Urine was collected pretreatment when animals had acclimatized to metabolism cages and was also collected at defined intervals throughout the course of each study. Blood was collected throughout the study (24 h, 48 h) and at termination (168 h). The overarching objective of COMET was to generate a metabonomic database of a diverse class of toxins and treatments that would present novel molecular pathology descriptors, with NMR spectroscopy as the principal analytical platform.

Multiple studies were carried out using the hepatotoxin hydrazine to assess species-specific differences and intersite reproducibility. The study of intersite reproducibility involved a split-set of urine samples from a study of acute hydrazine toxicity in the rat, with NMR data acquisition conducted at two sites with different ^{1}H spectrometer frequencies, 500 and 600 MHz.[34] The intersite urinary NMR spectral datasets generated were highly correlated and revealed near-identical patterns of hydrazine-induced metabolic change detected following data analysis at each site. A number of urinary metabolites including citrate, hippurate, and taurine displayed coefficients of variation (% CV) in the range of 4–8%, demonstrating excellent analytical reproducibility of the platform. The high analytical performance and reproducibility of NMR spectroscopy in the context of large-scale urinary metabolic profiling was an important outcome from the early work of the COMET consortium project and provided a stable base on which to develop the field. The COMET database also led to the generation of an expert system that could classify and predict the toxicity of compounds from ^{1}H NMR urinary spectral profiles. This approach utilized a density superposition approach, CLOUDS, based on probabilistic neural networks and showed the potential to correctly classify >85% of a subset ($n = 19$) of renal toxins and hepatotoxins.

20.5.2 COMET 2

The second phase of the COMET consortium project (COMET-2) involved the application of multiple analytical platforms, namely NMR spectroscopy and ultra-performance liquid chromatography coupled to mass spectrometry (UPLC-MS) for metabolic phenotyping in the toxicology setting. The work in COMET-2 involved the detailed and mechanistic study of two model toxins, a renal papillary toxin, bromoethanamine,[35] and a hepatotoxin, galN.[36]

GalN (2-amino-2-deoxy-D-galactose) is a simple amino sugar that has served as a model hepatotoxin since the 1960s. It was originally believed to represent a model of human viral hepatitis, yet this is now contested and it is more commonly used together with lipopolysaccharide (LPS) as a model of acute liver injury in mice. The hepatic metabolism of galN involves the production of UDP-galN, which effectively serves as a metabolic trap, as epimerization leads to UDP-glcN that cannot serve as a uridylate donor (unlike UDP-glucose/UDP-galactose). Ultimately, galN leads to a depletion of UDP-glucose and the hepatic uridine nucleotide pool, which results in hepatic inflammation and necrosis and inhibition of RNA and protein synthesis.

The COMET-2 project studied the endogenous metabolic consequences of galN administration in the rat in a number of NMR spectroscopic and UPLC-MS-based studies that included models of protection (glycine and uridine cotreatment) and that also encompassed the study of variability in response to treatment.[36]

The protective effect of glycine in galN hepatotoxicity had been attributed to glycine preventing activation of Kupffer cells and subsequent release of proinflammatory cytokines.[37] A metabonomic study[16] of the mechanism underlying glycine protection of galN hepatotoxicity involved treatment of rats with galN ($415\,mg\,kg^{-1}$ ip) or glycine (5% via the diet with a 6-day acclimatization period) or co-treatment with both galN and glycine. Urine, serum, and liver samples were collected and profiled using ^1H NMR and ^1H MAS NMR spectroscopy. Histopathological assessment revealed a protective effect of glycine in that moderate (5/8) or marked (3/8) hepatic necrosis was observed following galN treatment alone, in contrast to the co-administration group where predominantly minimal (5/8) hepatic necrosis was identified (with a degree of interanimal variability observed for three animals with mild (1/8), moderate (1/8), and marked (1/8) necrosis). Clinical chemistry supported the protective effect of glycine and also mirrored the inter-animal variability in that a lower serum alanine transaminase/aspartate transaminase (ALT/AST) activity was seen for all but one animal (marked histopathology necrosis score) in the co-treatment group (nonresponder to glycine protection) in comparison to the galN-alone treatment group. The NMR spectral profiles of biofluids and tissues revealed unique metabolic phenotypes for each treatment group and identified endogenous metabolites and pathways of relevance for understanding the protective effect of glycine on galN hepatotoxicity.

The NMR spectral profiles of urine and subsequent modeling revealed the metabolites that differentiated controls from galN treatment included galN, N-acetylglucosamine (glcNAc), and urocanic acid and decreased levels of 2-oxoglutarate, N-methylnicotinamide, 2′-deoxycytidine (dCyd),

and phenylacetylglycine. The discriminatory urinary metabolites identified when the co-treatment group (galN and glycine) was modeled against controls included elevated levels of galN, formate, and isovalerylglycine and decreased levels of N-methylnicotinamide and dCyd. In contrast to the galN-alone treatment group, glcNAc was not found to be significantly elevated in the urine of the co-treatment group. The level of glcNAc in the postdose urine was found to correlate strongly with the degree of galN-induced liver damage and was shown in further studies to represent a sensitive marker of the severity of galN hepatotoxicity and to reflect the extent of depletion of the uridine nucleotide pool. Furthermore, in the co-treatment group, urinary metabolic profiles were much closer to the control profile than the galN-alone treatment group, for example, urinary urocanic acid was not elevated and 2-oxoglutarate, D-3-hydroxybutyrate, and phenylacetylglycine were not depleted in the co-treatment group. The effect of treatment with glycine alone (5% in the diet) was also explored in the urinary metabolic profiles and revealed an increase in orotate, glycine, isovalerylglycine, isobutyrylglycine, 2-methylbutyrylglycine, 2-oxoglutarate, and fumarate together with reduced levels of dCyd, N-methylnicotinamide, and its catabolite N-methyl-4-pyridone-3-carboxamide (4-PY).

The serum NMR spectral profiles were also modeled to determine unique metabolic phenotypes for each treatment group. Discriminatory sera metabolites were identified between control and galN-alone treatment groups that included elevated tyrosine, phenylalanine, lactate, betaine, D-3-hydroxybutyrate, and dCyd together with reduced levels of the fatty acids of lipoproteins. The serum-based OPLS-DA model that discriminated between controls and the cotreatment group revealed the elevation of glycine and D-3-hydroxybutyrate in the cotreatment group and the absence of the changes described above for galN treatment alone, signifying a profile much closer to controls following cotreatment with galN and glycine.

A critical finding from the hepatic MAS NMR spectral data was the unexpected increase in hepatic levels of uridine, UDP-glucose, and UDP-galactose following treatment with glycine alone (5% via the diet) as shown in Figure 20.4. This data suggested that the protective role of glycine against galN toxicity was likely mediated by changes in the uridine nucleotide pool rather than through the prevention of Kupffer

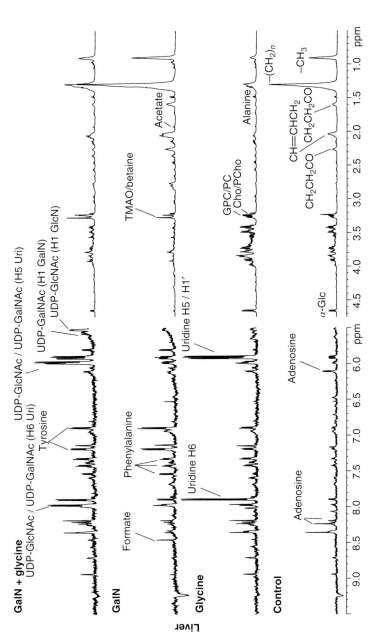

Figure 20.4. Representative ^{1}H MAS NMR liver tissue spectra of control, glycine-treated, galN-treated, and galN- and glycine-treated animals 24 h postdosing. Key: GalN, galactosamine; TMAO, trimethylamine-*N*-oxide; Cho/PCho, choline/phosphocholine; GPC/PC, glycerophosphatidylcholine/ phosphatidylcholine; UDP-glcNAc, UDP-*N*-acetylglucosamine; UDP-GalNAc, UDP-*N*-acetylgalactosamine; $-(CH_2)_n$, fatty acid methylene groups; $-CH_3$, fatty acid methyl groups. (Reprinted with permission from Coen, M.; Hong, Y. S.; Clayton, T. A.; Rohde, C. M.; Pearce, J. T.; Reily, M. D.; Robertson, D. G.; Holmes, E.; Lindon, J. C.; Nicholson, J. K., The mechanism of galactosamine toxicity revisited: a metabonomic study. Journal of Proteome Research 2007, 6 (7), 2711–9)

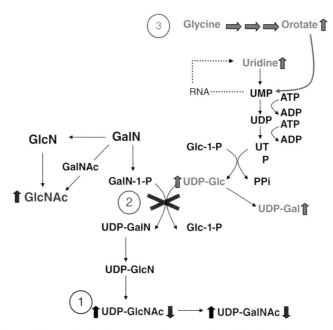

Figure 20.5. The hepatic metabolism of galN; 1 represents the production of UDP-aminosugars, 2 represents the inhibition of 1 and the ensuing increase in glcNAc when UDP-glc levels are depleted, 3 represents the proposed mode of glycine protection where pretreatment with glycine leads to increased levels of uridine nucleotides. The colored arrows indicate changes in the levels of metabolites as determined from the metabonomic data of urine/the liver and serum. (Reprinted with permission from Coen, M.; Hong, Y. S.; Clayton, T. A.; Rohde, C. M.; Pearce, J. T.; Reily, M. D.; Robertson, D. G.; Holmes, E.; Lindon, J. C.; Nicholson, J. K., The mechanism of galactosamine toxicity revisited; a metabonomic study. Journal of Proteome Research 2007, 6 (7), 2711–9)

cell activation (although in the absence of toxicity, the Kupffer cells would indeed not be activated, yet this was not the causal factor in the mechanism of protection). Increased levels of orotate in the urine following glycine treatment alone suggested that glycine increased the rate of carbamoyl phosphate synthesis, which would lead to increased UMP production and ultimately increased hepatic levels of uridine, UDP-glc, and UDP-gal. This study led to further studies of the effect of a range of levels of glycine supplementation on rat hepatic uridine levels together with a time-course assessment, which confirmed the results described above.

Hence, this study led to the generation of a novel hypothesis that administration of glycine increased the hepatic uridine nucleotide pool and that this metabolic effect alone counteracted the galN-induced depletion of these pools and facilitated complete metabolism of galN, as summarized in Figure 20.5. This study is an excellent exemplar for the ability of a non-hypothesis-driven strategy or 'untargeted'

approach to uncover novel data of mechanistic relevance.

20.6 TOXICO/PHARMACOMETABONOMICS

Inter-animal variability in response has been mentioned in the context of preclinical studies of toxins including galN and INH and represents an intriguing problem that has been addressed using metabolic phenotyping to shed light on the mechanistic basis for variability in response. In an ideal situation, it would be possible to predict therapeutic or toxic outcomes before drug administration via the metabolic phenotypes that are observed in the predose biofluid spectra of test organisms or patients. Recently, proof of concept metabonomic studies have used such predose spectral fingerprints of urine in order to construct models that can be used to predict the metabolic and toxic outcome of drug administration in the rat.[38]

These studies examined the fate of animals administered either galN or acetaminophen in terms of hepatotoxicity. In the case of galN, animals given the same dose of the toxin exhibited widely different responses such that they could effectively be classified as 'responders' or 'non-responders'. Interestingly, the variability in the response of the individual rats could be correlated with their predose urinary phenotypes. With acetaminophen, the predose urine profiles enabled models to be constructed that could be used to predict the amount of liver damage that resulted from drug administration, as determined from clinical chemistry and histopathological analysis. Analysis of the metabolic profile data from this study also involved determining the total urinary excretion of acetaminophen and its metabolites (obtained from the postdose spectra). The variations in drug and metabolite excretion were then used to construct a model relating these data to the metabolite profile determined from the predose urinary NMR spectra. The resulting models were shown to have high predictive ability and statistically significant variations in the predose spectral profiles that could be correlated with both postdose histopathology and xenobiotic metabolism. This demonstration that the metabolic fingerprint of an individual, which is sensitive to genetic, physiological, and environmental factors, can be used in a predictive way has led to the concept of pharmacometabonomics 'the prediction of the outcome (e.g., efficacy or toxicity) of a drug or xenobiotic intervention in an individual based on a mathematical model of preintervention metabolite signatures.' In contrast, pharmacogenomics relies on the individuals genetic makeup in order to predict factors such as therapeutic response to, and metabolism/toxicity of, drugs (or other xenobiotics). The pharmacometabonomic approach has recently been demonstrated in man in a study on the metabolic fate of acetaminophen in human volunteers.[39] As shown for rats, predicting metabolism from predose urinary metabolite profiles was demonstrated, based on the observation that individuals excreting comparatively high concentrations of p-cresol-*O*-sulfate generally excreted relatively less acetaminophen-*O*-sulfate, and larger amounts of acetaminophen-*O*-glucuronide, than subjects with low pre-dose urinary p-cresol-*O*-sulfate concentrations. It was thus hypothesized that acetaminophen and p-cresol as aromatic phenols may compete for sulfation, and given that sulfation is a relatively low capacity conjugation pathway compared to, e.g., glucuronidation, this is perhaps not

an unexpected result. The competing p-cresol is produced by gut-dwelling bacteria providing an additional source of variability between individuals to take into account in terms of ADME. Obviously, bacterially produced p-cresol will compete for sulfation with all phenolic drugs/metabolites and not only acetaminophen. This simple biomarker therefore may have pharmacometabonomic significance well beyond acetaminophen metabolism. Additional novel associations between pretreatment biofluid profiles and the effects of xenobiotics have been discovered and recently reviewed[40] and it is likely that pharmacometabonomics will play a major role in the development of personalized healthcare. Indeed, it may prove possible to predict which individuals are more likely to suffer adverse effects from particular therapeutic agents from pretreatment or early intervention metabolic profiles, with obvious benefit to the individual and the effective therapeutic use of the drug (see Chapter 25).

20.7 CONCLUSIONS

NMR spectroscopy, especially when combined with statistical methods of pattern recognition to aid biomarker discovery, represents a very useful method for examining both quantitative and qualitative changes in biofluid composition, offering information on both the location and severity of organ toxicity, as well as providing novel insights into the mechanisms of toxicity. The application of the technology to inform on the biochemical basis for the protective effect of glycine on galN hepatotoxicity represents an exemplar of the power of a nontargeted approach for unearthing novel mechanistic information. The methodology can also be used, via the pharmacometabonomic approach, to predict the likely response of an individual to a therapeutic intervention.

REFERENCES

1. J. K. Nicholson, J. C. Lindon, and E. Holmes, *Xenobiotica; the Fate of Foreign Compounds in Biological Systems*, 1999, **29**, 1181.

2. J. C. Lindon, J. K. Nicholson, E. Holmes, and J. R. Everett, *Concepts Magn. Reson.*, 2000, **12**, 289.

3. D. G. Robertson, M. D. Reily, and J. D. Baker, *Expert Opin. Drug Metab. Toxicol.*, 2005, **1**, 363.

4. A. C. Dona, B. Jimenez, H. Schafer, E. Humpfer, M. Spraul, M. R. Lewis, J. T. Pearce, E. Holmes, J. C. Lindon, and J. K. Nicholson, *Anal. Chem.*, 2014, **86**, 9887.

5. J. K. Nicholson, J. Connelly, J. C. Lindon, and E. Holmes, *Nat. Rev. Drug Discov.*, 2002, **1**, 153.

6. H. C. Keun, O. Beckonert, J. L. Griffin, C. Richter, D. Moskau, J. C. Lindon, and J. K. Nicholson, *Anal. Chem.*, 2002, **74**, 4588.

7. M. Coen, Y. S. Hong, O. Cloarec, C. M. Rhode, M. D. Reily, D. G. Robertson, E. Holmes, J. C. Lindon, and J. K. Nicholson, *Anal. Chem.*, 2007, **79**, 8956.

8. H. C. Keun, T. J. Athersuch, O. Beckonert, Y. Wang, J. Saric, J. P. Shockcor, J. C. Lindon, I. D. Wilson, E. Holmes, and J. K. Nicholson, *Anal. Chem.*, 2008, **80**, 1073.

9. N. E. Preece, J. K. Nicholson, and J. A. Timbrell, *Biochem. Pharmacol.*, 1991, **41**, 1319.

10. O. Beckonert, M. Coen, H. C. Keun, Y. Wang, T. M. Ebbels, E. Holmes, J. C. Lindon, and J. K. Nicholson, *Nat. Protoc.*, 2010, **5**, 1019.

11. R. Mirnezami, B. Jimenez, J. V. Li, J. M. Kinross, K. Veselkov, R. D. Goldin, E. Holmes, J. K. Nicholson, and A. Darzi, *Ann. Surg.*, 2014, **259**, 1138.

12. K. P. Gartland, C. R. Beddell, J. C. Lindon, and J. K. Nicholson, *Mol. Pharmacol.*, 1991, **39**, 629.

13. J. Trygg, E. Holmes, and T. Lundstedt, *J. Proteome Res.*, 2007, **6**, 469.

14. K. Cunningham, S. P. Claus, J. C. Lindon, E. Holmes, J. R. Everett, J. K. Nicholson, and M. Coen, *J. Proteome Res.*, 2012, **11**, 4630.

15. O. Cloarec, M. E. Dumas, J. Trygg, A. Craig, R. H. Barton, J. C. Lindon, J. K. Nicholson, and E. Holmes, *Anal. Chem.*, 2005, **77**, 517.

16. M. Coen, Y. S. Hong, T. A. Clayton, C. M. Rohde, J. T. Pearce, M. D. Reily, D. G. Robertson, E. Holmes, J. C. Lindon, and J. K. Nicholson, *J. Proteome Res.*, 2007, **6**, 2711.

17. M. Kyriakides, R. N. Hardwick, Z. Jin, M. J. Goedken, E. Holmes, N. J. Cherrington, and M. Coen, *Toxicol. Sci.: an Official Journal of the Society of Toxicology*, 2014, **142**, 105.

18. S. L. Robinette, J. C. Lindon, and J. K. Nicholson, *Anal. Chem.*, 2013, **85**, 5297.

19. E. Holmes, O. Cloarec, and J. K. Nicholson, *J. Proteome Res.*, 2006, **5**, 1313.

20. E. Holmes, R. L. Loo, O. Cloarec, M. Coen, H. Tang, E. Maibaum, S. Bruce, Q. Chan, P. Elliott, J. Stamler, I. D. Wilson, J. C. Lindon, and J. K. Nicholson, *Anal. Chem.*, 2007, **79**, 2629.

21. M. Coen, E. Holmes, J. C. Lindon, and J. K. Nicholson, *Chem. Res. Toxicol.*, 2008, **21**, 9.

22. E. Holmes, F. W. Bonner, B. C. Sweatman, J. C. Lindon, C. R. Beddell, E. Rahr, and J. K. Nicholson, *Mol. Pharmacol.*, 1992, **42**, 922.

23. L. B. Murgatroyd, R. J. Pickford, I. K. Smith, I. D. Wilson, and B. J. Middleton, *Hum. Exp. Toxicol.*, 1992, **11**, 35.

24. M. P. Harrison, D. V. Jones, R. J. Pickford, and I. D. Wilson, *Biochem. Pharmacol.*, 1991, **41**, 2045.

25. E. M. Lenz, J. Bright, R. Knight, F. R. Westwood, D. Davies, H. Major, and I. D. Wilson, *Biomarkers: Biochemical Indicators of Exposure, Response, and Susceptibility to Chemicals*, 2005, **10**, 173.

26. E. M. Lenz, J. Bright, R. Knight, I. D. Wilson, and H. Major, *Analyst* (Cambridge, U. K.), 2004, **129**, 535.

27. E. M. Lenz, J. Bright, R. Knight, I. D. Wilson, and H. Major, *J. Pharm. Biomed. Anal.*, 2004, **35**, 599.

28. J. A. Timbrell and C. J. Waterfield, *Adv. Exp. Med. Biol.*, 1996, **403**, 125.

29. B. M. Beckwith-Hall, J. K. Nicholson, A. W. Nicholls, P. J. Foxall, J. C. Lindon, S. C. Connor, M. Abdi, J. Connelly, and E. Holmes, *Chem. Res. Toxicol.*, 1998, **11**, 260.

30. N. J. Waters, E. Holmes, C. J. Waterfield, R. D. Farrant, and J. K. Nicholson, *Biochem. Pharmacol.*, 2002, **64**, 67.

31. N. J. Waters, E. Holmes, A. Williams, C. J. Waterfield, R. D. Farrant, and J. K. Nicholson, *Chem. Res. Toxicol.*, 2001, **14**, 1401.

32. J. C. Lindon, H. C. Keun, T. M. Ebbels, J. M. Pearce, E. Holmes, and J. K. Nicholson, *Pharmacogenomics*, 2005, **6**, 691.

33. J. C. Lindon, J. K. Nicholson, E. Holmes, H. Antti, M. E. Bollard, H. Keun, O. Beckonert, T. M. Ebbels, M. D. Reily, D. Robertson, G. J. Stevens, P. Luke, A. P. Breau, G. H. Cantor, R. H. Bible, U. Niederhauser, H. Senn, G. Schlotterbeck, U. G. Sidelmann, S. M. Laursen, A. Tymiak, B. D. Car, L. Lehman-McKeeman, J. M. Colet, A. Loukaci, and C. Thomas, *Toxicol. Appl. Pharmacol.*, 2003, **187**, 137.

34. H. C. Keun, T. M. Ebbels, H. Antti, M. E. Bollard, O. Beckonert, G. Schlotterbeck, H. Senn, U. Niederhauser, E. Holmes, J. C. Lindon, and J. K. Nicholson, *Chem. Res. Toxicol.*, 2002, **15**, 1380.

35. P. Shipkova, J. D. Vassallo, N. Aranibar, S. Hnatyshyn, H. Zhang, T. A. Clayton, G. H. Cantor, M. Sanders, M. Coen, J. C. Lindon, E. Holmes, J. K. Nicholson, and L. Lehman-McKeeman, *Xenobiotica; the Fate of Foreign Compounds in Biological Systems*, 2011, **41**, 144.

36. M. Coen, *Toxicology*, 2010, **278**, 326.

37. R. F. Stachlewitz, V. Seabra, B. Bradford, C. A. Bradham, I. Rusyn, D. Germolec, and R. G. Thurman, *Hepatology* (Baltimore, Md.), 1999, **29**, 737.

38. T. A. Clayton, J. C. Lindon, O. Cloarec, H. Antti, C. Charuel, G. Hanton, J. P. Provost, J. L. Le Net, D. Baker, R. J. Walley, J. R. Everett, and J. K. Nicholson, *Nature*, 2006, **440**, 1073.

39. T. A. Clayton, D. Baker, J. C. Lindon, J. R. Everett, and J. K. Nicholson, *Proc. Natl. Acad. Sci. U. S. A.*, 2009, **106**, 14728.

40. J. R. Everett, R. L. Loo, and F. S. Pullen, *Ann. Clin. Biochem.*, 2013, **50**, 523.

Chapter 21

Characterization of Pharmaceutical Compounds by Solid-state NMR

Frederick G. Vogt

Morgan, Lewis & Bockius, LLP, 1701 Market St., Philadelphia, PA 19103-2921, USA

21.1 INTRODUCTION

Solid-state nuclear magnetic resonance (SSNMR) is widely used in the characterization of solid and semisolid materials of importance in pharmaceutical research and development.[1-4] The use of SSNMR in studies of pharmaceutical materials has continued to expand in recent years, as new developments in the larger field of SSNMR experimentation have been adapted for pharmaceutical applications in both industrial and academic laboratories. Much of

the activity has focused on expanding the range of materials studied by SSNMR, exploring the use of computational methods to support spectral interpretation, increasing the use of 2D methods and other pulse sequence improvements (see Chapters 2 and 3), and broadening the use of nuclei outside of the traditional spin $1/2$ isotopes of ^{13}C, ^{15}N, ^{19}F, and ^{31}P. To an extent, the nontraditional nuclei include the ^1H nucleus, which has become increasingly accessible by direct and indirect observations because of higher static magnetic fields, ultrafast MAS, new 2D correlation pulse sequences, and increasingly efficient ^1H homonuclear dipolar decoupling pulse sequences. The nontraditional nuclei also include a number of important quadrupolar nuclei, such as the ^2H, ^{14}N, ^{17}O, ^{23}Na, and ^{35}Cl nuclei. Unlike SSNMR studies of biological solids, SSNMR studies of pharmaceutical materials typically rely on nuclei at natural abundance, with isotopic labeling normally confined to special studies wherein easily exchanged isotopes available in commercial solvents, such as ^2H, ^{17}O, and ^{13}C, are used to study a particular solvent environment in a hydrate or organic solvate.[1]

This chapter covers the major areas in drug development where SSNMR is currently applied to pharmaceutical characterization. Drug development currently represents the majority of the usage of SSNMR in this field, although increasing applications to drug discovery are underway and have been reviewed elsewhere.[1] In drug development, SSNMR is used to characterize three major types of material: drug substance (also referred to as active pharmaceutical ingredient or API),

NMR in Pharmaceutical Sciences. Edited by Jeremy R. Everett, John C. Lindon, Ian D. Wilson, and Robin K. Harris
© 2015 John Wiley & Sons, Ltd. ISBN: 978-1-118-66025-6
Also published in eMagRes (online edition)
DOI: 10.1002/9780470034590.emrstm1393

drug product (also referred to as a formulation or dosage form), and excipients. All three of these areas involve the use of SSNMR to characterize effects such as polymorphism, solvation (including hydration), and the formation of molecular complexes such as salts, cocrystals, and molecular dispersions. Polymorphs, salts, solvates, cocrystals, and amorphous forms are collectively referred to as phases, and most clinical and commercial drug substances are typically prepared as a single phase to ensure reproducible properties, including solubility and chemical and physical stability. In contrast, most drug products are phase mixtures, consisting of a number of excipient phases and at least one drug substance phase. The analysis of drug product, in particular, often faces sensitivity challenges when assessed by both SSNMR and other techniques, such as powder X-ray diffraction (PXRD), because the drug substance may only be present at low levels in the mixture. The pharmaceutical applications of SSNMR make use of experimental approaches that are particularly well suited to small- and mid-sized organic molecules with significant structural diversity, which exist in crystalline, disordered, and amorphous states, and which are generally not amenable to isotopic labeling. In the following sections, basic and emerging experimental SSNMR techniques of interest in pharmaceutical applications are first briefly reviewed. The applications of SSNMR to common pharmaceutical situations are then discussed in more detail.

21.2 BASIC EXPERIMENTAL METHODS

Many of the basic SSNMR experiments applied throughout the broader fields of chemistry and biochemistry also find use in studies of pharmaceuticals (see Chapter 2). The basic 1D MAS experiment with direct polarization created through a single pulse is often used in studies of ^{1}H, ^{19}F, and ^{31}P nuclei. The cross-polarization (CP) experiment is typically applied to enhance sensitivity when observing nuclei such as ^{13}C, ^{15}N, ^{19}F, and ^{31}P. Small molecule drug substances often contain ^{13}C environments that exhibit strong chemical shift anisotropy (CSA), such that MAS sidebands may interfere with peaks of interest. For example, aromatic carbons with isotropic ^{13}C chemical shifts in the range 110–160 ppm often exhibit higher CSA and, as a result, may yield spinning sidebands near to aliphatic carbons of interest using the magnetic field strengths and MAS rates commonly encountered in pharmaceutical applications. Because

of this, a total sideband suppression (TOSS) sequence may be applied before observation of the 1D ^{13}C CP-MAS signal to remove interferences and simplify analysis.[5] The combined sequence is often referred to as a CP-TOSS pulse sequence and is commonly applied in pharmaceutical studies of the ^{13}C nucleus. Basic quantitative methods are commonly employed in pharmaceutical applications using 1D spectra of nuclei such as ^{1}H, ^{13}C, ^{15}N, ^{19}F, and ^{31}P, and the differential effects of sidebands and CP can be accounted for with suitable calibration to achieve excellent detection limits for low levels of a given pharmaceutical phase present in a larger mixture.[6]

Nuclei such as ^{13}C, ^{15}N, ^{19}F, and ^{31}P in pharmaceutical materials are generally strongly coupled to nearby ^{1}H nuclei via heteronuclear dipolar interactions that are only partially averaged by MAS. Observation of these nuclei benefits from the use of heteronuclear decoupling methods. RF irradiation is generally applied to the ^{1}H spins while ^{13}C or other heteronuclear spins are observed. The two-pulse phase modulation (TPPM) and small phase incremental alternation (SPINAL) schemes are effective and popular methods for use in studies of pharmaceutical solids. In general, the best results are obtained using TPPM or SPINAL with ^{1}H RF fields equal to or greater than 100 kHz, which are readily achievable with modern probe designs with rotor outer diameters of 4 mm or less. Pharmaceutical materials often exhibit strong ^{1}H–^{13}C dipolar coupling, necessitating the use of these fields in most applications.

The observation of ^{1}H nuclei with resolution sufficient for use in most structural studies of pharmaceutical materials also requires special methods. The ^{1}H SSNMR spectra of organic solids lack the resolution of their solution-state counterparts because of the effects of strong ^{1}H–^{1}H dipolar coupling. Fast MAS spinning with rates of 30–70 kHz is increasingly utilized as probes using rotors with outer diameters of 1–2.5 mm become more widely accessible. Low resolution ^{1}H spectra obtained with single-pulse methods may still be used to measure relaxation times (such as ^{1}H T_1 or ^{1}H $T_{1\rho}$) or observe mobile species present in pharmaceutical materials. Both high-speed MAS and ^{1}H homonuclear decoupling pulse sequences have been developed to reduce ^{1}H–^{1}H dipolar coupling and enable ^{1}H SSNMR spectra. A popular sequence for this purpose is known as *decoupling using mind boggling optimization* (DUMBO), and is capable of improving resolution at MAS rates of up to 65 kHz.[7] Techniques such as DUMBO and high-speed MAS enable the experimental observation of ^{1}H SSNMR spectra in pharmaceutical solids with sufficient resolution to allow the study chemical shifts of key proton environments, although the resolution currently achievable is generally insufficient to observe subtle chemical shift differences and most J-couplings. These developments in experimental methods have allowed for better understanding of phenomena such as hydrogen bonding and aromatic π-stacking, as well as the routine use of ^{1}H dimensions in 2D experiments.

A number of spectral editing techniques have been developed for organic solids, based on both dipolar and J-coupling to nearby protons, some of which are used in pharmaceutical studies. The attached proton test (APT) experiment is often applied to assist with assignment of ^{13}C SSNMR spectra of small molecule drug substances. The APT experiment relies on ^{1}H–^{13}C one-bond J-coupling to distinguish carbon atoms through phase shifts that vary with the number of attached protons.[8] A similar result can be obtained for ^{15}N spectral editing, which is useful in assessing nitrogen protonation state in drug substance solid forms.[9] Closely related multiple-quantum filtering methods are also available and can be used, for example, to produce ^{13}C spectra exhibiting only CH or CH$_2$ signals.[10]

Perhaps the most frequently used spectral editing method in pharmaceutical applications is the ^{13}C or ^{15}N dipolar dephasing method (often referred to by its original name of 'nonquaternary suppression'), in which strong ^{1}H–^{13}C or ^{1}H–^{15}N dipolar coupling is reintroduced by removing ^{1}H decoupling during a period of the experiment, resulting in dephasing of ^{13}C or ^{15}N magnetization through T$_2$ relaxation processes.[11] This experiment normally produces signals from only quaternary carbon or nonprotonated nitrogen sites, which dephase more slowly than their protonated counterparts. The dipolar dephasing experiment also reveals signals from nuclei engaged in rapid molecular motion that averages ^{1}H–^{13}C or ^{1}H–^{15}N dipolar coupling (and hence results in slower dipolar dephasing), as is usually the case for, e.g., methyl carbons. An example of the use of ^{13}C dipolar dephasing, performed in conjunction with CP and TOSS, is shown in Figure 21.1(a) for Form A of cimetidine, a small molecule crystalline drug substance and histamine H$_2$ receptor antagonist.[9] Only the methyl and quaternary ^{13}C signals are observed in the dipolar dephasing experiment in Figure 21.1(a). The use of dipolar dephasing helps in the assignment of ^{13}C resonances with

similar frequencies, but different protonation states, such as distinguishing the resonances assigned to either the C1 or C2 positions, or those assigned to either the C6 or C9 positions. In Figure 21.1(b), the results of an ^{15}N version of the dipolar dephasing experiment are shown for Form A of cimetidine. The experiment was performed with CP and dephasing during a spin echo but without the use of TOSS (to avoid loss of sensitivity). For sensitivity reasons, the dephasing was performed for a shorter period, and thus small signals from dephased nitrogens N1, N3, and N4 remain, but the results still allow for unambiguous assignment of nitrogens with similar frequencies but different protonation states, such as N4 and N5.[9] Other relaxation-based editing methods that exploit differences between, e.g., ^{1}H T$_1$ and T$_{1rho}$ values are also employed in studies of pharmaceutical materials, particularly in analysis of complex mixtures such as drug products, where editing of the spectrum to observe drug substance or a particular excipient may be of interest.[12] Finally, methods based on ^{13}C T$_1$ and T$_{1\rho}$ values can be used as a filter to selectively observe or suppress fast relaxing carbons, such as methyl groups.[13] Specialized relaxation-based filtering methods are also applicable to other nuclei, such as ^{15}N or ^{19}F.

Relaxation time measurements also play an important role in many pharmaceutical SSNMR studies. Measurement of relaxation times is often accomplished using pseudo-2D experiments, in which a time dimension is incremented over a short number of values (e.g., 8–16 time points), followed by nonlinear fitting of the resulting data. The most common relaxation time measurement employed in pharmaceutical applications is the saturation recovery measurement of ^{1}H T$_1$ through heteronuclear (e.g., ^{13}C or ^{19}F) detection. The popularity of this experiment stems from its use to attempt to detect different ^{1}H T$_1$ values for different resonances in a ^{13}C spectrum. ^{1}H–^{1}H spin diffusion across a particle equalizes ^{1}H T$_1$ values so that a homogeneous ^{1}H T$_1$ is generally obtained within particles of a given phase.[14] As a result, the observation of two distinguishable sets of ^{1}H T$_1$ values across the resolved peaks in a ^{13}C spectrum can be used to demonstrate the presence of two separated phases in the sample. This type of result helps confirm that a sample of drug substance phase contains phase impurities, i.e., another physically separated phase that is present besides the desired phase. For example, another undesired polymorph present at the 10% w/w level in a batch of a desired drug substance polymorph

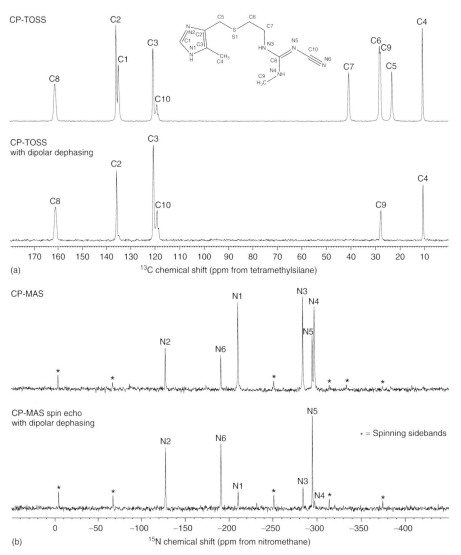

Figure 21.1. (a) ^{13}C CP-TOSS spectrum of cimetidine Form A obtained using a 2 ms contact time, compared to a dipolar dephasing CP-TOSS spectrum obtained with an added three rotor period shifted echo to obtain four total rotor periods of interrupted ^{1}H decoupling. Spectra were observed at 11.7 T and 273 K using a Bruker Avance II+ spectrometer equipped with a 4 mm HFX probe at an MAS rate of 8 kHz. The numbering scheme used is shown in the inset structure. (b) ^{15}N CP-MAS spectrum and dipolar dephasing CP-MAS spectrum of cimetidine Form A. Both spectra were obtained using a 5 ms contact time, with the dipolar dephasing spectrum obtained by adding a spin echo period of two rotor periods during which the ^{1}H decoupling was interrupted. Spectra were observed at 9.4 T and 298 K using a Bruker Avance spectrometer equipped with a 7 mm HX probe with an MAS rate of 5 kHz

is a phase impurity (particularly if its specific identity as a polymorph is not yet known), while in a salt of a drug substance, the presence of small amounts of a free base form is a phase impurity. In addition to heterogeneous ^{1}H T_1 values, phase impurities generally appear as small peaks in ^{13}C CP-MAS spectra relative to the larger peaks of the dominant phase, and may often be first assessed using only the 1D spectrum. This is possible because the intensity of peaks in a ^{13}C CP-MAS spectrum, while not quantitative, is sufficiently uniform that low level impurities are easily distinguished. While PXRD data can also be used to make the phase purity assessment, detailed powder indexing and refinement is typically required (and PXRD reflections inherently vary greatly in intensity). As a result, the more straightforward SSNMR techniques have become widely employed for this purpose.

^{1}H T_1 values can also be measured as a function of temperature to study dynamics. Heteronuclear detection is also widely employed for measurement of other relaxation values, including ^{1}H $T_{1\rho}$, which may also be used to assess phase purity and dynamics. Finally, measurements of heteronuclear T_1 or $T_{1\rho}$ values, such as CP experiments that determine ^{13}C T_1 or $T_{1\rho}$, are also employed to study dynamics in crystalline phases that exhibit significant molecular motion or in amorphous phases through study of the relaxation times of individual assigned resonances.

Finally, the measurement of the three principal components of chemical shift tensors (CSTs) is also widely employed in pharmaceutical applications, and in some cases, special experimental methods are utilized. While ^{19}F, ^{31}P, and ^{15}N CSTs are generally obtained from analysis of conventional MAS or CP-MAS experiments using methods such as the Herzfeld–Berger analysis, the spectral overlap encountered in ^{13}C spectra generally demands the use of special pulse sequences. Techniques such as magic-angle turning (MAT) and five-π replicated magic-angle turning (FIREMAT) are used to create 2D spectra wherein an isotropic chemical shift dimension is separated from an anisotropic chemical shift dimension, often in combination with specially developed convolution techniques and with CP. The phase-adjusted spinning sidebands (PASS) method, often combined with CP for ^{13}C studies (CP-PASS), is another commonly used method for analysis of CSTs in solid pharmaceuticals.[5] The CP-FIREMAT and CP-PASS methods can be performed at slow MAS rates of, e.g., 1.5 kHz on the standard 4 mm and

5 mm MAS probes often used for ^{13}C CP-MAS and CP-TOSS studies at higher MAS rates; alternatively, gas restrictors can be used to reduce spinning rates.

Single crystal X-ray diffraction (SCXRD) structures are available for many crystalline pharmaceutical phases of interest, including many drugs and small molecule excipients. The availability of an SCXRD structure allows for additional use of density functional theory (DFT) calculations to predict NMR parameters such as chemical shielding, electric field gradient (EFG), and J-coupling tensors of the crystalline phase, which can generally be used to interpret spectra in more detail than possible using empirical trends. Methods based on molecular clusters and methods that replicate the solid state using periodic boundary conditions, as well as hybrid methods incorporating aspects of both approaches, are currently in common use. DFT calculations on molecular clusters generally employ gas phase models where a drug molecule is surrounded by other molecules to replicate intermolecular interactions. Molecular cluster calculations are also useful in cases where periodicity is too large to be replicated by the DFT calculation, such as forcefield simulations of amorphous phases using large cubic cells, from which clusters can be extracted for NMR parameter computations. Finally, pure gas phase DFT calculations of NMR parameters are also useful in many cases, particularly in the prediction of the response of an NMR parameter to a structural change, such as the calculation of the principal components of a chemical shielding tensor as a torsional angle in a molecule is rotated, or the calculation of an isotropic chemical shielding value or quadrupolar parameter as a hydrogen bond or aromatic π-stacking interaction is geometrically varied.

21.3 EMERGING EXPERIMENTAL METHODS

The latest experimental SSNMR methods developed for use in other fields, such as materials science and biochemistry, are increasingly being adapted for use in pharmaceutical applications (see Chapter 2). Through these emerging experimental methods, the capabilities of an SSNMR instrument are more completely harnessed and more information can usually be obtained about materials of interest. The use of the full capabilities of an SSNMR instrument may be critical in an industrial setting, where such use can differentiate SSNMR from techniques often seen as less expensive competitors, including PXRD and vibrational spectroscopy.[15] Three general categories of emerging methods are discussed here: (i) 2D correlation methods based on dipolar coupling and J-coupling, (ii) methods that observe less accessible quadrupolar nuclei, and (iii) methods that improve sensitivity through dynamic nuclear polarization (DNP). 2D homonuclear correlation methods have been developed for use with many of the nuclei of interest in pharmaceutical solids, and include methods designed to detect $^1H-^1H$ proximity through dipolar coupling. One popular technique, known as *double-quantum back-to-back* (DQ-BABA) spectroscopy, is a robust method for obtaining ^1H, ^{19}F, and ^{31}P 2D homonuclear correlation spectra with a high-resolution double-quantum indirectly detected dimension. Most homonuclear correlation methods developed for ^{13}C and ^{15}N labeled biomolecules and based on dipolar and J-coupling are not commonly employed in studies of small molecule pharmaceuticals, where experiments rely on the presence of nuclei with naturally high abundance. However, a subset of homonuclear correlation methods, particularly those based on spin diffusion processes, are increasingly used in homonuclear correlation experiments involving ^1H and ^{19}F nuclei in pharmaceutical materials.

Various forms of heteronuclear correlation between spin $\frac{1}{2}$ nuclei, or between a spin $\frac{1}{2}$ nucleus and a quadrupolar nucleus, have also been developed. One of the most popular methods is a version of the 2D $^1H-^{13}C$ CP-HETCOR (cross-polarization heteronuclear correlation) experiment that includes frequency-switched Lee–Goldburg (FSLG) ^1H homonuclear decoupling (or a similar decoupling sequence) during ^1H evolution in the indirectly detected dimension, followed by CP and ^{13}C detection.[16] This experiment is also suitable for use for ^1H correlation with ^{15}N, ^{19}F, ^{23}Na, and ^{31}P nuclei, among other nuclei of interest in pharmaceutical materials, and is frequently used in this capacity. ^1H-detected versions of CP-HETCOR are also employed, and the CP step may be replaced with different heteronuclear dipolar mixing schemes that reintroduce dipolar coupling under MAS conditions. The results of CP-HETCOR experiments can be interpreted in conjunction with distances extracted from an SCXRD structure, if available, or in light of general atomic proximity arguments. In addition to interpretative structural analysis based on dipolar correlation, $^1H-^{13}C$ CP-HETCOR spectral results can be further combined with DFT

calculations of both ^1H and ^{13}C chemical shielding to assign crowded spectra.[17]

Correlation experiments based on homonuclear J-coupling are also available, although at present they are less frequently used in pharmaceutical applications. Heteronuclear correlation based on J-coupling transfer is more frequently used, particularly in the guise of the heteronuclear multiple-quantum coherence (HMQC) and heteronuclear single-quantum coherence (HSQC) experiments known as *MAS-J-HMQC* and *MAS-J-HSQC*. At present, the $^1H-^{13}C$ versions of these experiments are primarily used in pharmaceutical applications to assist with resonance assignments of ^1H and ^{13}C spectra.

There has been increasing interest in studies of quadrupolar nuclei in pharmaceuticals, including applications to nuclei that were previously difficult to study. The advent of higher magnetic field strengths and new pulse sequences has opened up the possibility of obtaining spectra of nuclei such as ^{14}N, ^{17}O, ^{35}Cl, and others. Nuclei with half-integral spin, such as ^{17}O, ^{23}Na, and ^{35}Cl, exhibit second-order quadrupolar central transition lineshapes that are reduced in breadth at higher field strengths. When combined with low to moderate magnitude EFG interactions and quadrupolar coupling parameters, half-integral spin nuclei can be useful in studies of crystalline pharmaceutical solids. For example, observation of ^{35}Cl in hydrochloride salts of drug substances is possible at higher fields using single-pulse MAS methods because of the low EFG magnitude generally encountered with chloride ions in organic crystals.[18] This has enabled ^{35}Cl MAS SSNMR studies of various drug substances that are hydrochloride salts and which exhibit polymorphism.[19–21] Nuclear environments experiencing greater quadrupolar coupling magnitudes, resulting from the EFG magnitude and/or from the properties of the nucleus, can produce extremely wide spectra not amenable to conventional methods and require specialized pulse sequences based on spin echoes to improve sensitivity, such as the quadrupolar Carr–Purcell–Meiboom–Gill (QCPMG) method. The QCPMG method can be used for ^{35}Cl SSNMR studies of aryl chloride groups, and also for observation of $^{79/81}Br$ and ^{127}I environments, which are found to a limited extent in pharmaceutical materials. Finally, spin 1 nuclei, which include ^2H and ^{14}N, require specialized experiments. The ^2H nucleus exhibits a low quadrupolar coupling magnitude and can be readily observed with traditional echo and MAS echo methods. Methods to address the observation of ^{14}N, which has a much larger quadrupolar coupling, have been developed more recently and applied to pharmaceutical materials.[9,22-24] These methods typically rely on indirect observation of a spin $^1/_2$ 'spy' nucleus using methods based on heteronuclear multiple quantum (HMQC) pulse sequences that are detected using the spin $^1/_2$ nucleus, such as the a 2D ^{13}C–^{14}N HMQC experiment. The transfer mechanism in these experiments at lower fields is the residual dipolar coupling between the spin $^1/_2$ nucleus and ^{14}N; at higher fields, the residual dipolar coupling decreases and J-coupling can also play a role. Furthermore, the use of dipolar recoupling sequences can be employed to reintroduce strong heteronuclear dipolar coupling and provide results similar to HETCOR experiments between spin $^1/_2$ nuclei. The ^{14}N–^{1}H HMQC experiment, which involves an ^{1}H–^{14}N–^{1}H transfer with ^{14}N evolution during t_1 and ^{1}H detection during t_2, is a prominent example of these approaches, and appears to be highly applicable in pharmaceutical applications because of the shorter experimental times required relative to 2D ^{13}C–^{14}N HMQC experiments Ref. 9.

The advent of commercial DNP experiments using microwave irradiation of biradicals inserted into a sample is an emerging technique that can greatly enhance the sensitivity of SSNMR experiments for less sensitive nuclei. The availability of gyrotron sources capable of reaching frequencies of 250 GHz and higher has allowed for the same strong static magnetic field to be used for both the electron and nuclear irradiation, resulting in improved, simplified systems that are more practical in pharmaceutical applications.[25,26] In the typical modern DNP experiment performed using MAS methods and SSNMR, signal enhancement is obtained for the spectrum obtained with continuous microwave irradiation using primarily the 'cross effect' for enhancement. The cross effect involves the use of a biradical with electron–electron dipolar coupling, which forms an eight-level system with the nuclear spin. The cross effect has advantages over other DNP mechanisms, including greater efficiency, ease of irradiation, and reduced loss of enhancement as static magnetic field increases (relative to other mechanisms, which fall off more rapidly at higher fields). Common biradicals useful in pharmaceutical SSNMR studies are based on 2,2,6,6-tetramethylpiperidin-1-yl)oxidanyl (TEMPO) and include bis-TEMPO-bis-ketal biradical (bTbK) and 1-(TEMPO-4-oxy)-3-(TEMPO-4-amino)propan-2-ol radical (TOTAPOL).

21.4 CRYSTALLINE PHASES OF DRUG SUBSTANCES AND EXCIPIENTS

Crystalline phases are commonly encountered in pharmaceutical materials, particularly when investigating small molecule drug substances and with small organic excipients. Crystalline phases may include both drug substances and excipients, although most excipients used in drug products are amorphous or semi-crystalline. Crystalline phases may also take the form of a salt, wherein the 'parent' molecule (usually the drug substance) is ionized and exists as a charge-balanced complex with a counterion, which itself may be an ion (e.g., chloride) or another ionized molecule (e.g., acetate). Zwitterionic crystalline phases of drugs also exist, e.g., where no counterion is present, and the drug itself is doubly ionized. Other crystalline phases of interest are nonionized molecular complexes, which include solvates and cocrystals. Solvates involve the cocrystallization of a drug or an excipient with organic solvent or water, such as trametinib dimethylsulfoxide or trehalose dihydrate. Cocrystals are generally defined as a complex between a parent molecule and a cocrystal former, wherein the two components, when crystallized as separate phases, are solids at room temperature. This definition differs only subtly from the definition of a solvate, in which at least one component is a liquid at room temperature, and thus solvates and cocrystals can both be treated as belonging to the same class of nonionized molecular complex.[27] Most crystalline phases have a sufficiently short ^{1}H T_1 value to allow for efficient CP experiments, although very small drug substances (such as aspirin) and excipients (such as mannitol) may exhibit very long ^{1}H T_1 values in some cases.

All of these crystalline phases are potentially subject to polymorphism, which may be generally defined as the existence of multiple crystalline phases for a single compound or complex.[28,29] Different polymorphs are often referred to as 'forms'. Most crystalline phases are screened for polymorphism and solvate formation during development. Screening has a number of benefits, such as allowing for identification of the optimal form, minimization of the risk of the unexpected appearance of a polymorph during large-scale crystallization, development of intellectual property, and identification of optimal solvents to use for crystallization (e.g., by selecting solvents that do not form solvates with the drug molecule during screening).

Figure 21.2 shows the ^{13}C CP-MAS spectra of two crystalline forms of azelastine HCl, an anhydrous

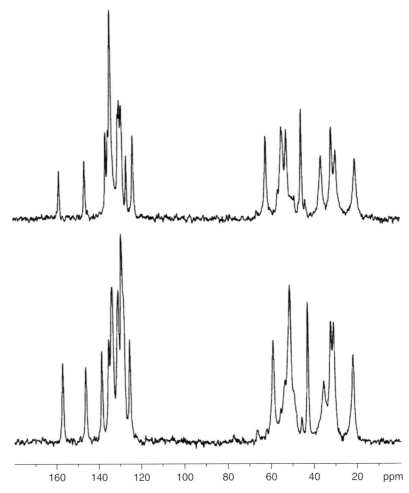

Figure 21.2. [13]C CP-MAS spectra of two crystalline forms of azelastine. The spectrum of azelastine HCl hydrate is shown at the top and the spectrum of anhydrous, nonsolvated azelastine HCl is shown at the bottom. Spectra were obtained at a static field strength of 9.4 T and at an MAS rate of 8 kHz. (Reprinted with permission from E. Maccaroni, E. Alberti, L. Malpezzi, G. Razzetti, C. Vladiskovic, and N. Masciocchi, Cryst. Growth & Des. 2009, 9, 517–524. Copyright 2009 American Chemical Society)

form and a hydrated form that was structurally related to two other organic solvates.[30] These spectra illustrate the typical resolution and information content of ^{13}C CP-MAS spectra of crystalline small molecule pharmaceuticals. Most small molecule organic drug substances contain carbonyl, aromatic, alkyl, and oxygen or nitrogen-containing heteroaromatic or heteroalkyl functional groups.[31] As a result, ^{13}C spectra of drug substances show resonances primarily in the ranges 200–110 ppm and 70–15 ppm. It is possible to assign many resonances found in ^{13}C CP spectra like those shown in Figure 21.2 using empirical trends established through extensive experience with ^{13}C solution-state NMR results. The ^{13}C spectra in Figure 21.2 were combined with SCXRD and PXRD results to examine structural features of the solvates in the absence of SCXRD data, including the likely space group and potential conformational effects involving a particular torsion angle.[30] The use of ^{13}C SSNMR to understand crystallographic symmetry is possible because of the potential occurrence of more than one independent molecule within the crystalline asymmetric unit, that is, more than one unique, nonsuperimposable molecular structure is present within the repeating cell of the crystal (using a simplified perspective that neglects mirror inversions in certain space groups as with racemic crystals). This situation is crystallographically referred to by a value designated Z′, wherein Z′ takes on a value of 1 if a single unique molecule is present, a value of 2 if a two unique molecules are present, and so on. Studies of pharmaceutical drug substances frequently evaluate Z′ using ^{13}C, ^{15}N, and ^{19}F spectra, and this can be used in combination with unit cell dimensions (available from PXRD analysis) to distinguish one potential space group from another.

Other common effects seen in ^{13}C SSNMR spectra of drug substances include the observation of residual dipolar-mediated quadrupolar coupling effects on spin $^1/_2$ resonances, such as ^{13}C nuclei covalently bonded to ^{14}N, $^{35/37}$Cl, and $^{79/81}$Br nuclei. These effects can help in the identification and assignment of spectra in some cases, and can interfere with analysis in others. Finally, motional effects are also observed in some crystalline drug substances, generally through broadening of a subset of resonances. When heteronuclear decoupling RF fields of 100 kHz or greater and modern heteronuclear sequences such as SPINAL or TPPM are employed, much of the residual linewidth in the ^{13}C spectra of small molecule pharmaceuticals is caused by either structural disorder or anisotropic bulk magnetic susceptibility (ABMS). The effects of ABMS can be confirmed through observation of linewidth changes in dilution experiments, in which a high-ABMS substance (such as many pharmaceutical drug substances) is physically mixed with a low-ABMS substance.

DFT methods are frequently used to assign the 1D spectra of complex crystalline drug substance phases, particularly the ^{13}C and ^{15}N spectra arising from small molecules with molecular masses of greater than 500 Da and in phases where Z′ ≥ 2. For example, two polymorphs of testosterone (the α- and β-phases) were assigned using the Gauge Including Projector Augmented Wave (GIPAW) DFT method for calculating CST, and individual molecules in a crystalline phase with Z′ = 2 were assigned using this calculation in conjunction with 2D ^{13}C–^{13}C correlation experiments.[32] The method has also been shown to be predictive of ^{19}F chemical shielding results for the fluorinated drug flubiprofen.[33] The use of the GIPAW method and similar DFT methods requires a crystal structure determined by SCXRD or other means. However, this method can be valuable even if crystal structures are not available for all crystalline phases in a family by allowing detailed assignment of at least

a few of the phases for which a crystal structure is available, which can then allow for comparative interpretation against the remaining phases. An example of the typical performance of the GIPAW method is shown in Figure 21.3, where a linear fit of calculated ^{13}C and ^{15}N isotropic chemical shielding values is given for Form A of cimetidine against previously reported experimental chemical shifts (which may be seen in the spectra of Figure 21.1).[9] The results show good agreement between calculation and experiment, and are typical of the results obtainable for rigid crystalline pharmaceutical phases. Linear fitting or linearization methods are frequently used to relate calculated chemical shieldings to measured chemical shifts and facilitate assignment, e.g., by allowing visualization of outliers that may correspond to incorrectly assigned positions or by allowing positions to be swapped to alternative assignments for assessment of the overall agreement between calculation and experiment.[34]

Experimental studies of polymorphism and solvation in drug substances often lead to situations where the SCXRD structure is not available for one or more members of a family of solid forms. This can arise from the inability to grow a single crystal of sufficient size or stability, e.g., for metastable or transient forms, or because of time restraints for growing crystals. Even if suitable crystals are available, SCXRD also has other limitations with respect to location of hydrogen atoms, and in cases where significant disorder is present within a structure (resulting in a poor SCXRD refinement). In situations where an SCXRD structure is not available for members of a family of polymorphs, SSNMR can provide structural information on aspects such as molecular conformation, hydrogen-bonding environment, and aromatic π-stacking interactions.[35] For example, a study of two polymorphs of sibenadet HCl made use of ^{1}H DQ-BABA techniques as well as related double-quantum techniques using ^{1}H homonuclear decoupling to observe correlations involving hydrogen-bonded NH and OH groups.[36] The results, depicted in Figure 21.4, confirm that the crystal structure of Form II (which is unavailable from SCXRD methods) is similar to the known crystal structure of Form I, at least with respect to the short ^{1}H–^{1}H contacts between the NH and OH positions. Double-quantum build-up rates were also analyzed to extract more detailed information about the relative ^{1}H environments in the two polymorphs.[36]

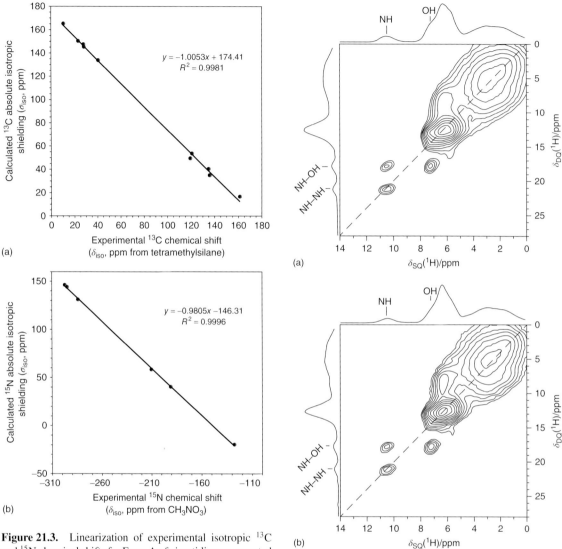

(a)

(b)

(a)

(b)

Figure 21.3. Linearization of experimental isotropic ^{13}C and ^{15}N chemical shifts for Form A of cimetidine, as reported in Ref. 9 and shown in Figure 21.1, against the isotropic shielding values obtained from a GIPAW DFT chemical shielding calculation. ^{13}C and ^{15}N results are shown in (a) and (b), respectively. The chemical shielding calculation was performed using the CASTEP module in the Materials Studio software package, version 6.0 (Accelrys, San Diego, CA, USA), with a Purdue–Burke–Ernzerhof functional, a plane wave cutoff energy of 440 eV, on-the-fly-generated pseudo-potentials, and a 2×1×1 k-space for Brillioun-zone integration, after optimization of hydrogen atom positions

Figure 21.4. DQ-BABA spectra of two polymorphs of sibenadet HCl, Form I (a) and Form II (b) obtained at 14.1 T and an MAS rate of 30 kHz using a 2.5 mm probe. An SCXRD structure was available for Form I, but not for Form II. One rotor period of BABA recoupling was used. (Reproduced with permission from Ref. 36. © John Wiley & Sons, Ltd., 2012)

SSNMR data can be used in some cases to assist with crystal structure determination in conjunction with established PXRD methods, such as Monte Carlo PXRD solution methods and Rietveld refinement.[35] This approach uses PXRD methods to solve the structure, which is adjusted to resolve ambiguities based on information obtained from SSNMR dipolar correlation methods such as ^{1}H DQ-BABA and [1]H–[13]C CP-HETCOR experiments, with verification using comparisons of calculated chemical shielding tensors to experimentally determined CSTs. The use of SSNMR data alone, without recourse to PXRD, can provide a crystal structure particularly in cases involving smaller molecules. These methods typically make use of ^{1}H spin diffusion build-up analysis and computational methods to use relative distance information to identify a structure, from which a space group and unit cell can then be obtained. Recently, the crystal structure of a moderate-sized drug molecule, 4-[4-(2-adamantylcarbamoyl)-5-*tert*-butyl-pyrazol-1-yl]benzoic acid, was solved using spin diffusion methods in conjunction with crystal structure prediction and GIPAW calculations, highlighting the ability of this approach to tackle more common drug substance phases.[37]

Another 2D SSNMR experiment that has been applied to crystalline pharmaceutical solids is the 2D [14]N–[1]H HMQC experiment. The results of this experiment applied to Form A of cimetidine are shown in Figure 21.5.[9] The spectrum was obtained using a recoupling time of 600 μs and primarily shows correlations produced by through-space [1]H–[14]N dipolar coupling. Internuclear distances obtained from the crystal structure are shown for several of these correlations. The experimental results shown in Figure 21.5 have the advantage of requiring just 6 h to obtain, relative to typically much longer acquisition times for [1]H–[15]N CP-HETCOR experiments. To obtain such results, higher MAS rates are generally necessary (a rate of 60 kHz was used in this study), and higher static field strengths are also helpful. The ^{14}N dimension exhibits second-order quadrupolar effects that are poorly resolved and lead to difficulties in determining the isotropic ^{14}N chemical shift, so an accompanying 1D ^{15}N CP-MAS experiment is generally useful in conjunction with the 2D [14]N–[1]H HMQC experiment. The R[3] dipolar recoupling scheme, which applies an ^{1}H RF field of twice the MAS rate and uses an $(x, -x)$ phase cycle, was used to obtain the spectrum shown in Figure 21.5.[24] The recoupled heteronuclear [1]H–[14]N dipolar interaction dominates the observed

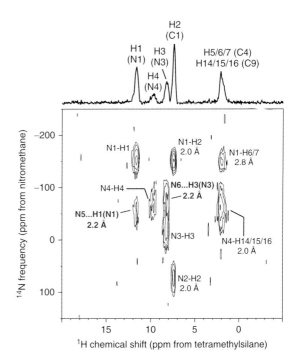

Figure 21.5. 2D [1]H-^{14}N HMQC spectrum of cimetidine Form A obtained at 850 MHz (B$_0$ = 20.0 T) using a 1.3 mm probe with a recoupling time of 600 μs and an MAS rate of 60 kHz. Correlations are labeled with respect to the heavy atom numbering scheme for cimetidine shown in Figure 21.1. Hydrogen atom labels are taken from the crystal structure (Cambridge Structural Database reference CIMETD) with attached heavy atoms shown in parenthesis. Distances are taken from the crystal structure after optimization of hydrogen positions using DFT (see Ref. 9). Correlations corresponding to intermolecular interactions are shown in red. Second-order quadrupolar shifts affect the frequencies of the ^{14}N resonances relative to the ^{15}N resonances shown in Figure 21.1(b). The F$_2$ projection (along the ^{1}H axis) is a skyline projection. (Reproduced from Ref. 9 with permission from The Royal Society of Chemistry)

correlations, while the effects of residual dipolar splitting and J-coupling are minimal.

Many marketed and developmental crystalline drug substances are delivered as salts.[38] Salts provide an alternative form in the event that a suitable crystalline phase of the free base or free acid cannot be identified, and also offer improved dissolution performance that can be critical to obtaining the needed in vivo bioavailability for drug with poor aqueous solubility.[38] Both organic and inorganic salts are common. Salt forms are different crystalline phases and thus usually yield distinct spectra for the commonly observed spin $^1/_2$ nuclei. The counterions of two of the most common inorganic salts used in pharmaceuticals, namely sodium and hydrochloride salts, may be directly addressed using ^{23}Na and ^{35}Cl (or ^{37}Cl) methods. For example, sodium naproxen and its hydrates have been studied using direct observation by ^{23}Na MAS methods as well as methods that narrow second-order quadrupolar lineshapes.[39,40] The ^{23}Na EFG was found to be sensitive to hydration in the sodium naproxen forms. This is consistent with the findings of earlier work involving a hydrated form of *N*-(3-(aminosulfonyl)-4-chloro-2-hydroxyphenyl)-*N'*-(2,3-dichlorophenyl) urea, a developmental CXCR2 antagonist, an IL8 inhibitor for inflammatory disorders, wherein the ^{23}Na MAS lineshape was strongly affected by dehydration.[41] Beyond 1D MAS experiments and the use of 1D or 2D methods to narrow second-order quadrupolar lineshapes, 2D ^{1}H–^{23}Na CP-HETCOR experiments provide a straightforward approach to linking the sodium ion resonances into the organic backbone of a drug substance crystal structure.[41] Hydrochloride salts have also been extensively studied by 1D ^{35}Cl MAS and static methods.[18–21] A variety of hydrochloride drug substances, including salts and their polymorphs and hydrates of more than a dozen commercially available compounds as well as several developmental compounds, have been studied to date.[19–21,42,43] The results have shown distinct spectra using MAS that reflect the subtle effects of the ^{35}Cl coordination sphere and hydrogen bonding to chloride on both EFGs and isotropic chemical shifts, using experiments that require several hours at higher fields. In many of these studies, ^{35}Cl spectra have also been calculated successfully using DFT methods. As with sodium salts, the ^{1}H-^{35}Cl CP-HETCOR experiment may be used to establish dipolar connectivity with the parent drug substance, while also probing structure beyond the ^{35}Cl environment.[42]

A large number of crystalline drug substance salts are organic acids or bases.[38] In addition, cocrystals between organic molecules and drug substance molecules are increasingly screened for potential development as alternative crystalline forms.[44] As the primary difference between a cocrystal and a salt is the transfer of a proton, the SSNMR techniques that can be applied are the same when the cocrystal or salt former is an organic molecule. Studies of cocrystals and organic salts by SSNMR generally involve confirmation of phase purity and analysis of interactions between the components along with any other species present (such as in a three-component system of drug, cocrystal former, and water of hydration).[45] 2D SSNMR methods based on dipolar correlation are particularly useful in studies of organic salts and cocrystals, because they can detect through-space interactions between the molecules and identify structural features of interest. For example, in Figure 21.6(a), the structure of a cocrystal of the nonsteroidal anti-inflammatory drug tenoxicam with succinic acid is depicted, with the corresponding ^{1}H–^{13}C CP-HETCOR spectrum shown in Figure 21.6(b).[46] The ^{1}H–^{13}C CP-HETCOR spectrum allows for observation of intermolecular correlations between tenoxicam and succinic acid as well as intramolecular correlations, which can be used in conjunction with other 1D and 2D SSNMR data to quickly evaluate the likely hydrogen-bonding network in the cocrystal in advance of a more definitive SCXRD study.

In addition to dipolar effects, chemical shifts can also provide sensitive indications of intermolecular interactions in pharmaceutical solids. Hydrogen-bonding trends are often observed from isotropic shift trends in ^{1}H spectra as well as from ^{13}C and other nuclei that participate or are near to donor and acceptor groups.[47] Hydrogen-bonding effects may also be probed through the analysis of the principal components of the CST either in relation to empirical trends or with the aid of DFT calculations. The measurement of CSTs is used for several purposes in crystalline pharmaceutical solids, including providing support for spectral assignments, analysis of molecular conformation (e.g., at a particular ^{13}C or ^{15}N site), detection of hydrogen-bonding effects, and analysis of ionization state, often in comparison to DFT calculations of the principal components of the chemical shielding tensor. For example, the principal components of ^{13}C CSTs measured by the CP-PASS experiment and ^{15}N CSTs measured by CP-MAS were used to analyze three polymorphs of 5-methyl-2-[(2-nitrophenyl)amino]-3-thiophenecarbonitrile and, with reference to DFT calculations, and determine that the phenyl-thiophene ring coplanar angle was empirically related to CST changes between the forms.[48] ^{13}C CP-PASS experiments for analysis of CST principal components have been used in the studies of other pharmaceutical polymorphs, and CP-FIREMAT experiments have been applied to the structural analysis of cocrystals.[35,45]

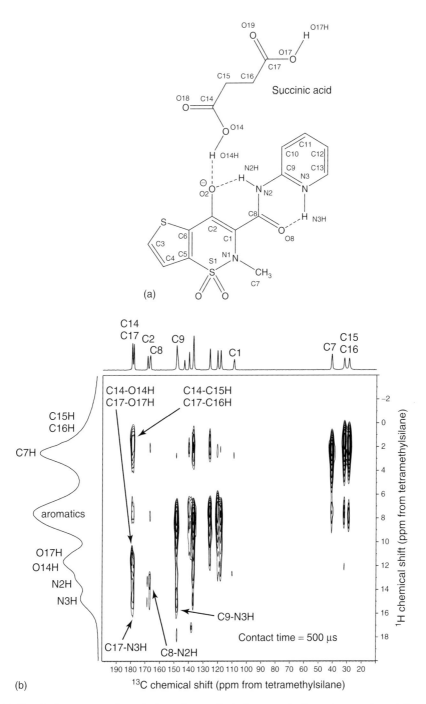

Figure 21.6. (a) Chemical structure and numbering scheme of a 1 : 1 cocrystal of tenoxicam and succinic acid. (b) ^{1}H-^{13}C CP-HETCOR spectrum of this crystalline phase, obtained at an MAS rate of 12.5 kHz and at a field strength of 9.4 T, showing key correlations including intermolecular correlations. The ^{1}H MAS and ^{13}C CP-TOSS spectra are displayed along the F_1 (vertical) and F_2 (horizontal) dimensions, respectively. (Reproduced with permission from Ref. 46. © Elsevier, 2012)

DNP methods have been used to study crystalline pharmaceutical solids, including polymorphs of the drug substances acetaminophen (paracetamol) and sulfathiazole.[49] It was found that the longer ^{1}H T$_1$ times (on the order of 200 s) of these drugs assisted with DNP enhancement through spin diffusion over long distances. An enhancement factor of about 60 (over conventional CP-MAS results) was obtained for microcrystalline sulfathiazole at 9.4 T and 105 K. In addition to enhancing ^{13}C CP-MAS spectral acquisition, this result also increased sensitivity such that insensitive natural abundance 2D ^{13}C–^{13}C J-coupling correlation experiments became much more feasible. While most drug substances have much shorter ^{1}H T$_1$ times even at low temperatures and may not perform as well in this type of approach, the ability to use DNP methods on materials with long ^{1}H T$_1$ times, which includes a small percentage of developmental drugs and a number of crystalline excipients, opens up new possibilities for the study of these difficult materials.

Many crystalline phases exhibit molecular motion, with a common case being the motion of water in hydrated drug substances or excipients. The water molecule is known to undergo rapid jump motion in many hydrates, which can be readily observed using spin labeling with D$_2$O and/or H$_2$^{17}O. Spin labeling of hydrated drug substances can often be achieved through simple vapor exchange or by small-scale slurry experiments. An ^{2}H and ^{17}O study of a hydrated form of N-(3-(aminosulfonyl)-4-chloro-2-hydroxyphenyl)-N'-(2,3-dichlorophenyl) urea, a developmental CXCR2 antagonist and IL8 inhibitor for inflammatory disorders, was performed using wideline methods and vapor exchange with D$_2$O and H$_2$^{17}O.[41] This allowed for observation of jump-motion behavior of the exchanged water molecules. Well-defined group or atomic motion may also occur in other functional groups in crystalline phases, and has been observed for groups such as small aliphatic sidechains and cycloalkyl rings.[29]

Disordered crystalline phases represent a particularly challenging area of application for SSNMR. Disorder can appear within the unit cell of a crystal, where atoms are distributed between multiple positions in a crystal structure. This effect can lead to broadening of resonances, but can still provide information about the nature of disorder in such a phase and its relationship with other phases. For example, a study of the hydrochloride salt of a crystalline developmental drug identified two structurally similar polymorphs; the first polymorph was studied using SCXRD and found to exhibit disorder involving two functional groups, whereas the second polymorph, which was not amenable to SCXRD analysis, was found to exhibit even greater disorder by spectroscopic techniques including ^{13}C and ^{35}Cl SSNMR.[20] The SSNMR study allowed for inferences to be drawn about the structural elements responsible for the disorder and the relationship between the forms. Disorder effects can also appear at larger scale, such as in composite crystals, where nanoscale layers comprising different crystal structures are fused or similar structural features are present. An ^{1}H and ^{13}C SSNMR and SCXRD study of a pleuromutilin-derived drug substance exhibiting this phenomenon demonstrated the ability of SSNMR techniques, including ^{1}H-^{13}C CP-HETCOR experiments, to provide structural information in these complex systems that can be challenging to understand using SCXRD methods alone.[50]

21.5 AMORPHOUS PHASES OF DRUG SUBSTANCES AND EXCIPIENTS

Amorphous phases lack long range order, and are generally not amenable to conventional PXRD analysis of Bragg reflections, although pair distribution function (PDF) analysis of diffuse PXRD scattering can be applied to obtain more limited structural information.[51] Instead, most amorphous phases of pharmaceutical materials are investigated using spectroscopic methods, particularly SSNMR and vibrational spectroscopy. Of these techniques, SSNMR can offer the most structural insight through investigations of multiple nuclei using 1D and 2D experiments. A number of important materials used in pharmaceutical development are amorphous, disordered, or semicrystalline, including many excipients, most drug–polymer dispersions, most mesoporous dispersions (which involve a small molecule drug substance residing within a mesoporous host, such as silica), and most biopharmaceutical, oligonucleotide, and oligopeptide drugs. Amorphous phases consisting of only drug substance are frequently prepared for experimental studies, to provide a high-energy state to improve aqueous solubility, for solid form screening studies and for use as a reference in analytical method development. However, because amorphous drug substance is often chemically and physically unstable and can have poor physical properties (such as excessive hygroscopicity), it is rarely chosen as a form for development.[29,30] Although 1D and 2D SSNMR studies of these amorphous forms exhibit broad lineshapes, they can still provide structural information not available from other techniques and are useful in probing the relationship between an amorphous form and crystalline forms (e.g., if an amorphous form results from dehydration of a crystalline hydrate, or milling of a crystalline form).

Amorphous phases are often encountered in studies of amorphous drug–polymer dispersions, which are widely employed in pharmaceutical development to improve the dissolution performance of drugs with poor water solubility.[52] Amorphous dispersions straddle the definition of drug substance and drug product, as they can be employed in some cases as the final drug product by direct filling into capsules. Most amorphous dispersions can be described as a solid solution, in that the drug substance is dispersed at the molecular level within the polymer. Interactions between the drug and polymer, such as hydrogen bonds or van der Waals forces, can help stabilize amorphous dispersions. SSNMR methods can detect such interactions, and furthermore can demonstrate that the drug substance and polymer have formed a solid solution using 2D HETCOR methods. Phase separation in a dispersion or the formation of nanodomains enriched in drug substance and polymer can be observed using heteronuclear-detected ^{1}H T_1 and ^{1}H $T_{1\rho}$ measurements.[52,53] SSNMR has numerous advantages in the role of detection of nanodomain formation or the formation of a true solid solution in relation to the two other primary techniques of modulated differential scanning calorimetry and PDF PXRD analysis, particularly in more complex dispersions containing more than two components.[52,53] The ^{14}N–^{1}H HMQC experiment has also been successfully applied to the study of a two-component amorphous dispersion.[54] This experiment may be particularly useful in studies of dispersions because of its combined sensitivity to nitrogen chemical shift and EFG (as nitrogen is a common hydrogen bond donor and acceptor in drug substances) and to ^{1}H chemical shifts.

Amorphous excipients are also frequently studied by SSNMR, including 'functional' excipients that affect the performance of drug product. Many polymeric excipients are commonly found as amorphous phases, including polyvinylpyrrolidone (PVP) and many starch-based and cellulosic polymers. For example, a ^{13}C SSNMR study of the amorphous excipient sodium alginate, a swelling agent, binder, and disintegrant used in controlled release formulations, analyzed monomer content using ^{13}C spectral deconvolution methods.[55] Correlations were observed among intrinsic viscosity, water content, and ^{1}H T_1 relaxation time, which offered the possibility of examining these effects directly in a formulation (wherein sodium alginate may be a minor component of a mixture); most alternative techniques require pure substances for analysis.

Amorphous organic solids may also be readily analyzed by DNP methods. In contrast to crystalline phases, a biradical may be added to many of these amorphous materials by simple redissolution or soaking in organic solvents, or by initial preparation by conventional methods (e.g., spray-drying, melt extrusion, or lyophilization) using an added biradical, without significant deleterious effects on the obtained solid phase. This straightforward sample preparation allows for direct study of important materials such as amorphous dispersions and amorphous drug substances and excipients. Furthermore, even the simplest preparation approach of direct soaking of biradical solutions into the amorphous material may be able to achieve better biradical dispersion and better enhancements than in crystalline materials. For example, DNP analysis of amorphous *ortho*-terphenyl led to greater enhancements compared to DNP analysis of the crystalline phase of this small molecule, an effect that was attributed to the better dispersion of the radical in the amorphous solid.[56] A similar result can be obtained for typical amorphous dispersions, where significant ^{13}C CP enhancements were observed in a dispersion of 16% w/w ezetimibe in mesoporous silica, and also in a dispersion of 30% w/w diflunisal in hydroxypropylmethylcellulose acetate succinate (HPMC-AS), using a 9.4 T static magnetic field strength and a 263 GHz gyrotron at a sample temperature of 100 K.[52]

21.6 DRUG PRODUCTS

From a regulatory perspective, the most critical stage of solid form control occurs with respect to the drug product ultimately delivered to patients. Thus, while control over the drug substance form to be used as input to drug product manufacturing is important, any form changes that might occur during manufacturing or stability of drug product are of greater importance to regulatory agencies. SSNMR is frequently employed to support this effort, and has many advantages in the direct analysis of drug products relative to other techniques. First, it offers high spectral resolution, and because of the organic functional groups present in drug substances and excipients, often allows for observation of key drug substance resonances (e.g., ^{13}C aromatic resonances) with little overlap with excipients. Second, SSNMR allows for analysis of the most common dosage forms, such as oral tablets, with minimal sample preparation. Normally, a small section of a tablet is punched or cut and inserted (still as an intact fragment) directly into an MAS rotor. (Most techniques require some form of sample preparation – PXRD normally requires tablets to be crushed to a fine powder, whereas vibrational microscopy usually requires cross-sectioning). Third, SSNMR offers good sensitivity for drug substance at typical levels in dosage forms, which generally range from 5% to 50% by weight although many exceptions, e.g., for highly potent drugs, are known. A large number of applications of SSNMR to various drug products have appeared, investigating formulations that range from tablets to suspensions, and are too numerous to review here although more detailed reviews are available.[2,4]

Reasonable sensitivity is possible for ^{13}C detection of drug substance form within most drug products, typically using longer experimental times of 24 h or greater. For example, a study of bambuterol hydrochloride and terbutaline sulfate investigated drug products containing up to 5% w/w of each drug.[57] As an example, a limit of detection (LOD) of 0.5% w/w and a limit of quantitation (LOQ) of 1.0% w/w were established for the desired form of bambuterol hydrochloride in these formulations. For quantitative purposes in drug product, it is often necessary (from a regulatory perspective) to provide an LOQ and/or LOD for undesired forms that is as low as possible; for example, in a drug product containing 20% w/w of drug substance, providing a method that can detect undesired forms to a level of 5% w/w yields an absolute target LOD of 1% w/w. As demonstrated by the bambuterol hydrochloride example, this is feasible for drug products containing higher levels of drug substance. Additional sensitivity can be gained using more sensitive nuclei, if available in the drug substance. For example, the much greater sensitivity of ^{19}F detection relative to ^{13}C detection was exploited by showing that an amorphous form (with broader resonances that are inherently less sensitive) could be detected by ^{19}F SSNMR in tablets containing only 2% w/w of a selective M3 neuronal receptor antagonist.[58] Although many drugs are fluorinated, the availability of an ^{1}H-based approach would help to broaden high sensitivity to nearly all drug substances. Because of much more significant spectral overlap in the ^{1}H spectrum of a drug product relative to the ^{13}C spectrum, even under fast MAS and homonuclear decoupling, novel approaches are needed. An ^{1}H 2D DQ experiment using ^{1}H homonuclear decoupling was used to demonstrate one possibility, allowing detection of a single form of a drug substance within a formulation.[59] Finally, DNP methods also offer a solution to sensitivity challenges and have been applied to the study of different commercial drug products containing 4.8–8.7% by weight of the antihistamine cetirizine dihydrochloride by soaking of biradical solutions into the formulations.[60] DNP enhancements on the order of 40–90 were observed, which allowed for the use of the normally insensitive ^{1}H–^{15}N CP-HETCOR experiment directly on the formulated samples. It was also possible to obtain domain size information within the formulations from detailed spin diffusion analysis.[60]

21.7 CONCLUSION

SSNMR methods for the study of pharmaceutical materials continue to advance as new approaches are harnessed for the unique applications in this field. Experimental techniques that rely primarily on natural abundance nuclei and are robust enough for routine industrial application have tended to be the most widely adopted methods for pharmaceutical applications. The use of SSNMR in the studies of pharmaceutical materials is also frequently characterized by a multidisciplinary approach, often combining SSNMR with other spectroscopic, diffractometric, and physical techniques. The multidisciplinary nature of solid form studies is driven by scientific factors as well as the need to provide regulatory agencies with as complete of a picture as possible for drug products seeking marketing approval.

In the future, the use of 2D SSNMR methods is likely to continue to advance given the need to study increasingly complex pharmaceutical materials and the inherent flexibility of current systems. Studies of drug products using SSNMR, including the use of the sensitive ^{19}F nucleus, are also expected to increase in importance. The traditional spin $\frac{1}{2}$ nuclei observed in pharmaceutical SSNMR studies are now frequently supplemented using nuclei such as ^{1}H, ^{14}N, and ^{35}Cl, a trend that is also likely to continue. Finally, significant advances in sensitivity enhancement methods based on DNP that have been recently achieved in academic laboratories are likely to find increasing application in industrial laboratories.

ACKNOWLEDGMENTS

The author thanks his colleagues and collaborators at GlaxoSmithKline plc., Bruker Biospin, Inc., and many academic and industrial institutions for helpful discussions about pharmaceutical SSNMR analysis.

RELATED ARTICLES IN EMAGRES

Magic Angle Spinning

Rotating Solids

Cross Polarization in Rotating Solids: Spin-1/2 Nuclei

Heteronuclear Decoupling in Solids

Fast Magic-Angle Spinning: Implications

Spectral Editing Techniques: Hydrocarbon Solids

Through-Bond Experiments in Solids

Spin Diffusion in Solids

Spin diffusion for NMR crystallography

Chemical Shift Tensor Measurement in Solids

Sideband Analysis in Magic Angle Spinning NMR of Solids

Magic-Angle Turning Spectra of Solids Involving 5π-Pulse Sequences

Shielding Tensor Calculations

Quadrupolar Coupling: An Introduction and Crystallographic Aspects

Double-Quantum NMR Spectroscopy of Dipolar-Coupled Spins Under Fast Magic-angle Spinning

Dipolar Recoupling: Homonuclear Experiments

Dipolar Recoupling: Heteronuclear

Indirect Coupling and Connectivity

Quadrupolar Nuclei in Solids

Chlorine, Bromine, and Iodine Solid-State NMR

Nitrogen-14 NMR Studies of Biological Systems

Nitrogen-Proton Correlation Experiments of Organic Solids at Natural Isotopic Abundance

High-Frequency Dynamic Nuclear Polarization

Magic Angle Spinning: Effects of Quadrupolar Nuclei on Spin-1/2 Spectra

Magic Angle Spinning Carbon-13 Lineshapes: Effect of Nitrogen-14

Magnetic Susceptibility and High Resolution NMR of Liquids and Solids

Crystallography and NMR: An Overview

Fundamental Principles of NMR Crystallography

Quadrupolar NMR to Investigate Dynamics in Solid Materials

REFERENCES

1. F. G. Vogt, in *New Applications of NMR in Drug Discovery and Development*, eds L. Garrido and N. Beckmann, The Royal Society of Chemistry: London, 2013, Chap. Solid-state NMR in drug discovery and development, 43.

2. M. Geppi, G. Mollica, S. Borsacchi, and C. A. Veracini, *Appl. Spectrosc. Rev.*, 2008, **43**, 202.

3. R. K. Harris, *Analyst*, 2006, **131**, 351.

4. F. G. Vogt, J. S. Clawson, M. Strohmeier, T. N. Pham, S. A. Watson, and A. J. Edwards, in *Pharmaceutical Sciences Encyclopedia: Drug Discovery, Development, and Manufacturing*, ed S. C. Gad, John Wiley & Sons, Inc.: New York, 2011, 1.

5. O. N. Antzutkin, *Prog. NMR Spectrosc.*, 1999, **35**, 203.

6. T. J. Offerdahl, J. S. Salsbury, Z. Dong, D. J. W. Grant, S. A. Schroeder, I. Prakash, E. M. Gorman, D. H. Barich, and E. J. Munson, *J. Pharm. Sci.*, 2005, **94**, 2591.

7. E. Salager, R. S. Stein, S. Steuernagel, A. Lesage, B. Elena, and L. Emsley, *Chem. Phys. Lett.*, 2009, **469**, 336.

8. A. Lesage, S. Steuernagel, and L. Emsley, *J. Am. Chem. Soc.*, 1998, **120**, 7095.

9. A. S. Tatton, T. N. Pham, F. G. Vogt, D. Iuga, A. J. Edwards, and S. P. Brown, *CrystEngComm*, 2012, **14**, 2654.

10. D. Sakellariou, A. Lesage, and L. Emsley, *J. Magn. Reson.*, 2001, **151**, 40.

11. S. J. Opella and M. H. Frey, *J. Am. Chem. Soc.*, 1979, **101**, 5854.

12. M. Asada, T. Nemoto, H. Mimura, and K. Sako, *Anal. Chem.*, 2014, **86**, 10091.

13. P. D. Murphy, *J. Magn. Reson.*, 1985, **62**, 303.

14. N. Zumbulyadis, B. Antalek, W. Windig, R. P. Scaringe, A. M. Lanzafame, T. Blanton, and M. Helber, *J. Am. Chem. Soc.*, 1999, **121**, 11554.

15. F. G. Vogt, *Future Med. Chem.*, 2010, **2**, 915.

16. B. J. van Rossum, H. Förster, and H. J. M. de Groot, *J. Magn. Reson.*, 1997, **124**, 516.

17. R. K. Harris, P. Hodgkinson, T. Larsson, A. Muruganantham, I. Ymen, D. S. Yufit, and V. Zorin, *Cryst. Growth Des.*, 2008, **8**, 80.

18. D. L. Bryce, M. Gee, and R. E. Wasylishen, *J. Phys. Chem. A*, 2001, **105**, 10413.

19. H. Hamaed, J. M. Pawlowski, B. F. T. Cooper, R. Fu, S. H. Eichhorn, and R. W. Schurko, *J. Am. Chem. Soc.*, 2008, **130**, 11056.

20. F. G. Vogt, G. R. Williams, M. N. Johnson, and R. C. B. Copley, *Cryst. Growth Des.*, 2013, **13**, 5353.

21. M. Hildebrand, H. Hamaed, A. M. Namespetra, J. M. Donohue, R. Fu, I. Hung, Z. Gan, and R. W. Schurko, *CrystEngComm*, 2014, **16**, 7334.

22. S. Cavadini, *Prog. NMR Spectrosc.*, 2010, **56**, 46.

23. L. A. O'Dell, *Prog. NMR Spectrosc.*, 2011, **59**, 295.

24. Z. Gan, J. P. Amoureux, and J. Trébosc, *Chem. Phys. Lett.*, 2007, **435**, 163.

25. M. Rosay, L. Tometich, S. Pawsey, R. Bader, R. Schauwecker, M. Blank, P. M. Borchard, S. R. Cauffman, K. L. Felch, R. T. Weber, R. J. Temkin, R. G. Griffin, and W. E. Maas, *Phys. Chem. Chem. Phys.*, 2010, **12**, 5850.

26. A. B. Barnes, G. D. Paëpe, P. C.van der Wel, K. N. Hu, C. G. Joo, V. S. Bajaj, M. L. Mak-Jurkauskas, J. R. Sirigiri, J. Herzfeld, R. J. Temkin, and R. G. Griffin, *Appl. Magn. Reson.*, 2008, **34**, 237.

27. G. R. Desiraju, *CrystEngComm*, 2003, **5**, 466.

28. J. Bernstein, *Polymorphism in Molecular Crystals*, Oxford University Press: New York, 2002.

29. S. R. Byrn, R. R. Pfeiffer, and J. G. Stowell, *Solid-state Chemistry of Drugs*, SSCI: West Lafayette, 1999.

30. E. Maccaroni, E. Alberti, L. Malpezzi, G. Razzetti, C. Vladiskovic, and N. Masciocchi, *Cryst. Growth Des.*, 2009, **9**, 517.

31. E. H. Kerns and L. Di, *Drug-like Properties: Concepts, Structure Design and Methods*, Elsevier: Burlington, 2008.

32. R. K. Harris, S. Joyce, C. J. Pickard, S. Cadars, and L. Emsley, *Phys. Chem. Chem. Phys.*, 2006, **6**, 137.

33. J. R. Yates, S. E. Dobbins, C. J. Pickard, F. Mauri, P. Y. Ghi, and R. K. Harris, *Phys. Chem. Chem. Phys.*, 2005, **7**, 1402.

34. R. K. Harris, P. Hodgkinson, C. J. Pickard, J. R. Yates, and V. Zorin, *Magn. Reson. Chem.*, 2007, **45**, S174.

35. F. G. Vogt, L. M. Katrincic, S. T. Long, R. L. Mueller, R. A. Carlton, Y. T. Sun, M. N. Johnson, R. C. B. Copley, and M. E. Light, *J. Pharm. Sci.*, 2008, **97**, 4756.

36. J. P. Bradley, C. J. Pickard, J. C. Burley, D. R. Martin, L. P. Hughes, S. D. Cosgrove, and S. P. Brown, *J. Pharm. Sci.*, 2012, **101**, 1821.

37. M. Baias, J.-N. Dumez, P. H. Svensson, S. Schantz, G. M. Day, and L. Emsley, *J. Am. Chem. Soc.*, 2013, **135**, 17501.

38. P. H. Stahl and C. G. Wermuth (eds), *Handbook of Pharmaceutical Salts: Properties, Selection, and Use*, Wiley-VCH: New York, 2002.

39. K. M. Burgess, F. A. Perras, A. Lebrun, E. Messner-Henning, I. Korobkov, and D. L. Bryce, *J. Pharm. Sci.*, 2012, **101**, 2930.

40. A. D. Bond, C. Cornett, F. H. Larsen, H. Qu, D. Raijada, and J. Rantanen, *IUCrJ*, 2014, **1**, 328.

41. F. G. Vogt, J. Brum, L. M. Katrincic, A. Flach, J. M. Socha, R. M. Goodman, and R. C. Haltiwanger, *Cryst. Growth Des.*, 2006, **6**, 2333.

42. F. G. Vogt, G. R. Williams, M. Strohmeier, M. N. Johnson, and R. C. B. Copley, *J. Phys. Chem. B*, 2014, **118**, 10266.

43. F. G. Vogt, G. R. Williams, and R. C. B. Copley, *J. Pharm. Sci.*, 2013, **102**, 3705.

44. I. Miroshnyk, S. Mirza, and N. Sandler, *Expert Opin. Drug Deliv.*, 2009, **6**, 333.

45. F. G. Vogt, J. S. Clawson, M. Strohmeier, A. J. Edwards, T. N. Pham, and S. A. Watson, *Cryst. Growth Des.*, 2009, **9**, 921.

46. J. R. Patel, R. A. Carlton, T. E. Needham, C. O. Chichester, and F. G. Vogt, *Int. J. Pharm.*, 2012, **436**, 685.

47. G. A. Jeffrey, *An Introduction to Hydrogen Bonding*, Oxford University Press: New York, 1997.

48. J. R. Smith, W. Xu, and D. Raftery, *J. Phys. Chem.*, 2006, **110**, 7766.

49. A. J. Rossini, A. Zagdoun, F. Hegner, M. Schwarzwälder, D. Gajan, C. Copéret, A. Lesage, and L. Emsley, *J. Am. Chem. Soc.*, 2012, **134**, 16899.

50. J. S. Clawson, F. G. Vogt, J. Brum, J. Sisko, D. B. Patience, W. Dai, S. Sharpe, A. D. Jones, T. N. Pham, M. N. Johnson, and R. C. B. Copley, *Cryst. Growth Des.*, 2008, **8**, 4120.

51. S. J. L. Billinge and M. G. Kanatzidis, *Chem. Commun.*, 2004, 749.

52. F. G. Vogt, in *Solid state Characterization of Amorphous Dispersions*, ed A. Newman, John Wiley & Sons: Hoboken, New Jersey, USA, 2015.

53. T. N. Pham, S. A. Watson, A. J. Edwards, M. Chavda, J. S. Clawson, M. Strohmeier, and F. G. Vogt, *Mol. Pharm.*, 2010, **7**, 1667.

54. A. S. Tatton, T. N. Pham, F. G. Vogt, D. Iuga, A. J. Edwards, and S. P. Brown, *Mol. Pharm.*, 2013, **10**, 999.

55. D. M. Sperger, S. Fu, L. H. Block, and E. J. Munson, *J. Pharm. Sci.*, 2011, **100**, 3441.

56. T. C. Ong, M. L. Mak-Jurkauskas, J. J. Walish, V. K. Michaelis, B. Corzilius, A. A. Smith, A. M. Clausen, J. C. Cheetham, T. M. Swager, and R. G. Griffin, *J. Phys. Chem. B*, 2013, **117**, 3040.

57. R. K. Harris, P. Hodgkinson, T. Larsson, and A. Muruganantham, *J. Pharm. Biomed. Anal.*, 2005, **38**, 858.

58. R. M. Wenslow, *Drug Dev. Ind. Pharm.*, 2002, **28**, 555.

59. J. M. Griffin, D. R. Martin, and S. P. Brown, *Angew. Chem. Int. Ed. Engl.*, 2007, **119**, 8182.

60. A. J. Rossini, C. M. Widdifield, A. Zagdoun, M. Lelli, M. Schwarzwälder, C. Copéret, A. Lesage, and L. Emsley, *J. Am. Chem. Soc.*, 2014, **136**, 2324.

Chapter 22

Structure-based Drug Design Using NMR

Mark Jeeves, Lee Quill, and Michael Overduin

School of Cancer Sciences, University of Birmingham, Birmingham B15 2TT, UK

22.1 INTRODUCTION

NMR is a sensitive and versatile tool that has proven useful throughout the drug discovery process. It provides information at the atomic level and gives fine details of molecular interaction that can be used to drive drug design and help to validate candidates from high throughput screening. Each atom of a target provides a potential reporter for interactions through the chemical shift and relaxation properties of its nucleus. Its frequency is dependent on the environment of the nucleus in question, and as such provides a highly sensitive measure of small molecule binding. Given assignment of the chemical shifts of both protein and ligand, each molecule's respective binding moieties can be determined.

NMR in Pharmaceutical Sciences. Edited by Jeremy R. Everett, John C. Lindon, Ian D. Wilson, and Robin K. Harris
© 2015 John Wiley & Sons, Ltd. ISBN: 978-1-118-66025-6
Also published in eMagRes (online edition)
DOI: 10.1002/9780470034590.emrstm1430

Modern drug discovery begins with the identification of efficient starting points, for which NMR provides several screening assays. Common experiments involve detection of (i) the change in spin lattice relaxation of the ligand upon protein binding via $T_{1\rho}$[1] values, (ii) the transfer of magnetization from the protein to the ligand by Saturation Transfer Difference-NMR (STD-NMR) experiments,[2] and (iii) transfer of bulk water magnetization to the ligand by water–Ligand Observed via Gradient Spectroscopy (waterLOGSY)[3] (see Chapter 3 for more extensive discussion of NMR methods used in pharmaceutical R&D). These methods perform well with larger protein targets, and are widely used. Other ligand-observed methods include experiments using paramagnetic probes,[4] heteronuclear detection using ^{31}P and ^{19}F nuclei,[5–7] and target immobilized NMR screening.[8] The major advantages of such ligand-observed NMR methods include the requirement for comparatively low protein concentration ($1–10\,\mu M$), detection of even weakly binding compounds ($100\,nM < K_D < 2\,mM$), identification of compounds that bind virtually anywhere on the protein (including allosteric sites), and mapping of binding epitopes on the ligand. For detection of high affinity ligands the technique of reporter screening can be used where a stronger ligand is used to displace the test compound in a concentration-dependent manner.[9] The binding site can also be divulged in a similar manner by using a competitor with a known binding location. These techniques are particularly useful in the field of fragment-based drug design, an approach that provides a promising alternative to conventional high-throughput screening. These techniques also allow identification of a single binding molecule in a cocktail of compounds, thus reducing the time required to screen fragment libraries. NMR works well as a complementary screen in parallel with orthogonal methods such as surface plasmon resonance (SPR) to weed out false positives and validate hits from activity screens.

A complementary NMR screening approach involves observing the changes in the protein resonances upon ligand interaction. This requires significantly higher concentrations of soluble protein (<50 μM), typically isotopically labeled with ^{15}N, ^{13}C, or both ^{15}N and ^{13}C. With this approach, the maximum protein molecular weight limit is ~35 kDa owing to the increased relaxation and complexity of the spectra. With advances including perdeuteration, specific labeling of the protein, and use of transverse relaxation optimized spectroscopy (TROSY[10]), this upper size limit can be increased to 100 kDa in favorable cases. The binding of a ligand induces progressive chemical shift perturbations (CSPs) in the protein resonances, thus yielding ligand binding sites and affinities. The use of fast methods such as the SOFAST-HMQC experiments,[11] high throughput autosamplers, and low volume cryogenic probes mean that small molecule libraries can now be cost-effectively screened within several days. With protein resonance assignments in hand the results are more powerful as the perturbed signals can then be used to identify residues that are at any potential site of interaction in the protein.[12] Rapid structure determination of protein–ligand complexes is feasible, particularly for relatively small proteins (<15 kDa) or by using an existing 3D structure of the protein. This yields intimate details of the binding site, affords comparison of differential binding properties of ligands, and facilitates modeling using programs such as HADDOCK.[13] NMR is currently the most widely used technique for probing the molecular structure of biological molecules in solution and is uniquely able to characterize the dynamics and flexible regions of biological molecules in solution (see Chapter 1 for general discussion of the role of NMR in drug discovery and development). Finding the most relevant target protein state for therapeutic intervention is critical for rational drug design, and NMR is well placed to give insights into the multiple conformations and multidomain orientations adopted during activation and signaling. To illustrate the potential within the field we discuss three examples of proteins where NMR-based screening and design has yielded novel inhibitors and drug-like molecules with demonstrated efficacy.

22.2 PROTEIN KINASES

Protein tyrosine phosphorylation is of paramount importance in modulating intracellular signaling. This intrinsically reversible process is governed by the coordinated activities of protein tyrosine kinase (PTK) and protein tyrosine phosphatase (PTP) enzymes. These signaling enzymes play important roles in determining the fate of cellular events controlling proliferation, differentiation, migration, survival, and apoptosis.[14] Dysregulation of these networks is a key factor in many diseases including cancers, diabetes, and inflammatory conditions. For these reasons, protein kinases have long been seen as excellent targets for the development of drugs because of their key roles in signaling networks. The 518 protein kinases encoded within the human genome[15] all share a catalytic domain that has a conserved sequence and structure. However, the regulation of the catalytic activity of this domain is profoundly different. Owing to their key roles in cellular signaling and implication in disease pathways, around 30% of the proteins being targeted by pharmaceutical companies for drug development are either protein or lipid kinases, with several inhibitors already on the market.[16] The first site targeted for drug development has been the ATP-binding pocket that is found between the two lobes of the kinase fold. The conformation of the activation loop is key to activity of the kinase domain. Efficient catalysis requires the motif to adopt a conformation in which the DFG motif orients in.[17] The alternative DFG-out conformation results in an inactive state and has been seen in a number of kinase structures.[18] The dynamics of loop movement in kinases is important in assessing compound binding and can give insights into new avenues for compound development. However, lack of selectivity and the limited number of available chemotypes that bind the ATP binding site can become an issue, leading to the targeting of less conserved surrounding pockets, affording selective interactions while retaining the affinity generated by occupying the ATP site.

22.2.1 NMR Studies of Kinases

Protein kinase catalytic domains have been challenging to study by NMR owing to the relatively large size of the catalytic domain (>32 kDa), difficulties in expression of sufficient quantities of isotopically labeled protein, and their complex internal dynamics.

However, a range of technological advances are making these drug targets more amenable for exploitation by NMR-based methods. Historically, kinase domains have been difficult to study by NMR owing to the protein size and the internal dynamics of the apo form around the hinge region. To overcome the dynamics problem, the binding of an inhibitor to the ATP site can be employed to lock the kinase into a single conformation. To date, several kinases have had their NMR resonances assigned, usually in complex with an inhibitor, including cAMP dependent protein kinase both in complex with AMP-PNP and the free form,[19] the p38 protein,[20] Src in complex with imatinib,[21] the FGFR1 protein,[22] Abl with imatinib,[23] the Eph receptor,[24] ERK2,[25] and VRK1[26,27] allowing in-depth studies of their ligand binding modes in solution. The use of NMR in conjunction with X-ray crystallography integrates the complementary strengths of both techniques to rationally design effective drug molecules for kinase targets.

22.2.2 Abelson Tyrosine Kinase (Abl) as a Target

The Abl kinase is an important target because of its role in chronic myelogenous leukemia (CML), and represents a paradigm for target-based therapeutic intervention. CML is caused by the $t(9,22)$ chromosomal translocation that gives rise to the Philadelphia chromosome[28] and the subsequent production of BCR-Abl protein which is a fusion of the breakpoint cluster region protein with the catalytic domain of Abl kinase. The resulting expression of a constitutively activated tyrosine kinase causes uncontrolled replication and proliferation of progenitor cells, and provides a unique molecular target for design of selective inhibitors.

The structure of c-Abl kinase domain in complex with the inhibitor imatinib has been solved by X-ray crystallography and shows the classical kinase fold with two lobes and two possible sites for drug development (Figure 22.1). Imatinib was shown to be bound to the ATP site, contacting both the activation loop and the C-helix (Figure 22.2a).[29] The discovery of imatinib as a specific inhibitor of Abl has led to its use as a front line drug in targeted cancer therapy for CML. However, over time, cancer cells can become resistant to this drug owing to mutations acquired within the kinase domain of BCR-Abl. The second generation drugs, dasatinib and ponatinib, are able to overcome

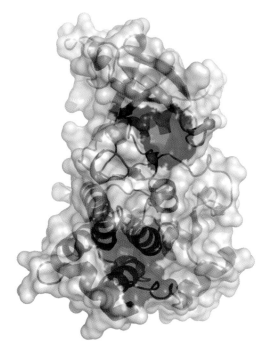

Figure 22.1.   The structure of Abl kinase domain showing the ATP (blue) and myristate (Red) binding sites that are the key sites for designing inhibitors directed toward the BCR-Abl target[34]

most of these mutations but not T315I. Threonine 315 has become known as the *gatekeeper residue* as it maps to the periphery of the nucleotide binding site in Abl and regulates access to a deep hydrophobic pocket in the active site.[30] Mutations in T315 result in very poor prognosis for the patient.[30] By monitoring titrations of fully and specifically labeled Abl kinase domain with various inhibitors by heteronuclear single quantum coherence (HSQC) NMR experiments, the different modes of binding of second generation drugs for CML could be revealed. This yielded key insights into possible routes to design improved drugs that bypass the gatekeeper mutation at the core of the hinge region.[31]

22.2.3   NMR, Abl Kinase, and Drug Design

Assignment of 96% of the backbone resonances of the imatinib-bound form of Abl kinase was achieved using

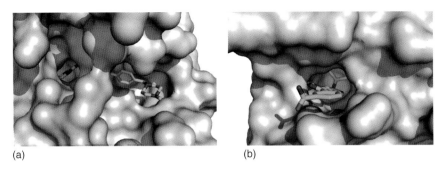

Figure 22.2. The binding of imatinib to the deep ATP pocket (a) and GNF-2 to the shallower allosteric myristate site (b) of the Abl kinase structure[32]

uniform labeling with ^{13}C/^{15}N in baculovirus Sf9 insect cells. Residue type identification was achieved using selective labeling techniques.[23] Residue-specific assignment allowed structure–activity relationship (SAR) screening using ^{15}N-HSQC experiments to be performed, and has shown that imatinib and nilotinib bind to an inactive form of the kinase domain with the activation loop in the out conformation (DFG-out). In contrast, dasatinib (an Abl and Src inhibitor) interacts with the DFG-in structure (i.e., it binds to the active form) as does AFN941, a generic kinase inhibitor that is a staurosporine derivative.[31] Knowledge of the different ways in which these inhibitors function is vital for the discovery of improved drugs for the treatment of resistant BCR-Abl. Chemical shift changes observed in ^{15}N-HSQC spectra, in combination with X-ray crystallography, mutagenesis, and mass spectrometry, show that the inhibitor GNF-2 and its derivative GNF-5 bind not to the ATP site but instead to the allosteric myristate site of Abl (Figure 22.2b), thus opening up a new avenue for the design of new drug molecules with high specificity to be used in conjunction with active site inhibitors.[32,33] An NMR fragment-based screen, using waterLOGSY and $T_{1\rho}$, on Abl kinase complexed with imatinib to occlude the active site, was used to identify compounds binding to allosteric sites. Conformational changes in the myristate site of those binders were identified using CSPs of a ^{15}N-Val selectively labeled protein. Whether compounds binding to the myristate site act as inhibitory antagonists or as activating agonists was determined by monitoring the length and bending of helix I.[34] This was achieved by observing the resonances of Val525, which is highly flexible in the uninhibited state and thus gives sharp NMR lines, but forms an α-helix in the

inhibited form yielding a correspondingly broader signal. These agonists, while not useful in treating CML patients with the Philadelphia chromosome, could be useful in treating patients after radiation therapy owing to the beneficial effects of c-Abl in DNA damage repair. The potential benefits of kinase activators have been subjected to limited study, and remain an unexploited avenue for therapeutic intervention.

The mechanism by which myristate is bound has been explored using insights from NMR relaxation measurements and residual dipolar couplings as well as small angle X-ray scattering studies. Binding of a second drug molecule to this allosteric site for myristate, in conjunction with imatinib, can restore the closed conformation of Abl.[35] Imatinib induces an open inhibited state of the protein but the binding of compounds to the myristate site can recreate the closed conformation seen in the apo protein. This has the effect of hiding a vital phosphorylation site key to full activation of BCR-Abl and so suggests a way to overcome the T334I gatekeeper mutant that is immune to imatinib, dasatinib, and nilotinib.

An NMR study of the intramolecular dynamics within the N-terminal regulatory domains revealed that c-Abl may be targeted to the cellular membrane leading to possible new target sites in Abl[36] that would control its intracellular location and thereby its activity. Thus, NMR is one of the several key tools that provide understanding into how regulatory and drug molecules bind to and affect the structure and activity of Abl kinase. This in turn is opening up new avenues to explore for the design of next generation agents to combat forms of CML that are resistant to existing drug therapies.

22.3 PROTEIN TYROSINE PHOSPHATASE 1B (PTP1B)

The focus on the development of small molecule inhibitors (SMIs) that specifically abrogate PTK function has left other equally important components of phosphorylation-dependent signaling processes, such as PTPs, relatively unexploited. Given the wealth of successful drug discovery initiatives targeting PTKs and the notion that accurate maintenance of cellular phosphotyrosine levels are reciprocally controlled by the actions of kinases and phosphatases, PTPs have been posited as a class of drug targets with significant untapped potential.[37–39]

PTPs constitute a family of 107 enzymes responsible for catalyzing the removal of phosphate groups from phosphotyrosine-containing target substrates.[38] Conventional methods of targeting the active sites of PTPs have presented technical challenges arising largely from the conserved, charged nature of this pocket, and the catalytically critical cysteine. As such, the development of efficacious SMIs has been fraught with complications as the charges and reactivity of the ligands that have emerged display poor cell permeability and selectivity. These obstacles have prompted suggestions that although a potentially promising reservoir of drug targets, phosphatases may be an undruggable class of enzymes that are refractory to conventional active site-targeted inhibition.[40,41] This in turn has motivated a search for novel pockets and allosteric sites that could prove more druggable.

22.3.1 PTP1B Signaling Mechanism

Since being discovered 25 years ago, PTP1B has been an enzyme of significant importance in evolving our understanding of phosphatase mechanisms that contribute to normal cellular function and the etiology of human disease.[42] In a cellular context, PTP1B plays a central role in the dephosphorylation and subsequent downregulation of the insulin receptor in response to elevated blood glucose levels, leading to restoration of blood sugar homeostasis.[43] Maintenance of insulin signaling achieved through selective modulation with PTP1B-targeted inhibition may hold therapeutic potential for the treatment of type 2 diabetes and obesity. It is anticipated that treatment with a PTP1B inhibitor may increase the half-life of the phosphorylated state of the insulin receptor, thereby extending the time frame through which the receptor is optimally responsive to insulin.[44,45]

22.3.2 NMR Studies and Rational Drug Design

Drug discovery initiatives focusing on the development of potent, novel, and highly selective PTP1B-directed inhibitors have greatly benefited from NMR-based insights into conformational dynamics and novel druggable sites. Indeed, favorable strategies have emerged that utilize NMR to elucidate the binding modes of hits and lead molecules. More specifically, these strategies have centered on the use of combinatorial approaches, where NMR-based screening methods coupled with rational drug design have been adopted in order to identify SMIs with high affinity and selectivity for PTP1B.[46] These combinatorial approaches have represented a unique modular approach to SMI design, and demonstrate the systematic development of a lead inhibitor harnessing a novel synthetic chemistry-based approach. Owing to the versatility and wide range of ligand affinities that can be detected, NMR has informed multiple levels of the drug design process, including the identification of novel scaffolds and the design of inhibitors. In particular, NMR has been useful for hit validation in multiple PTP1B drug discovery programs (hit and lead validation using NMR are discussed at greater length in Chapter 11). PTP1B-directed screening strategies comprising libraries of 10 000 fragments have relied heavily on the elucidation of hits by HSQC NMR spectroscopy. In this case, initial hit validation of test compounds binding to either uniformly labeled ^{15}N PTP1B or PTP1B selectively labeled with δ-^{13}CH$_3$ Ile was confirmed through monitoring CSPs in ^{1}H,^{15}N, or ^{1}H,^{13}C-resolved HSQC spectra. Significant CSPs were subsequently assigned and mapped to key residues Val49, Gly228, and Gly218 that surround the catalytic site. The use of NMR in the initial hit verification stage is particularly advantageous in that fragments that bind with relatively weak affinities can be readily detected, while information concerning the fragment binding mode can also be obtained. The results of the initial screens revealed a compound with phosphotyrosine-mimicking propensity ($K_D = 100\ \mu$M), which was optimized in an attempt to gain increased occupancy of the active site. Corroborative X-ray structures revealed why the optimized compound possessed markedly increased affinity for the PTP1B active site. Owing to the highly conserved nature of phosphatase architecture and active site chemistry, the design of a selective inhibitor that is exclusively active-site directed is unlikely without additional contacts to nonhomologous regions outside the active site cavity. Circumventing these difficulties therefore requires a redirected approach focusing on the development of bivalent inhibitors offering multiple contact points each with differential specificity to ultimately boost cellular potency. In this way, synthesizing a compound that binds not only to the active site but also to a second binding site positioned nearby provides an excellent strategy to improve both inhibitor selectivity and affinity in an in vivo setting.

Screening for a second compound was established by selective labeling of PTP1B with ^{13}C-methionine, and initial hits were specifically monitored by observing the ^{13}C-methionine resonance shifts in corresponding ^{1}H,^{13}C-HSQC spectra. This yielded a ligand for the secondary binding site, which was subsequently joined to the initial active site binder using a chemical linker synthesized on the basis of the predetermined co-crystal structure with this compound. The new bivalent inhibitor exhibited markedly improved binding parameters with an experimentally determined K_D = 20 nM, derived from 4-nitrophenol phosphate hydrolysis activity assays. When screened against the same panel of phosphatases, the selectivity profile of the linked compound against LAR, SHP-2, CD-45, and calcineurin ranged from 36-fold to 10 000-fold, with moderate twofold selectively observed over T-cell tyrosine phosphatase TCPTP. A direct comparison of the selectivity profiles of the two compounds indicated that the linked compound compared favorably with that of the active site binding compound, and confirmed that the additional contacts provided by the second-site ligand accounted for the large difference in both specificity and selectivity of the two compounds. This established the principle that the development of a ligand into a second site can yield benefits in both affinity and specificity, thereby demonstrating the validity of this approach for the rational design of therapeutic agents for clinically significant phosphatases.[12]

Despite being highly potent with promising selectivity profiles, bivalent oxamic acid-based PTP1B inhibitors (Figure 22.3) elicit unfavorable physicochemical properties, and resultantly lack obvious cell permeability.[46] Additional screens, utilizing selective ^{13}C labeling at the methyl groups of Ile219 and monitoring subsequent chemical shifts, but focusing on screening monocarboxylic acid and noncarboxylic

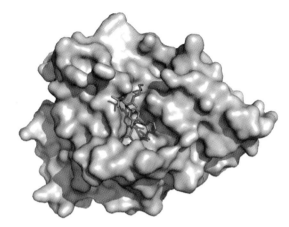

Figure 22.3. The structure of PTP1B catalytic domain complexed with a bivalent inhibitor (turquoise).[46] The nonselective oxamic acid bound in the active site is linked to a second site ligand identified by NMR-based fragment screening leading to a highly specific and potent inhibitor of PTP1B

acid-based fragments as the initial chemotypes for potential catalytic site ligands, led to inhibitors with enhanced cell permeability.[47] Selecting starting chemicals with a low charge density, such as mono and noncarboxylic acid-based fragments not only maximizes the possibility of cell permeability but also increases the likelihood of obtaining a more tractable inhibitor. In this case, the structure-based design and further chemical modification of the weak affinity ligands for both the catalytic site and a second phosphotyrosine site subsequently delivered a potent, selective, and cellular active inhibitor against PTP1B that superseded the cell impermeable oxamic acid-based inhibitors (Figure 22.4). In addition, the identified PTP1B-targeted inhibitor possessed micromolar potency and 30-fold selectivity over the highly homologous TCPTP enzyme.[47]

PTP1B has become a validated therapeutic target for obesity and diabetes, and as such drug discovery efforts directed toward PTP1B inhibition have direct therapeutic and translational relevance. In addition to well-characterized roles in metabolic regulation and insulin sensitivity, PTP1B has shown an increased propensity to become overexpressed in breast tumors, along with the HER2 receptor.[48,49] Mice engineered with oncogenic activating mutations in the HER2 receptor in mammary glands developed multiple tumors and chronic metastases to the lung. More significantly, when these mice were crossed with wild type PTP1B mice, tumor development was reduced and the

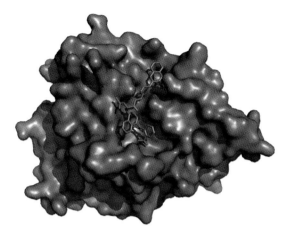

Figure 22.4. Close up of the PTP1B catalytic domain complexed with a cell-permeable oxalylarylaminobenzoic acid-based inhibitor (turquoise)[47]

induction of incipient lung metastases decreased. In the same study, stimulating overexpression of PTP1B was sufficient for the transformation of mammary cells into an oncogenic phenotype.[49] These observations suggest that PTP1B may have a positive role in the signaling events underlying mammary tumorigenesis and highlight the potential for pharmacological inhibition of PTP1B as a target for breast cancer. However, the development of an appropriate chemotherapeutic strategy is confounded by the difficulties associated with targeting the phosphatase active site with SMIs. Consequently, novel approaches are required that circumvent these obstacles and allow for a new avenue for developing SMIs with unique binding modes, occupancies, and specificities for phosphatase targets.

Recently, one such strategy providing a robust challenge to the proposed undruggable nature of phosphatases has demonstrated an alternative approach to the inhibition of PTP1B by exploiting the conformational flexibility of the C-terminal region. In this instance, two-dimensional $^1H,^{15}N$-HSQC and three-dimensional experiments were recorded on $^2H/^{13}C/^{15}N$-labeled protein using TROSY experiments. NMR was used to probe the inherent flexibility of amino acid residues 300–393 comprising the C-terminal region of PTP1B, and thus highlighted the key residues defining the binding site for the inhibitor MSI-1436. Chemical shifts corresponding to residues 300–393 of PTP1B in the ^{15}N-HSQC-TROSY spectrum were observed at approximately fivefold greater intensity compared with those from PTP1B(1–301)

and were found clustered in the center of the spectrum indicative of flexible, unstructured regions. Overlaying of the two-dimensional TROSY spectra of the PTP1B constructs PTP1B(1–393) and PTP1B(1–301) enabled the successful assignment of the disordered region, providing key molecular details for establishing the mode of inhibition of MSI-1436. Indeed, NMR analysis combined with complementary DiFMUP (6,8-difluoro-4-methylumbelliferyl phosphate) enzymatic assays and further biophysical characterization established MSI-1436 (more widely known as *trodusquemine*[50] in the clinic) as a bivalent, allosteric inhibitor of PTP1B that binds reversibly and selectively to both the disordered C-terminus of PTP1B and a unique pocket situated in close proximity to the active site. The therapeutic potential of such an approach to PTP1B inhibition has been validated by successful translation of these findings into in vivo models. For example, targeted inhibition of PTP1B with MSI-1436 resulted in the downregulation of HER2 signaling, marked reduction in tumor growth, and attenuation of lung metastasis in HER2-positive animal models of breast cancer.[51] This study illustrates the utility of NMR spectroscopy in probing and resolving molecular details of the interaction between SMIs and full length PTP1B. Perhaps even more importantly, it represents a novel approach for inhibiting phosphatases through intrinsically disordered regions that offer specific binding sites for SMIs.

NMR is now being used in conjunction with other complementary approaches to expand our knowledge of the structure, function, and drug development potential of a range of other clinically and therapeutically significant phosphatases. These include the tyrosine-protein phosphatase nonreceptor type 11 (PTPN11) that has been implicated in a wide array of developmental disorders and in tumorigenesis[52–54] and the striatal-enriched protein tyrosine phosphatase (STEP) that has been implicated in neurological disorders. With indications of these targets in Alzheimer's disease,[55] schizophrenia[56], and Huntington's disease,[57] NMR-based phosphatase inhibitor design is positioned to have broad clinical benefits.

22.4 RAS SUPERFAMILY

The Ras superfamilies of small guanosine triphosphatases (GTPases) drive cell proliferation and migration within eukaryotic cells, and are critically involved in cancer progression. The human genome contains over 150 members that are divided into five major classes based on sequence divergences, which are concentrated in their membrane-interactive termini, while all share a highly conserved 19 kDa catalytic domain.[58] The most advanced drug target is Ras, which is one of the most genetically deregulated oncogenes, while members of the similar Rho, Rab, Ran, and Arf subfamilies are also of growing interest for therapeutic intervention. Nonetheless, this superfamily of proteins has, until recently, been considered undruggable because of their flexible nature and the absence of obvious deep pockets to offer footholds for drug design. This has stimulated an effort among NMR spectroscopists and others to generate new tools and creative approaches to tackle these valuable targets.

22.4.1 NMR Studies

Being comprised of a relatively small single domain, the Ras family is particularly amenable to NMR analysis. Indeed, the chemical shifts of multiple Ras forms as well as RalB, Rheb, RhoA, Rac1, Cdc42, and Arf1 GTPases have been reported, although loops in the activated catalytic site and the membrane binding elements are generally dynamic and exhibit broad lines. Nonetheless, complete and partial backbone assignments of H-Ras in GTP analog- and GDP-bound states, respectively, have been made and show significant perturbations from those of oncogenic mutants.[59] All the backbone and many side chain resonances of the K-Ras catalytic domain bound to GDP were assigned at physiological pH using ^{2}H/^{13}C/^{15}N labeled protein and TROSY methods.[60] The full-length nonprocessed and farnesylated forms of cysteine-substituted H-Ras are also assigned, revealing a network of dynamic and chemical shift changes that are induced in the structural domain by extension of the flexible terminus.[61] However, no Ras family member structure has been determined in the membrane-bound form, despite this state being particularly relevant for oncogenic signaling and drug design. Nonetheless, a large number of NMR groups have contributed to a growing body of chemical shift, structural, and dynamic information about these proteins that can now be exploited.

22.4.2 Structures and Dynamics

The structure of the catalytic mechanism of Ras was first elucidated 25 years ago,[62] and since then several hundred structures of this and related GTPases including of Arf, Cdc42, Rab, RAD, Ran, Rap, Rheb, and Rho have been solved, largely by X-ray crystallography. They reveal a remarkably conserved fold. The enzymes all comprise a single, compact domain with a central, six-stranded β sheet flanked by five α helices. Two proximal elements exhibit conformational changes upon GTP binding: switch I interacts with downstream effectors, while switch II interacts with a guanine nucleotide exchange factor (GEF). To stabilize critical target states Ras active site mutations such as substitutions of Thr35 for a serine and Pro40 for an Asp were designed based on ^{31}P NMR studies,[63] yielding NMR-based and crystallographic structural insights into GTP-binding transitions and novel druggable pockets.[64,65] Beyond the catalytic domain, the farnesyl group is seen by CSPs to be directly recognized by part of the downstream effector, Raf, and may also interact with residues near switch I, suggesting functionally significant coupling between these elements.[61] Altogether, the application of NMR and X-ray crystallography to wild-type and mutant forms of Ras using substrate analogs has exposed a range of inducible sites of interest for drug design.

22.4.3 Signaling Mechanism

The intrinsic catalytic activities of Ras superfamily members are generally very weak, but are dramatically enhanced when a GTP activating protein (GAP) binds. For example, RasGAP enhances Ras activity by positioning Gln61 for nucleophilic attack and stabilizing the transition state. The subsequent exchange of the bound GDP molecule for GTP is normally very slow, unless a GEF is recruited. As revealed by the structure of the Ras complex, members of the Cdc25 family such as Son-of-Sevenless (SOS) bind between residues 32 and 67, sterically occluding magnesium and inducing GDP release.[66] This frees the site for binding to GTP, for which it has picomolar affinity, and the GEF becomes dislodged. Thus regenerated, the activated state can then bind downstream effectors, notably Raf to stimulate MEK/ERK kinase pathway signaling. Interaction with Raf uniquely increases the intrinsic rate of GTP hydrolysis by Ras, thus freeing the GTPase from GAP-dependent activation.[67] The Raf interaction is the strongest amongst a diverse set of competing Ras effectors, as shown by comparative analysis of how multiple effector combinations perturb the HSQC spectra of activated ^{15}N-labeled Ras.[68] Moreover, the effectors are seen to compete directly with both GAP and SOS proteins for Ras binding, thus constituting a complex system of controlled access with feedback cycles that can be monitored in real-time by NMR. The oncogenic mutations of Ras can either accelerate or decelerate nucleotide exchange and hydrolysis,[59] and dramatically alter the profile of binding affinities for the range of effector molecules, consistent with their wide ranging perturbations of signaling outputs in cells.

22.4.4   Role of the Membrane

Ras proteins have been found to signal on endosomal, ER, Golgi, mitochondrial, and plasma membranes. This presents challenges for structural characterization of the target states that drive signaling or disease progression, as these active forms are farnesylated, proteolytically cleaved and methylated at a C-terminal cysteine, and palmitoylated at additional nearby Cys residues. Together these modifications along with basic linker sequence motifs and catalytic domain surfaces mediate selective insertion into lipid raft and nonraft domains, thus influencing regulatory and effector protein interactions. GDP-bound Ras is attracted to rafts by its lipid anchors, while the GTP-bound state is recruited to disordered regions of membrane through its adjacent linker sequence, although the mechanisms remains elusive.[69] The bilayer bound states have not been experimentally resolved because of the complexity and dynamics of the lipid interactions and the requirement for full length, lipid-modified proteins. Thus, the mechanisms of the most critical target states including coupling between GDP/GTP exchange, protein–protein interactions, and membrane–protein association remain elusive. Computational modeling of the full-length H-Ras protein on a bilayer suggests that the catalytic domain orients on the membrane via α4 residues Arg128 and Arg135 binding to phospholipid head groups, influencing the accessibility of elements that interact with protein partners, and influencing downstream signaling.[70] This offers a novel potentially druggable site that has yet to be exploited.

22.4.5   Therapeutic Significance

The hyperactivity of Ras family members contributes to cancer progression, with 20–30% of tumors expressing mutant forms. The incidence is particularly high in pancreatic cancer, where 60% of cases exhibit mutant forms of K-ras. Most frequently mutated are residues 12, 13, and 61, which are involved in GTP hydrolysis. The lower GTPase activity leads to constitutive Ras signaling and tumorigenesis. Mutations or overexpression of upstream receptors such as the epidermal growth factor receptor (EGFR) also contribute to cancer causation, as does mutations of downstream effectors such as Raf. The upregulation or activation of Ras is a frequent unwanted consequence of treatment of patients with EGFR or Raf inhibitors, underscoring the need for Ras inhibitors to combat the drug resistant mutants that are selected for and emerge during the course of treatment. Many other targets are emerging, for example, the overexpression of Rho family members and their regulators contributes to invasion and metastasis in several cancers as well as cardiac hypertrophy.

22.4.6   Target State

Given the therapeutic importance of several Ras superfamily members, several strategies have been used to design inhibitors. NMR-based methods have been used to identify compounds that bind directly to Ras, including one of the two GTP-bound states that are in dynamic equilibrium. Another approach to inhibit aberrant Ras signaling involves screening for compounds that bind to the GEF interface, thus perturbing the recycling of the enzyme's active state. Alternatively, the formation of complexes with downstream effectors or membrane surfaces can be targeted. Up to eight allosteric states have been proposed, with four Ras conformers being detected by high pressure and ^{31}P-NMR studies.[71] Thus, despite the absence of obvious druggable pockets, several functional sites and conformation states as well as a growing set of probe molecules useful for chemical biology experiments are now available to aid in drug design.

22.4.7   Ligand Screening and Optimization

The presence of several protein and membrane binding surfaces[72] rather than clear druggable pockets (Figure 22.5) renders Ras proteins most suitable for NMR-based fragment screening to identify efficient starting points for drug design. Several groups have initiated campaigns to discover novel Ras inhibitors using this approach. The results of fragment screening

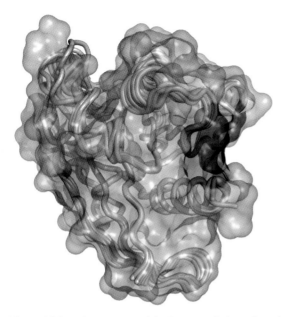

Figure 22.5. The structure of the Ras target in its activated state is shown with the candidate binding sites highlighted.[72] Sites identified include the nucleotide-binding site (red) and new potential binding sites p1, p2, and p3 (in pink, green, and blue, respectively)

campaigns focused on K-Ras have been reported by several groups. A team at Genentech screened 3285 drug-like fragments by 1D NMR methods, generating 266 fragments as confirmed ligands of the protein.[73] The binding sites were mapped by ^{15}N-HSQC spectroscopy using the GDP-loaded GTPase domain, yielding 25 hits that interacted with an inducible pocket normally occupied by the side chain of Tyr71, which also forms hydrogen bonds with the SOS protein. Those ligand binding affinities in the upper micromolar range could be estimated in titration experiments based on progressive chemical shift changes. This provided a basis for testing series of similar compounds and improving the affinity and activity of the ligands. Crystal structures of several hits revealed a shared binding pocket lined by Lys5, Leu6, Val7, Ile55, Leu56, and Thr74, consistent with the NMR studies. Plasticity within the site is also evident, expanding by up to 50% upon occupancy with a small molecule, and being proximal to the Ras–SOS interface.[74] The most promising hits included DCAI, which inhibits nucleotide release from K-Ras by blocking the SOS interaction, providing a basis for lead optimization based on competition with GEF docking.

A different approach was taken by Fesik and colleagues, who relied on ^{15}N-HSQC screening of the GDP-bound K-Ras protein using a library of 11 000 fragments.[75] The resulting 140 hits mainly bound near the SOS site with millimolar affinities, while others interact with an adjacent site beside switch I. Two hits were covalently linked using structural-based design to generate a bivalent ligand having an affinity of 190 μM and being an inhibitor of nucleotide exchange and SOS interactions. This approach was extended by preparing a S39C mutant of K-Ras which was then modified by a thiol-reactive compound to occlude the proximal site for a first ligand molecule.[76] This covalent complex was used in a NMR-screening campaign to identify second site binders, thus avoiding the need for saturating concentrations of a first ligand to identify complementary ligands for linking.

A team at Kobe University has exploited a Ras mutant that adopts a critical GTP-bound conformation. Virtual ligand screening of a library of over 2 million compounds yielded hits for validation by a Raf binding assay.[77] The solution structures of ligand complexes were determined using intermolecular NOEs and CSPs from the GTP analog-loaded form of H-Ras, as co-crystals could not be generated. The drug binding pocket involves conserved Lys5, Leu56, Met67, Gln70, and Tyr 74 residues near switch I. Relaxation edited 1D proton NMR studies revealed binding to the GDP-loaded form as well as a number of other Ras family members, indicating a need to design in further target specificity and potency.

A metal-cyclen that binds weakly to another pocket near the dynamic GTP binding site of Ras has been identified by Kalbitzer's group.[78] This interaction was identified by monitoring ligand effects on the ^{31}P NMR signals of a GTP analog bound to full-length and mutant forms of H-Ras. This led to the discovery that Zn^{2+}-cyclen binding shifts the Ras-GTP complex toward an inactive conformation that has reduced affinity for downstream effectors.

One of the earlier Ras inhibitors to be designed was from the Schering-Plough Research Institute. The design of the compound, SCH-54292, was based on a GDP analog, and exhibits an IC_{50} of 700 nM. It binds in a flexible pocket that forms under switch II upon nucleotide exchange. It could be docked to the H-Ras structure using NMR chemical shifts and intermolecular NOE distance restraints,[79] explaining its selectivity for the GDP-bound state. A related analog displays moderate inhibitory activity in PC-12 cells with an EC_{50} of 10–20 μM,[80] thus constituting a potential lead for further optimization. Together this illustrates the profusion of novel pockets and starting points that can now be identified for the design of next generation inhibitors of this high value target.

22.5 CONCLUSIONS

NMR is a key technique in structure-based drug design. It occupies a unique position in that it can provide atomic details of structure, dynamics, and ligand interactions, which when combined with complementary techniques such as X-ray crystallography, small angle X-ray scattering, and SPR, enables the expansion and refinement of basic initial hits into highly specific leads and eventually drug molecules capable of treating previously incurable conditions. NMR is also an ideal tool for searching for new binding sites in target proteins previously described as undruggable. There are many other examples of the use of NMR in drug discovery including for the design of Bcl-2,[81] BACE1,[82] and Hsp90[83] inhibitors that have progressed toward clinical trials. We have presented three established targets, PTP1B, Abl, and Ras, which have been exploited to find new pockets for drug design and to elaborate previously discovered drug molecules. This has led to a deeper understanding of their mechanisms of action and reinforced the promise of fragment-based NMR approaches, providing further opportunities for drug discovery for emerging targets with these major superfamilies of proteins.

ACKNOWLEDGMENTS

The authors would like to thank the Biotechnology and Biological Sciences Research Council, Cancer Research UK and Leukaemia & Lymphoma Research for funding.

REFERENCES

1. P. J. Hajduk, E. T. Olejniczak, and S. W. Fesik, *J. Am. Chem. Soc.*, 1997, **119**, 12257.

2. B. Meyer, J. Klein, M. Mayer, R. Meinecke, H. Moller, A. Neffe, O. Schuster, J. Wulfken, Y. Ding, O. Knaie, J. Labbe, M. M. Palcic, O. Hindsgaul, B. Wagner, and B. Ernst, *Ernst Schering Res. Found. Workshop*, 2004, 149.

3. C. Dalvit, G. Fogliatto, A. Stewart, M. Veronesi, and B. Stockman, *J. Biomol. NMR*, 2001, **21**, 349.

4. W. Jahnke, *Chembiochem.*, 2002, **3**, 167.

5. C. Dalvit, M. Flocco, M. Veronesi, and B. J. Stockman, *Comb. Chem. High Throughput Screen.*, 2002, **5**, 605.

6. T. Tengel, T. Fex, H. Emtenas, F. Almqvist, I. Sethson, and J. Kihlberg, *Org. Biomol. Chem.*, 2004, **2**, 725.

7. F. Manzenrieder, A. O. Frank, and H. Kessler, *Angew. Chem. Int. Ed. Engl.*, 2008, **47**, 2608.

8. S. Vanwetswinkel, R. J. Heetebrij, J.van Duynhoven, J. G. Hollander, D. V. Filippov, P. J. Hajduk, and G. Siegal, *Chem. Biol.*, 2005, **12**, 207.

9. X. Zhang, A. Sanger, R. Hemmig, and W. Jahnke, *Angew. Chem. Int. Ed. Engl.*, 2009, **48**, 6691.

10. K. Pervushin, R. Riek, G. Wider, and K. Wuthrich, *Proc. Natl. Acad. Sci. U. S. A.*, 1997, **94**, 12366.

11. P. Schanda, E. Kupce, and B. Brutscher, *J. Biomol. NMR*, 2005, **33**, 199.

12. S. B. Shuker, P. J. Hajduk, R. P. Meadows, and S. W. Fesik, *Science*, 1996, **274**, 1531.

13. C. Dominguez, R. Boelens, and A. M. Bonvin, *J. Am. Chem. Soc.*, 2003, **125**, 1731.

14. T. Hunter, *Cell*, 1995, **80**, 225.

15. G. Manning, D. B. Whyte, R. Martinez, T. Hunter, and S. Sudarsanam, *Science*, 2002, **298**, 1912.

16. M. Vieth, J. J. Sutherland, D. H. Robertson, and R. M. Campbell, *Drug Discov. Today*, 2005, **10**, 839.

17. P. Badrinarayan and G. N. Sastry, *Curr. Pharm. Des.*, 2013, **19**, 4714.

18. L. Garuti, M. Roberti, and G. Bottegoni, *Curr. Med. Chem.*, 2010, **17**, 2804.

19. T. Langer, M. Vogtherr, B. Elshorst, M. Betz, U. Schieborr, K. Saxena, and H. Schwalbe, *Chembiochem*, 2004, **5**, 1508.

20. M. Vogtherr, K. Saxena, S. Grimme, M. Betz, U. Schieborr, B. Pescatore, T. Langer, and H. Schwalbe, *J. Biomol. NMR*, 2005, **32**, 175.

21. R. Campos-Olivas, M. Marenchino, L. Scapozza, and F. L. Gervasio, *Biomol. NMR Assign.*, 2011, **5**, 221.

22. N. Vajpai, A. K. Schott, M. Vogtherr, and A. L. Breeze, *Biomol. NMR Assign.*, 2014, **8**, 85.

23. N. Vajpai, A. Strauss, G. Fendrich, S. W. Cowan-Jacob, P. W. Manley, W. Jahnke, and S. Grzesiek, *Biomol. NMR Assign.*, 2008, **2**, 41.

24. S. Wiesner, L. E. Wybenga-Groot, N. Warner, H. Lin, T. Pawson, J. D. Forman-Kay, and F. Sicheri, *EMBO J.*, 2006, **25**, 4686.

25. A. Piserchio, K. N. Dalby, and R. Ghose, *Methods Mol. Biol.*, 2012, **831**, 359.

26. J. Shin, G. Chakraborty, N. Bharatham, C. Kang, N. Tochio, S. Koshiba, T. Kigawa, W. Kim, K. T. Kim, and H. S. Yoon, *J. Biol. Chem.*, 2011, **286**, 22131.

27. J. Shin, G. Chakraborty, and H. S. Yoon, *Biomol. NMR Assign.*, 2014, **8**, 29.

28. A. de Klein, A. G.van Kessel, G. Grosveld, C. R. Bartram, A. Hagemeijer, D. Bootsma, N. K. Spurr, N. Heisterkamp, J. Groffen, and J. R. Stephenson, *Nature*, 1982, **300**, 765.

29. B. Nagar, W. G. Bornmann, P. Pellicena, T. Schindler, D. R. Veach, W. T. Miller, B. Clarkson, and J. Kuriyan, *Cancer Res.*, 2002, **62**, 4236.

30. T. Zhou, L. Parillon, F. Li, Y. Wang, J. Keats, S. Lamore, Q. Xu, W. Shakespeare, D. Dalgarno, and X. Zhu, *Chem. Biol. Drug Des.*, 2007, **70**, 171.

31. N. Vajpai, A. Strauss, G. Fendrich, S. W. Cowan-Jacob, P. W. Manley, S. Grzesiek, and W. Jahnke, *J. Biol. Chem.*, 2008, **283**, 18292.

32. J. Zhang, F. J. Adrian, W. Jahnke, S. W. Cowan-Jacob, A. G. Li, R. E. Iacob, T. Sim, J. Powers, C. Dierks, F. Sun, G. R. Guo, Q. Ding, B. Okram, Y. Choi, A. Wojciechowski, X. Deng, G. Liu, G. Fendrich, A. Strauss, N. Vajpai, S. Grzesiek, T. Tuntland, Y. Liu, B. Bursulaya, M. Azam, P. W. Manley, J. R. Engen, G. Q. Daley, M. Warmuth, and N. S. Gray, *Nature*, 2010, **463**, 501.

33. D. Fabbro, P. W. Manley, W. Jahnke, J. Liebetanz, A. Szyttenholm, G. Fendrich, A. Strauss, J. Zhang, N. S. Gray, F. Adrian, M. Warmuth, X. Pelle, R. Grotzfeld, F. Berst, A. Marzinzik, S. W. Cowan-Jacob, P. Furet, and J. Mestan, *Biochim. Biophys. Acta*, 2010, **1804**, 454.

34. W. Jahnke, R. M. Grotzfeld, X. Pelle, A. Strauss, G. Fendrich, S. W. Cowan-Jacob, S. Cotesta, D. Fabbro, P. Furet, J. Mestan, and A. L. Marzinzik, *J. Am. Chem. Soc.*, 2010, **132**, 7043.

35. L. Skora, J. Mestan, D. Fabbro, W. Jahnke, and S. Grzesiek, *Proc. Natl. Acad. Sci. U. S. A.*, 2013, **110**, E4437.

36. G. A. de Oliveira, E. G. Pereira, G. D. Ferretti, A. P. Valente, Y. Cordeiro, and J. L. Silva, *J. Biol. Chem.*, 2013, **288**, 28331.

37. L. Bialy and H. Waldmann, *Angew. Chem. Int. Ed. Engl.*, 2005, **44**, 3814.

38. N. K. Tonks, *Nat. Rev. Mol. Cell Biol.*, 2006, **7**, 833.

39. N. K. Tonks and B. G. Neel, *Curr. Opin. Cell Biol.*, 2001, **13**, 182.

40. J. N. Andersen and N. K. Tonks, *Top. Curr. Genet.*, 2004, **5**, 201.

41. X. Li, K. A. Oghi, J. Zhang, A. Krones, K. T. Bush, C. K. Glass, S. K. Nigam, A. K. Aggarwal, R. Maas, D. W. Rose, and M. G. Rosenfeld, *Nature*, 2003, **426**, 247.

42. N. K. Tonks, *FEBS J.*, 2013, **280**, 346.

43. S. Walchli, M. L. Curchod, R. P. Gobert, S. Arkinstall, and R. Hooft van Huijsduijnen, *J. Biol. Chem.*, 2000, **275**, 9792.

44. M. Elchebly, P. Payette, E. Michaliszyn, W. Cromlish, S. Collins, A. L. Loy, D. Normandin, A. Cheng, J. Himms-Hagen, C. C. Chan, C. Ramachandran, M. J. Gresser, M. L. Tremblay, and B. P. Kennedy, *Science*, 1999, **283**, 1544.

45. L. D. Klaman, O. Boss, O. D. Peroni, J. K. Kim, J. L. Martino, J. M. Zabolotny, N. Moghal, M. Lubkin, Y. B. Kim, A. H. Sharpe, A. Stricker-Krongrad, G. I. Shulman, B. G. Neel, and B. B. Kahn, *Mol. Cell Biol.*, 2000, **20**, 5479.

46. B. G. Szczepankiewicz, G. Liu, P. J. Hajduk, C. Abad-Zapatero, Z. Pei, Z. Xin, T. H. Lubben, J. M. Trevillyan, M. A. Stashko, S. J. Ballaron, H. Liang, F. Huang, C. W. Hutchins, S. W. Fesik, and M. R. Jirousek, *J. Am. Chem. Soc.*, 2003, **125**, 4087.

47. G. Liu, Z. Xin, Z. Pei, P. J. Hajduk, C. Abad-Zapatero, C. W. Hutchins, H. Zhao, T. H. Lubben, S. J. Ballaron, D. L. Haasch, W. Kaszubska, C. M. Rondinone, J. M. Trevillyan, and M. R. Jirousek, *J. Med. Chem.*, 2003, **46**, 4232.

48. M. Bentires-Alj and B. G. Neel, *Cancer Res.*, 2007, **67**, 2420.

49. S. G. Julien, N. Dube, M. Read, J. Penney, M. Paquet, Y. Han, B. P. Kennedy, W. J. Muller, and M. L. Tremblay, *Nat. Genet.*, 2007, **39**, 338.

50. K. A. Lantz, S. G. Hart, S. L. Planey, M. F. Roitman, I. A. Ruiz-White, H. R. Wolfe, and M. P. McLane, *Obesity (Silver Spring)*, 2010, **18**, 1516.

51. N. Krishnan, D. Koveal, D. H. Miller, B. Xue, S. D. Akshinthala, J. Kragelj, M. R. Jensen, C. M. Gauss, R. Page, M. Blackledge, S. K. Muthuswamy, W. Peti, and N. K. Tonks, *Nat. Chem. Biol.*, 2014, **10**, 558.

52. G. Chan, D. Kalaitzidis, and B. G. Neel, *Cancer Metastasis Rev.*, 2008, **27**, 179.

53. R. J. Chan and G. S. Feng, *Blood*, 2007, **109**, 862.

54. M. Tartaglia, C. M. Niemeyer, A. Fragale, X. Song, J. Buechner, A. Jung, K. Hahlen, H. Hasle, J. D. Licht, and B. D. Gelb, *Nat. Genet.*, 2003, **34**, 148.

55. Y. Zhang, P. Kurup, J. Xu, N. Carty, S. M. Fernandez, H. B. Nygaard, C. Pittenger, P. Greengard, S. M. Strittmatter, A. C. Nairn, and P. Lombroso, *J. Proc. Natl. Acad. Sci. U. S. A.*, 2010, **107**, 19014.

56. N. C. Carty, J. Xu, P. Kurup, J. Brouillette, S. M. Goebel-Goody, D. R. Austin, P. Yuan, G. Chen, P. R. Correa, V. Haroutunian, C. Pittenger, and P. J. Lombroso, *Transl. Psychiatry*, 2012, **2**, e137.

57. A. Saavedra, A. Giralt, L. Rue, X. Xifro, J. Xu, Z. Ortega, J. J. Lucas, P. J. Lombroso, J. Alberch, and E. Perez-Navarro, *J. Neurosci.*, 2011, **31**, 8150.

58. K. Wennerberg, K. L. Rossman, and C. J. Der, *J. Cell Sci.*, 2005, **118**, 843.

59. M. J. Smith, B. G. Neel, and M. Ikura, *Proc. Natl. Acad. Sci. U. S. A.*, 2013, **110**, 4574.

60. U. Vo, K. J. Embrey, A. L. Breeze, and A. P. Golovanov, *Biomol. NMR Assign.*, 2013, **7**, 215.

61. R. Thapar, J. G. Williams, and S. L. Campbell, *J. Mol. Biol.*, 2004, **343**, 1391.

62. E. F. Pai, U. Krengel, G. A. Petsko, R. S. Goody, W. Kabsch, and A. Wittinghofer, *EMBO J.*, 1990, **9**, 2351.

63. M. Spoerner, C. Herrmann, I. R. Vetter, H. R. Kalbitzer, and A. Wittinghofer, *Proc. Natl. Acad. Sci. U. S. A.*, 2001, **98**, 4944.

64. F. Shima, Y. Ijiri, S. Muraoka, J. Liao, M. Ye, M. Araki, K. Matsumoto, N. Yamamoto, T. Sugimoto, Y. Yoshikawa, T. Kumasaka, M. Yamamoto, A. Tamura, and T. Kataoka, *J. Biol. Chem.*, 2010, **285**, 22696.

65. M. Araki, F. Shima, Y. Yoshikawa, S. Muraoka, Y. Ijiri, Y. Nagahara, T. Shirono, T. Kataoka, and A. Tamura, *J. Biol. Chem.*, 2011, **286**, 39644.

66. H. Sondermann, S. M. Soisson, S. Boykevisch, S. S. Yang, D. Bar-Sagi, and J. Kuriyan, *Cell*, 2004, **119**, 393.

67. G. Buhrman, G. Holzapfel, S. Fetics, and C. Mattos, *Proc. Natl. Acad. Sci. U. S. A.*, 2010, **107**, 4931.

68. M. J. Smith and M. Ikura, *Nat. Chem. Biol.*, 2014, **10**, 223.

69. B. Rotblat, I. A. Prior, C. Muncke, R. G. Parton, Y. Kloog, Y. I. Henis, and J. F. Hancock, *Mol. Cell Biol.*, 2004, **24**, 6799.

70. A. A. Gorfe, M. Hanzal-Bayer, D. Abankwa, J. F. Hancock, and J. A. McCammon, *J. Med. Chem.*, 2007, **50**, 674.

71. H. R. Kalbitzer, I. C. Rosnizeck, C. E. Munte, S. P. Narayanan, V. Kropf, and M. Spoerner, *Angew. Chem. Int. Ed. Engl.*, 2013, **52**, 14242.

72. B. J. Grant, S. Lukman, H. J. Hocker, J. Sayyah, J. H. Brown, J. A. McCammon, and A. A. Gorfe, *PLoS One*, 2011, **6**, e25711.

73. T. Maurer and W. Wang, *Enzymes*, 2013, **33**, 15.

74. T. Maurer, L. S. Garrenton, A. Oh, K. Pitts, D. J. Anderson, N. J. Skelton, B. P. Fauber, B. Pan, S. Malek, D. Stokoe, M. J. Ludlam, K. K. Bowman, J. Wu, A. M. Giannetti, M. A. Starovasnik, I. Mellman, P. K. Jackson, J. Rudolph, W. Wang, and G. Fang, *Proc. Natl. Acad. Sci. U. S. A.*, 2012, **109**, 5299.

75. Q. Sun, J. P. Burke, J. Phan, M. C. Burns, E. T. Olejniczak, A. G. Waterson, T. Lee, O. W. Rossanese, and S. W. Fesik, *Angew. Chem. Int. Ed. Engl.*, 2012, **51**, 6140.

76. Q. Sun, J. Phan, A. R. Friberg, D. V. Camper, E. T. Olejniczak, and S. W. Fesik, *J. Biomol. NMR*, 2014, **60**, 11.

77. F. Shima, Y. Yoshikawa, S. Matsumoto, and T. Kataoka, *Enzymes*, 2013, 1.

78. I. C. Rosnizeck, T. Graf, M. Spoerner, J. Trankle, D. Filchtinski, C. Herrmann, L. Gremer, I. R. Vetter, A. Wittinghofer, B. Konig, and H. R. Kalbitzer, *Angew. Chem. Int. Ed. Engl.*, 2010, **49**, 3830.

79. A. K. Ganguly, Y. S. Wang, B. N. Pramanik, R. J. Doll, M. E. Snow, A. G. Taveras, S. Remiszewski, D. Cesarz, J.del Rosario, B. Vibulbhan, J. E. Brown, P. Kirschmeier, E. C. Huang, L. Heimark, A. Tsarbopoulos, V. M. Girijavallabhan, R. M. Aust, E. L. Brown, D. M. DeLisle, S. A. Fuhrman, T. F. Hendrickson, C. R. Kissinger, R. A. Love, W. A. Sisson, S. E. Webber, *et al.*, *Biochemistry*, 1998, **37**, 15631.

80. A. G. Taveras, S. W. Remiszewski, R. J. Doll, D. Cesarz, E. C. Huang, P. Kirschmeier, B. N. Pramanik, M. E. Snow, Y. S. Wang, J. D.del Rosario, B. Vibulbhan, B. B. Bauer, J. E. Brown, D. Carr, J. Catino, C. A. Evans, V. Girijavallabhan, L. Heimark, L. James, S. Liberles, C. Nash, L. Perkins, M. M. Senior, A. Tsarbopoulos, S. E. Webber, *et al.*, *Bioorg. Med. Chem.*, 1997, **5**, 125.

81. A. M. Petros, J. R. Huth, T. Oost, C. M. Park, H. Ding, X. Wang, H. Zhang, P. Nimmer, R. Mendoza, C. Sun, J. Mack, K. Walter, S. Dorwin, E. Gramling, U. Ladror, S. H. Rosenberg, S. W. Elmore, S. W. Fesik, and P. J. Hajduk, *Bioorg. Med. Chem. Lett.*, 2010, **20**, 6587.

82. A. Stamford and C. Strickland, *Curr. Opin. Chem. Biol.*, 2013, **17**, 320.

83. S. Roughley, L. Wright, P. Brough, A. Massey, and R. E. Hubbard, *Top. Curr. Chem.*, 2012, **317**, 61.

Chapter 23

Pharmaceutical Technology Studied by MRI

David G. Reid[1] and Stephen J. Byard[2]

[1]*Department of Chemistry, University of Cambridge, Lensfield Road, Cambridge CB2 1EW, UK*
[2]*Covance Laboratories, Alnwick, UK*

23.1 INTRODUCTION

MRI is best known as one of the foremost techniques for noninvasive scanning in medical diagnoses and procedures. Its great advantages in these applications are also well known. MRI directly determines the distribution of water or fat and is, therefore, excellent for scanning normal and diseased soft tissue. Importantly, MRI can be 'tailored' to emphasize zones of water according to their physicochemical environment using multiple 'weighting' or 'sensitization' techniques, based on familiar NMR properties ($T1$,[a] $T2$, $T2*$, magnetization transfer, magnetic susceptibility, paramagnetism, flow, and molecular diffusion) and produce image contrast between tissues with similar water content. The basis and applications of these contrast mechanisms are thoroughly summarized elsewhere[1] (see

Further Reading) and in some cases with specific focus on pharmaceutical applications.[2] MRI is equally versatile, however, across a range of disciplines including material and plant science, and, as was recognized early, pharmaceutical technology.[3] In the case of pharmaceutical technology, MRI is uniquely capable of exploring the behavior of drug delivery devices and products as they interact with artificial or biological media to deliver their active pharmaceutical ingredients (APIs). There are good recent reviews of MRI applications in pharmaceutics, including thorough accounts of MRI theory and methodology.[4,5]

Owing to the requirements in the treatment of many frequently encountered indications and the inherent properties of drug substances, there is an increasingly intense effort directed toward development of materials and devices that will deliver the active compounds over sustained periods of time with both predictable and controllable kinetics.[6] Design and delivery principles of such delivery devices or products are comprehensively treated in standard texts (see Further Reading). In this article, the general categories of products or delivery devices are described briefly and include a variety of modified release forms: delayed (drug release after a time lag), prolonged (drug presented for absorption over a period of time, possibly after an initial delay), sustained (some drug released in an initial burst to provide a therapeutic concentration followed by continued slower release), extended or controlled (drug continuously released to maintain therapeutic drug levels), and repeated (controlled doses released at

NMR in Pharmaceutical Sciences. Edited by Jeremy R. Everett, John C. Lindon, Ian D. Wilson, and Robin K. Harris
© 2015 John Wiley & Sons, Ltd. ISBN: 978-1-118-66025-6

intervals). Some common abbreviations, which are frequently encountered in the technical literature without definition, are also introduced.

Design strategies depend on numerous factors: the desired pharmacokinetics of the active ingredient, its physicochemical properties, polymorphic form, solubility, stability in the different pH environments of the gastrointestinal tract (GIT), hydrophilicity, protonation states, interactions with food and GIT lipids and macromolecules, salt form and counter-ion, whether it is absorbed by passive or active transport, and where this occurs in the GIT. Skilful formulation strategies can significantly enhance the effectiveness of a drug molecule and, by implication, substantially improve benefits to patients. In some cases, alternative indications or treatments can be addressed through improved understanding and implementation of controlled delivery. This can also be of commercial value through intellectual property.

Excipients are generally pharmacologically inert substances used as drug carriers or vehicles. They are universally employed in solid dose formulations to provide a homogeneous product of appropriate volume with good uniformity and, sometimes quite critically, to influence release kinetics. While the function of the excipient in facilitating the formulation process itself is beyond the scope of this article, it should not be overlooked completely as the choice of excipient blend with a focus solely on the ease of manufacture does not cover design space adequately with respect to product performance. Thus, for example, increased lubricant to aid tablet manufacture may seriously impact on bioavailability through altered dissolution profile. Broadly, most solid dose formulations will include binders (e.g., lactose), glidants (e.g., colloidal silica), lubricants (e.g., magnesium stearate), fillers/diluents (e.g., dibasic calcium phosphate, mannitol, or lactose monohydrate), often chosen through extensive chemometric experimental design. However, a wide variety of excipient materials are available that can be used to engineer the behavior of the formulated product, and an extensive knowledge of these is important because their properties in a hydrated state, in vivo, influence their mode of action. NMR parameters and MRI acquisition protocols have to be optimized accordingly. Numerous functionalized derivatives of starch (sodium starch glycolate), crosslinked sodium carboxymethylcellulose (Croscarmellose), and crosslinked polyvinylpyrrolidone (Cros-povidone) promote fast tablet disintegration when rapid drug release is desired. For sustained release over longer periods one frequently uses ethyl, hydroxyethyl, hydroxypropyl or mixed hydroxypropyl methyl cellulose ethers (EC, HEC, HPC, and HPMC, respectively), polymethacrylate (PMA), polymers or copolymers of ethyl acrylate, methyl methacrylate (MMA), or trimethyl ammonium ethacrylate, polyvinyl pyrollidone/polyvinyl acetate dispersion, pregelatinized starch, gelatine, hydrophobic waxes or fatty acid glycerides, or polyethylene glycol (PEG). Acid-labile APIs are protected in their passage through the stomach by 'enteric coating' with acid-inert materials such as cellulose acetate phthalate (CAP) or hydroxypropyl methyl cellulose acetate phthalate (HPMCAP), which become permeable or disintegrate when the tablet passes from the stomach into the more alkaline lower regions of the GIT. A growing area is the design of implantable or injectable slow release vehicles to release active agents over weeks or months. These are usually based on hydrolyzable biocompatible polyesters of glycolic acid (PGA), polyesters of lactic acid (PLA), or polyesters of mixed glycolic-lactic acid (PGLA or PLGA).

The drug release process may occur through a combination of different mechanisms. At a high level, these include diffusion of water into the excipient matrix, swelling or erosion of the matrix, and diffusion of the dissolved API through and out of the hydrated matrix. The relative importance of these different mechanisms for a specific product is of great interest to formulators, and, of course, MRI is ideal for studying all in detail. Tablet or capsule performance is traditionally tested by measuring drug-release rates in a variety of standardized United States Pharmacopoeia/European Pharmacopoeia (USP/EP) devices using defined dissolution media such as hydrochloric acid 'simulated gastric fluid' (SGF) or near-neutral phosphate buffered saline 'simulated intestinal fluid' (SIF). Unfortunately, these methods give little opportunity for characterizing the macroscopic and microscopic behavior of a solid-dosage form in terms of solvent ingress, porosity, swelling, and erosion. This is where MRI adds significant value to the development processes.

23.2 PRACTICALITIES

23.2.1 MRI Technology

High-field superconducting magnet microscopy or in vivo scanning instruments, with either vertical or horizontal magnets, are widely used in pharmaceutical MRI. Commercial microimaging RF resonators between about 10 and 25 mm in diameter are ideal probes as they can accommodate a cell in which a typical tablet can be analyzed in the relevant medium. Also, flow-through cells may be accommodated in which the solid-dosage form can be analyzed in situ under different conditions of dissolution. There are detailed designs of suitable cells in the literature.[7,8] Good sensitivity at high field allows high temporal and spatial resolution, and acquisition of 3D isotropic images that can be interrogated by 'slicing' the data matrix in any plane (for example, equatorial, pole-to-pole, on center, off center, oblique). This approach can be especially useful for complex formulations based on multilayer, shaped, coated, or membrane-controlled technologies. Usually, established standard microscopy pulse sequences work well to highlight critical aspects of solid-dose performance. $T1$ weighted ($T1$W, short echo time TE of a few milliseconds, short repetition time TR of tens to hundreds of milliseconds), $T2$ weighted ($T2$W, longer TE of tens to hundreds of milliseconds, longer TR of hundreds of milliseconds to a few seconds), and spin or proton density (short TE, long TR) weighted pulse sequences are all commonly used. Figure 23.1 shows a slice from spin echo (SE) datasets acquired at various times from an HPMC-sustained release tablet; the standard approach used enables solvent, swollen and solvent-penetrated HPMC, and dry core to be clearly resolved and quantified. Rapid release formulations are best captured with faster gradient echo methods, and their usefulness in characterizing the more challenging rapidly occurring events is shown in Figure 23.2.

High-field instrumentation is expensive, generally requires skilled operators, and is often committed much of the time to other purposes. The latter two factors, especially, have deterred widespread use (or at least published use) of MRI in the pharmaceutical and drug delivery industries. However, affordable lower-field permanent magnet 'benchtop' systems[9] operable by non-MRI experts are proving useful for elucidating tablet performance despite the inevitable sensitivity compromise. These instruments can be designed around a standard USP testing cell allowing the dissolution medium to be passed through a temperature-controlled probe in a defined way. With this configuration, the active substance release can be measured simultaneously 'downstream' by conventional methods such as ultraviolet (UV) absorption,

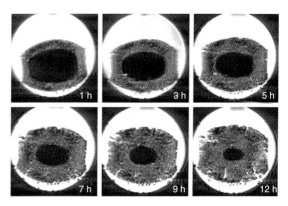

Figure 23.1. $T2$W images, acquired at the times shown on each panel, through a monolithic HPMC tablet designed for sustained delivery of an antibiotic. Images (7 T) were acquired from a purpose-built cylindrical flow-through plastic cell designed to fit into a 15 mm i.d. resonator (SIF at 10 ml min^{-1}). The perfusion medium appears bright, the core of the tablet, unpenetrated by solvent, is black, and the penetrated outer zone is intermediate in intensity. Note how the tablet swells and erodes as the solvent front advances. (Bruker AMX300, 3D $T2$W RARE, 'Rapid Acquisition with Relaxation Enhancement', TE/TR 30/1000 ms, Data matrix size 128^3, FOV = $4 \times 4 \times 4$ cm, RARE factor 8, TE$_{eff}$ 120 ms)

while MRI images are acquired of the tablet itself. Correlation between physical changes observed by MRI and release of API as monitored by UV absorption usually provides a wealth of information concerning release mechanisms and helps downstream problem solving. Another advantage is that the flow-through arrangement effectively provides sink conditions and may give a better approximation of in vivo behavior. However, it has to be emphasized that model conditions (nonbiologically relevant dissolution medium and restricted volume MRI system) are often required to achieve the desired product discrimination, and the use of static high-field systems is highly relevant. In addition, it should be noted that the inherent nature of the experiments, where temporal aspects are always critical, compromises the image quality compared with many other forms of MR microscopy where no significant changes are observed on a time scale of hours. Despite this, the information content of the data is invariably high. Moreover, in principle, all procedures can be compliant with good manufacturing practice (GMP) analytical requirements. Such instrumentation should assume a significant role in pharmaceutical development laboratories.

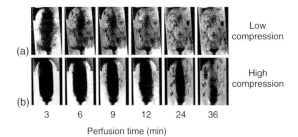

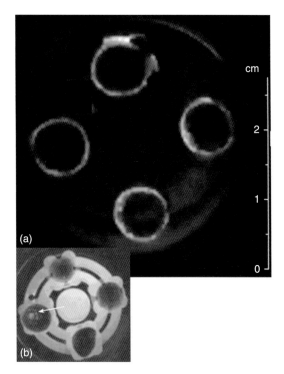

Figure 23.2. A comparison using fast acquisition techniques (flow compensated GE, TE/TR 2/200 ms, flip angle, α, 30°, 128^2 matrix, 13 slices 1 mm thick, separated by 1 mm) of the solvent penetration and disintegration characteristics of rapid polymethacrylate (PMA)-based tablets prepared under low (a) and high compression (b). Imaging was performed in 2 M acetic acid, flow rate 15 ml min^{-1}, at 7 T. The harder, more compressed, tablets clearly absorb solvent and disintegrate more slowly in accordance with their active substance release kinetics

23.2.2 When to Use MRI?

Applying a time- and effort-intensive technique such as MRI to routine formulation design processes is not appropriate under normal circumstances. However, it provides unique and potentially crucial information when new excipients or devices are being developed, especially where more complex designs are involved. A practical example appears in Figure 23.3, which compares images of the penetration of neutral phosphate buffered SIF into the enteric coat of a commercial formulation of an acid-sensitive compound (a) after preincubation in SGF, with images (b) from an experimental enteric coat that is clearly failing.

The MRI approach is frequently overlooked but, usually and rapidly, provides valuable insights when established formulation processes start to show unexpected or out-of-specification drug release kinetics. Typically, this may be because of physical defects in the device leading, for instance, to solvent ingress along defects between layers in a multilayer, or between the membrane and interior of a membrane-controlled system. It may be because of the changes in excipient properties or formulation conditions, possibly unsuspected, leading to changes in matrix erodibility or penetrability, or diffusion rates of dissolution medium or active substance. Issues of this nature can occur at any stage in the development process, including late stage manufacturing trials,

Figure 23.3. Enteric coat penetration. Experimental tablets, designed to deliver an acid-sensitive active molecule, were immersed in 0.1 M HCl simulated gastric fluid for 2 h. The four tablets portrayed in panel (a) were imaged following a further 30 min in simulated intestinal fluid (Tris buffer), pH 7.4. Penetration of the enteric coat by solvent is clearly visible in (a); no penetration was observed for this batch when the tablets were imaged before transfer to pH 7.4 buffer, so the enteric coat is behaving as intended. However, high image intensity solvent 'hot spots', arrowed in panel (b), are clearly visible in image slices positioned across the surface of a different batch after 2 hours of immersion in 0.1 M HCl. In this batch the enteric coat is clearly failing (same acquisition parameters as Figure 23.1)

and may be of significant importance for ensuring quality and performance of the finished products. Answers to such questions can be of great urgency and commercial importance; MRI can provide many of them.

23.3 APPLICATIONS

23.3.1 Controlled Release

Many published studies describe controlled-release products because they place the greatest demands on the pharmaceutical engineering and often require a multidisciplinary approach involving formulation scientists, analysts, and chemists. Feedback of data derived from MRI studies to the different contributors is required for optimization. Typical solvent-density images in Figure 23.4 summarize the behavior of a wax-based matrix tablet in flowing SIF. The slices shown are reformatted from a 3D dataset enabling slicing in any desired plane. The images show the development and propagation of cracks in the wetted zone that potentially provide unexpected routes for drug egress, and a partially hydrated transition zone between core and hydrated exterior.

MRI is particularly useful for gaining insights about the behavior of complex formulations. Figure 23.5 shows three different reformatted views over time through a product in which drug release from an HPMC matrix is controlled by holes in an impermeable membrane. The ability to reformat the 3D datasets is particularly useful here because it provides the opportunity of interrogating events in any arbitrary image plane.

$T1$ and $T2$ are closely related to solvent molecular motion, and hence interactions with macromolecules. This gives useful insights into solvent–excipient binding,[10] as in the case of formulations of pentoxyphylline with xanthan gum at different pHs and gum–water ratios. Here, pH and composition strongly influence $T2$. Additives can affect the pharmaceutical properties of a bulk matrix material and these influences, such as the effects of mannitol and calcium phosphate, on the swelling and erosion of HPMC can be studied in detail.[11] HPMC matrices loaded with different amounts of tetracycline[12] in HCl undergo a transition from a glassy amorphous state to a gel state under hydration. Drug release is regulated by solvent-front advance that occurs at a constant rate, and subsequent rapid diffusion of drug through the swollen matrix. Gel layer thickness increased at a greater rate with increased tetracycline loading in the HPMC, implying the drug was reducing the integrity or increasing the molecular disorder of the HPMC matrix. Depositing EC coatings by aqueous or organic solvent processes has contrasting effects on release from pulsatile capsules (chronopharmaceuticals).[13] Aqueous coating processes produce more rapid and erratic release, which is clearly shown by MRI to be associated with penetration of the medium through irregularities or defects in the enteric coating. This is a good example of MRI explaining undesirable release

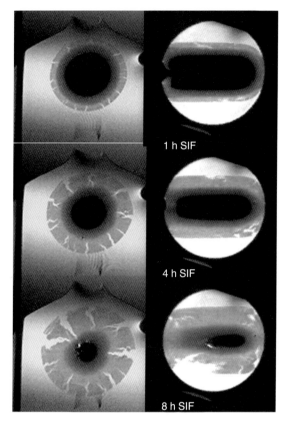

Figure 23.4. Slices through 3D proton density MR images of a wax-based sustained-release device composed of fatty acid glycerides bound to polyethylene glycol. The left-hand column is a slice selected orthogonal to the long axis of the cylindrical tablet. The right-hand column is a slice parallel to this axis. MRI clearly shows solvent penetration, swelling, cracking and erosion, and the existence of a 'transition zone' between fully hydrated and unhydrated zones (same acquisition parameters as for Figure 23.1)

kinetics by the physical behavior of the capsule and the tablet it encloses.

Implantable delivery devices are becoming more attractive for treatment of certain indications where controlled drug release is desired. PGA is a useful matrix for such products, and slow hydrolysis of the ester linkages is an integral part of drug release mechanisms. MRI supports a 'four-stage degradation' model[14]: stage I – very small amounts of water adsorb to the surface; stage II – molecular weight falls due to hydrolysis, but little drug (theophylline) substance is lost; stage III – a highly hydrated reaction-erosion front begins to move through the sample, with rapid

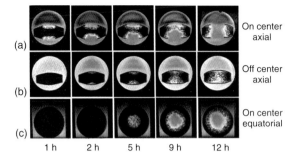

Figure 23.5. MRI of the behavior of a sustained release design based on an inert outer shell with holes drilled into each face; drug release kinetics can be finely controlled by changing the size of the hole. The excipient can be a release-retarding polymer such as HPMC or polymethacrylate. By using a true 3D acquisition protocol, isotropic resolution is achieved and the 3D image matrix can be sliced in any direction, an important consideration when dealing with complex shaped or layered devices. (a, b, c) The behavior of the tablet by illustrating a slice from 'pole to pole' (axial) positioned through the center of the tablet, a parallel axial slice but off center, and a slice through the equator. Thus, (a) shows the immediate penetration of SIF through each central hole, (b) depicts the delay before the solvent reaches the more outer regions of the core, and (c) shows the delay in reaching the equatorial slice. The images show that solvent 'creep' along the shell-filler junction is minimal, an important prerequisite for predictable active compound release

drug diffusion from the porous regions behind the front (measured by UV spectroscopy); and stage IV – 100% drug release is achieved when the advancing fronts meet at the center of the sample. The entire process takes many days, and individual samples were imaged after removal from the dissolution medium at different time points. Gravimetric and MRI analyses of water uptake correlate well. The tendency of the PGA ester linkages to hydrolyze more slowly in D_2O than in H_2O provides insights into the chemistry underlying drug release.[15] Combination with deuterium-specific scanning microbeam nuclear reaction analysis (NRA) was used to support these observations.

A modified flow-through USP 4 device[8] with metal parts replaced by plastic has been employed as a sample holder for MRI and used to study HPMC tablets, a ranitidine-HCl–lactose–HPC formulation, and a simple salbutamol osmotic pump extended release device. The latter uses sodium chloride as the osmotic agent and a small hole in a semipermeable membrane as a release orifice to produce drug release, which is linear

with time. Tablet disintegration is faster in flowing than in static media (SGF or SGF doped with 5 mM $CuSO_4$ for relaxation enhancement).

Wang *et al.*[16] studied the effects of pH (to model the contrasts between the acid of the stomach and the more alkaline conditions of the lower GIT) and ionic strength on interactions with solvent of variously derivatized chitosan and polysaccharides with tailored mean pK_a values. Using the antibiotic tetracycline as a model compound, Kowalczuk and Trit-Goc[17] explored the behavior of HPMC matrices – swelling, drug release, solvent ingress, and the effects of solvent pH.

In an instructive study of the relationship of matrix and drug properties with drug release kinetics, HPMC formulations loaded with two different model drugs[18] have been studied. The advantages of fluorine in both compounds were exploited to track the fate of the drug itself by spin density ^{19}F NMR. Solvent $T2$ provides a measure of the HPMC content of the penetrated zone. However, bubbles from air entrapped during the compression process can be a significant source of error in quantifying the dimensions of the infiltrated zone. The drugs, one charged but hydrophobic (triflupromazine-HCl), the other neutral but hydrophilic (5-fluorouracil), were chosen as model compounds because they undergo quite different interactions with the HPMC matrix. $T1$, $T2$, and self-diffusion coefficient decrease with increasing HPMC content. Movement of triflupromazine-HCl through the matrix follows matrix hydration, and the drug release depends on HPMC erosion kinetics. On the other hand, 5-fluorouracil diffuses rapidly at all matrix hydration levels. The presence of NMR-active fluorine-19 in the lipid-lowering fluvastatin has also been exploited to simultaneously monitor HPMC-based delivery device performance (using proton MRI) and drug mobilization and release (by ^{19}F spectroscopy).[19,20] Chemical shift imaging (CSI) can be used to monitor the release of drug and polymer constituents from hydrophobically modified polyacrylic acid (HMPAA) matrices; the CSI technique enables visualization of drug molecules that do not contain a heteronuclear NMR 'reporter' atom such as ^{19}F.[21]

The use of MRI to study the behavior of a multicomponent 'regulated release' device is summarized in Figure 23.6. The device consists of a 'sandwich' of two circular plastic discs connected by a central and three peripheral pillars; the inner surfaces of each disc are shaped so that when the gap between them is packed with formulation, drug release takes place from the cylindrical surface, the area of which can be controlled

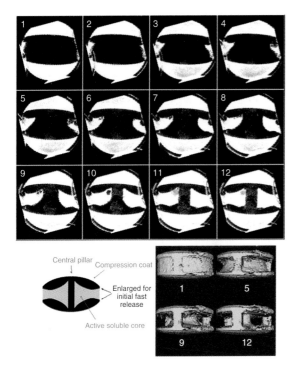

Figure 23.6. Slices through multiecho 3D datasets acquired at the time points shown (in hours) in flowing SGF (1, 2 h) and SIF (subsequent images) of a shaped controlled-release device consisting of an inert plastic support (see cross-sectional diagram at bottom left) packed with formulation. (The flow-through cell enables switch-over between different perfusion media without disturbing the tablet or interrupting scanning.) The formulation surface area for drug release that the device presents to solvent as it erodes is controlled by the cross-sectional profile of the support, thereby providing a simple means of controlling drug release kinetics. Because the MRI was performed using high-resolution 3D imaging, it is possible, using commercially available image processing software, to not only measure standard parameters of interest such as solvent-exposed surface or residual volume but also 'surface render' the images to produce 3D representations such as those shown at bottom right. 3D rendering is a powerful tool for understanding the performance of complex-shaped devices

by the shape of the inner surfaces of the sandwiching discs. This makes precise control of release kinetics possible. Acquisition of 3D isotropic images not only enables data to be interrogated in any oblique plane but also allows 'rendering' of the solvent-exposed faces to give a more comprehensive impression of the behavior of the device than is usually possible from 2D slices.

There is an interesting study on a similar shaped device to release ranitidine-HCl at a constant rate from a lactose-HPC formulation.[22] MR microscopy showed SGF penetration through and around the shaped plastic coat, explaining anomalously fast drug-release kinetics. This early example is instructive in showing how MRI can highlight the factors underlying unexpected product performance, in a way that is difficult to envisage using other techniques.

HPMC formulations of the atypical antipsychotic quetiapine have been studied for the progression of glassy to gel regions under USP 4 flow conditions.[23] Amorphous or molecular dispersions of drug molecules are comparatively rare although they offer advantages in terms of assisting dissolution of sparingly soluble compounds. In a multinuclear study using one-dimensional spatial localization methods and multinuclear detection (^{19}F of the nonsteroidal anti-androgen flutamide, ^{1}H of HPMC excipient, and ^{2}H of solvent D_2O) to differentiate the components of the formulation, Dahlberg *et al.*[24] showed that $T1$ can be exploited to study drug crystalline/amrphous transformations and nanoparticle coalescence, and how they are related to solvent and behavior. Using ^{2}H and ^{1}H imaging to follow formulation hydration (using D_2O) and polymer mobilization, respectively, the same group related drug release kinetics to mode of preparation (milling or rotary evaporation) and pore size; released compound (the soluble model compound pyrazine) was measured directly by ^{1}H NMR spectroscopy facilitated by the D_2O medium.[25]

A potentially general empirical equation has been proposed relating drug release from HPMC matrices to core and hydrogel area (measured by MRI), and excipient viscosity and level of hydroxymethylation.[26]

23.3.2 Rapid Release

MRI of rapid release formulations is equally useful in highlighting structural reasons underlying batch performance differences or anomalies. Figure 23.7 compares the response to dissolution medium of four different batches of a fast-release formulation, enabling the correlation of unexpected interbatch differences in drug release with solvent ingress as measured by the decrease in the size of the nonpenetrated tablet core.

Rapid Snapshot FLASH (fast low angle shot) with 500 ms image acquisition enables real-time interrogation of drug release mechanisms from 'immediate release' acetaminophen tablets formulated with

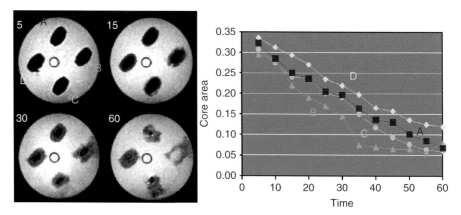

Figure 23.7. A comparison of four different rapid release antibiotic formulations (labeled A to D) immersed in 0.01 M HCl (pH 2) solution before imaging at 7 T using a standard multislice proton density SE (TE/TR 7/2400 ms, matrix size 128^2, FOV 4 × 4 cm, slice thickness 2 mm, 8 slices per time point acquired in 5 min blocks). Selected time points are shown beside the image to which each corresponds). The central cross-sectional surface area of the nonpenetrated core of each tablet was measured, and mean core areas (in cm^2) plotted against immersion time. The rate of surface erosion correlated well with formulation methodology, and with compound release kinetics measured separately

starch, lactose, or polyvinylpyrrolidone over several minutes.[27] Drug release correlates with tablet disintegration and depends on excipient, preparation process, and drug content. 'Solid-friendly' single point imaging (SPI) can be used on samples with very short relaxation times, and shows that the distribution of the drug in dry tablets is uniform. FLASH imaging in 2.5 minute windows[28] has been applied to rapid release tablets of the mucolytic agent bromhexine, comprising PEG of various chain lengths, lactose, and potato starch. Again, constituents and tabletting processes have definite effects on dissolution and disintegration. For instance, slow solidification and increasing PEG chain length caused slower disintegration and overall drug release.

Food in the GIT can be an important factor in delivery device performance and it is important to anticipate and understand these effects. For instance, a high-fat nutritional drink delays disintegration of an HPMC fosamprenavir immediate-release tablet.[29] The study used a commercial 'artificial gut' designed to mimic the pH and digestive enzyme content of the various compartments of the GIT. Complete rapid disintegration occurs in SGF, but only swelling and little disintegration in the nutritional drink. Effectively, food transforms the release kinetics from immediate to delayed release with potential significant effects on therapy.

23.3.3 Diffusion

Controlled drug release is a function of diffusion of the dissolution medium into the drug product and drug substance out of it, in conjunction with swelling, erosion, and porosity changes of the product itself. Drug released is often analyzed using a form of the semiempirical equation

$$Q/Q_{inf} = kt^n$$

in which Q/Q_{inf} is the fraction of total drug released after time t, and k is a system-dependent constant. Two extremes of contrasting behaviors are recognized. The case where $n = {}^1/_2$ is often referred to as 'Fickian', 'case I', or diffusion controlled, release, governed by the rate of solvent diffusion into the matrix and diffusion of drug substance out of it. This behavior is usually characteristic of hydrophobic matrices such as EC. The situation where $n = 1$ is referred to as 'case II' behavior with release controlled by matrix swelling, and is usually found in hydrophilic or erodible matrices such as HPMC and PGLA.

Simple diffusion-sensitized imaging is often sufficient to distinguish between these two extremes. Figure 23.8 shows diffusion-weighted (DW) SE image slices through a trilayer tablet, in which one of the outer layers is designed to release active substance by an erosion-based mechanism and the other outer layer by diffusion. NonDW SE images show little difference

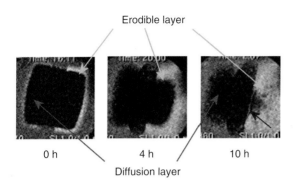

Erodible layer

0 h   4 h   10 h

Diffusion layer

Figure 23.8. Three DW images acquired from a trilayer tablet designed to present erosion- and diffusion-governed layers to solvent. In the erodible layer solvent diffusion is inhibited and so there is minimal signal loss during the diffusion-sensitization period, and this layer hence appears bright although it swells noticeably. Without diffusion sensitization both erosion- and diffusion-governed layers appear equally intense. MRI also shows an unexpected influx of solvent between the erodible layer and the inert core layer, which would compromise predicted release kinetics

between the layers at any time. However, with DW the solvent-penetrated diffusion layer appears dark because freely diffusing solvent signal is destroyed during the DW (case I), but the solvent-penetrated erodible layer appears bright as diffusion-inhibited solvent signal is not destroyed by the sensitization procedure (case II). This represents a simple approach to establishing that different layers are performing as intended.

More detailed quantitative information such as apparent diffusion coefficients (ADC) can be obtained by using a series of diffusion-sensitization gradients.[30] Fast quantitative methods used on HPMC matrix swelling, erosion, and solvent front penetration give new insights into the glassy transition zone lying between the dry core and the edge of the fully hydrated gel layer. Swelling and water mobility of low-substituted HPC[31] loaded with two different model drugs, theophylline (poorly water soluble) and procainamide-HCl (highly water soluble), have been compared with more hydrophilic HPMC and HPC. Less hydrophilic low-substituted HPC hydrated much less than the other materials, which showed a gradual increase in solvent $T2$ and ADC, from core to exterior of the swollen layer. Incorporation of either drug increased the extent of the gel layer and swelling.

Solvent penetration and matrix degradation of drug-loaded PGLA implantable slow-release cylinders have been studied with $T2$ mapping;[32] samples were removed from the incubation medium for imaging at intervals of a few days. Inhomogeneous solvent ingress is explained by autocatalytic degradation of the PGLA polyester by its own acidic degradation products. The solvent ADC decreases toward the interior of the cylinder, but then increases again in the autocatalytically degraded center. MRI thus provides significant information for the design of implantable products with predictable drug release characteristics. Implantable PGLAs are attractive for chronic release of polypeptides and other macromolecules that would be either degraded in or not absorbed from the GIT after oral administration.[33] Polypeptide incorporation strongly influences implant performance. In the drug-free product (placebo), PBS ingress is rapid and typical of 'case II' behavior, i.e., solvent diffusion into the device is limited by the rate at which the polymer matrix hydrates and relaxes. In the drug-loaded product, ingress was much slower and followed either case I or case II kinetics in different regions of the device. This article gives excellent accounts of MR microscopy and of the different diffusion regimes in hydrated polymers.

In hydrophilic HPMC, HEC, and HPC matrices[34] containing the model hydrophilic drug sodium salicylate, pulsed-field gradient spin echo (PGSE) spectroscopy (not strictly speaking MRI) has been used to measure both salicylate and water diffusibility to determine whether drug transport is diffusion controlled (Fickian) in the swollen hydrogel matrix. All three polymers exert very similar effects on solute diffusivity, which decreases as weight percent polymer increases. High amylose starch tablets, gelatinized with NaOH and crosslinked with epichlorhydrin in a highly swollen state, have been studied. Uncharacteristically, crosslinking actually increases water uptake. Hydrogel tablet swelling was followed in water over about 20 hours.[35] Predominantly, case II kinetics are related to conversion of the noncrystalline dry matrix on hydration to a highly crystalline 'B' double helical form of amylose.

HPMC behavior can be pH dependent.[36] In case I diffusion, solvent advances in proportion to the square root of time of exposure to solvent, there is a solvent gradient from exterior to interior, and $T2$ is constant throughout the swollen region. In case II, diffusion distance increases linearly with time, solvent

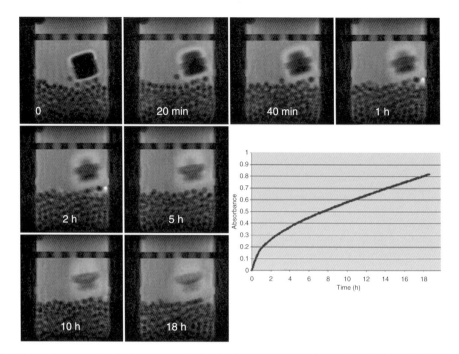

Figure 23.9. Spin-echo proton density images acquired at the times shown in each panel on a 0.5 T Oxford Instruments Maran DI permanent magnet scanner (TE/TR 6/1800 ms, FOV 32 mm, Slice thickness 2 mm, Matrix size 128²). The experiment interrogated the behavior of an experimental trilayer tablet, in which each layer has been designed with different release characteristics to produce a specific overall release profile. The scanner is 'hyphenated' with a UV cell for measurement of released drug concentrations, plotted against time in the bottom right panel

concentration is uniform throughout the swollen region, and $T2$ decreases from swollen exterior toward tablet core (amorphous polymers below glassy state). At pH 2, initial solvent diffusion follows case I, and subsequently becomes case II. At pH 6, solvent diffusion follows case II behavior throughout. At pH 2, the HPMC takes up less solvent, more slowly, with increased solvent-matrix hydrogen bonding causing progressive decrease in $T2$, and multicomponent $T2$ relaxation processes. These findings could have important implications for drug release kinetics in acidic (stomach) and less acidic (lower gut) zones of the GIT.

23.3.4 Benchtop Approaches

Benchtop MRI systems are attractive because they are affordable, have modest space requirements, and can be simple to use. Good quality data are achievable, as exemplified in Figure 23.9, which shows images from a trilayer tablet. Each layer of the tablet is designed

to release drug at a different rate to achieve an overall desirable release profile. MR images of a hydrating bilayer HPMC tablet clearly show solvent penetrating between the bilayer junctions.[37] The bilayer design is used to, for instance, separate incompatible drug substances, confer separate release kinetics on two different drugs, or to provide biphasic release characteristics such as a burst followed by slow release.

A benchtop instrument has also been used to examine a bilayer 'osmotic pump'-controlled release tablet.[38] The delivery mechanism involves controlled water diffusion through a semipermeable membrane, with drug release through an orifice. Floating tablets delay passage out of the stomach and into the lower GIT and are amenable to benchtop imaging.[39,40] In the first study, tablets were removed periodically from a standard USP paddle dissolution apparatus and imaged in 0.1 M HCl. Swelling of polymer was found to be only marginally dependent on the proportions of the hydrophilic model drug, propranolol. In the second study, the same matrix polymer and model drug were used, with addition of sodium bicarbonate to the tablets to generate CO_2 gas on contact with stomach acid. Shape changes, solvent penetration, and gas bubble generation were clearly observed in the MRI experiment. A benchtop 0.5 T magnet system has been successfully used to correlate polyethylene oxide/PEG matrix behavior with release of a test compound acetaminophen.[41]

23.3.5 Combination Techniques

It is useful to combine MRI with other modalities such as X-ray microtomography[42] exemplified by a study of an HPMC/MCC/lactose hydrophilic swelling formulation in PBS.[43] MRI offers clear advantages when investigating physicochemical events in situ, providing direct information about dynamics and interactions as well as morphology and distribution. However, the attainable resolution is less than that of X-ray microtomography, making the two techniques highly complementary. Movement of nonhydrated MR invisible parts of the tablet was measured by X-ray detection of implanted glass microspheres. NMR parameters were correlated with compositional changes during hydration, as in previous studies.[44,45] Under certain circumstances, MRI and electron paramagnetic resonance (EPR) can be a powerful combination. Stable paramagnetic nitroxide 'spin probe' molecules have been invoked to model drug behavior.[46] Nitroxides incorporated into biodegradable acid anhydride implant materials mimic hydrophilic drug release. EPR distinguishes the spin probe in the implant from the free probe molecules because they give different EPR signals due to their differing molecular motions. Solvent ingress, erosion, and shape changes are followed by MRI in subcutaneous implants in rats, as well as in vitro; nitroxide release kinetics are followed from the EPR signal of the free nitroxide. The same combination has been used[47] to study anhydride polymer degradable implant products; in a sandwich design, one layer contained ^{14}N- and the other ^{15}N-nitroxide. The different EPR properties of these two labels allow the rate of nitroxide release from the different layers of the sandwich to be separately monitored. In an interesting multidisciplinary approach,[48] electric current density imaging (EDI) has been used to measure release of model ionic drug model substances (organic acids) from rapid oral disintegration tablets, simultaneously with MRI.

23.3.6 Pharmaceutical MRI In Vivo

Most studies, including MRI, of delivery device performance are carried out under controlled, nonbiological conditions. This is ideal and indeed necessary for understanding the reproducibility of a manufacturing process. USP devices provide reproducible conditions under which variation in drug release kinetics can be ascribed confidently to variation between delivery devices, which is a prerequisite for product development and quality control. Such approaches, however, cannot give a complete picture of how a tablet will perform in a real biological environment such as the GIT. Here, different conditions of, for instance, peristaltic motions, pH, ionic strength, and food content are encountered. Thus, methods to elucidate the behavior and passage of a device in the GIT itself are extremely valuable. This is challenging because the GIT moves and delivery devices are small and usually MRI invisible. However, experimental formulations have been 'labeled' with small quantities of superparamagnetic magnetite (Fe_3O_4) and tracked using $T2^*$ enhancement. This creates a signal 'void' in the vicinity of the tablet.[49-51] At the same time, incorporation of a gadolinium $T1$ enhancement agent tracks 'drug' release by signal enhancement in $T1$-weighted images. Subjects were seated upright in an open 0.5 T magnet clinical scanner. Gelatine capsules containing magnetite have been tracked in the stomach after a meal, in a clinical 1.5 T scanner.[52]

A significant in vivo application used a 0.5 T benchtop instrument with a 23 mm usable sample diameter, maintained at 37 °C, to study the performance of PGLA implants for slow drug release in an anesthetized mouse.[53] The solidification of the PGLA after injection, subsequent body fluid interactions, and erosion of the implant were studied in a series of imaging sessions over several months. The PEG vehicle appears bright and the precipitated PGLA polymer dark on $T1$-weighted images. The attractions of using MRI to track the performance of implantable devices are obvious; the subject can be scanned repeatedly so the behavior of the same device can be observed longitudinally, effectively acting as its own control. Another area of likely future importance is that of implantable 'scaffolds' for in vivo tissue replacement and regeneration.[54] Composite hollow hydroxyapatite-collagen-chitosan scaffolds have been studied in a 0.5 T benchtop MRI system. Mass transport into the scaffold was observed using the $T1$-enhancing contrast agent gadolinium (III) diethylenetriamine pentaacetate and compared for different scaffold types. Cells labeled with superparamagnetic iron oxide particles can also be tracked as they migrate into the scaffold material. It is likely that noninvasive tracking of device performance in vivo will become a significant area of pharmaceutical MRI.

23.4 FUTURE DEVELOPMENTS

MRI is already a valuable tool in pharmaceutical technology. It will continue to provide unique insights into the performance of new drug-delivery products and devices, the influence of manufacturing processes, raw materials, and storage conditions such as temperature and relative humidity. It has a valuable role in 'troubleshooting' when established processes fail to produce expected outcomes, either for in vitro tests or in volunteers and patients. These activities will continue to use and develop the methods discussed in this article. MRI's unique ability to observe and quantify solvent flow is being used to study the performance of delivery devices under USP testing conditions and their possible influence on drug release. Thus, a detailed MRI study of hydrodynamics inside a USP 4 flow cell device operating under standard flow conditions found that the 'flow field' is heterogeneous and not necessarily laminar.[55] Much wider application in industry is anticipated, facilitated by the availability of space-friendly instruments designed for nonexpert users. Pharmaceutical MRI could become an attractive offering from contract research organizations to their pharmaceutical and biotechnology industry customers. We expect that MRI data will be used in new drug-delivery device patent applications and submissions for approval to regulatory agencies. Because MRI data provide a refined and detailed method of comparing and analyzing the performance of different devices, including those offered by commercial competitors, MRI will not only continue to play its current part in internal decision making but also could assume a significant role in intellectual property defense.

Extant studies already cited have shown how useful MRI is in tracking the performance of delivery devices in vivo, and in following the progress of tablets through the GIT. With current technology, tablet tracking requires labeling with small quantities of paramagnetic marker substances. If appropriate equivalence studies prove that the labeled device supports the same pharmacokinetics as the unlabeled one, a strong case can be made for using labeled tablets in trials in both preclinical species and volunteers, removing potential uncertainties from conventional drug absorption studies. Alternative methods for in vivo tablet tracking require exposing subjects to ionizing radiation and their use is limited by safety issues; in principle, there is no limitation on the frequency with which a patient or clinical trials volunteer can have MRI scans.

MRI also has great potential as a method of monitoring the performance of implanted materials such as PGLA reservoirs for chronic drug release. The field of tissue replacement using organic materials such as collagen and polysaccharides is an area of intense activity. MRI is a unique tool for observing the incorporation of such materials into native tissue. Again, patients or volunteers can be scanned as often as diagnostically necessary, or as a clinical trial protocol dictates.

23.5 END NOTE

The notation for relaxation parameters such as $T1$ conforms to the normal usage for MRI work. In contrast to the usual situation for more general NMR articles, the specific indications for the parameter involved such as the '1' in '$T1$' are not subscripted. The T will be italicized except when it is part of a label such as '$T1$-weighted'.

ACKNOWLEDGMENTS

DGR acknowledges numerous GlaxoSmithKline former colleagues: In the MRI group: Keith Brooks, Albert Busza, Simon Campbell, Kumar Changani, John Hare, Paul Hockings, Amy Nicholson, Dick Payne, Torsten Reese, Sean Smart, and Sarah Squires; in Pharmaceutical Development, Chi Li, Gino Martini, Tracey Naylor, Vin Re, Laurence Robinson, David Valentini, Peter Coles, and Helen Willey. SJB acknowledges Estelle Chouin and Gareth Lewis in the Pharmaceutical Sciences Group of Sanofi-aventis (France), Richard Bowtell (University of Nottingham), and Kevin Nott (Oxford Instruments).

REFERENCES

1. M. M. Britton, *Chem. Soc. Rev.*, 2010, **39**, 4036.
2. C. D. Melia, A. R. Rajabi-Siahboomi, and R. W. Bowtell, *Pharm. Sci. Tech. Today*, 1998, **1**, 32.

3. M. Ashraf, V. L. Iuorno, D. Coffinbeach, C. A. Evans, and L. L. Augsburger, *Pharm. Res.*, 1994, **11**, 733.

4. M. D. Mantle, *Int. J. Pharm.*, 2011, **417**, 173.

5. M. D. Mantle, *Curr. Opin. Colloid Interf. Sci.*, 2013, **18**, 214.

6. J. C. Richardson, R. W. Bowtell, K. Mader, and C. D. Melia, *Adv. Drug Deliv. Rev.*, 2005, **57**, 1191.

7. S. Abrahmsen-Alami, A. Korner, I. Nilsson, and A. Larsson, *Int. J. Pharmaceut.*, 2007, **342**, 105.

8. C. A. Fyfe, H. Grondey, A. I. Blazek-Welsh, S. K. Chopra, and B. J. Fahie, *J. Controlled Release*, 2000, **68**, 73.

9. K. P. Nott, *Eur. J. Pharmaceut. Biopharmaceut.*, 2010, **74**, 78.

10. U. Mikac, A. Sepe, J. Kristl, and S. Baumgartner, *J. Controlled Release*, 2010, **145**, 247.

11. F. Tajarobi, S. Abrahmsen-Alami, A. S. Carlsson, and A. Larsson, *Eur. J. Pharm. Sci.*, 2009, **37**, 89.

12. J. Kowalczuk, J. Tritt-Goc, and N. Pislewski, *Solid State Nucl. Magnet. Reson.*, 2004, **25**, 35.

13. J. C. D. Sutch, A. C. Ross, W. Kockenberger, R. W. Bowtell, R. J. MacRae, H. N. E. Stevens, and C. D. Melia, *J. Controlled Release*, 2003, **92**, 341.

14. G. E. Milroy, R. E. Cameron, M. D. Mantle, L. F. Gladden, and H. Huatan, *J. Mater. Sci. Mater. Med.*, 2003, **14**, 465.

15. G. E. Milroy, R. W. Smith, R. Hollands, A. S. Clough, M. D. Mantle, L. F. Gladden, H. Huatan, and R. E. Cameron, *Polymer*, 2003, **44**, 1425.

16. Y. J. Wang, E. Assaad, P. Ispas-Szabo, M. A. Mateescu, and X. X. Zhu, *Int. J. Pharm.*, 2011, **419**, 215.

17. J. Kowalczuk and J. Tritt-Goc, *Eur. J. Pharm. Sci.*, 2011, **42**, 354.

18. C. A. Fyfe and A. I. Blazek-Welsh, *J. Controlled Release*, 2000, **68**, 313.

19. C. Chen, L. F. Gladden, and M. D. Mantle, *Molec. Pharmaceut.*, 2014, **11**, 630.

20. Q. L. Zhang, L. Gladden, P. Avalle, and M. Mantle, *J. Controlled Release*, 2011, **156**, 345.

21. P. Knoos, D. Topgaard, M. Wahlgren, S. Ulvenlund, and L. Piculell, *Langmuir*, 2013, **29**, 13898.

22. B. J. Fahie, A. Nangia, S. K. Chopra, C. A. Fyfe, H. Grondey, and A. Blazek, *J. Controlled Release*, 1998, **51**, 179.

23. P. Kulinowski, P. Dorozynski, A. Mlynarczyk, and W. P. Weglarz, *Pharm. Res.*, 2011, **28**, 1065.

24. C. Dahlberg, S. V. Dvinskikh, M. Schuleit, and I. Furo, *Molec. Pharmaceut.*, 2011, **8**, 1247.

25. C. Dahlberg, A. Millqvist-Fureby, M. Schuleit, and I. Furo, *Eur. J. Pharmaceut. Biopharmaceut.*, 2010, **76**, 311.

26. P. P. Dorozynski, P. Kulinowski, A. Mendyk, A. Mlynarczyk, and R. Jachowicz, *AAPS Pharm. Sci. Tech.*, 2010, **11**, 588.

27. J. Tritt-Goc and J. Kowalczuk, *Eur. J. Pharm. Sci.*, 2002, **15**, 341.

28. S. Kwiecinski, M. Weychert, A. Jasinski, P. Kulinowski, I. Wawer, and E. Sieradzki, *App. Magnet. Reson.*, 2002, **22**, 23.

29. J. Brouwers, B. Anneveld, G. J. Goudappel, G. Duchateau, P. Annaert, P. Augustijns, and E. Zeijdner, *Eur. J. Pharmaceut. Biopharmaceut.*, 2011, **77**, 313.

30. Y. Y. Chen, L. P. Hughes, L. F. Gladden, and M. D. Mantle, *J. Pharm. Sci.*, 2010, **99**, 3462.

31. M. Kojima, S. Ando, K. Kataoka, T. Hirota, K. Aoyagi, and H. Nakagami, *Chem. Pharm. Bull.*, 1998, **46**, 324.

32. A. Djemai, L. F. Gladden, J. Booth, R. S. Kittlety, and P. R. Gellert, *Magnet. Reson. Imag.*, 2001, **19**, 521.

33. T. M. Hyde, L. F. Gladden, and R. Payne, *J. Controlled Release*, 1995, **36**, 261.

34. C. Ferrero, D. Massuelle, D. Jeannerat, and E. Doelker, *J. Controlled Release*, 2008, **128**, 71.

35. W. E. Baille, C. Malveau, X. X. Zhu, and R. H. Marchessault, *Biomacromolecules*, 2002, **3**, 214.

36. J. Tritt-Goc and N. Pislewski, *J. Controlled Release*, 2002, **80**, 79.

37. H. Metz and K. Mader, *Int. J. Pharmaceut.*, 2008, **364**, 170.

38. V. Malaterre, H. Metz, J. Ogorka, R. Gurny, N. Loggia, and K. Mader, *J. Controlled Release*, 2009, **133**, 31.

39. S. Strubing, T. Abboud, R. V. Contri, H. Metz, and K. Mader, *Eur. J. Pharmaceut. Biopharmaceut.*, 2008, **69**, 708.

40. S. Strubing, H. Metz, and K. Mader, *J. Controlled Release*, 2008, **126**, 149.

41. T. Tajiri, S. Morita, R. Sakamoto, M. Suzuki, S. Yamanashi, Y. Ozaki, and S. Kitamura, *Int. J. Pharm.*, 2010, **395**, 147.

42. P. P. Dorozynski, P. Kulinowski, A. Mlynarczyk, and G. J. Stanisz, *Drug Discov. Today*, 2012, **17**, 110.

43. P. R. Laity, M. D. Mantle, L. F. Gladden, and R. E. Cameron, *Eur. J. Pharmaceut. Biopharmaceut.*, 2010, **74**, 109.

44. S. Baumgartner, G. Lahajnar, A. Sepe, and J. Kristl, *Eur. J. Pharmaceut. Biopharmaceut.*, 2005, **59**, 299.

45. C. A. Fyfe and A. I. Blazek, *Macromolecules*, 1997, **30**, 6230.

46. K. Mader, Y. Cremmilleux, A. J. Domb, J. F. Dunn, and H. M. Swartz, *Pharm. Res.*, 1997, **14**, 820.

47. K. Mader, G. Bacic, A. Domb, O. Elmalak, R. Langer, and H. M. Swartz, *J. Pharm. Sci.*, 1997, **86**, 126.

48. U. Mikac, A. Demsar, F. Demsar, and I. Sersa, *J. Magnet. Reson.*, 2007, **185**, 103.

49. A. Steingoetter, P. Kunz, D. Weishaupt, K. Mader, H. Lengsfeld, M. Thumshirn, P. Boesiger, M. Fried, and W. Schwizer, *Alimentary Pharmacol. Ther.*, 2003, **18**, 713.

50. A. Steingoetter, D. Weishaupt, P. Kunz, K. Mader, H. Lengsfeld, M. Thumshirn, P. Boesiger, M. Fried, and W. Schwizer, *Pharm. Res.*, 2003, **20**, 2001.

51. A. Steingoetter, D. Weishaupt, P. Kunz, K. Mader, H. Lengsfeld, M. Thumshirn, M. Fried, P. Boesiger, and W. Schwizer, *Gastroenterology*, 2002, **122**, A337.

52. M. Knorgen, R. P. Spielmann, A. Abdalla, H. Metz, and K. Mader, *Eur. J. Pharmaceut. Biopharmaceut.*, 2010, **74**, 120.

53. S. Kempe, H. Metz, P. G. C. Pereira, and K. Mader, *Eur. J. Pharmaceut. Biopharmaceut.*, 2010, **74**, 102.

54. H. Nitzsche, H. Metz, A. Lochmann, A. Bernstein, G. Hause, T. Groth, and K. Mader, *Tissue Eng. C Methods*, 2009, **15**, 513.

55. G. Shiko, L. F. Gladden, A. J. Sederman, P. C. Connolly, and J. M. Butler, *J. Pharm. Sci.*, 2011, **100**, 976.

FURTHER READING

Pharmaceutics

M. E. Aulton, Pharmaceutics: The Design and Manufacture of Medicines, 3rd edn, Churchill Livingstone: Edinburgh, 2007.

D. Jones, Pharmaceutics – Dosage Form and Design, Pharmaceutical Press: London, Chicago, 2008.

C.-J. Kim, Advanced Pharmaceutics: Physicochemical Principles, Boca Raton: CRC Press, 2004.

Y. Perrie and T. Rades, Pharmaceutics – Drug Delivery and Targeting, Pharmaceutical Press: London, Chicago, 2010.

MRI

M. A. Brown and R. C. Semelka, MRI: Basic Principles and Applications, 4th edn, Wiley-Blackwell: Oxford, 2010.

S. C. Bushong, Magnetic Resonance Imaging: Physical and Biological Principles, St. Louis: Mosby, 2003.

P. T. Callaghan, *Principles of NMR Microscopy*, Oxford University Press: Oxford, 1994.

D. W. McRobbie, E. A. Moore, M. J. Graves, and M. R. Prince, MRI: From Picture to Proton, Cambridge University Press: Cambridge, 2007.

D. G. Mitchell and M. S. Cohen, MRI Principles, 2nd edn, Saunders: London, 2004.

D. Weishaupt, V. D. Kochli, and B. Marincek, How Does MRI Work?, Springer: Berlin, London, 2003.

PART E
Clinical Development

Chapter 24

NMR-based Metabolic Phenotyping for Disease Diagnosis and Stratification

Beatriz Jiménez

Clinical Phenotyping Centre, Division of Computational and Systems Medicine, Department of Surgery and Cancer, Imperial College London, London SW7 2AZ, UK

24.1 INTRODUCTION

There exists a dynamic network of metabolic pathways in the human body that are interconnected to an individual's genome and proteome. Metabolite concentrations in tissues and biofluids are largely determined by enzymes and other regulatory molecules that are themselves interlinked with various cellular processes including transcription, translation, and protein–protein interactions. Factors such as diet, lifestyle, host–microbe interaction, and disease pathology greatly influence metabolism. Deviation from a normal, healthy phenotype to a disease state will similarly alter local and system biochemistry. Thus, metabolites, either as intermediates or products, can provide important insight into the functional biochemical processes of a living system and its phenotype. Accurate and detailed analysis of the metabolite composition of tissues and biofluids can be used for the classification of an individual's physiological state using an approach termed *metabonomics*. Formally, metabonomics is defined as the 'the quantitative measurement of the dynamic multiparametric metabolic response of living systems to pathophysiological stimuli or genetic modification'.[1] The term *metabolomics* was also coined in the field of plant biology[2] and in relation with the term *metabolome* that was defined as 'the total complement of metabolites in a cell, to look at global shifts in cellular function under different growth conditions'.[3] Nowadays, the two concepts are used interchangeably; however, the main difference between them is that metabonomics has an intrinsic time-related and holistic component as it deals with identifying and quantifying time-related metabolic changes in a biological system instead of characterizing an individual cell at a set time point. NMR-based metabolic phenotyping has been used for a wide range of clinical applications including the study of disease mechanism and as a powerful tool for the diagnosis of cancer,[4,5] coronary heart disease,[6] liver disease,[7] and diseases of the nervous system.[8] Moreover, it has been successfully utilized for examining interacting systems such as between humans, their nutrition, and their symbiotic gut microflora,[9] prediction of drug response,[10] and large-scale epidemiological studies.[11] While various platforms permit the analysis of metabolites, NMR

NMR in Pharmaceutical Sciences. Edited by Jeremy R. Everett, John C. Lindon, Ian D. Wilson, and Robin K. Harris
© 2015 John Wiley & Sons, Ltd. ISBN: 978-1-118-66025-6
Also published in eMagRes (online edition)
DOI: 10.1002/9780470034590.emrstm1394

spectroscopy and mass spectrometry (MS) have emerged as the analytical tools of choice for metabolic phenotyping. NMR spectroscopy is particularly advantageous owing to its speed, high reproducibility, and capacity to simultaneously analyze and quantify a wide variety of small chemical compounds without the need of extraction and separation techniques. As a result, NMR-based 'metabolic phenotyping' or 'metabotyping'[12] is being increasingly used as a tool for improving understanding of the human disease process, for stratifying patient populations on the basis of their likelihood to respond to a therapeutic strategy or not, as well as for patient diagnostic and prognostic purposes. This chapter explores the conceptual and practical aspects of experimental design, data analysis, and technological requirements particularly relevant to NMR spectroscopy-based metabotyping. Recent examples of NMR-based metabotyping in a clinical setting are provided to showcase how this technique is contributing to the field of precision medicine.

24.2 NMR SPECTROSCOPY AND METABOLIC PHENOTYPING

24.2.1 Human Metabolism and the Metabotype

Cellular metabolism results in the production of biochemical endpoints that are typically excreted from the cell into the surrounding tissue and the circulatory system. Changes in energy substrate availability or oxygen supply are two obvious examples that would precipitate a shift in intracellular metabolic pathways and subsequently shift the metabolic profile of a cell and its surrounding tissue. However, human metabolism is influenced by both static and dynamic factors including the host genome, diet, drugs, and other environmental factors as well as host–microbe interactions.[9] Moreover, metabolic outputs are closely interfaced with circadian physiology and the central nervous system.[13] Therefore, understanding and defining basic human metabolic homeostasis is a challenging, yet critical component of patient metabotyping as it permits the detection of dysfunctional metabolism that accompanies pathophysiology. For example, tumor cells require enhanced energy support for their development and proliferation. This involves metabolic changes within the tumor cell itself that are reflected in its local tissue environment and, ultimately, circulating biofluids. Consistent with this, decreasing the availability

of glucose and biosynthetic precursors to tumor cells through enzymatic modulation of tumor metabolism reduces cell proliferation and tumor expansion rates.[14] The capacity to detect and distinguish such metabolic deviations from the norm in tissue and biofluid samples and relate them to pathophysiology and phenotype lies at the heart of metabonomics and metabotyping.

24.2.2 Experimental Design and Sampling Considerations for Metabotyping

An individual's metabotype essentially consists of a unique chemical signature measurable in a clinical sample by one or more analytical techniques. The general workflow followed for NMR-based metabotyping involves patient recruitment, sample collection and preparation, spectral acquisition, data processing, chemometric modeling of the data, and biomarker identification. However, before delving deeper into the practical and technical aspects of NMR spectroscopy as used for metabotyping, it is worthwhile considering the type of biochemical information that can be derived from specific clinical samples, and how this information can be harnessed for defining health status and subsequently informing the clinical decision-making process.

24.2.2.1 Sample Types, Collection, and Preparation

Normal cellular metabolism is in a constant state of flux modulated by signaling and transport molecules often functioning in an autocrine or paracrine manner. Utilization of energy substrates and the end products of metabolic reactions is collectively reflected in the surrounding tissue. As such, metabonomic analysis of tissue biopsies provides localized information that can be used for elucidating disease mechanism as well as for diagnostic and prognostic applications.[15] Tissue biopsies are typically obtained during surgery or as part of routine clinical practice; yet practical and ethical limitations mean that appropriate samples are not always readily available for analysis. Blood and urine represent the two most accessible and widely used clinical samples for metabolic phenotyping. However, the biological information contained within each differs considerably. Blood samples – analyzed as serum or plasma preparations – provide 'snap-shot' data of systemic metabolism. In contrast, urine is stored for periods of time in the bladder before being sampled and therefore provides time-averaged data on metabolism. In some cases, urine sampling is performed across a 24 h collection time period to account for circadian events. The biochemical properties of the two fluids also differ markedly. While blood has relatively stable pH and ionic strength and contains both small and large molecular weight molecules, urine samples exhibit variable pH and ionic strength and typically contain only small molecules and compounds <2 kDa. NMR spectroscopy-based metabotyping can also be readily undertaken using other biofluids (e.g., saliva, cerebrospinal fluid, amniotic fluid, gastric fluids, and bile) or even 'artificial' biofluids (e.g., bronchiolar lavage fluid, fecal water, and cell extracts). Although these samples can reveal important information regarding local biochemistry, practical and ethical issues often hamper their clinical accessibility. As these samples are not routinely assessed in metabolic phenotyping studies, focus will remain on tissue, blood, and urine analyses.

Metabolic phenotyping is based on the assumption that the metabotype defined for a particular clinical sample is representative of its in vivo state. As tissue samples and most biofluids contain proteins and enzymes that remain active after collection, care must be taken to prevent post-sampling metabolic activity. Depending on sample type, this is usually achieved by quenching metabolic activity using rapid deep freezing. For this reason, sample collection and processing should follow defined standard operating procedures (SOPs). Detailed protocols for biofluid (urine, serum/plasma)[16] and tissue[17] sample collection and preparation, including the extraction of polar and lipophilic metabolites from tissues and accompanying ^{1}H NMR spectroscopic techniques have been published. An updated set of protocols for human biofluid (blood/urine) has been recently presented.[18] These revised SOPs consider key aspects of NMR spectroscopy-based metabolic phenotyping including sample preparation, spectrometer parameters, NMR pulse sequences, throughput, reproducibility, and quality control measures.

24.2.3 NMR Metabotyping Workflow

High-resolution NMR spectroscopy is a powerful analytical platform for phenotyping. Although the sensitivity of NMR spectroscopy, with detection limits in the low micromolar range, is less than that of MS (femtomolar), NMR spectroscopy is advantageous;

in that, it is an inherently quantitative technique, is nondestructive, permits the characterization of organic molecules regardless of their physicochemical properties, and provides detailed information on solution-state molecular structures. Of importance and relevance to clinical metabotyping applications, it is highly reproducible and requires small sample volumes (10–600 μl) and minimal sample preparation. As samples are placed into a sample holder (glass tube for biofluids and a ZrO_2 rotor in the case of tissues), they never come into direct contact with any part of the spectrometer meaning there is no problem with instrument contamination and different sample types can be sequentially assessed. NMR spectroscopic approaches can also be used to investigate metabolite molecular dynamics and mobility through the use of NMR spin relaxation and molecular diffusion experiments.[19] Technological advances in probe design, as well as software development, have led to increased sensitivity of ^{1}H and ^{13}C acquisition. This coupled with improved automation of setup routines (e.g., tuning, matching, and shimming) means accurate metabolic profiles for biofluids can now be acquired within 4–5 min of data acquisition time. The range of detectable metabolic features is broad and includes amino acids, carbohydrates, Krebs' cycle metabolites, low-molecular-weight organic acids, organic amines, fatty acids, eicosanoids, steroids, bile acids, phospholipids, ceramides, triglycerides, peptides, and vitamins, among others.

A typical ^{1}H NMR spectrum of a biofluid contains thousands of resonances predominantly low molecular weight metabolites.[20,21] Examples of representative metabolic profiles for clinical samples are provided in Figure 24.1. Blood plasma and serum samples contain both low and high molecular weight compounds, the latter seen as broader width signals owing to their faster transverse (R_2) relaxation and overlap of peaks. Further details of specific NMR experiments used for metabotyping are provided below in the relevant sections.

24.2.3.1 *Statistical Modeling of NMR-derived Metabolic Data and Metabolite Identification*

Once a full set of NMR experiments has been obtained for a patient population, the spectra undergo automated processing (Fourier transformation, phasing, baseline correction, and calibration) and quality control assessment before then being converted into

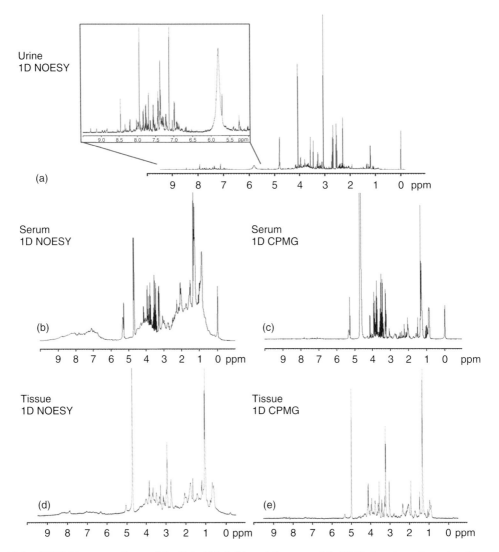

Figure 24.1. ^{1}H NMR spectra of accessible clinical biofluids and tissues used for NMR-based metabotyping. Representative ^{1}H NMR spectra of (a) human urine acquired using a 1D NOESY pulse sequence, (b) human plasma using a 1D NOESY pulse sequence, and (c) a 1D CPMG pulse sequence to filter out signals coming from fast relaxing protons derived from macromolecules. (d) ^{1}H NMR spectra from human colorectal mucosa tissue obtained using a NOESY pulse sequence and an HR-MAS NMR probe and (e) using a 1D CPMG pulse sequence. All the pulse sequences include a presaturation pulse during the relaxation period to suppress the water signal

numerical values, typically by integration of resonance peaks, converting the spectra into specific regions using a process known as *bucketing* or *binning*, or even nowadays using the full-resolution spectral data points. Multivariate statistical approaches, often referred to as chemometrics analyses, can then be used for data reduction, pattern recognition, and the construction

of predictive models[22] (although the rigorous use of univariate statistics is also possible and is popular in epidemiological studies). This approach enables the identification of metabolite profile patterns (sometime referred to as fingerprints) that describe a phenotype. A combination of unsupervised (e.g., principal components analysis) and supervised statistical approaches (e.g., orthogonal partial least squares) is typically employed. Parameters describing the goodness of 'fit' of the predictive models (e.g., Q^2 value) generated by the statistical analysis are often presented. However, the model validation is best achieved using an independent set of samples to test the model built from the training set. Specificity and selectivity can then be easily calculated by statistical assessment of the number of blind samples correctly classified or not by the devised model.[23]

In order to provide some insight about the disease mechanism and reveal biomarkers, it is desirable and often essential to unambiguously identify the signals in the data that are characteristic of the disease metabotype (biomarkers). The NMR spectrum contains important structural information that resides in both the chemical shifts of the target molecule, its coupling constants, and peak integrals. Although complex, many resonances in ^{1}H NMR spectra of biofluids can be directly assigned based on their chemical shifts and signal multiplicities, and further information can be obtained by using spectral editing techniques and 2D NMR spectroscopy. Extensive online databases also exist that provide NMR spectral data acquired under different conditions for thousands of metabolites often observed in human tissue and biofluid samples. Such databases are useful starting points for the metabolite identification process and some of the more recognized ones include the 'Human Metabolome Database' (http://www.hmdb.ca/) and the 'Biological Magnetic Resonance Data Bank' (http://www.bmrb.wisc.edu/). Many laboratories also utilize commercially available databases such as S-BASE (Bruker Biospin) and Chenomx (Chenomx Inc) and it is increasingly common to construct bespoke in-house spectral libraries.

24.2.3.2 NMR-based Metabotyping of Biofluids

As touched upon earlier, metabotyping studies by NMR spectroscopy have largely been conducted on urine and plasma samples and accordingly, the majority of methods for NMR spectroscopy-based metabonomics have been developed for these two biofluids. Sample preparation procedures for metabolic profiling studies were first described in 2007 as part of the Standard Metabolomic Reporting Structures working group.[16] Since then hundreds of studies have shown that by adhering to those guidelines, it is possible to reduce sampling artifacts and maintain consistency with other metabonomics studies

in the field. Following technological advancements in both NMR machinery and software, alongside the establishment of 'Phenome Centers' that increasingly rely on automation of sample processing, data acquisition, and analysis, updated guidelines have recently been developed for standardizing experimental procedures for high-throughput NMR-based metabotyping where interstudy and interlaboratory comparisons are of critical importance.[18] These guidelines have been specifically developed for human urine and blood-derived samples but still utilize the classical cassette of NMR experiments used for metabonomics. Firstly, 1D NOESY-presat experiments and J-res 2D experiments are run in automation at 300 K for urine and 310 K for plasma or serum in a Bruker Avance III 600 spectrometer working at 14.1 T equipped with a BBI probe (these are inverse broadband high-resolution probes, the inner coil of the probe is optimized to observe ^{1}H, and the outer coil is tunable to acquire or decouple a wide range of nuclei from ^{31}P to ^{109}Ag). The number of scans has been decreased to 32 free-induction decays (FID) to ensure a quicker turnover per sample, while the majority of parameters have been kept the same as in the previous method.[16] For plasma samples, it is necessary to acquire the additional relaxation-edited Carr–Purcell–Meiboom–Gill (CPMG) spin-echo sequence experiment to eliminate signals belonging to proteins and other macromolecules in the ^{1}H NMR spectrum and to be able to observe mainly signals belonging to metabolites and small molecules. In addition, a diffusion-edited experiment to highlight macromolecule peaks can be acquired for serum and plasma samples, but this depends on the information required for the data set. Technological improvements to both spectrometer hardware and software enable processing of the spectra (Fourier transformation, phasing, baseline correction, and calibration) to be performed automatically using TopSpin 3.2 (Bruker Coorporation, Germany). Using these experimental guidelines, each urine sample will spend around 15 min inside the magnet, whereas each blood sample will spend 19 min. A third of this time involves temperature equilibration, locking, tuning, and matching as well as shimming. With this procedure, 90 urine samples and 70 blood samples can be analyzed per day.

The main challenge faced by NMR spectroscopists working with biological samples is the removal or suppression of the water signal without affecting any other signals in the sample. By using a 1D NOESY-presat pulse sequence for the acquisition of a simple 1D ^{1}H spectrum, is possible to include in the mixing time of the pulse sequence a low power pulse that is applied on the water resonance to improve suppression of the water signal. A pulse sequence version with gradients is used so this mixing time can be kept very short (10 ms) and the loss of signal due to relaxation can be decreased as much as possible. The residual water signal in this experiment should have a line width of about 0.1 ppm and less intensity than a signal derived from a metabolite containing an equivalent proton concentration of around 10 mM.

It is increasingly recognized that successful high-throughput NMR spectroscopy for metabotyping applications relies heavily on an exhaustive experimental setup designed to assess the performance of the spectrometer before the analysis of each sample set. Calibration of temperature, water suppression, and the inclusion of an external reference are essential for obtaining consistent, reproducible, spectroscopic data and for accurate quantitative measurements of biomarkers from the NMR spectra. The use of internal reference compounds can be dangerous because they often interact with other molecules in the sample or add new signals in an already crowded NMR spectrum. An external reference sample for quantification provides the best solution for the quantification issue by enabling a regression curve to be created permitting more accurate estimation of target metabolite concentrations without disturbing the metabolic profile of the origin sample. Specifically, the ERETIC method (electronic reference to access in vivo concentrations) introduces a reference signal, synthesized by an electronic device, which can be used for the determination of absolute concentrations.[24] The identification and quantification of metabolite biomarkers in metabotyping studies is of key importance as it facilitates translation to the clinic. Once a true biomarker is determined, similar measurements can be obtained using different field strength spectrometers or even other analytical platforms.

Comprehensive metabolic phenotyping of biofluids by ^{1}H NMR spectroscopy is increasingly being used for biomarker discovery and for patient diagnosis and prognosis in a wide variety of biomedical fields. For example, in the field of oncology, ^{1}H NMR spectroscopic metabolic profiling of serum samples from patients suffering metastatic colorectal cancer (mCRC) can be used not only as a predictor of the disease but also as a marker for overall survival among these patients. Serum metabolic profiles of

mCRC patients were shown to have characteristically lower levels of lactate and polyunsaturated lipids and increased levels of 3-hydroxybutyrate characteristic of an altered energetic pathway and an inflammatory status.[25] Serum metabolic profiles of patients suffering early breast cancer have also been recently studied by ^{1}H NMR spectroscopy and shown to be markedly different to those suffering metastatic tumors.[26] The authors identified nine discriminatory metabolites including histidine, acetoacetate, glycerol, pyruvate, glycoproteins (N-acetyl), mannose, glutamate, phenylalanine, and 3-hydroxybutyrate that could be related with the high energetic demand of cells due to tumor proliferation. The NMR spectroscopic serum metabolic profile has also proven to be useful in the stratification of patients suffering different molecular subtypes of leukemia.[4] Serum NMR spectral metabolic phenotyping has also shown great promise in clinical cardiovascular applications. The proven relationship existing between cardiovascular disease and lipoproteins contained in blood can be readily assessed and analyzed quantitatively from the ^{1}H NMR spectroscopic serum metabolic profile.[27]

24.2.3.3 NMR-based Tissue Metabotyping

High-resolution magic angle spinning (HR-MAS) NMR spectroscopy is a technique that enables metabolic profiles of intact, solid tissue samples to be acquired within minutes. By spinning tissue samples in a specially designed rotor at an angle of 54.7° relative to the applied magnetic field (the term 'magic angle'), line-broadening effects that would ordinarily obscure ^{1}H spectra are substantially reduced.[28] As such, this platform is receiving increasing attention for its applicability to biochemical characterization of tissue samples in a clinical setting. To this extent, it has been recently shown how HR-MAS NMR spectroscopic metabolic profiling of tumor biopsies obtained from colorectal cancer patients can be used for rapid tumor classification and prognostication in a hospital environment.[15,29]

As touched upon in Section 24.2.3.2 in order to produce high-throughput and high-quality spectroscopic data useful for metabotyping, it is necessary to have highly automated protocols for experiment acquisition. Efforts to address this issue have begun with the development of a robot for sample exchanging (Sample Pro, Bruker Biospin GmbH) that enables bulk loading of up to 48 tissue samples that are kept at −20 °C

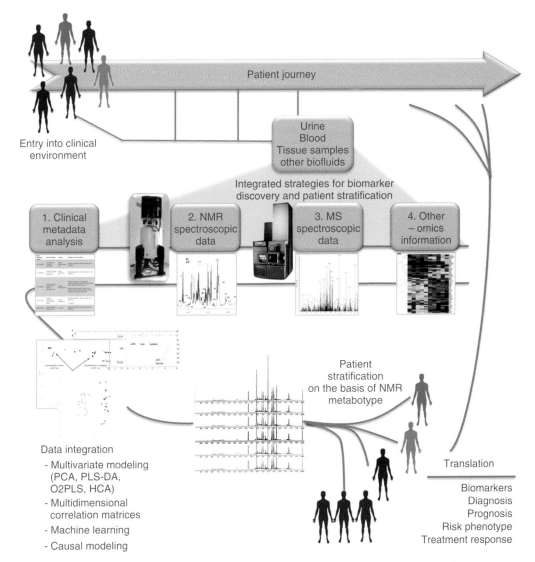

Figure 24.2. Schematic describing the patient population journey and how the implementation of metabonomics into clinical practice can help in decision-making, diagnosis, prognosis, and patient stratification. Metabolic phenotypes or 'metabotypes' of patients are obtained using biofluids and tissue samples collected at different time points of the clinical journey. Samples can be analyzed using different analytical platforms (NMR, MS, and any other suitable omics technique) and the spectra will be modeled on the basis of available metadata. Once the chemometrics analysis is performed, different metabotypes can be used for patient stratification. This process aims to be iterative and patient follow-up would be performed to assess disease evolution or treatment response

to ensure the integrity of the sample. The use of disposable Teflon inserts permits immediate storage of the tissue after measurement and facilitates the cleaning of the rotors. However, sample exchanging remains the only process sufficiently automatized for HR-MAS NMR spectroscopy. Automated setup routines for tuning, matching, and shimming have yet to be developed and implemented for the high-resolution solid probes. As such, the high standards set for the metabotyping analyses of biofluids cannot yet be extended to the analysis of intact tissues. The reader is encouraged to refer to published protocols for further experimental and methodological details for HR-MAS analysis of tissue samples.[17]

The optimal sample size for a 4 mm HR-MAS rotor is around 10 mg of tissue. However, in many cases, this may be challenging to obtain when biopsy tissue is limited and the available tissue required examination via different techniques (e.g., histopathological assessment and targeted clinical assays). This exposes a potential limitation of HR-MAS NMR spectroscopy-based metabotyping. This has begun to be addressed through the use of a magic-angle coil spinning (MACS) microcoil that permits metabonomics analysis of as little as 500 ng of muscle tissue.[30] The MACS microcoil is a small resonant solenoid coil that can be remotely excited and detected using resonant-inductive coupling to commercial HR-MAS probes. Owing to the fact that the microcoil rotates together with the sample at the magic angle inside the MAS rotor, high-resolution spectra of weight-limited tissues can be acquired. As such it is envisaged that this approach will provide a new avenue for single cell metabonomic analysis, improve analysis of heterogeneous tissue samples, and enable targeted metabolic mapping of tissue biopsy regions.

24.2.4 The Role of Phenome Centers and Datasharing in Metabotyping

Metabotyping of accessible tissue and biofluids provides an objective assessment of the biochemical interactions among host genes, lifestyle, environmental factors, and critically, how these biochemical interactions change in response to disease processes. As a result, there is increasing interest in metabotyping approaches for population-based disease risk studies as well as for patient stratification strategies (Figure 24.2).[31] Such investigations require the analysis of large sample cohorts using appropriate technology platforms that follow validated, standardized procedures and protocols ensuring robust and reproducible data acquisition. Dedicated Phenome Centers provide an optimal environment and infrastructure for such large-scale studies.

The UK MRC-NIHR National Phenome Centre (NPC) is the first of these centers in the world to be established and provides a potential model for similar centers. Opened in 2013, the NPC is funded to Imperial College London and Kings College London

by the Medical Research Council (MRC), the National Institute for Health Research (NIHR) as well as leading NMR (Bruker Biospin GmbH) and MS (The Waters Corporation) manufacturers. The remit of the NPC is to conduct metabolic phenotyping studies on large cohorts of samples derived from epidemiological studies to assess metabolic biomarkers of disease risk in populations. It complements the Imperial Clinical Phenotyping Centre (CPC) located at St. Mary's Hospital Paddington, London, which undertakes studies directed toward metabolic phenotyping individual patient journeys with the aim of improving patient diagnostics and prognostics and therapy management.[32] These centers will also play important roles in the development of improved protocols for fully automated high-throughput NMR and MS experimentation. By establishing SOP-based methods and robust experimental procedures, large-scale data cohorts collected within different laboratories will be compatible and comparable. Such efforts are critical for the formation of worldwide resources designed to support metabolic phenotype data sharing. In Europe, the European Institute for Bioinformatics (EBI), which is part of the European Molecular Biology Laboratory (EMBL), has recently launched MetaboLights – an open source database designed for uploading metabonomics data sets and linked metadata.[33] A similar online resource funded by the US Common Fund of the National Institutes of Health entitled Metabolomics Workbench has also been recently launched.[34] Such databases are valuable resources for future metabotyping studies but care must be taken to ensure deposited data follow established standardize protocols for sample preparation and experimental acquisition.

24.2.5 The Future, New Avenues, and Challenges

NMR-based metabotyping is a powerful tool that has wide reaching clinical applications; however, future challenges remain to ensure its full potential is realized. This can be categorized into two main areas; firstly, the challenge of introducing NMR metabotyping into routine clinical practice and secondly, addressing technical shortcomings that limit the amount of biochemical information that can be detected using current methodologies. NMR spectroscopy-based metabotyping offers the opportunity to introduce a more objective clinical decision-making process that is driven by relevant biochemical and molecular information. Undoubtedly, a future challenge will be to convert the opinions and practices of clinicians. One way that this can be approached is by engaging with, and working alongside, clinical researchers and medical practitioners in a complementary manner that promotes transfer of knowledge and by undertaking studies designed by clinicians that are driven from a research perspective as to provide scientifically relevant results that can be then fed back and translated into general clinical practices.

A number of technical issues still hamper NMR spectroscopy-based profiling techniques. The most obvious of these is sensitivity and resolution. Current NMR spectroscopy-based phenotyping is generally conducted using 14.1 T magnets, i.e., at 600 MHz for ^{1}H observation. It is feasible that future phenotyping will be performed using higher field magnets that would improve sensitivity. Present NMR spectroscopy-based metabotyping approaches are limited to the analysis of one-dimensional ^{1}H spectra for the creation of metabolic profiles or fingerprints even though important metabolite information can also be obtained from ^{13}C spectra.[35] Other fields of NMR spectroscopy have addressed this issue of limited resolution by utilizing multidimensional NMR experiments. There exists great potential for establishing multidimensional NMR spectral methodologies for NMR-phenotyping approaches that would ultimately permit the generation of multidimensional metabolic fingerprints. This would not only address the issue of resolution but also enhance structural information obtained during data acquisition useful for metabolite identification. Identification of discriminatory metabolites remains a bottleneck in NMR spectroscopy-based phenotyping studies; however, recently developed multivariate statistical approaches such as STORM (subset optimization by reference matching) may help address this by permitting the recovery of metabolite structure information from multiple ^{1}H NMR spectra in population sample sets.[36]

Apart from the issue of sensitivity and resolution, the fact remains that NMR spectra harbor much relevant biological information that current methodologies are unable to extract and utilize. For example, pH and ionic strength variation between samples is largely removed during sample preparation through the addition of buffers. While this reduces chemical shift variation among spectra and thus facilitates downstream peak alignment and signal pattern recognition, it may mask physiochemical properties that

could reflect biologically meaningful differences between samples. For example, it is well known that tumor cells, because of their high glucose intake, acidify the intracellular environment through the excretion and accumulation of lactate.[37]

Metabolic flux analysis is essential to understand mechanisms involved in regulation of metabolic activity. Stable isotopes (e.g., ^{13}C or ^{15}N) can be used to label target substrates such as glucose or amino acids. Metabolism of the labeled compound can then be traced using NMR spectroscopy. This approach has been used in cell culture models to study cancer cell metabolism[38] and for examining brain metabolism of glucose and glutamine using in vivo MRI.[39] Historically, a limiting factor of studying natural abundance ^{13}C by NMR is sensitivity. Hyperpolarization techniques have helped overcome this using dynamic nuclear polarization (DNP) to generate hyperpolarized ^{13}C substrates. Moreover, the development of a dissolution process that facilitates injection of such hyperpolarized ^{13}C materials into living subjects has advanced the clinical potential of in vivo magnetic resonance spectroscopy, although to date the number of useful reagents is limited by the requirement of long relaxation times, otherwise the hyperpolarization is too short lived. The fact that flux analysis requires an intervention before the sample collection has likely hindered its applicability to metabotyping, yet it undoubtedly provides mechanistic information that in the future could be utilized to help define metabolic pathway activity related to phenotype.

24.3 CONCLUSION

NMR spectroscopy-based metabotyping is a powerful tool for examining detailed interactions among an individual's genome, their lifestyle, and response to environmental stimuli. Using NMR spectroscopy, detailed biochemical information can be acquired in a high-throughput manner from clinically attainable samples that can help inform the clinical decision-making process. It has opened new avenues for rapid diagnostics and prognostics and for assessing population-wide disease risks and the stratification of patients on the basis of disease etiology and potential treatment response. We are now at the stage of developing robust, standardized protocols for high-throughput metabotyping that will be translatable across laboratories ensuring meaningful data sharing and ultimately translation into a clinical setting.

ACKNOWLEDGMENTS

The author wishes to acknowledge members and colleagues from the Imperial Clinical Phenotyping Centre and the MRC-NIHR National Phenome Centre and the Division of Computational and Systems Medicine. Dr David A MacIntyre (Imperial College London) is also acknowledged for invaluable critique and feedback provided during the preparation of this article. The research was supported by the National Institute for Health Research (NIHR) Biomedical Research Centre based at Imperial College Healthcare NHS Trust and Imperial College London. The views expressed are those of the author and not necessarily those of the NHS, the NIHR, or the Department of Health.

REFERENCES

1. J. K. Nicholson, J. C. Lindon, and E. Holmes, *Xenobiotica; the Fate of Foreign Compounds in Biological Systems*, 1999, **29**, 1181.

2. O. Fiehn, *Plant Mol. Biol.*, 2002, **48**, 155.

3. H. Tweeddale, L. Notley-McRobb, and T. Ferenci, *J. Bacteriol.*, 1998, **180**, 5109.

4. D. A. MacIntyre, B. Jiménez, E. J. Lewintre, C. R. Martin, H. Schafer, C. G. Ballesteros, J. R. Mayans, M. Spraul, J. Garcia-Conde, and A. Pineda-Lucena, *Leukemia*, 2010, **24**, 788.

5. J. L. Spratlin, N. J. Serkova, and S. G. Eckhardt, *Clin. Cancer Res. Off. J. Am. Assoc. Cancer Res.*, 2009, **15**, 431.

6. J. T. Brindle, H. Antti, E. Holmes, G. Tranter, J. K. Nicholson, H. W. Bethell, S. Clarke, P. M. Schofield, E. McKilligin, D. E. Mosedale, and D. J. Grainger, *Nat. Med.*, 2002, **8**, 1439.

7. B. Jiménez, C. Montoliu, D. A. MacIntyre, M. A. Serra, A. Wassel, M. Jover, M. Romero-Gomez, J. M. Rodrigo, A. Pineda-Lucena, and V. Felipo, *J. Proteome Res.*, 2010, **9**, 5180.

8. R. Kaddurah-Daouk and K. R. Krishnan, *Neuropsychopharmacol.: Off. Publ. Am. Coll. Neuropsychopharmacol.*, 2009, **34**, 173.

9. J. K. Nicholson, E. Holmes, and I. D. Wilson, *Nat. Rev. Microbiol.*, 2005, **3**, 431.

10. T. A. Clayton, J. C. Lindon, O. Cloarec, H. Antti, C. Charuel, G. Hanton, J. P. Provost, J. L. Le Net, D. Baker, R. J. Walley, J. R. Everett, and J. K. Nicholson, *Nature*, 2006, **440**, 1073.

11. J. C. Lindon, H. C. Keun, T. M. Ebbels, J. M. Pearce, E. Holmes, and J. K. Nicholson, *Pharmacogenomics*, 2005, **6**, 691.

12. C. L. Gavaghan, I. D. Wilson, and J. K. Nicholson, *FEBS Lett.*, 2002, **530**, 191.

13. K. L. Eckel-Mahan, V. R. Patel, R. P. Mohney, K. S. Vignola, P. Baldi, and P. Sassone-Corsi, *Proc. Natl. Acad. Sci. U. S. A.*, 2012, **109**, 5541.

14. D. Anastasiou, Y. Yu, W. J. Israelsen, J. K. Jiang, M. B. Boxer, B. S. Hong, W. Tempel, S. Dimov, M. Shen, A. Jha, H. Yang, K. R. Mattaini, C. M. Metallo, B. P. Fiske, K. D. Courtney, S. Malstrom, T. M. Khan, C. Kung, A. P. Skoumbourdis, H. Veith, N. Southall, M. J. Walsh, K. R. Brimacombe, W. Leister, S. Y. Lunt, Z. R. Johnson, K. E. Yen, K. Kunii, S. M. Davidson, H. R. Christofk, C. P. Austin, J. Inglese, M. H. Harris, J. M. Asara, G. Stephanopoulos, F. G. Salituro, S. Jin, L. Dang, D. S. Auld, H. W. Park, L. C. Cantley, C. J. Thomas, and M. G. Vander Heiden, *Nat. Chem. Biol.*, 2012, **8**, 839.

15. B. Jiménez, R. Mirnezami, J. Kinross, O. Cloarec, H. C. Keun, E. Holmes, R. D. Goldin, P. Ziprin, A. Darzi, and J. K. Nicholson, *J. Proteome Res.*, 2013, **12**, 959.

16. O. Beckonert, H. C. Keun, T. M. Ebbels, J. Bundy, E. Holmes, J. C. Lindon, and J. K. Nicholson, *Nat. Protoc.*, 2007, **2**, 2692.

17. O. Beckonert, M. Coen, H. C. Keun, Y. Wang, T. M. Ebbels, E. Holmes, J. C. Lindon, and J. K. Nicholson, *Nat. Protoc.*, 2010, **5**, 1019.

18. A. C. Dona, B. Jiménez, H. Schäefer, E. Humpfer, M. Spraul, M. R. Lewis, J. T. M. Pearce, E. Holmes, J. C. Lindon, and J. K. Nicholson, *Anal. Chem.*, 2014, **46**, 9887.

19. M. Liu, J. K. Nicholson, and J. C. Lindon, *Anal. Chem.*, 1996, **68**, 3370.

20. J. K. Nicholson, P. J. Foxall, M. Spraul, R. D. Farrant, and J. C. Lindon, *Anal. Chem.*, 1995, **67**, 793.

21. S. Bouatra, F. Aziat, R. Mandal, A. C. Guo, M. R. Wilson, C. Knox, T. C. Bjorndahl, R. Krishnamurthy, F. Saleem, P. Liu, Z. T. Dame, J. Poelzer, J. Huynh, F. S. Yallou, N. Psychogios, E. Dong, R. Bogumil, C. Roehring, and D. S. Wishart, *PLoS One*, 2013, **8**, e73076.

22. J. Trygg, E. Holmes, and T. Lundstedt, *J. Proteome Res.*, 2007, **6**, 469.

23. E. Saccenti, H. C. J. Hoefsloot, A. K. Smilde, J. A. Westerhuis, and M. M. W. B. Hendriks, *Metabolomics Off. J. Metabolomic Soc.*, 2014, **10**, 361.

24. L. Dreier and G. Wider, *Magn. Reson. Chem.*, 2006, **44**, S206.

25. I. Bertini, S. Cacciatore, B. V. Jensen, J. V. Schou, J. S. Johansen, M. Kruhoffer, C. Luchinat, D. L. Nielsen, and P. Turano, *Cancer Res.*, 2012, **72**, 356.

26. E. Jobard, C. Pontoizeau, B. J. Blaise, T. Bachelot, B. Elena-Herrmann, and O. Tredan, *Cancer Lett.*, 2014, **343**, 33.

27. N. J. Rankin, D. Preiss, P. Welsh, K. E. Burgess, S. M. Nelson, D. A. Lawlor, and N. Sattar, *Atherosclerosis*, 2014, **237**, 287.

28. J. P. Shockcor and E. Holmes, *Curr. Top. Med. Chem.*, 2002, **2**, 35.

29. R. Mirnezami, B. Jiménez, J. V. Li, J. M. Kinross, K. Veselkov, R. D. Goldin, E. Holmes, J. K. Nicholson, and A. Darzi, *Ann. Surg.*, 2014, **259**, 1138.

30. A. Wong, B. Jiménez, X. Li, E. Holmes, J. K. Nicholson, J. C. Lindon, and D. Sakellariou, *Anal. Chem.*, 2012, **84**, 3843.

31. J. C. Lindon and J. K. Nicholson, *Expert Opin. Drug Metab. Toxicol.*, 2014, **10**, 915.

32. J. K. Nicholson, E. Holmes, J. M. Kinross, A. W. Darzi, Z. Takats, and J. C. Lindon, *Nature*, 2012, **491**, 384.

33. K. Haug, R. M. Salek, P. Conesa, J. Hastings, P. de Matos, M. Rijnbeek, T. Mahendraker, M. Williams, S. Neumann, P. Rocca-Serra, E. Maguire, A. Gonzalez-Beltran, S. A. Sansone, J. L. Griffin, and C. Steinbeck, *Nucleic Acids Res.*, 2013, **41**, D781.

34. http://www.metabolomicsworkbench.org.

35. H. C. Keun, O. Beckonert, J. L. Griffin, C. Richter, D. Moskau, J. C. Lindon, and J. K. Nicholson, *Anal. Chem.*, 2002, **74**, 4588.

36. J. M. Posma, I. Garcia-Perez, M. De Iorio, J. C. Lindon, P. Elliott, E. Holmes, T. M. Ebbels, and J. K. Nicholson, *Anal. Chem.*, 2012, **84**, 10694.

37. A. Schulze and A. L. Harris, *Nature*, 2012, **491**, 364.

38. C. M. Metallo, J. L. Walther, and G. Stephanopoulos, *J. Biotechnol.*, 2009, **144**, 167.

39. J. Kurhanewicz, R. Bok, S. J. Nelson, and D. B. Vigneron, *J. Nucl. Med. Off. Publ. Soc. Nucl. Med.*, 2008, **49**, 341.

Chapter 25

NMR-based Pharmacometabonomics: A New Approach to Personalized Medicine

Jeremy R. Everett

Medway Metabonomics Research Group, University of Greenwich, Chatham Maritime, Kent ME4 4TB, UK

25.1 INTRODUCTION TO PERSONALIZED MEDICINE

Many medicines are ineffective, or even unsafe, in a significant proportion of the patient population. Personalized medicine (also known as individualized medicine or precision medicine) seeks to avoid this issue by delivering the right treatment, including the right drug, to the right patient group at the right time. It

NMR in Pharmaceutical Sciences. Edited by Jeremy R. Everett, John C. Lindon, Ian D. Wilson, and Robin K. Harris.
© 2015 John Wiley & Sons, Ltd. ISBN: 978-1-118-66025-6
Also published in eMagRes (online edition)
DOI: 10.1002/9780470034590.emrstm1395

is important to note that from a drug development perspective, medicine will always be *personalized* rather than fully *personal* or individual. This is because with current technologies and processes, it is impossible to deliver different new drugs to individual patients. All that can be aimed for presently is to define and then target subgroups of the general population for treatment with drug X rather than drug Y, on the basis that that subgroup would enjoy improved efficacy and safety from treatment with that choice. To date, this endeavor has been underpinned by genomic information, which has been used both to link gene variation to disease risk and to the different responses of individuals to drugs. Some would argue that medicine has always been individualized, with clinicians attempting to deliver the best possible treatment for individuals through an assessment of their medical history, environment, weight, age, diet, and genetic factors. For instance, the dosage of a given drug is frequently tailored to the age or weight of a patient. Until recently, however, the genetic input was limited to an assessment of the family history of the patient. Nowadays, with access to cheap, high-throughput genotyping, genetic information can be obtained with unprecedented detail.[1]

Pharmacogenomics (also now interchangeably known as pharmacogenetics) seeks to relate variation in patients' genes with their responses to drug treatment. This science is now 50 years old and is starting to have an impact on clinical practice.[2] For instance, a clinical study of close to 2000 patients with HIV showed that a human lymphocyte antigen (HLA) genetic variant called *HLA-B*5701* was associated with adverse hypersensitivity reactions to the AIDS drug abacavir and that alternatives should be used: this is now included in US treatment guidelines. Another HLA gene variation was found to be responsible for a carbamazepine-induced skin reaction called *Stevens–Johnson syndrome*. In a study of patients experiencing this adverse drug reaction, the carrier frequency of the *HLA-B*1502* genetic variant was 100%, but it was only 3% in carbamazepine-tolerant patients and 9% in the general population. Thus, this is a sensitive and reasonably specific genetic marker. Indeed, this test is now in routine use in hospitals in Taiwan. The selective prescription of carbamazepine, over costlier alternative drugs without the skin reaction problem, to non-*HLA-B*1502* carriers, has been estimated to save their health service around $1 billion per year.[2] Thus, there are some examples of the effective use of pharmacogenomics emerging. However, difficulties remain with the use of genetic information in complex, multifactorial diseases, and a recent analysis[3] of cardiovascular pharmacogenomics studies that were published between 2005 and 2011 showed that of 289 studies reported, the majority (229) were at the basic biomedical research stage [candidate gene or genome-wide association study (GWAS)]. Most of the 229 studies reported positive findings and resulted in a total of 220 unique single-nucleotide polymorphism (SNP) to drug-effect associations. However, of these 220, only 19 were confirmed by the criteria of the analysis[3] and at the time of publication, none had been recommended for inclusion into clinical practice guidelines. This is a sobering statistic and likely reflects the fact that much variation between patients comprises environmental factors such as diet, the taking of other medicines, and particularly, the composition and activity of the person's microbiome.

It is unrealistic that a technology such as pharmacogenomics relying solely on human genetic information will be able to predict drug outcomes for patients in all circumstances, as both the disease state and the interaction between the patient and the drug are likely to involve both genetic and environmental factors. In this situation, it is expected that technologies reporting on both genetic and environmental factors will be able to make a contribution to the delivery of personalized medicine. In the rest of this chapter, we will review what pharmacometabonomics is, what has been achieved to date (with a focus on NMR spectroscopy-led studies), and what role pharmacometabonomics may play in personalized medicine in the future.

25.2 INTRODUCTION TO METABONOMICS AND PHARMACOMETABONOMICS

25.2.1 Metabonomics

Metabonomics is defined as 'the study of the metabolic response of organisms to disease, environmental change, or genetic modification'.[4] The notion was conceived in 1999 by Jeremy Everett and Jeremy Nicholson in the course of a collaboration between Pfizer and Imperial College London on the use of biofluid analysis by NMR spectroscopy to better understand and predict pre-clinical drug safety. The term was defined in order to provide a holistic framework for a series of studies integrating genomics, proteomics, metabolite profiling, and clinical information. The concept behind the definition is that an intervention to an organism, such as giving a patient a dose of a drug, has an effect on that person's metabolic profile. The metabonomics study would then be to determine the changes in the patient's metabolite profile from before the intervention to after the intervention. This can also include the concept of a metabolic trajectory over time, as the patient responds to the drug administration.[4,5] This differentiates metabonomics studies from simple observational studies in the absence of any intervention.

Soon after metabonomics was defined, an alternative definition was published for the similar term *metabolomics*. This was defined as follows: 'Metabolomics is a postgenomic technology which seeks to provide a comprehensive profile of all the metabolites present in a biological sample'.[6] This definition is observational in nature and will be very difficult, if not impossible, to achieve, as the number of metabolites in human urine, for instance, is astonishingly high. The uses of the terms *metabonomics* and *metabolomics* used to be differentiated by various technological or sample type factors, but they are now used interchangeably. We use the original term *metabonomics*, as the definition is not only more precise and useful but also etymologically compelling.[7]

25.2.2 The Discovery of Pharmacometabonomics

Metabonomics is a critical technology for drug discovery and development with applications ranging from target validation to pre-clinical and clinical efficacy and safety studies.[8] In metabonomics, the objective is generally to study changes that occur as the result of some intervention, which could be, for instance, the administration of a drug. In a number of pre-clinical drug administration studies conducted in Pfizer in the mid-1990s and before that in Beecham Pharmaceuticals, in the 1980s, it was common to observe a wide variety of responses in the animals under study. This was typically loosely ascribed to 'biological variation' but was sometimes so extreme that doubts were expressed as to whether the drug dosing had been conducted properly. This issue surfaced in the course of a Pfizer/Imperial College metabonomics collaboration meeting in Amboise, France on 18th October 2000. Inconsistent results had been observed again among subgroups of animals dosed with either galactosamine or isoniazid. The hypothesis was proposed at the meeting that differences in metabolic response between the animals might be related to differences in their pre-dose metabolic status, and that the latter should be reflected in the pre-dose NMR spectral profiles. This was a radical proposal as the animals in the experiments were matched for age, gender, and weight, and treated identically. However, a series of experiments were then proposed that culminated in a large study of the metabolism and safety of paracetamol in rats. The experiment was designed to test the hypothesis that there was a relationship between pre-dose metabolic profiles and postdose outcomes. The result of this experiment was that the hypothesis was confirmed: the pre-dose urine profiles of the rats did give a statistically significant prediction of the post-dose outcomes in terms of paracetamol metabolic pathways and toxicity. This was the birth of pharmacometabonomics.[9]

Pharmacometabonomics is defined as the 'prediction of the outcome (e.g., efficacy or toxicity) of a drug or xenobiotic intervention in an individual based on a mathematical model of pre-intervention metabolite signatures'.[9] It should be noted that the notion of a mathematical model is very broad and could include, for instance, the concentration of just a single, pre-dose biomarker molecule.

Pharmacometabonomics is thus related to pharmacogenomics and pharmacoproteomics, which are concerned with the prediction of drug effects, particularly safety and efficacy, based on an analysis of a patient's genome or proteome. The reader should be aware however, that there are some confusing uses of the word pharmacoproteomics in the literature, with some groups using the term to denote changes induced in proteomes due to drug administration: that is just proteomics.[10] A similar phenomenon is starting to happen with misuse of the term *pharmacometabonomics* as well: readers beware.

25.3 THE FIRST DEMONSTRATIONS OF PHARMACOMETABONOMICS

The first demonstration of pharmacometabonomics came in a study of the effects of paracetamol (acetaminophen) administration in Sprague–Dawley rats by a Pfizer/Imperial College research team. The aim of this study was to determine if pre-dose urine metabolite profiles could be interpreted to predict post-dose liver histopathology and drug metabolism results (Figure 25.1).

Representative 600 MHz ^{1}H NMR spectra of pre- and post-dose rat urine from this study are shown in Figure 25.2. The resonances of a number of endogenous, microbiome-derived, and drug-related metabolites are observed.

Unbiased multivariate analysis of the *pre-dose* urine NMR spectra using principal components analysis (PCA) showed that there was a partial separation of the animals according to their *post-dose* mean liver histopathology score (MHS) indicating that a relationship had been established between pre-dose metabolite profiles and post-dose safety outcomes (Figure 25.3).[9]

The PCA results in Figure 25.3 showed that weak but statistically significant correlations existed between the *pre-dose* urine metabolite profiles and the *post-dose* liver histopathology. Other results from this study demonstrated that it was possible to get statistically significant prediction of the metabolism of the drug as well, in terms of the ratio of glucuronide metabolite to parent drug (G/P). It was interesting to note that one of the metabolites found to be discriminating in terms of predicting the post-dose G/P ratio was a glucuronide signal in the pre-dose urine spectra.

Following this encouraging result in animals, an ambitious, ethically approved clinical trial was conducted in 100 human volunteers in 2003 to see if the methodology could be extended from the pre-clinical to the clinical setting.[11] Obviously, it would not be ethical

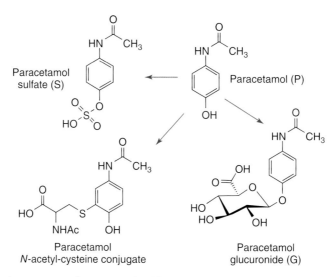

Figure 25.1. The molecular structures of paracetamol and its major metabolites

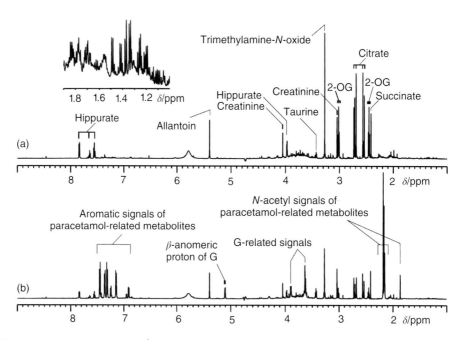

Figure 25.2. Representative 600 MHz 1D ^{1}H NMR spectra of rat urine: (a) before dosing and (b) after doing with paracetamol (600 mg kg^{-1}). 2-OG = 2-ketoglutaric acid. G = glucuronide metabolite of paracetamol. (Reprinted by permission from Macmillan Publishers Ltd: T. Clayton, J. Lindon, O. Cloarec, H. Antti, C. Charuel, G. Hanton, J. Provost, J. Le Net, D. Baker, R. Walley, J. Everett and J. Nicholson, Nature, 2006, 440, 1073–1077. copyright 2006)

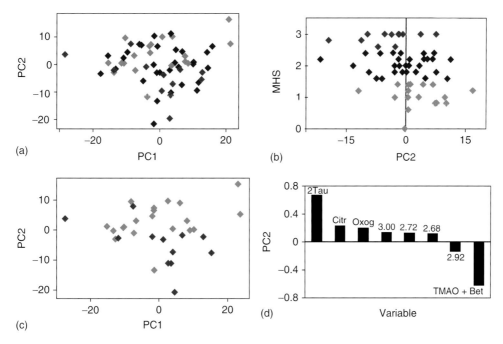

Figure 25.3. Unbiased, multivariate PCA of the *pre-dose* urine spectra from rats dosed with paracetamol at 600 mg kg^{-1}. (a) Scores plot of a PCA of the multivariate *pre-dose* NMR data. Each diamond represents a different animal in the study, color coded by *post-dose* histopathology classification: green: Class 1 (no or minimal necrosis), blue: Class 2 (mild necrosis), and red: Class 3 (moderate necrosis). There is a partial separation across PC2 between histology Classes 1 and 3. (b) A plot of mean histopathology score (MHS) postdose against the pre-dose PC2 score from (a). A weak but significant correlation is observed. (c) Scores plot of a PCA with the same color coding as above, for the animals in histopathology Classes 1 and 3 only. (d) A loadings plot corresponding to the PCA in (c), showing the variables in the pre-dose NMR spectra that most influenced the differentiation of the animals' pre-dose urine scores across PC2, and the direction of that influence. Variables are listed by the midpoint of the chemical shift range of the spectral region they represent, or by their names: Tau, taurine; Citr, citrate; Oxog, 2-ketoglutarate; TMAO + Bet, trimethylamine-*N*-oxide, and betaine. (Reprinted by permission from Macmillan Publishers Ltd: T. Clayton, J. Lindon, O. Cloarec, H. Antti, C. Charuel, G. Hanton, J. Provost, J. Le Net, D. Baker, R. Walley, J. Everett and J. Nicholson, Nature, 2006, 440, 1073–1077. copyright 2006)

to try to predict safety outcomes in human volunteers. The aim of this study was, therefore, to determine if *pre-dose* urine metabolite profiles could predict the metabolic fate of paracetamol (acetaminophen) after a normal, 1 g oral dose of the drug. The protocol for the trial did not specify a standard diet but placed certain restrictions on the diet and on alcohol consumption. Volunteers were only eligible if not taking drugs, herbal medicines, or dietary supplements before sample collection. Pre-dose, 0–3 h, and 3–6 h post-dose urines were then collected from each volunteer and analyzed by 600 MHz ^{1}H NMR spectroscopy. Figure 25.4 shows representative pre-dose and post-dose NMR spectra from two of the volunteers, all of whom were fit and healthy males aged 18–64.

With hindsight, the analysis of these spectra is trivial. However, at the time, over a year was spent in the investigation of sophisticated pattern recognition and multivariate statistical analysis methods, in order to determine if there were correlations between the pre-dose urinary metabolite profiles and the post-dose paracetamol metabolism patterns. In the end, the analysis was completed by visual inspection: all the computational methods failed. This is an unusual case that we have not experienced since. The significant metabolite differences are already clearly visible in Figure 25.4. The volunteer with a higher *pre-dose* urine level of metabolite 4 has a lower *post-dose* ratio of paracetamol sulfate to paracetamol glucuronide (S/G) metabolites (labeled 7 and 8, respectively, in Figure 25.4a and b). On the

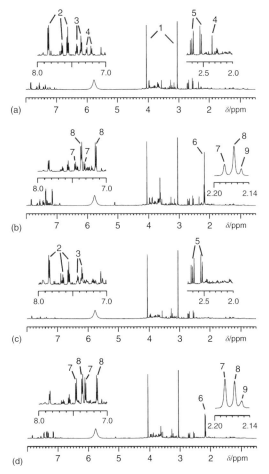

Figure 25.4. Representative 600 MHz flow-mode ^{1}H NMR spectra from two different human volunteers. Spectra (a) and (b): pre-dose and 0–3 h post-dose urines, respectively, from a volunteer excreting a relatively low amount of paracetamol sulfate (S, peak 7) relative to paracetamol glucuronide (G, peak 8). Spectra (c) and (d): corresponding pre-dose and 0–3 h post-dose urines from a second volunteer whose excreted ratio of S/G was much higher. Note that the key difference between the pre-dose and post-dose spectra (a vs b and c vs d) is due to the presence of the acetyl and aromatic signals for paracetamol and its metabolites in the postdose spectra. The insets are expansions of those regions of the NMR spectra. Key to peak numbers: 1, creatinine; 2, hippurate; 3, phenylacetylglutamine; 4, metabolite 4 (see text); 5, citrate; 6, cluster of *N*-acetyl groups from paracetamol-related compounds; 7, paracetamol sulfate; 8, paracetamol glucuronide; 9, other paracetamol-related compounds. (Reproduced from T. A. Clayton, D. Baker, J. C. Lindon, J. R. Everett and J. K. Nicholson, Proceedings of the National Academy of Sciences of the United States of America, 2009, 106, 14728–14733)

other hand, the volunteer with no obviously visible signal from metabolite 4 in their *pre-dose* urine has a much higher ratio of S/G in their *post-dose* urine (Figure 25.4c and d). When examined over the whole cohort of volunteers, these results held generally true (Figures 25.5 and 25.6).

What is immediately clear from inspection of Figure 25.5 is that there is a significant trend toward lower *postdose* ratios of S/G metabolites if the pre-dose levels of metabolite 4 are high. This is most clearly seen by inspection of these data for all of the volunteers (Figure 25.6).

Inspection of Figure 25.6 shows that for all those volunteers with a *pre-dose* urinary ratio of metabolite 4 to creatinine >0.06, then the *post-dose* ratio of S/G was invariably low and less than 0.8. On the other hand, if the *pre-dose* ratio of metabolite 4/creatinine was less than 0.06, then the *post-dose* ratio of S/G was highly variable and not predictable. It thus became very important to determine the identity of metabolite 4.

Statistical total correlation spectroscopy (STOCSY)[12] analysis had shown that metabolite 4 was associated with a singlet peak at about 2.35 ppm, interpreted as an sp^2C–CH$_3$ group, and two aromatic signals at about 7.21 and 7.29 ppm. These aromatic signals showed the characteristic second-order, pseudo-doublet patterns expected for a *para*-di-substituted phenyl ring. The metabolite was identified by two methods. Firstly, treating a sample of pre-dose urine with sulfatase enzyme decreased the signals from metabolite 4 and produced new signals identified as *para*-cresol by comparison with an authentic standard sample. It was thus hypothesized that metabolite 4 was *para*-cresol sulfate. Secondly, and in confirmation of this hypothesis, a sample of *para*-cresol sulfate was synthesized from *para*-cresol and was shown to have identical spectra to metabolite 4 (see Figure 25.7 for a spectrum of an authentic sample of *para*-cresol sulfate).

Thus, metabolite 4 was confirmed as *para*-cresol sulfate. This finding shocked us as *para*-cresol sulfate is not human in origin: it derives from the sulfation of *para*-cresol, a bacterial metabolite, known to be excreted into the human gut by *Clostridium* species in particular. In order to gain confidence in the data, the entire analysis was repeated on the original urine samples in 2007, 4 years after the NMR spectroscopy experiments were completed, but this time using conventional 5 mm NMR tubes, rather than a flow probe. No significant changes in the results were found. In addition, ultra-performance liquid

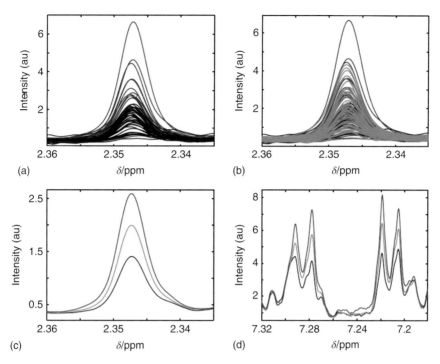

Figure 25.5. (a) Expansions of the 600 MHz flow-mode ^{1}H NMR spectra of the *pre-dose* urines of human volunteers subsequently dosed with a standard 1 g oral dose of paracetamol, in the region of the singlet signal of metabolite 4, for the 25 volunteers with the highest ratio of S/G in their *post-dose* urine (colored blue), and also the 25 volunteers with the lowest ratio of S/G in their *post-dose* urine (colored red). (b) As for (a) but with additional traces in green for the 49 volunteers with intermediate S/G ratios in their post-dose urines. (c) The averaged 600 MHz ^{1}H NMR spectra of the volunteers' *pre-dose* urines in the region of the singlet signal of metabolite 4: color coding according to *post-dose* outcome as above. (d) Corresponding plot in the region of the aromatic signals of metabolite 4 with the same color coding. (Reproduced from T. A. Clayton, D. Baker, J. C. Lindon, J. R. Everett and J. K. Nicholson, Proceedings of the National Academy of Sciences of the United States of America, 2009, 106, 14728–14733)

chromatography–mass spectrometry (UPLC-MS) methods were used to provide additional validation of results. The 2007 NMR-based, *pre-dose para*-cresol sulfate/creatinine results were also confirmed by UPLC-MS combined with NMR in 2007, with a correlation coefficient between the two sets of data of 0.927. Finally, 5 years after the original NMR analyses, the 2003 NMR-based, 3–6 h post-dose S/G ratio results were confirmed by UPLC-MS with a correlation coefficient of 0.991 and with no outliers. A variety of other checks were performed in order to gain confidence in the data before publication in 2009.[11]

So, finally, it turned out that not only was it possible to achieve a statistically significant prediction of the ratio of S/G metabolites in *post-dose* human urine, after a 1 g dose of paracetamol, from an analysis of *pre-dose* levels of urinary *para*-cresol sulfate, i.e.,

pharmacometabonomics was demonstrated to work in humans for the first time, but that the key biomarker was of microbial and not human origin. This was a particularly startling result given the hundreds of papers published on the metabolism of paracetamol, with no indication of any bacterial involvement in the metabolism. This result is, however, perhaps less surprising when we compare the molecular structures of *para*-cresol and paracetamol (Figure 25.8).

Both paracetamol and *para*-cresol possess a *para*-disubstituted benzene ring with one of the substituents being a hydroxyl group. Both molecules are sulfated by the same human sulfotransferase enzymes e.g. SULT1A1 and will therefore be in competition with one another for both the enzyme and the sulfonate donor, 3'-phosphoadenosine 5'-phosphosulfate (PAPS). In contrast to rats and mice, humans

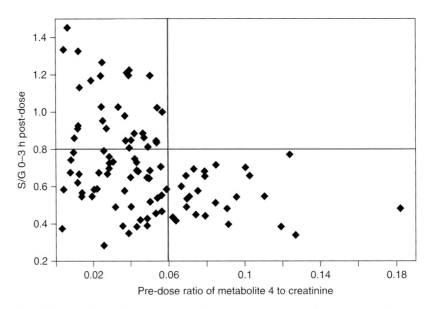

Figure 25.6. A plot of the post-dose ratio of paracetamol sulfate to paracetamol glucuronide metabolites (S/G) against the pre-dose ratio of metabolite 4 normalized to creatinine. The vertical red and horizontal blue lines are at critical cut-off values. (Reproduced from T. A. Clayton, D. Baker, J. C. Lindon, J. R. Everett and J. K. Nicholson, Proceedings of the National Academy of Sciences of the United States of America, 2009, 106, 14728–14733)

metabolize *para*-cresol largely by sulfation with relatively little metabolic flux via glucuronidation. Thus, significant exposure to *para*-cresol will place a correspondingly significant demand for sulfation on a human. A 1 g dose of paracetamol also seems to place a significant demand on the sulfation capacity of the human body as, for every volunteer, it was found that the post-dose S/G ratio was lower in the 3–6 h urine than in the 0–3 h urine, implying that the sulfation capacity was being diminished over this time. The average values for the post-dose S/G ratios were 0.71 (0–3 h) and 0.53 (3–6 h).

The hypothesis is, therefore, that an individual who is exposed to significant levels of microbiome-excreted *para*-cresol will have his/her sulfation capacity depleted to such an extent that a subsequent dose of 1 g of paracetamol will cause the body's metabolic pathways to switch significantly from sulfation to glucuronidation, as observed. Conversely, in the absence of a *para*-cresol load, the human body will metabolize paracetamol across both sulfation and glucuronidation pathways in a manner dependent on a number of genetic and environmental factors and this is not predictable, at least at the current time. It is important to note that this finding should apply equally to *any drug requiring sulfation* to assist its elimination from

the body and is not just of importance for paracetamol. Equally, this discovery may have implications for the sulfation of endogenous molecules in the human body such as proteins, glycans, hormones, and neurotransmitters and may have even broader implications.[11] Finally, it is worth noting that a number of disease states are associated either with increased *para*-cresol levels or decreased S/G ratios after paracetamol dosing. These diseases include childhood hyperactivity, multiple sclerosis, Parkinson's disease, and childhood autism. Thus, impaired sulfation may be expected in each of these cases and it is likely that there will be a significant microbiome involvement causing these changes.

A commentary[13] by Ian Wilson accompanied the publication of this study.[11]

25.4 THE CURRENT STATUS OF PHARMACOMETABONOMICS

There have been several reviews of pharmacometabonomics and all pre-clinical and clinical studies up to April 2013 were recently reviewed.[10] In addition, a review of pharmacometabonomics and its implications for clinical pharmacology and systems

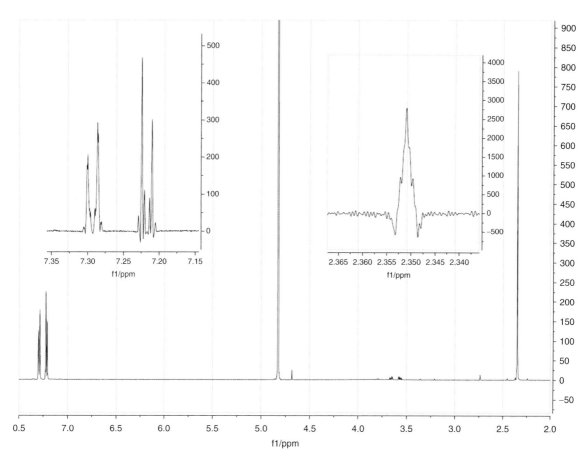

Figure 25.7. A 600 MHz ^{1}H NMR spectrum of a sample of authentic *para*-cresol sulfate. The inset expansions are resolution enhanced by Gaussian multiplication of the free induction decay and show clearly the second-order nature of the aromatic pseudo-doublets and fine structure due to long-range coupling from the aromatic protons to the methyl protons. In high-resolution COSY spectra of biofluids containing *para*-cresol sulfate, cross peaks between the methyl protons and the aromatic protons at about 7.29 are clearly observed, and the higher frequency aromatic signal is clearly of lower intensity than the lower frequency signal due to a 4-bond coupling

pharmacology appeared recently.[14] In the latter review, pharmacometabolomics/pharmacometabonomics was described as involving 'determination of the metabolic state of an individual as affected by environmental, genetic, and gut microbiome influences – the so-called metabotype – to define signatures both at baseline (before) and after drug exposure that might inform treatment outcomes.' This contrasts with the original and precise definition of 'prediction of the outcome (e.g., efficacy or toxicity) of a drug or xenobiotic intervention in an individual based on a mathematical model of *pre-intervention* metabolite signatures'.[9]

In this study, we cover the key NMR-based contributions to this emerging field in the pre-clinical and clinical arenas, but work on cell-based systems is not covered here. There has been much progress across many groups since the original demonstrations of pharmacometabonomics, detailed earlier, were published. Similarly to the use of the two terms metabonomics and metabolomics, there is now the joint and interchangeable use of the two terms pharmacometabonomics and pharmacometabolomics. Herein, we use the original term and definition[9] of pharmacometabonomics throughout for clarity and precision.

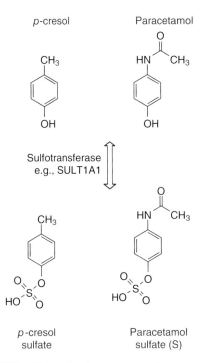

p-cresol   Paracetamol

Sulfotransferase
e.g., SULT1A1

p-cresol   Paracetamol
sulfate   sulfate (S)

Figure 25.8. The molecular structures of *para*-cresol and paracetamol and their sulfation products. Both molecules are sulfated in humans by identical sulfotransferase enzymes such as SULT1A1

25.4.1 Progress in Pre-clinical Pharmacometabonomics

Kwon *et al.*[15] showed that ¹H NMR analysis of pre-dose rat urine could differentiate between nephrotoxicity-prone animals (higher 2-ketoglutarate) and nephrotoxicity-resistant animals (higher pre-dose allantoin, creatinine, and succinate), when subsequently dosed with the oncology drug cisplatin. This is prediction of safety.

The Pfizer/Imperial College group recently published one of the studies giving rise to the notion of pharmacometabonomics in 2000[16] involving the antituberculosis drug isoniazid. This drug is associated with a number of side effects in humans including liver toxicity and central nervous system (CNS) effects. This study showed that pre-dose urine metabolite profiles in Sprague–Dawley rats clearly differentiated between high-dose animals having CNS effects and those with no CNS response. The high-dose CNS responders had elevated pre-dose urine ¹H NMR signals at 5.11 and 6.72 ppm, ascribed

to a 1,2,4-trisubstituted phenolic ether glucuronide and a decreased ratio of acetylisoniazid to parent drug, isoniazid, in the post-dose urines. Remarkably, although no CNS effects were seen in low-dose animals, the *pre-dose* levels of the signals at 5.11 and 6.72 ppm were negatively correlated with their *post-dose* levels of acetylisoniazid metabolite excretion. Thus, this study demonstrated both prediction of safety and prediction of metabolism of the drug.[16]

Coen *et al.*[17] have reported a new ¹H NMR study on variability of hepatotoxic responses to galactosamine in male Sprague–Dawley rats. Following an initial (primary) dose with galactosamine on day 1, animals were separated into responders and nonresponders on the basis of their clinical findings. Four primary responders and nine primary nonresponders received a second dose of galactosamine at day 12. All the primary nonresponders now had a clinical response to the drug, but the primary responders now had a reduced response relative to these induced/secondary responders. Pharmacometabonomic analyses enabled discrimination of the *pre-dose* urinary and fecal extract metabolic profiles of primary nonresponders relative to induced responders, but not of primary nonresponders from primary responders. This is the first demonstration of differentiation between a nonresponder and an induced-responder phenotype and represents a prediction of safety. A key fecal extract metabolite discriminating between the primary nonresponder and the induced responders groups was γ-aminobutyrate (GABA), which is synthesized from glutamate in the gut microbiome by colonic bacterial decarboxylase. It was concluded that gut microbiome differences existed between the nonresponders and the induced responders and that these were responsible for the GABA fecal extract differentiation.

In addition, some mass spectrometry (MS)-based publications have demonstrated the use of pharmacometabonomics to predict preclinical drug PK and disease susceptibility.[18,19]

25.4.2 Progress in Clinical Pharmacometabonomics

Weight gain is a serious issue for women undergoing breast cancer chemotherapy. In the first validated pharmacometabonomics study with patients, as opposed to volunteers, Keun *et al.*[20] used ¹H NMR to show that predose serum concentrations of lactate, alanine, and percentage body fat were predictive of weight gain in the patients.

Paracetamol is hepatotoxic at high doses. The group of Watkins *et al.*[21] used ^{1}H NMR to show, in a group of 71 adult male and female volunteers, that administration of paracetamol at a high dose of 4 g per day led to reversible liver injury in some of the volunteers (responders) as judged by serum alanine aminotransferase (ALT) levels. The analysis of pre-dose urine profiles was *not* able to produce a predictive model for the responders vs the nonresponders. However, analysis of the urine metabolite profiles shortly after drug administration, and before any changes in ALT levels, led to statistically significant discrimination, based on the levels of the toxic metabolite *N*-acetyl paraquinone imine (NAPQI) plus some other endogenous metabolites. The group labeled this 'early intervention pharmacometabonomics'.

Keun *et al.*[22] have used ^{1}H NMR to demonstrate that pre-dose serum metabolite profiles can predict toxicity in cancer patients treated with capecitabine. Higher toxicity was associated with higher pre-dose levels of low-density lipoprotein-derived lipids including choline phospholipids and polyunsaturated fatty acids.

A prediction of the induction of cytochrome P450 3A4 (CYP3A4) has been published by Kemsley *et al.*[23] In this study, 301 subjects from the TwinsUK registry were dosed with 300 mg of a formulation of 0.03% St John's Wort extract, three times a day for 14 days. On the evening of day 14, each subject was dosed with quinine (Q) sulfate (300 mg). Q is metabolized, mainly by CYP3A4 to 3-hydroxyquinine (3-OHQ), and the ratio of 3-OHQ to Q is used to measure CYP3A4 activity. UPLC-MS was used to quantify post-dose Q and 3-OHQ levels. ^{1}H NMR spectroscopy was used to measure pre-dose endogenous metabolite levels in a spectral binning approach. Multiple linear regression was used to establish relationships between the pre-dose metabolite levels and the post-dose 3-OHQ : Q ratios. Nine pre-dose metabolite bins including those with signals from indoxyl sulfate, guanidine acetate, glycine, *scyllo*-inositol, proline betaine, and alanine were used to construct a predictive model for the post-dose 3-OHQ : Q ratio based on the pre-dose bins, but no rationale for linking these metabolites to CYP3A4 induction was found. Further models included BMI and UPLC-MS batch numbers as predictive variables; however, the significance of using these is not understood.

Finally, several MS-based clinical pharmacometabonomics publications have appeared. These have been concerned with the prediction of the PK of the immunosuppressant Tacrolimus,[24] the efficacy of the lipid-lowering drug simvastatin,[25–27] the efficacy of some schizophrenia treatments,[28] the efficacy of selective serotonin reuptake inhibitors (SSRIs) for major depressive disorder (including the ability to predict placebo response),[29–32] the efficacy of aspirin,[33,34] and finally, the activity of a human cytochrome P450 of importance in drug metabolism, CYP3A4.[35] A particularly important feature of the work on efficacy prediction for SSRIs[29] was the development of a pharmacometabonomics-led pharmacogenomics approach. This will undoubtedly be an important paradigm in the future.

25.5 FUTURE DEVELOPMENTS AND PREDICTIVE METABONOMICS

Pharmacometabonomics was discovered in 2000 and the first publication on it appeared in 2006. By contrast, pharmacogenomics is a science that is now over 50 years old. Already, pharmacometabonomics has demonstrated a remarkable ability to predict drug efficacy, placebo efficacy, drug metabolism, drug toxicity, drug pharmacokinetics, and enzyme induction and also to be able to inform pharmacogenomics studies. In the future, we envisage that pharmacometabonomics and pharmacogenomics will increasingly be used together in this synergistic manner. A further concept that has emerged is that of longitudinal pharmacometabonomics, particularly for patients, where metabolic profiling is undertaken before, during, and repeatedly after an intervention, in order to create a metabolic trajectory that is predictive of future patient treatment outcomes.[36]

It should also be recognized that pharmacometabonomics is just one example of a broader branch of study that is called *predictive metabonomics*. Pharmacometabonomics is defined as the 'prediction of the outcome (e.g., efficacy or toxicity) of a drug or xenobiotic intervention in an individual based on a mathematical model of pre-intervention metabolite signatures'. Predictive metabonomics is correspondingly defined as the 'prediction of the outcome (e.g., disease onset) of an intervention in an individual based on a mathematical model of pre-intervention metabolite signatures'. Predictive metabonomics thus has a broader coverage than pharmacometabonomics, where the intervention is specifically drug treatment. In predictive metabonomics, the intervention could be exercise, environmental change, diet change, or simply the passage or a defined period of time.

Examples of the successful implementation of predictive metabonomics have already emerged. A major predictive metabonomics study of pre-diabetes was conducted by Wang-Sattler *et al.*[37] in a large cohort of hundreds of subjects [population-based Cooperative Health Research in the Region of Augsburg cohort (KORA)]. This showed that low levels of glycine and lysophosphatidylcholine (LPC) (18 : 2) at baseline predicted the risks of developing impaired glucose tolerance (IGT) and/or type-2 diabetes (T2D). Glycine and LPC were also shown to be predictors of glucose tolerance as much as 7 years before disease onset. Although this study was MS based, it is a very important indicator that it will be possible to predict disease progression from baseline metabolite profiles in at least some cases. This opens up the prospect of keeping people healthier for longer by advising lifestyle changes before the onset of the disease and any drug treatments!

The area of predictive metabonomics is likely to expand considerably in the near future. There have been a series of landmark studies published recently that relate GWAS with metabolic profiles in human blood, serum, and urine (determined by NMR- and MS-based metabonomics), on a very large scale.[38-41] These exciting studies provide information on the metabolic variance associated with genetic as opposed to environmental factors in human biofluids. In addition, they can provide new metabolic insights into disease progression, approaches to disease treatment, and the prediction of drug efficacy and toxicity, via previous GWAS associations with these important factors. A key feature mentioned in several of these studies is the stability, precision, and low variance of NMR-based studies leading to improved study outcomes.[42]

One recent example of how this information can be used came in an NMR-based GWAS-metabonomics study of metabolic traits associated with SNPs found in 835 Caucasians in the CoLaus study.[43] One finding from this study was an association between SNP rs492602 in the *FUT2* gene (encodes a fucosyltransferase enzyme) and the urine concentrations of fucose. Importantly, fucose is known to impact human microbiome composition and, therefore, gut health. As the *FUT2* gene has been associated with Crohn's disease (CD), it was proposed that elevated urinary fucose levels may be an early indicator for the dysbiosis associated with CD: if validated, this would be a predictive metabonomics finding. This study also employed a 'metabomatching' method that generates a pseudo-NMR spectrum of the metabolite correlating with the SNP, thus facilitating the identification of candidate metabolites. In addition, proposals for the identities of unknown metabolites can be made on the basis of the genetic associations established in these kinds of studies.

It is not often that NMR spectroscopy opens up an entirely new area of science, but it did so in pharmacometabonomics. No doubt many other unanticipated discoveries will be made in this area in the near future. The need for better personalized medicine is significant: we hope that NMR-based pharmacometabonomics will help to deliver it.

REFERENCES

1. K. Salari, H. Watkins, and E. A. Ashley, *Eur. Heart J.*, 2012, **33**, 1564–1570.

2. T. J. Urban and D. B. Goldstein, *Sci. Transl. Med.*, 2014, **6**, 221–229.

3. G. D. Kitsios and D. M. Kent, *Br. Med. J.*, 2012, **344**, e2161.

4. J. Lindon, J. Nicholson, E. Holmes, and J. Everett, *Concepts Magn. Resonan.*, 2000, **12**, 289.

5. J. K. Nicholson, J. C. Lindon, and E. Holmes, *Xenobiotica*, 1999, **29**, 1181.

6. J. Taylor, R. D. King, T. Altmann, and O. Fiehn, *Bioinformatics*, 2002, **18** (Suppl 2), S241.

7. J. K. Nicholson, *Mol. Syst. Biol.*, 2006, **2**, 52.

8. J. C. Lindon, J. K. Nicholson, and E. Holmes, The Handbook of Metabonomics and Metabolomics, Elsevier: Amsterdam, 2007.

9. T. Clayton, J. Lindon, O. Cloarec, H. Antti, C. Charuel, G. Hanton, J. Provost, J. Le Net, D. Baker, R. Walley, J. Everett, and J. Nicholson, *Nature*, 2006, **440**, 1073.

10. J. R. Everett, R. L. Loo, and F. S. Pullen, *Ann. Clin. Biochem.*, 2013, **50**, 523.

11. T. A. Clayton, D. Baker, J. C. Lindon, J. R. Everett, and J. K. Nicholson, *Proc. Natl. Acad. Sci. U. S. A.*, 2009, **106**, 14728.

12. S. L. Robinette, J. C. Lindon, and J. K. Nicholson, *Anal. Chem.*, 2013, **85**, 5297.

13. I. D. Wilson, *Proc. Natl. Acad. Sci. U. S. A.*, 2009, **106**, 14187.

14. R. Kaddurah-Daouk, R. M. Weinshilboum, and Pharmacometabolomics Research Network, *Clin. Pharmacol. Ther.*, 2014, **95**, 154.

15. H. N. Kwon, M. Kim, H. Wen, S. Kang, H. J. Yang, M. J. Choi, H. S. Lee, D. Choi, I. S. Park, Y. J. Suh, S. S. Hong, and S. Park, *Kidney Int.*, 2011, **79**, 529.

16. K. Cunningham, S. P. Claus, J. C. Lindon, E. Holmes, J. R. Everett, J. K. Nicholson, and M. Coen, *J. Proteome Res.*, 2012, **11**, 4630.

17. M. Coen, F. Goldfain-Blanc, G. Rolland-Valognes, B. Walther, D. G. Robertson, E. Holmes, J. C. Lindon, and J. K. Nicholson, *J. Proteome Res.*, 2012, **11**, 2427.

18. L. Liu, B. Cao, J. Aa, T. Zheng, J. Shi, M. Li, X. Wang, C. Zhao, W. Xiao, X. Yu, R. Sun, R. Gu, J. Zhou, L. Wu, G. Hao, X. Zhu, and G. Wang, *PLoS One*, 2012, **7**, e43389.

19. H. Li, Y. Ni, M. Su, Y. Qiu, M. Zhou, M. Qiu, A. Zhao, L. Zhao, and W. Jia, *J. Proteome Res.*, 2007, **6**, 1364.

20. H. C. Keun, J. Sidhu, D. Pchejetski, J. S. Lewis, H. Marconell, M. Patterson, S. R. Bloom, V. Amber, R. C. Coombes, and J. Stebbing, *Clin. Cancer Res.*, 2009, **15**, 6716.

21. J. H. Winnike, Z. Li, F. A. Wright, J. M. Macdonald, T. M. O'Connell, and P. B. Watkins, *Clin. Pharmacol. Ther.*, 2010, **88**, 45.

22. A. Backshall, R. Sharma, S. J. Clarke, and H. C. Keun, *Clin. Cancer Res.*, 2011, **17**, 3019.

23. N. Rahmioglu, G. Le Gall, J. Heaton, K. L. Kay, N. W. Smith, I. J. Colquhoun, K. R. Ahmadi, and E. K. Kemsley, *J. Proteome Res.*, 2011, **10**, 2807.

24. P. B. Phapale, S. D. Kim, H. W. Lee, M. Lim, D. D. Kale, Y. L. Kim, J. H. Cho, D. Hwang, and Y. R. Yoon, *Clin. Pharmacol. Ther.*, 2010, **87**, 426.

25. R. Kaddurah-Daouk, R. A. Baillie, H. J. Zhu, Z. B. Zeng, M. M. Wiest, U. T. Nguyen, S. M. Watkins, and R. M. Krauss, *Metabolomics*, 2010, **6**, 191.

26. R. Kaddurah-Daouk, R. A. Baillie, H. Zhu, Z. B. Zeng, M. M. Wiest, U. T. Nguyen, K. Wojnoonski, S. M. Watkins, M. Trupp, and R. M. Krauss, *PLoS One*, 2011, **6**, e25482.

27. M. Trupp, H. Zhu, W. R. Wikoff, R. A. Baillie, Z. B. Zeng, P. D. Karp, O. Fiehn, R. M. Krauss, and R. Kaddurah-Daouk, *PLoS One*, 2012, **7**, e38386.

28. R. Condray, G. G. Dougherty, M. S. Keshavan, R. D. Reddy, G. L. Haas, D. M. Montrose, W. R. Matson, J. McEvoy, R. Kaddurah-Daouk, and J. K. Yao, *Int. J. Neuropsychopharmacol.*, 2011, **14**, 756.

29. Y. Ji, S. Hebbring, H. Zhu, G. D. Jenkins, J. Biernacka, K. Snyder, M. Drews, O. Fiehn, Z. Zeng, D. Schaid, D. A. Mrazek, R. Kaddurah-Daouk, and R. M. Weinshilboum, *Clin. Pharmacol. Ther.*, 2011, **89**, 97.

30. R. Abo, S. Hebbring, Y. Ji, H. Zhu, Z. B. Zeng, A. Batzler, G. D. Jenkins, J. Biernacka, K. Snyder, M. Drews, O. Fiehn, B. Fridley, D. Schaid, N. Kamatani, Y. Nakamura, M. Kubo, T. Mushiroda, R. Kaddurah-Daouk, D. A. Mrazek, and R. M. Weinshilboum, *Pharmacogenet. Genomics*, 2012, **22**, 247.

31. R. Kaddurah-Daouk, S. H. Boyle, W. Matson, S. Sharma, S. Matson, H. Zhu, M. B. Bogdanov, E. Churchill, R. R. Krishnan, A. J. Rush, E. Pickering, and M. Delnomdedieu, *Transl. Psychiatry*, 2011, **1**, 1.

32. H. Zhu, M. B. Bogdanov, S. H. Boyle, W. Matson, S. Sharma, S. Matson, E. Churchill, O. Fiehn, J. A. Rush, R. R. Krishnan, E. Pickering, M. Delnomdedieu, R. Kaddurah-Daouk, and Pharmacometabolomics Research Network, *PLoS One*, 2013, **8**, e68283.

33. L. M. Yerges-Armstrong, S. Ellero-Simatos, A. Georgiades, H. Zhu, J. P. Lewis, R. B. Horenstein, A. L. Beitelshees, A. Dane, T. Reijmers, T. Hankemeier, O. Fiehn, A. R. Shuldiner, R. Kaddurah-Daouk, and Pharmacometabolomics Research Network, *Clin. Pharmacol. Ther.*, 2013, **94**, 525.

34. S. Ellero-Simatos, J. P. Lewis, A. Georgiades, L. M. Yerges-Armstrong, A. L. Beitelshees, R. B. Horenstein, A. Dane, A. C. Harms, R. Ramaker, R. J. Vreeken, C. G. Perry, H. Zhu, C. L. Sanchez, C. Kuhn, T. L. Ortel, A. R. Shuldiner, T. Hankemeier, and R. Kaddurah-Daouk, *CPT Pharmacometrics Syst. Pharmacol.*, 2014, **3**, e125.

35. K. H. Shin, M. H. Choi, K. S. Lim, K. S. Yu, I. J. Jang, and J. Y. Cho, *Clin. Pharmacol. Ther.*, 2013, **94**, 601.

36. J. K. Nicholson, J. R. Everett, and J. C. Lindon, *Expert Opin. Drug Metab. Toxicol.*, 2012, **8**, 135.

37. R. Wang-Sattler, Z. Yu, C. Herder, A. C. Messias, A. Floegel, Y. He, K. Heim, M. Campillos, C. Holzapfel, B. Thorand, H. Grallert, T. Xu, E. Bader, C. Huth, K. Mittelstrass, A. Doering, C. Meisinger, C. Gieger, C. Prehn, W. Roemisch-Margl, M. Carstensen, L. Xie, H. Yamanaka-Okumura, G. Xing, U. Ceglarek, J. Thiery, G. Giani, H. Lickert, X. Lin, Y. Li, H. Boeing, H.-G. Joost, M. H.de Angelis, W. Rathmann, K. Suhre, H. Prokisch, A. Peters, T. Meitinger, M. Roden, H. E. Wichmann, T. Pischon, J. Adamski, and T. Illig, *Mol. Syst. Biol.*, 2012, **8**, 615.

38. K. Suhre, S.-Y. Shin, A.-K. Petersen, R. P. Mohney, D. Meredith, B. Waegele, E. Altmaier, P. Deloukas, J. Erdmann, E. Grundberg, C. J. Hammond, M. Hrabe

de Angelis, G. Kastenmueller, A. Koettgen, F. Kronenberg, M. Mangino, C. Meisinger, T. Meitinger, H.-W. Mewes, M. V. Milburn, C. Prehn, J. Raffler, J. S. Ried, W. Roemisch-Margl, N. J. Samani, K. S. Small, H. E. Wichmann, G. Zhai, T. Illig, T. D. Spector, J. Adamski, N. Soranzo, C. Gieger, and CardioGram, *Nature*, 2011, **477**, 54.

39. K. Suhre, H. Wallaschofski, J. Raffler, N. Friedrich, R. Haring, K. Michael, C. Wasner, A. Krebs, F. Kronenberg, D. Chang, C. Meisinger, H. E. Wichmann, W. Hoffmann, H. Voelzke, U. Voelker, A. Teumer, R. Biffar, T. Kocher, S. B. Felix, T. Illig, H. K. Kroemer, C. Gieger, W. Roemisch-Margl, and M. Nauck, *Nat. Genet.*, 2011, **43**, 565.

40. J. Kettunen, T. Tukiainen, A.-P. Sarin, A. Ortega-Alonso, E. Tikkanen, L.-P. Lyytikainen, A. J. Kangas, P. Soininen, P. Wuertz, K. Silander, D. M. Dick, R. J. Rose, M. J. Savolainen, J. Viikari, M. Kahonen, T. Lehtimaki, K. H. Pietilainen, M. Inouye, M. I. McCarthy, A. Jula, J. Eriksson, O. T. Raitakari, V. Salomaa, J. Kaprio, M.-R. Jarvelin, L. Peltonen, M. Perola, N. B. Freimer, M. Ala-Korpela, A. Palotie, and S. Ripatti, *Nat. Genet.*, 2012, **44**, 269.

41. S.-Y. Shin, E. B. Fauman, A.-K. Petersen, J. Krumsiek, R. Santos, J. Huang, M. Arnold, I. Erte, V. Forgetta, T.-P. Yang, K. Walter, C. Menni, L. Chen, L. Vasquez, A. M. Valdes, C. L. Hyde, V. Wang, D. Ziemek, P. Roberts, L. Xi, E. Grundberg, M. Waldenberger, J. B. Richards, R. P. Mohney, M. V. Milburn, S. L. John, J. Trimmer, F. J. Theis, J. P. Overington, K. Suhre, M. J. Brosnan, C. Gieger, G. Kastenmueller, T. D. Spector, N. Soranzo, and Multiple Tissue Human Expression Resource (MuTHER) Consortium, *Nat. Genet.*, 2014, **46**, 543.

42. R. B. Schnabel, *Circ. Cardiovasc. Genet.*, 2012, **5**, 479.

43. R. Rueedi, M. Ledda, A. W. Nicholls, R. M. Salek, P. Marques-Vidal, E. Morya, K. Sameshima, I. Montoliu, L. Da Silva, S. Collino, F.-P. Martin, S. Rezzi, C. Steinbeck, D. M. Waterworth, G. Waeber, P. Vollenweider, J. S. Beckmann, J. Le Coutre, V. Mooser, S. Bergmann, U. K. Genick, and Z. Kutalik, *PLoS Genet.*, 2014, **10**, e1004132.

Chapter 26

Clinical MRI Studies of Drug Efficacy and Safety

David G. Reid[1], Paul D. Hockings[2], and Nadeem Saeed[3]

[1]*Department of Chemistry, University of Cambridge, Lensfield Road, Cambridge CB2 1EW, UK*
[2]*Antaros Medical, BioVenture Hub, 43183 Mölndal, Sweden*
[3]*BioClinica, 72 Hammersmith Road, London W14 8UD, UK*

26.1 INTRODUCTION

Magnetic resonance imaging (MRI) is one of the foremost techniques for noninvasive scanning in medical diagnoses and procedures. Indeed, it is the most familiar face of NMR to the general public. This chapter cannot give a detailed account of the principles of MRI, image contrast, and the best techniques to identify and diagnose different pathologies, which are thoroughly treated in numerous standard texts and online resources (see Further Reading). MRI is certainly the method of choice for diseased (and normal) soft tissue. Importantly, MRI can be 'tailored' to emphasize zones of water according to their physicochemical environment using multiple 'weighting' or 'sensitization' techniques and can thus distinguish between tissue types and organs even though their basic water contents are very similar. Many of these are based on familiar NMR properties: $T1$ and $T2$ contrast arises from differences in relaxivity of different tissues, which is a function of water–macromolecule interactions and cellularity; $T2^*$ contrast is based on differences between tissue magnetic susceptibilities and is the basis of BOLD (blood oxygenation level determination) methods that distinguish oxygenated (diamagnetic) from deoxygenated (paramagnetic) hemoglobin; magnetization transfer contrast (MTC) directly reports the extent of water–macromolecule interactions and detects edema; a number of synthetic paramagnetic reagents are licensed as 'contrast agents' for use in measuring tissue perfusion using 'dynamic contrast enhancement' (DCE); arterial flow can be depicted using 'bright blood' 'time of flight' (ToF), or 'dark blood', techniques producing magnetic resonance angiograms (MRAs). MRI can be sensitized to water diffusion and diffusibility (DWI – diffusion weighted MRI), particularly useful in detecting cellular edema in, for instance, stroke, where the decrease in water diffusibility as cells absorb water and swell is

NMR in Pharmaceutical Sciences. Edited by Jeremy R. Everett, John C. Lindon, Ian D. Wilson, and Robin K. Harris
© 2015 John Wiley & Sons, Ltd. ISBN: 978-1-118-66025-6
Also published in eMagRes (online edition)
DOI: 10.1002/9780470034590.emrstm1420

the earliest imaging marker of 'area at risk'; DWI and has been used in nerve 'tractography.'

The advantages of noninvasive techniques such as MRI for characterizing pathology-specific anatomical and functional markers have been frequently discussed. Briefly, patients/volunteers can be scanned as often as necessary and resources permit to plot the course of disease progression and response to treatment; pretreatment scans enable each volunteer (normal subject or patient) to act as their own control, and can be used before the treatment commences to minimize accidental bias in assignment to treatment groups. Because MRI is generally accepted as essentially totally safe (provided standard precautions are taken against hazards from ferromagnetic objects), there is no limit in principle to how often or frequently volunteers can be scanned. Besides these real scientific advantages, offering a scanning technique to volunteers, which does not involve exposure to ionizing radiation or radioactivity, enhances volunteer perceptions of safety and can facilitate recruitment to a study.

The advantages of a costly and resource-intensive technique such as MRI in a clinical trial of months- or years-long duration need to be clear and compelling. Significant questions are: Will MRI provide unique decision-influencing data unobtainable by other routes? Can equivalent information be obtained using other, likely less expensive imaging techniques such as ultrasonography or planar X-radiography? Will MRI offer higher quality information, whether unique or not, and thereby a stronger treatment effect 'signal' producing robust data using fewer subjects, shorter study durations, and more 'volunteer friendly' procedures? Will the patient groups tolerate (e.g., cardiovascular and asthmatic) and cooperate both during the course of scanning (e.g., ability to keep still, or do controlled breath holds) and in terms of attending scanning sessions consistent with the trial protocol (e.g., demented and Alzheimer's, or psychiatric, subjects) with extended scanning? Ultimately, medical imaging enhances the ability to quantify the impact of drugs on human health compared to conventional or invasive clinical assessments. The increased power of the experiments can reduce patient numbers and/or duration of exposure to test compounds and hence accelerate drug trials and render them safer.

This chapter will not comprehensively review all pharmacological interventional clinical trials. Instead, we concentrate on studies illustrating fundamental principles of clinical drug trial conduct (as well as the requisite MRI methodology for specific

pathologies), and which have resulted in citeable journal publications and discussion rather than mere uncritical data tabulation in databases.

26.2 BIOMARKERS AND SURROGATE MARKERS

Conventional markers of drug efficacy are clinical outcomes: does the treatment make the patient get better, feel better or suffer less pain, enjoy better quality of life, or simply survive longer. In many chronic degenerative diseases (e.g., arthritis, Alzheimer's, and atherosclerosis), a positive drug effect 'signal' can only be reliably discerned after months or even years of treatment. A 'biomarker' is a readily measureable parameter that reflects aspects of disease severity but which in itself is not fully predictive of outcome, e.g., blood protein markers of inflammation in rheumatoid arthritis (RA). A few biomarkers may eventually be accepted as robust predictors of clinical outcome, such as hypertension in occlusive vascular disease or elevated low-density lipoprotein (LDL) cholesterol in dyslipidemia, and are referred to as *surrogate biomarkers* and accepted as alternatives to clinical outcomes. Surrogate markers favorably affected by therapy can then replace conventional outcome measures as indicators of efficacy. Imaging in clinical trials is usually used to provide a biomarker of disease[1-3], e.g., carotid artery atherosclerosis as a surrogate imaging marker,[4] not yet shown to relate to clinical outcome, e.g., stroke or survival, but used by pharmaceutical companies for internal decision making on progression of a drug project to the next phase of clinical development. In some instances, imaging is used as a surrogate biomarker for new drug registration, e.g., change in tumor size for certain types of cancer, or reduction in the number of multiple sclerosis lesions, after treatment (see latter sections).

The FDA's Guidance for Industry Standards for Clinical Trial Imaging Endpoints[5] represents the FDA's current thinking on the use of imaging endpoints in confirmatory clinical trials, e.g., phase 3 trials intended to confirm a drug's clinical efficacy conclusively. Imaging acquisition and interpretation standards can be divided into a *medical practice* standard or a *clinical trial* standard. Briefly, the medical practice imaging standard may rely on an investigator's response to a clinical question such as 'What is a patient's cardiac ejection fraction' determined by any available medical practice method. The clinical trial imaging standard specifies the standardization of image acquisition and analysis in order to enhance the ability to detect a drug effect and to verify data integrity. Clinical trial imaging standards exceed those used in medical practice.

Imaging has long been used in phase 1 and phase 2 trials in addition to its use in confirmatory clinical trials. Danhof and colleagues[6] present the definition of a biomarker as 'a measure that characterizes, in a strictly quantitative manner, a process, which is on the causal path between drug administration and effect', of which they have classified seven classes:

Type 0: Genotype or phenotype, e.g., Is the pharmacological target (receptor, enzyme, transporter, etc.) expressed, where, at what level, and are there isoforms?

Type 1: Concentration of drug and/or metabolite, i.e., Does the active compound (and any possibly active metabolites) actually reach its pharmacological target?

Type 2: Target occupancy, e.g., What proportion of the pharmacological target population does the drug bind to, and what proportion of the drug binds?

Type 3: Target activation, determined by the intrinsic efficacy of the drug and the level of expression of the receptor in the target tissue.

Type 4: Physiological measures or laboratory tests, in integrated biological systems, e.g., hemodynamic measurements in testing antihypertensives.

Type 5: Disease progression, e.g., inflammatory protein markers in RA.

Type 6: Clinical scales, e.g., cognitive tests in neurodegenerative diseases.

MRI has been used to phenotype patients to provide either a prognostic biomarker that characterizes patients by degree of risk for disease occurrence or progression, or a predictive biomarker that characterizes patients by their likelihood of response to a particular treatment, an example of the use of MRI outputs as a Type 0 biomarker. However, its main clinical application to date has been to measure physiology (Type 4 biomarker) and disease progression (Type 5 biomarker) in phase 2 trials.

26.3 PHASES OF CLINICAL TRIALS

When preclinical safety and efficacy results are satisfactory a new compound or biological will be progressed into clinical trials, broadly categorizable into four different phases:

Phase 1 includes the 'first time in man' administration to human volunteers, usually healthy, principally to assess safety rather than efficacy. Initial low doses are gradually escalated and absorption, metabolism, duration in the body, and any safety issues will be assessed.

Phase 2 progresses the therapy into small groups of patients (up to a few hundred in total) who display the target disease indication, to assess efficacious dose, delivery route and frequency, and safety as patient responses may differ from those of healthy volunteers.

It is in Phases 1 and, especially, 2 that MRI has so far been most effectively applied.

Phase 3 trials assess efficacy in large groups of patients (several hundred to several thousand) and are often conducted across numerous centers and in more than one country. Successful outcomes result in submission of an NDA (new drug application) to relevant registration agencies such as the FDA and European Medicines Agency before commercial launch.

Phase 4 studies are conducted 'post marketing' after launch to identify any long-term side effects and benefits, and optimize use, and can span many years and involve thousands of subjects.

Phase 1 and 2 studies will normally apply clinical trial imaging standards and use imaging endpoints for internal decision making. Phase 3 studies with imaging as a primary endpoint will often use medical practice imaging standards and form part of the submission package to regulators.

26.4 THERAPEUTIC AREAS

There are a number of therapeutic areas in which the application of imaging techniques, including, and often especially, MRI, is particularly compelling. These include disorders involving degeneration of soft and connective tissue, changes in soft tissue morphology, pathological deposition or modification of tissue, and changes in tissue vascularity and perfusibility. A few organizations may be fortunate enough to have access to preclinical in vivo MRI facilities and in vivo disease models, in which case the best scanning protocols to highlight the pathology can, to some extent, be worked out in advance of committing to a study in man (see Chapter 20).

Practically all clinical trials and relevant essentials, e.g., purpose, treatment, sponsor, partners, and collaborating centers, volunteer numbers and demographics, exclusion criteria, endpoints, results and analyses, and relevant journal publications (if any), are captured in websites such as clinicaltrials.gov and clinicaltrialsregister.eu. A search using the term *MRI* returns hundreds of hits, many dealing with optimization of imaging methodology for specific pathologies, and with characterizing MRI markers of pathologies and the 'natural history' of their progression. These can be essential to subsequent drug intervention trials as providing the baseline against which a drug effect must be assessed, and quantifying test–retest and interindividual variation, and hence patient numbers necessary for sufficient statistical power in subsequent interventional trials. It appears the case that citeable journal publications tend to arise from trials showing positive treatment effects, so the journal literature tends to be skewed toward reports of positive drug trials. In addition to government agency websites, many pharmaceutical companies now publish more or less detailed accounts of all their clinical trials and outcomes in open access websites, and may consider requests from researchers for anonymized patient-level data for research. In this article, we have used citeable journal publications from industry sponsored pharmaceutical interventional trials, wherever possible, to exemplify the incorporation of MRI.

26.4.1 Arthritis

MRI has been extensively applied in RA and osteoarthritis (OA), where it is a frontline diagnostic technique for evaluation of cartilage erosion, synovitis, pannus formation, bone marrow edema, and synovial permeability using DCE imaging. MRI in drug trials is particularly compelling because progression tends to be slow so disease modifying therapy will only gradually become apparent above the natural course of degeneration. Nonimaging disease and progression biomarkers are imprecise (e.g., joint swelling and tenderness, pain, disability scores) or indirect (e.g., circulating inflammatory marker proteins) so, while indispensable, more objective and precise

MRI markers of actual disease progression have considerable potential to shorten trials and decrease volunteer numbers required for statistical power.

Semiquantitative reading of images by blinded expert readers is particularly important in this disease area and assuring consistency between repeat reads by the same reader, and between readers, is a significant factor in designing a trial (see later). There are also several alternative scoring systems, for instance, WORMS (Whole Organ MR Scoring) and BLOKS (Boston-Leeds OA Knee Scoring) which are comparably robust.[7] A study exemplifying the principles of MRI in arthritis treatment trials shows therapy slowing the degeneration of MRI bone erosion score assessed using RAMRIS (Rheumatoid Arthritis Magnetic Resonance Image Scoring System) and OMERACT (Outcome Measures in Rheumatoid Arthritis Clinical Trials) scoring systems.[8]

26.4.2 Alzheimer's

As in arthritis, inclusion of MRI in neurodegenerative diseases trials such as Alzheimer's is particularly compelling. Conventional measures such as behavioral and cognitive outcomes are imprecise and subjective. Neurodegeneration measured by MRI correlates with conventional symptomatic deterioration so it is likely that disease modifying treatments will give a more precise MRI 'signal', again potentially shortening trials and requisite volunteer numbers. The small number of trials posting journal publications is a reflection not on this hypothesis but on the failure until now of treatments to ameliorate disease. MRI can play a significant role in pre-trial screening to exclude certain patients, who, for instance, show signs of mini stroke. MRI here has a dual role: measuring brain atrophy but also looking at safety in the drug trial. The EMA has advised that low hippocampal MRI volume can be used to select patients with early stage cognitive impairment for Alzheimer's disease clinical trials.[9]

Total brain volume (Figure 26.1) and volume of specific structures such as the hippocampus are typical neurodegenerative markers. In a one-year study of the effects of the acetylcholine esterase inhibitor memantine on total brain atrophy (TBA), this was shown to correlate well with cognitive and behavioral measures.[10] Although no positive disease modifying effect of treatment was demonstrated this work exemplifies the incorporation of MRI into drug trials in this therapeutic area.

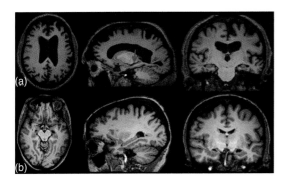

Figure 26.1. *Alzheimer's disease*: Slices from multislice T1-weighted spin echo images of the brains of (a) an Alzheimer's disease patient and (b) an age-matched healthy control. Image orientations are (from left to right) axial, sagittal, and coronal. The imaging method has been chosen to render brain tissue bright against a dark background of cerebrospinal fluid. Brain and gray matter shrinkage, extreme enlargement of the ventricular spaces, and atrophy (blue arrow) of the hippocampus (red arrows), in the Alzheimer's patient are readily obvious. It is straightforward from such image sets to extract and calculate neurodegeneration marker parameters such as total brain, gray matter, and ventricular volumes. (Acknowledgements: Dr. Paul Mullins, Bangor University Imaging Unit, and the NeuroSKILL project)

26.4.3 Cardiovascular Disease (CVD) and Cardiac Remodeling

Cardiac anatomical, functional, and myocardial perfusion changes are potentially highly significant consequences of several common disease states and also of some drug treatments. Cine MRI triggered in phase with the cardiac cycle is the method of choice for measurement of a number of critical disease marker parameters including ejection fraction, stroke volume, cardiac output, filling and ejection rates, ventricular hypertrophy, myocardial perfusion (using DCE), and, with image guided spectroscopy, myocardial triglyceride fat. Cardiac magnetic resonance imaging (cMRI) provides excellent interstudy reproducibility, resulting in reliable estimates of observed changes and reduced participant numbers in clinical trials[11] with MRI preferred for volume and left ventricular ejection fraction (LVEF) estimation because of its 3D approach, and superior image quality.[12] Application of cMRI in trials is well illustrated by a safety study of rosiglitazone effects on CV performance and cardiac function.[13,14]

The Cardiac Safety Research Consortium (CSRC) fosters collaborations between academics, industry, and regulators in the assessment of cardiac and vascular safety of new medical products, for instance, in drug-induced myocardial toxicity.[15] The report was issued because it has been demonstrated that, for example, the anthracycline-induced myocyte lesion is a predictable result of cumulative dosing, and imaging tools used to detect LV dysfunction could be used to guide the subsequent dosing of doxorubicin.

26.4.4 Atherosclerosis

Atherosclerosis is another chronic condition where MRI is uniquely suitable for observing and quantifying pathological lesions directly and precisely,[16] (Figure 26.2) and monitoring their progression or regression under therapy.[17] Zhao *et al.*[18–20] give a good MRI methodology description, describing measurement of arterial (aorta, carotids) thickness, plaque evolution, plaque inflammation (by increasing K_{trans} measured by DCE), presence of lipid rich necrotic core, plaque composition, and the effects of various treatments. The benefits of lipid lowering therapy on carotid artery plaque lipid content are demonstrable by a combination of different contrast techniques (bright blood ToF, PDW, T1W, T2W, and contrast-enhanced T1W). Cilostazol has been shown using cerebral MR

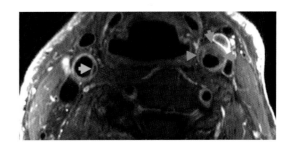

Figure 26.2. *Atherosclerosis*: Representative slice from multislice T2-weighted spin echo cardiac and respiratory synchronized dataset from the upper chest of an atherosclerotic patient. Various quantifiable imaging disease biomarkers in both of the common carotid arteries are arrowed: Red – Vascular calcification (Low signal); green – fibrous cap; blue – lipid rich necrotic core. Volumes of all three biomarker deposits are readily measured from the full multislice dataset, from which slowing or reversal of disease progression under therapy can be quantified. (Acknowledgements: Dr. Lars Johansson, University of Uppsala)

angiography (MRA) to reduce progression of intracranial arterial stenosis by decreasing remnant lipoprotein cholesterol.[21] 3D dark blood MRI of atherosclerosis in thoracic and abdominal aorta was used to evaluate the effects of the renin–angiotensin–aldosterone system (RAAS) inhibitor aliskiren on vessel wall thickness.[22,23]

26.4.5   Metabolic Diseases

In metabolic conditions such as diabetes and obesity MRI is well suited to measure disease-specific markers such as adipose tissue distribution, both organ specific or whole-body. Thus, MRI can distinguish between and quantify hepatic, visceral, and peripheral fat and inter- and intramyocellular fat, each of which pools carries specific risks. This potential is well illustrated by a study of the effects of the anti-obesity treatment orlistat, measuring and quantifying fat distribution.[24]

26.4.6   Psychiatry

Psychiatric disorders are a particularly attractive, although challenging, field for application of functional magnetic resonance imaging (fMRI) techniques, because they promise precise, localized measurements of abnormal brain function, brain activation, and response of brain activity to therapy. As in neurodegeneration, fMRI offers objectivity, quantitation, and precision, in comparison with conventional psychological testing. Image-guided spectroscopy can quantify high energy and other pathology-linked metabolites (e.g., phosphocreatine, ATP, *N*-acetyl aspartate, lactate, glutamate, and choline). Additionally, MRI can confirm that the test treatment is engaging receptors at the site and structures in the brain known to be associated with the disorder. However, applications are fraught with complications, particularly controlling for effects of normal, and drug-induced, changes in blood pressure that confound the origin of the fMRI signal.

A study of the effects of sertraline in geriatric depression (NCT00245557, no publications at time of writing) uses MRS (magnetic resonance spectroscopy) to measure critical brain metabolites in geriatric depression, and compare and correlate these with depression and geriatric depression rating scales. A study has been reported (NCT01051466) describing the use of fMRI (BOLD) in the amygdala brain structures of participants with major depressive disorder (MDD) to explore the mode and site of action of the SSRI sertraline.

fMRI has also been used to show target activation (Type 3 biomarker) in anti-obesity studies. Fletcher *et al.*[25] showed that a sibutramine-induced modulation of the hypothalamic response was correlated with the drug's impact on both weight and subsequently measured ad libitum eating. The satiety-induced modulation of striatal response also correlated with subsequent ad libitum eating.

26.4.7   Cancer

MRI methodologies are at the forefront of detection and characterization of many types of solid tumors and MRI has the potential to precisely measure solid tumor regression, and vital properties such as tumor vascularity (DCE-MRI). This potential is acknowledged by publication of recommendations for the application of MRI in clinical trials of antiangiogenic and antivascular therapies.[26,27] The need to qualify novel biomarkers and to standardize the way they are applied are questions ideally suited to precompetitive research and public–private partnerships. A good example is the QuIC-ConCePT (Quantitative Imaging in Oncology: Connecting Cellular Processes to Therapy) consortium initiated by the Innovative Medicines Initiative (IMI) jointly between the European Union and the European Federation of Pharmaceutical Industries and Associations.[28,29] They work toward standardization of acquisition and analysis, image-pathology correlation, cross-sectional clinical–biomarker correlations, and correlation with outcome. The aim is to qualify imaging biomarkers of tumor cell proliferation, apoptosis, and necrosis, which will allow drug developers to reliably demonstrate the modulation of these pathologic processes in tumors in cancer trials.

RECIST[30] is a set of published rules that define when cancer patients improve ('respond'), stay the same ('stable'), or deteriorate ('progression') during treatments. The criteria were evolved by an international collaboration including the European Organization for Research and Treatment of Cancer (EORTC), National Cancer Institute (NCI) of the United States, and the National Cancer Institute of Canada Clinical Trials Group. With the importance of MRI in solid tumor detection and evaluation, RECIST includes guidelines on acquisition and measurement methodology with a principal goal being consistency between repeat patient scans, and between collaborating sites.

26.4.8 Multiple Sclerosis (MS)

Although a relatively small market, MRI has found intensive use in MS trials, probably because MRI gives more objective, precise, and reproducible measures of drug efficacy than conventional clinical scores.[31] A wide variety of MRI disease markers have been applied in MS trials, including the presence and number of Gd-enhancing T1-lesions, T2-lesion volumes, and magnetization transfer as a measure of demyelination or remyelination.[32] All these measures can be used to assign volunteers to treatment groups.[33] Recommendations for improving the use of MRI in MS trials: (i) images should be acquired using 3D pulse sequences, with near-isotropic spatial resolution and multiple image contrasts to allow more comprehensive analyses of lesion load, and atrophy, across timepoints. Image artifacts need special attention given their effects on image analysis results. (ii) Automated image segmentation methods integrating the assessment of lesion load and atrophy are desirable. (iii) A standard dataset with benchmark results should be set up to facilitate development, calibration, and objective evaluation of image analysis methods for MS.[34]

26.4.9 Stroke

Stroke is another potentially rewarding, but challenging, area of application. DWI methods can delineate 'area at risk' shortly (hours) after an ischemic event, and correlation between these volumes and those of later T1-, T2-, and DCE-visible infarct zones can indicate the efficacy of 'acute' interventions such as early thrombolytic treatment, or blockade of excitatory neurotransmitters or ion channel activity. While this area remains a major application of preclinical MRI, practical problems in clinical trials are considerable: scanning patients shortly after the ischemic event to establish area at risk is often impractical and/or undesirable clinically, and subsequent scans to assess efficacy of chronic therapy can be confounded by inability of severely compromised patients to cooperate with the procedure. As for Alzheimer's the relatively low number of journal publications may reflect these challenges, and possibly the lack of positive outcomes to current therapy.

A study of the effects of cilostazol in intracranial arterial stenosis[29] used frequency of recurrence, and of new, lesions relative to baseline scans, as outcome measures. A study of the efficacy of a synthetic peptide in iatrogenic stroke after endovascular aneurysm repair shows a beneficial effect using DWI and fluid-attenuated inversion recovery (IR) MRI on number of infarcts.[35]

26.5 CLINICAL TRIALS CONDUCT: SPECIFIC CONSIDERATIONS

Most MRI scans are performed with the aim of obtaining qualitative information to assist diagnosis. The aims and therefore conduct of a clinical trial incorporating imaging are fundamentally different, and the NMR scientist involved in a trial needs to pay particular attention to these differences. The processes of setting up and conducting an imaging clinical drug trial are in some respects similar to those followed in a preclinical drug efficacy or safety MRI study (see Chapter 20). Indeed, the preclinical activities can be an informative predictive platform for translating imaging biomarkers into clinical research. The clinical lead scientist or physician (principal investigator, PI) for the therapeutic area, and imaging specialists and statisticians, who may be members of the sponsoring company or a partner contract research organization (CRO), review the disease target and the role imaging can play in enhancing decision making (efficacy, safety, and patient stratification-inclusion/exclusion). If imaging is a 'go-forward' decision it is incorporated in the clinical protocol. Regulators have issued guidance on the Qualification Process for Drug Development Tools,[36] and on defining Best Practice for clinical trial imaging endpoints.[37,38]

The sponsor or CRO imaging specialists then submit an imaging capability questionnaire (ICQ) to potential clinical sites to assess their capabilities; the overwhelming majority, if not all, trials using imaging are carried out in collaborating clinical radiology departments. Areas of particular significance are: Imaging hardware, covering field strength, e.g., 1.5 vs 3 T, availability and performance of coils specific to the anatomical area of interest, e.g., head, hand, hip, knee, and chest, scanner age, maintenance and calibration procedures, qualifications and experience of lead site technologists, previous experience in imaging clinical trials, and image storage, backup, and transmission capabilities. Given satisfactory answers to the ICQ, an

MRI manual is developed specifying: Details of MRI methodology including acquisition software and version, coils, positioning of volunteers and the anatomical area of interest, and the field of view (FOV) needed to cover the target anatomy. The need for reproducibility within a site and consistency between sites can be contradictory to the needs of diagnostic imaging, which is concerned with producing the best quality image to visualize a specific pathology in a given individual. This frequently involves 'tweaks' to pulse sequence parameters, FOV positioning and image slice orientation, and of course upgrading hardware and software to the latest and best affordable. Such changes in scanning protocols and hard/software may be unacceptable or only acceptable within narrow specified limits in a clinical trial, and may be a serious consideration for a center planning to participate in a long term trial. All the sites have to use the same MRI protocol for a specific scanner (GE, Siemens, and Philips). A site has to then 'lock' the MRI protocol for the full duration of the study. No changes are allowed. All the three major manufacturers provide images in the acceptable universal DICOM (Digital Imaging and Communications in Medicine) format; labeling volunteer information and images, including a site ID, patient ID, a visit code, and an examination date, must be decided and standardized, as well as the means of image transmission to the Central/Core Imaging Lab (CIL), whether on 'hard' storage media such as CD/DVD, or web upload or ftp, and ensuring that no Personal Identifiable Information (PII) is transmitted to the CIL.

All participating sites must be trained to implement a standardized imaging protocol, followed by submission of test data acquired on a standard 'phantom' set to confirm instrument performance and reproducibility against specifications for review by the CIL. Mixing scanner manufacturers is a quite common necessity in multicenter trials, but the imaging CRO has to test MRI methodologies on all manufacturers' machines to ensure that scan appearance is as consistent as possible. A single field strength, currently 1.5 T, is preferable but if a sponsor insists, then 3 T could also be brought in. It is best not to use any other field as there is not enough information in the literature for comparison with 1.5 and 3 T. The goal is to minimize variability by keeping the systems as homogenous as possible. Once test data is approved, the clinical site is authorized by a formal authorization letter to start scanning patients for the clinical trial.

Before images are loaded into the CIL database, volunteer demographics (subject ID, visit ID, exam date, age, date of birth, gender, etc.) must be reviewed against information provided by the sponsor, and discrepancies, inconsistencies, and incomplete data queried with the relevant site and all images must pass a quality control examination. Time-point blinding or unblinding of the datasets may be required. In OA and RA trials time-point blinding is extensively used, so that cartilage volume, for instance, is computed independently for each time-point. Time-point unblinding is generally limited to oncology studies where disease state such as progressive disease, stable disease, partial response, or complete response is generally continuously reported. A rigorous image 'reading system', using volunteer test data must be developed and analyzed for interoperator and intersite consistency with specifications. Electronic Case Report Forms (CRFs), whereby the sponsor collects data from the participating sites, will be customized to the scoring methodology. A validation report showing that the reading system consistently produces the expected results may be required for regulatory compliance. Image reader training and qualification depends on whether semiquantitative MRI analysis and visual readouts are employed, such as the formalized RAMRIS in RA, and WORMS or BLOKS in OA, or whether quantitative MRI analysis is possible and desirable, such as cartilage segmentation in OA, bone marrow edema and bone erosion in RA, or brain volume changes in Alzheimer's disease.

Protocols for test and trial data export need to be programmed. Data Transfer to the sponsor will address issues such as: details of the data structure and format and the data variables to be transferred, and whether data transfer will be cumulative or individual. 'Data' usually consists of scores, brain volumes, cartilage thickness, and so on. Some sponsors also request transfer of images for additional analysis, by themselves or an independent key opinion leader (KOL). Typically, the data transfer file contains variables, e.g., 'H-vol' for hippocampus volume, with a value attached, e.g., '5 ml.' A database audit is necessary before the final transfer, data encryption and delivery is finalized, and data storage and archiving methodology agreed.

Patient welfare, and nonimaging endpoint (e.g., blood markers and cognitive testing) and safety data gathering, throughout the trial, is the responsibility of the site and sponsor. However, during the trial if the imaging CRO's reader notices any adverse findings in the scans (tumor or stroke in an Alzheimer's disease patient for instance), then this is reported to the site and sponsor, and the PI is asked to investigate further.

26.6 FUTURE DEVELOPMENTS

Multimodality imaging such as combined MRI–positron emission tomography (PET) is becoming available in commercial integrated clinical scanners in which the high image resolution and soft tissue definition of MRI is combined with the high sensitivity of PET to track metabolic activation (using ^{18}F-fluorodeoxyglucose, FDG) and target organ drug levels and drug-receptor binding using PET isotope labeled drugs or probe molecules. In respect of specific strengths already mentioned, and weaknesses, MRI and PET are fully complementary. MR is unable to directly detect molecules (metabolites or drugs) present at tissue concentrations less the millimolar level, and even at these levels detection can be compromised by line broadening from, for instance, interactions with macromolecules. PET offers no direct information about anatomical location of tracer molecules, apart from that available from externally positioned phantoms, and apart from some commercially available metabolic (FDG) and receptor probe molecules, requires synthetic route development and on-site synthesis of tracer molecules owing to the short half-lives of positron emitting biologically relevant isotopes (e.g., ^{18}F ca 110 min and ^{11}C ca 20 min). Although studies to date in the clinical trials registries have been of a validatory or method development nature, enormous potential clearly exists for combination techniques in drug trials. Any of the MRI approaches described earlier in this chapter could be combined with appropriate tracer PET imaging to prove, for instance, that a drug accesses its target site (representative of a Type 1 biomarker) or engages its receptor (Type 2 biomarker).

It is certain that the uses of MRI in clinical trials that have already been described in this chapter will increase and begin to encompass more multicenter phase 3 activities. There is an expectation that pharmaceutical companies, CROs, CILs, and regulatory authorities will need to work together in order to harmonize the endpoints used in clinical trials so that regulators will be able to relate imaging biomarkers to clinical outcome. The overall effect will be a major contribution to shorter, smaller, and more powerful, clinical trials, reduction of drug development attrition rates and time scales, and generally progressing new therapies into patients more rapidly and safely and at lower overall cost.

ACKNOWLEDGMENTS

DGR acknowledges numerous GlaxoSmithKline former colleagues: In the MRI group (apart from the coauthors) Albert Busza, Simon Campbell, Kumar Changani, John Hare, and Paul Mullins, colleagues in GSK's clinical groups, in particular Jim Semple, and among collaborators especially Professor David Doddrell and his colleagues at the Centre for Magnetic Resonance at University of Queensland. NS would like to acknowledge ex-colleagues at Imperial College London (UK), GSK and Pfizer, and also colleagues at BioClinica.

REFERENCES

1. H. H. Pien, A. J. Fischman, J. H. Thrall, and A. G. Sorensen, *Drug Discov. Today*, 2005, **10**, 259.

2. Y. X. Wang, *Clin. Radiol.*, 2005, **60**, 1051.

3. Y. X. Wang and M. Deng, *J. Thorac. Dis.*, 2010, **2**, 245.

4. R. Duivenvoorden, E.de Groot, E. S. Stroes, and J. J. Kastelein, *Atherosclerosis*, 2009, **206**, 8.

5. Website of the Food and Drug Administration (FDA), Draft guidance for industry standards for clinical trial imaging endpoints, 2011. http://www.fda.gov/downloads/Drugs/GuidanceComplianceRegulatory Information/Guidances/UCM268555.pdf

6. M. Danhof, G. Alvan, S. G. Dahl, J. Kuhlmann, and G. Paintaud, *Pharm. Res.*, 2005, **22**, 1432.

7. J. A. Lynch, F. W. Roemer, M. C. Nevitt, D. T. Felson, J. Niu, C. B. Eaton, and A. Guermazi, *Osteoarthr. Cartilage*, 2010, **18**, 1393.

8. S. B. Cohen, R. K. Dore, N. E. Lane, P. A. Ory, C. G. Peterfy, J. T. Sharp, D.van der Heijde, L. Zhou, W. Tsuji, and R. Newmark, *Arthritis Rheum.*, 2008, **58**, 1299.

9. Website of the European Medicines Agency (EMA), Qualification opinion of low hippocampal volume (atrophy) by MRI for use in clinical trials for regulatory purpose – in pre-dementia stage of Alzheimer's disease, 2011. http://www.ema.europa.eu/docs/en_GB/document_library/Regulatory_and_procedural_guideline/2011/12/WC500118737.pdf.

10. D. Wilkinson, N. C. Fox, F. Barkhof, R. Phul, O. Lemming, and P. Scheltens, *J. Alzheimers Dis.*, 2012, **29**, 459.

11. F. Grothues, G. C. Smith, J. C. Moon, N. G. Bellenger, P. Collins, H. U. Klein, and D. J. Pennell, *Am. J. Cardiol.*, 2002, **90**, 29.

12. N. G. Bellenger, M. I. Burgess, S. G. Ray, A. Lahiri, A. J. Coats, J. G. Cleland, and D. J. Pennell, *Eur. Heart J.*, 2000, **21**, 1387.

13. D. K. McGuire, S. M. Abdullah, R. See, P. G. Snell, J. McGavock, L. S. Szczepaniak, C. R. Ayers, M. H. Drazner, A. Khera, and J. A. de Lemos, *Eur. Heart J.*, 2010, **31**, 2262.

14. D. K. McGuire, R. See, S. M. Abdullah, P. G. Snell, J. M. McGavock, C. R. Ayers, and L. S. Szczepaniak, *Diab. Vasc. Dis. Res.*, 2009, **6**, 43.

15. J. B. Christian, J. K. Finkle, B. Ky, P. S. Douglas, D. E. Gutstein, P. D. Hockings, P. Lainee, D. J. Lenihan, J. W. Mason, P. T. Sager, T. G. Todaro, K. A. Hicks, R. C. Kane, H. S. Ko, J. Lindenfeld, E. L. Michelson, J. Milligan, J. Y. Munley, J. S. Raichlen, A. Shahlaee, C. Strnadova, B. Ye, and J. R. Turner, *Am. Heart J.*, 2012, **164**, 846.

16. R. D. Santos and K. Nasir, *Atherosclerosis*, 2009, **205**, 349.

17. C. Yuan, W. S. Kerwin, V. L. Yarnykh, J. Cai, T. Saam, B. Chu, N. Takaya, M. S. Ferguson, H. Underhill, D. Xu, F. Liu, and T. S. Hatsukami, *NMR Biomed.*, 2006, **19**, 636.

18. L. Dong, W. S. Kerwin, H. Chen, B. Chu, H. R. Underhill, M. B. Neradilek, T. S. Hatsukami, C. Yuan, and X. Q. Zhao, *Radiology*, 2011, **260**, 224.

19. X. Q. Zhao, L. Dong, T. Hatsukami, B. A. Phan, B. Chu, A. Moore, T. Lane, M. B. Neradilek, N. Polissar, D. Monick, C. Lee, H. Underhill, and C. Yuan, *JACC Cardiovasc. Imaging*, 2011, **4**, 977.

20. X. Q. Zhao, B. A. Phan, B. Chu, F. Bray, A. B. Moore, N. L. Polissar, J. T.Dodge Jr, C. D. Lee, T. S. Hatsukami, and C. Yuan, *Am. Heart J.*, 2007, **154**, 239.

21. D. E. Kim, J. Y. Kim, S. W. Jeong, Y. J. Cho, J. M. Park, J. H. Lee, D. W. Kang, K. H. Yu, H. J. Bae, K. S. Hong, J. S. Koo, S. H. Lee, B. C. Lee, M. K. Han, J. H. Rha, Y. S. Lee, G. M. Kim, S. L. Chae, J. S. Kim, and S. U. Kwon, *Stroke*, 2012, **43**, 1824.

22. J. Deiuliis, G. Mihai, J. Zhang, C. Taslim, J. J. Varghese, A. Maiseyeu, K. Huang, and S. Rajagopalan, *J. Hum. Hypertens.*, 2014, **28**, 251.

23. G. Mihai, J. Varghese, T. Kampfrath, L. Gushchina, L. Hafer, J. Deiuliis, A. Maiseyeu, O. P. Simonetti, B. Lu, and S. Rajagopalan, *J. Am. Heart Assoc.*, 2013, 2.

24. E. L. Thomas, A. Makwana, R. Newbould, A. W. Rao, G. Gambarota, G. Frost, B. Delafont, R. G. Mishra, P. M. Matthews, E. S. Berk, S. M. Schwartz, J. D. Bell, and J. D. Beaver, *Eur. J. Clin. Nutr.*, 2011, **65**, 1256.

25. P. C. Fletcher, A. Napolitano, A. Skeggs, S. R. Miller, B. Delafont, V. C. Cambridge, S.de Wit, P. J. Nathan, A. Brooke, S. O'Rahilly, I. S. Farooqi, and E. T. Bullmore, *J. Neurosci.*, 2010, **30**, 14346.

26. M. O. Leach, K. M. Brindle, J. L. Evelhoch, J. R. Griffiths, M. R. Horsman, A. Jackson, G. Jayson, I. R. Judson, M. V. Knopp, R. J. Maxwell, D. McIntyre, A. R. Padhani, P. Price, R. Rathbone, G. Rustin, P. S. Tofts, G. M. Tozer, W. Vennart, J. C. Waterton, and S. R. Williams, *Br. J. Radiol.*, 2003, **76**, S87.

27. M. O. Leach, K. M. Brindle, J. L. Evelhoch, J. R. Griffiths, M. R. Horsman, A. Jackson, G. C. Jayson, I. R. Judson, M. V. Knopp, R. J. Maxwell, D. McIntyre, A. R. Padhani, P. Price, R. Rathbone, G. J. Rustin, P. S. Tofts, G. M. Tozer, W. Vennart, J. C. Waterton, S. R. Williams, and P. Workmanw, *Br. J. Cancer*, 2005, **92**, 1599.

28. J. C. Waterton and L. Pylkkanen, *Eur. J. Cancer*, 2012, **48**, 409.

29. J. M. Jung, D. W. Kang, K. H. Yu, J. S. Koo, J. H. Lee, J. M. Park, K. S. Hong, Y. J. Cho, J. S. Kim, S. U. Kwon, and Toss-Investigators, *Stroke*, 2012, **43**, 2785.

30. Website of the European Organization for Research and Treatment of Cancer (EORTC), Response evaluation criteria in solid tumors (RECIST), 2014. http://www.eortc.org/investigators-area/recist.

31. A. C. Evans, J. A. Frank, J. Antel, and D. H. Miller, *Ann. Neurol.*, 1997, **41**, 125.

32. R. Zivadinov, M. G. Dwyer, S. Markovic-Plese, C. Kennedy, N. Bergsland, D. P. Ramasamy, J. Durfee, D. Hojnacki, B. Hayward, F. Dangond, and B. Weinstock-Guttman, *PLoS One*, 2014, **9**, e91098.

33. B. M. Segal, C. S. Constantinescu, A. Raychaudhuri, L. Kim, R. Fidelus-Gort, L. H. Kasper, and Ustekinumab MS Investigators, *Lancet Neurol.*, 2008, **7**, 796.

34. H. Vrenken, M. Jenkinson, M. A. Horsfield, M. Battaglini, R. A.van Schijndel, E. Rostrup, J. J. Geurts, E. Fisher, A. Zijdenbos, J. Ashburner, D. H. Miller, M. Filippi, F. Fazekas, M. Rovaris, A. Rovira, F. Barkhof, N. de Stefano, and Magnims Study Group, *J. Neurol.*, 2013, **260**, 2458.

35. M. D. Hill, R. H. Martin, D. Mikulis, J. H. Wong, F. L. Silver, K. G. Terbrugge, G. Milot, W. M. Clark, R. L. Macdonald, M. E. Kelly, M. Boulton, I. Fleetwood, C. McDougall, T. Gunnarsson, M. Chow, C. Lum, R. Dodd, J. Poublanc, T. Krings, A. M. Demchuk, M.

Goyal, R. Anderson, J. Bishop, D. Garman, M. Tymianski, and Enact Trial Investigators, *Lancet Neurol.*, 2012, **11**, 942.

36. Website of the Food and Drug Administration (FDA) qualification process for drug development tools http://www.fda.gov/downloads/Drugs/GuidanceCompliance RegulatoryInformation/Guidances/UCM230597.pdf.

37. R. Ford and P. D. Mozley, *Drug Info. J.*, 2008, **42**, 515.

38. Website of the European Society of Radiology, 2013. http://www.ncbi.nlm.nih.gov/pmc/articles/PMC3609959/.

FURTHER READING

Books

MRI theory

P. T.Callaghan, *Principles of NMR Microscopy*, Oxford University Press: Oxford, 1994.

MRI methodology

D. W. MacRobbie, E. A. Moore, M. J. Graves, and M. R. Prince, *MRI from Picture to Proton*, Cambridge University Press: Cambridge, 2006.

D. Weishaupt, V. D. Kochli, and B. Marincek, *How Does MRI Work? An Introduction to the Physics and Function of Magnetic Resonance Imaging*, 2nd edn, Springer: Berlin, 2008.

C. Westbrook, *Handbook of MRI Technique*, Blackwell Publishing: Oxford, 2008.

C. Westbrook, C. Kaut Roth, and J. Talbot, *MRI in Practice*, 4th edn, Blackwell-Wiley: Oxford, 2011.

Diagnostic MRI

A. Bright, *Planning and Positioning in MRI*, Churchill Livingstone Elsevier: Chatswood, NSW, 2011.

F. A. Burgener, S. P. Meyers, R. K. Tan, and W. Zaunbauer, *Differential Diagnosis in Magnetic Resonance Imaging*, Thieme: Stuttgart, 2002.

G. Burghart and C. A. Finn, *Handbook of MRI Scanning*; Mosby-Elsevier: St, Louis, MO, 2011.

E. C. Lin, E. J. Escott, K. D. Garg, A. G. Bleicher, and D. Alexander, *Practical Differential Diagnosis for CT and MRI*, 1st edn., Thieme: Stuttgart, 2008.

MRI in drug discovery and development

N. Beckmann (ed), *In Vivo MR Techniques in Drug Discovery and Development*, Taylor and Francis, New York, 2006.

Imaging in clinical trials specifically

C. Miller, J, Krasnow, L. H. Schwartz (eds), *Medical Imaging in Clinical Trials*, Springer-Verlag: London, 2014.

Useful web resources

MRI principles
http://www.fda.gov/drugs/developmentapprovalprocess/developmentresources/ucm092895.htm.

FDA Medical Imaging and Drug Development
http://www.fda.gov/drugs/developmentapprovalprocess/developmentresources/ucm092895.htm.

Chapter 27

The Role of NMR in the Protection of Intellectual Property in Pharmaceutical R&D

Frederick G. Vogt

Morgan, Lewis & Bockius, LLP, 1701 Market St., Philadelphia, PA 19103-2921, USA

27.1 INTRODUCTION

The pharmaceutical and biopharmaceutical industries rely on strong intellectual property (IP) protections perhaps more than any other industry because of the need to protect costly and risky investments in the development of treatments for human disease.[1] Although IP protections vary by nation and by jurisdiction within nations, in general four primary categories of IP are recognized and are of significance in the pharmaceutical and biotechnology fields: patents, trademarks, copyrights, and trade secrets.[2] Of these categories of IP, patent protection remains the most visible form

NMR in Pharmaceutical Sciences. Edited by Jeremy R. Everett, John C. Lindon, Ian D. Wilson, and Robin K. Harris.
© 2015 John Wiley & Sons, Ltd. ISBN: 978-1-118-66025-6
Also published in eMagRes (online edition)
DOI: 10.1002/9780470034590.emrstm1396

used by pharmaceutical and biotechnology research organizations and academic institutions to protect potential therapies. A granted patent generally confers on the owner or exclusive licensee the right to exclude others from making, using, selling, or offering to sell an invention. In essence, these rights grant the owner or exclusive licensee the right to prevent competitors from copying and selling products that infringe on their patent rights, generally for a period of approximately 20 years from the filing date, subject to extensions and adjustments in many cases. Trade secret protections are also employed in the pharmaceutical and biopharmaceutical industries, particularly with respect to manufacturing processes ranging from chemical syntheses to bioprocesses for antibody production. However, because trade secrets cannot protect a product from reverse engineering using modern analytical methods, they are generally not employed to protect the actual marketed form of a pharmaceutical or biopharmaceutical therapy. Other forms of exclusivity that are beyond traditional IP, such as data exclusivity (the right to reference a competitor's regulatory data), also offer protection from competitors for some products.[3] However, the longer duration of exclusive rights available through patents has ensured that they remain the principal form of protection for the majority of commercial pharmaceutical and biopharmaceutical products and thus an intense area of competitive activity.

This chapter focuses primarily on the role of the NMR and MRI techniques in patent prosecution and in the enforcement of patent rights. In this role, NMR and MRI are applied in a manner similar to many other common analytical and bioanalytical techniques, including mass spectrometry (MS), microscopy and imaging methods, liquid chromatography (LC), and methods for sequencing polynucleotides and polypeptides. The widespread laboratory use of NMR and MRI methods is an advantage in this role, as a rare or unusual analytical method runs the risk of being difficult to reproduce when the need may arise in the decades after the patent application containing the data is filed. A brief review of patent rights is first given in the present chapter, followed by examples of the typical usage of solution-state NMR, solid-state nuclear magnetic resonance (SSNMR), and MRI in patent prosecution and enforcement. This article is focused on patents involving pharmaceutical and biopharmaceutical products and only briefly discusses the important role of patents that claim NMR or MRI techniques or hardware, such as RF and microwave electronics, data processing methods, and experimental NMR or MRI methods. This article also covers the use of NMR and MRI techniques in applications relating to small organic pharmaceutical molecules (typically made through chemical or biosynthetic methods, with a molecular weight <1 kDa), oligopeptides and oligonucleotides (with molecular weights ranging from about 500 Da to 20 kDa), and larger protein-based molecules such as monoclonal antibodies (mAbs) and proteins (typically made by biosynthetic methods, with molecular weights ranging from about 20 to 150 kDa).

27.2 PATENT RIGHTS

Patents are granted as the result of successful patent applications. Applications are examined to determine if they meet the criteria set by each country or region for patentability. Patent examination is performed by the governmental authorities of each country, such as the United States Patent and Trademark Office (USPTO), with the exception of cases where regional authorities, such as the European Patent Office (EPO) and the Eurasian Patent Organization (EAPO), examine patents on behalf of member countries. The examination process generally requires multiple rounds of formal discussion between the applicants and the national or regional authority (e.g., the USPTO or EPO), which is often referred to as prosecution. Patent applications are usually members of families that descend from

an initial parent with a filing date that is of importance in the determination of priority over other applicants. In many pharmaceutical and biopharmaceutical patent applications, a provisional patent application filed in the United States begins the process at an initial, earliest, date and is followed a year later by the filing of an international patent application under the Patent Cooperation Treaty (PCT), to which most countries in the world are currently signatories. If the patent is successfully granted, the filing date of the international patent application (also known as a PCT application) typically starts the period of exclusivity, which may run approximately 20 years but may be extended under various situations.[2] The international patent application is then 'entered' into each nation or region approximately 18–20 months after its filing date, depending on jurisdiction (and thus 30–32 months after the filing of the US provisional application). At the entry point (referred to as *national phase entry*), examination commences in each country or region and patents may issue from this process, typically over the course of several years. A schematic depiction of this strategy, which is the most common approach used for patent applications in the pharmaceutical and biopharmaceutical fields, is shown in Figure 27.1. Many variations on this process are possible, including filing of concurrent national phase and international phase applications on the same date, as well as entry into the international phase from a national phase application filed in a country that does not allow provisional applications and direct entry from a US provisional into a national phase application.

Patent applications and patents contain claims, which define the property right being sought and which are the main focus of the examination. From a legal perspective, the claims are the most critical aspect of a patent.[2] In pharmaceutical and biopharmaceutical applications, patent claims are typically directed to a particular chemical or biological composition or to a method of using such a composition. Patent claims contain words that function as limitations in that they confine the scope of the claim to a specific composition or method. In some cases described in later sections, the claims of patents may directly include or be supported by NMR, SSNMR, MRI, and related analytical data. The claims appear as a numbered list, normally at the end of the patent document, and may either stand alone (as independent claims) or depend on

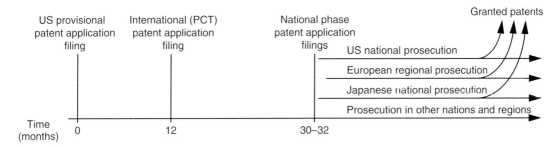

Figure 27.1. Simplified diagram of the major steps in the typical patent application strategy used in the pharmaceutical and biopharmaceutical fields. A US provisional patent application is followed after a year by an international (PCT) application, which is followed by national phase applications that prosecute in each country or region until patents are issued or the applications are abandoned. In most countries, the national phase application must be filed not later than 30 months after the US provisional application, as shown in the figure. However, in the European region, the application may be filed not later than 31 months, and other countries allow filing of the national phase application not later than 32 months. The applications shown are all connected through a 'claim of priority' or a 'priority chain' such that they are all given the benefit of the filing date of the US provisional application with respect to the qualification of prior art. Prosecution may continue for extended periods in each nation or region, and multiple patents may issue in each nation or region

an earlier claim and add limitations (as dependent claims).

In most countries, basic criteria must be met for a patent to be granted with a particular set of claims. In the United States, for example, patentability is set by statute and the law of the federal courts and includes requirements that (i) the invention must be useful, (ii) the invention must be directed to subject matter that is not excluded from patent protection, (iii) the invention must be novel (as judged against the 'prior art' previously known to the world before the patent application filing date), and (iv) the invention must not be obvious.[2] For example, a small-molecule anticancer drug and its family of related compounds might satisfy criterion (i) by showing that it can cause apoptosis of a cancerous cell line relative to a control, criterion (iii) by being a previously unknown chemical structure, and criterion (iv) by not being an obvious variant of some previously known chemical structure. Criterion (ii) is normally not an issue for patent applications claiming compositions of matter, but in some countries, certain categories of method patent claims (e.g., methods of treating a human disease) and compositions (e.g., cell therapies or materials isolated from nature), which are of importance in the pharmaceutical and biopharmaceutical industries, may be excluded from patent protection.

Other steps are also required to complete the process of obtaining patent rights. In the United States, these include providing a patent application that contains an adequate description of the claimed subject matter and providing a disclosure that enables a worker in the field to make and use the invention. For example, a newly synthesized small-molecule candidate drug might be adequately described by drawing its chemical structure and providing its standardized chemical name and might be further described by its NMR spectra, mass spectra, and other spectral properties. The synthesis of the compound might be enabled by providing an experimental write-up describing the synthetic steps, structures of intermediates, spectra of intermediates (including NMR spectra), purification steps (e.g., LC or column chromatography methods), and crystallization steps. For patent applications that seek to claim a 'genus' of many compounds, it may be necessary to provide synthetic guidance for making a large subset of these compounds, so that the average worker in the field of chemistry could readily determine how to make any compound in the genus.

As noted earlier, patents provide their owner with the right to exclude others (i.e., restrict competitors) from activities related to the claims of a patent, which define the scope of the property right conferred, and the claims of patents for pharmaceutical and biopharmaceutical products tend to focus on two main areas: compositions of matter (such as a chemical or biological substance) and methods of using a composition of matter. In the case of a small-molecule drug, in many cases, an initial patent application is filed to seek claims to the genus of molecules (typically thousands of compounds) and also to specific members of the genus that are of interest as potential drug candidates. If granted, this initial application may provide broad protection over compositions and methods of using the compositions (e.g., in an initial set of therapeutic areas). This initial application may be followed later by separate, more specific patent applications directed to a pharmaceutical formulation (e.g., a specialized oral formulation), a polymorph or other solid phase (e.g., a crystalline, amorphous, or disordered phase), or a method of using a drug to treat a particular disease (e.g., a method of treating a form of cancer with a known compound that previously was not known to be useful in this disease). Many pharmaceutical and biopharmaceutical compounds of commercial interested are thus protected by multiple patents.

Patent enforcement occurs primarily through civil litigation brought in countries where patent rights exist. For example, a US patent owner might bring a civil lawsuit in a US federal district court to enforce the patent, while an owner of a European Patent validated in Germany might bring suit for infringement in the German district courts. The defendant in a patent infringement suit is likely to challenge the patent's validity. In the United States, for example, this normally occurs in the same district court where the civil suit was brought, or before the USPTO; in Germany, invalidity is determined by a separate Federal Patent Court. In either case, the analytical data in the patent may play a key role in the litigation, whether relevant to the claims directly or by providing support for the claims.

27.3 SOLUTION-STATE NMR

Solution-state NMR is widely used in the pharmaceutical industry for structural characterization of both small molecules and biomolecules (see Chapter 14). In the common case of a small organic molecule that is a potential candidate for clinical studies, solution-state NMR is commonly combined with MS techniques, including MS^n and accurate mass analysis, to establish the covalent chemical structure of the molecule of pharmaceutical interest. Common experiments for small molecule analysis are established and include a range of 1D and 2D methods making use of ^1H, ^{13}C, and in some cases other accessible natural-abundance nuclei such as ^{15}N,

^{19}F, and ^{31}P (see Chapters 2 and 3).[4] Small biomolecules with a molecular weight <1 kDa, including small peptides, oligonucleotides, and oligosaccharides, are typically amenable to the approaches used with small organic molecules without the need for spin labeling or other special considerations. The importance of solution-state NMR in structural determination of small-molecule drugs results in the extensive use of NMR data in most modern patent applications directed to these molecules.

In the case of biomolecular structures, ranging from oligonucleotides and oligopeptides up to proteins and mAbs, solution-state NMR may play a role in structural determination.[5] However, in studies of larger proteins and mAbs, NMR is usually less utilized than methods that determine the primary amino acid sequence including peptide mapping (digestion using enzymes followed by reversed-phase LC-MS analysis) and amino acid sequencing using MS fragmentation methods. Specialized techniques are also used to characterize glycans attached to mAbs and to analyze disulfide and sulfhydryl groups, among other structural features. In addition, techniques such as electronic circular dichroism and deuterium exchange MS methods are used to characterize secondary and tertiary structure. Solution-state NMR, primarily using isotopically labeled analogs, can play a key role in the characterization of these materials but is usually applied as a secondary technique, particularly in cases when detailed aspects of the 3D structure are of interest. As a result, solution-state NMR data is not as commonly found in patent applications for proteins and mAbs as is the case for small molecules, although 3D biomolecular structures are included in some cases.

Patent applications on potential small molecule therapeutics normally involve synthesis and testing of numerous structures with a chemical genus. For example, the patent application that disclosed and claimed the free acid of atorvastatin included a genus containing multiple substitutable groups R_1–R_4 and a bridging group X, along with structures corresponding to the opening of the lactone ring:[6]

and is intended to encompass the species of atorvastatin:

which is also named in its ring-closed form in the patent application. ^{1}H NMR data along with elemental analysis (which has today largely been replaced by accurate MS methods) and infrared spectra are reported for compounds within the genus. A typical pharmaceutical composition of matter patent at the present time may include hundreds of examples of compounds within the genus, as well as intermediates, with ^{1}H and ^{13}C NMR data provided along with MS data for each example.

It is also possible to claim a chemical structure using solution-state NMR data alone. For example, a claim could be directed to a composition with a characteristic ^{1}H NMR spectrum in D_2O at a particular temperature and concentration (and if needed, pH value or buffer) that exhibits a selection of signals chosen from the ^{1}H NMR spectrum. The could take the form of a partial list of distinctive chemical shift values and J-coupling constants. While this is less common, particularly for small molecules, it may offer some advantages in the detection of infringement by allowing a more direct comparison of a test product and a claim with less reliance on expert interpretation. If this route is taken, attention should be given to the identification and selection of ranges for individual NMR signals too so that these ranges are broad enough to allow for some variability but are narrow enough to avoid potentially capturing similar, previously known compounds of the prior art. Claiming a structure or structures through an NMR spectrum may be of particular interest when working with complex mixtures, such as those of natural products, where the full structures may be too complex to definitively assign or the product is characterized by a range of structures related to branching and substitution, polymerization or stereoisomerization. Examples of complex structures marketed as drugs include heparin sulfate, enoxaparin and related glycosaminoglycans, and glatiramer acetate. In patent applications directed to complex mixtures, a greater variety of analytical data may be included to help illustrate the properties of the invention, such as more detailed LC-MS analysis, additional chromatographic and electrophoretic results, and other spectroscopic analyses. Additional NMR data beyond the conventional ^{1}H and ^{13}C results may also be included, such as DOSY results, which are useful in analysis of drug mixtures.[7]

Small organic pharmaceutical molecules often contain one or more chiral centers. In these cases, the confident determination of absolute stereochemistry or relative stereochemistry is an important aspect of the structural characterization included in patent applications with claims to compositions of matter. Solution-state NMR is widely used for both absolute and relative stereochemical determinations in small organic molecules using a range of techniques that rely on chemical shift, J-coupling, and dipolar coupling (see Chapters 16 and 17).[8] NMR methods are particularly useful in cases where a reliable X-ray diffraction structure is not yet available, which may be the case at the relatively early time in drug development at which initial patent applications are drafted.

Data from solution-state NMR methods for studies of drug binding for use in drug discovery may also be included in patent applications and the methods have themselves been the subject of patent applications and granted patents.[9–12] These methods generally involve detection of an interaction of a small molecule or fragment of a molecule with a target biomolecule (e.g., a protein) in solution, based on chemical shift effects, diffusion, relaxation, saturation or NOE transfer, and/or double-quantum relaxation effects. The claims are often directed toward methods of screening rather than compositions, such as the use of the NMR technique to identify candidate drugs or fragments that bind to the biomolecule of interest.

Calibration of solution-state NMR instruments before acquisition of NMR data for use in a patent application is an important consideration. Established procedures for standardization of measurements and calibration of instruments are available.[13] The calibration of chemical shifts against a reference and the optimization of lineshape quality are two particularly important steps, as they can directly affect the spectral output in a manner that, if performed improperly, can make it difficult for others to reproduce the result. An adequate description of the experimental procedures and parameters used within the patent is also an important consideration. If NMR spectral properties are used to support a patent claim, the inability to reproduce a measurement because of lack of disclosure of experimental parameters (such as sample temperature, sample concentration, or solvent) in the patent application or granted patent may result in problems during patent prosecution or in later litigation. For example, under US law, a patent claim may be found to be invalid (i.e., the patent right is lost) because it is 'indefinite', in that it does not inform the public of its scope because the measurement cannot be definitively reproduced by others. Thus, a claim that is based on an NMR spectrum that varies strongly with temperature but is presented without information about the temperature used might be found indefinite because others cannot determine when they might infringe. A claim that uses NMR data that has been incorrectly measured may also simply be unenforceable against a product that should infringe, because measurement of that product's characteristics by commonly accepted standards might not produce a result that falls within the claims. A simple example is a claim that includes a limitation drawn to 'a ^{13}C resonance with a maximum at 23.0 ppm ± 0.1 ppm relative to tetramethylsilane', which may not ensnare a competitor's product if the instrument was miscalibrated such that chemical shifts differ from their true values by 0.2 ppm. Similar considerations may apply if solvent, concentration, or other parameters or procedures are not adequately specified in the patent or patent application.

If quantitative NMR methods are used (see Chapter 5), it may be necessary to provide the details of the quantitative analysis including any standards used or validation performed and T_1 relaxation time analysis. Quantitative NMR analysis generally refers to analyses where an accurate and precise quantitative value (e.g., the concentration of a component) is calculated directly from the NMR data, as opposed to the more approximate quantitative integration normally performed with ^{1}H solution-state NMR spectra. Quantitative NMR results can play a critical role in patent applications. Basic measurements such as salt or solvent stoichiometry are often reported, and assay results against a standard are also applied in cases where more common techniques such as LC are not suitable. Quantitative solution-state NMR methods accessing nuclei such as ^{1}H, ^{19}F, and ^{31}P can be used to characterize complex mixtures in pharmaceutical preparations, excipients, and complex preservatives through the quantitative ratio of resonances of interest.[14]

27.4 SOLID-STATE NMR

SSNMR is widely used in pharmaceutical R&D, primarily for the characterization of solid drug substances, excipients, drug product intermediates, and drug products (see Chapter 21).[15] Many of these solid materials are also of interest to an organization's patent protection efforts. The most common examples of interest for patent protection include novel crystalline and amorphous phases of a drug substance, formulations of a drug, and materials that are intermediate between formulations and drug substance (such as amorphous dispersions of a drug in a polymer or a formulation intermediate). In this role, SSNMR is often teamed with techniques such as powder X-ray diffraction (PXRD), single crystal X-ray diffraction (SCXRD), and vibrational spectroscopic techniques including Raman spectroscopy and infrared (IR) spectroscopy. Typical nuclei explored in SSNMR studies of pharmaceutical solids include ^{13}C, ^{15}N, ^{1}H, ^{19}F, and ^{31}P. While efforts are most commonly focused on analysis of the drug, novel formulations may also be characterized by SSNMR to characterize excipients or excipient–drug interactions.

The crystalline phases commonly encountered in pharmaceutical science include such materials as polymorphs, solvates, salts, and cocrystals.[16–18] These are often claimed in patent applications as unique crystalline phases through characteristic spectroscopic or diffractometric properties. Polymorphism and solvation (the latter including hydration) are frequent occurrences in pharmaceutical development, and screens are normally performed on candidate drug molecules to identify novel polymorphs and solvates, primarily to enable effective process development, and satisfy regulatory authorities. However, the identification of a novel crystalline form of a drug, whether a polymorph, a solvate, or a hydrate, may also provide an opportunity for patent protection. This is of special interest in cases where the crystalline form was difficult to identify and/or has unexpectedly useful properties, as the chemical structure of the compound may have already been disclosed and may serve as prior art against a later-filed application claiming a polymorph or a solvate. SSNMR is frequently used to characterize polymorphs and solvates, which normally yield unique ^{13}C spectra as well as spectra of other nuclei suitable for use in patent applications. For example, in Figure 27.2, the ^{13}C spectra of two polymorphic forms of foretinib drug substance are shown.[19] These spectra were

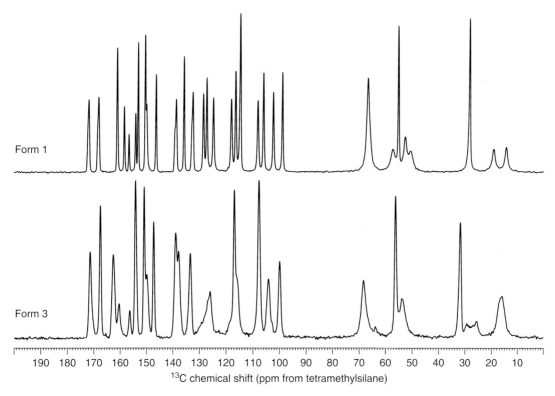

Figure 27.2. [13]C spectra of two polymorphic forms of foretinib obtained with cross polarization and sideband suppression (of sidebands produced by MAS), after Ref. 19. The spectra show distinctive ^{13}C resonances that can be used to draft claims for each of the polymorphic forms

obtained using cross polarization and MAS methods with sideband suppression.[19] The chemical structure of foretinib, shown below, exhibits the typical complexity of a small-molecule pharmaceutical drug substance:

The spectra in Figure 27.2 are sufficiently resolved to allow the forms to be distinguished and claimed using selected resonances in their ^{13}C spectra.[19] Good resolution for polymorphs of a particular drug substance is typically observed with ^{13}C SSNMR techniques and in SSNMR studies of other nuclei.[15]

Salts and cocrystals are two classes of molecular complexes that are also frequently encountered during drug development and are similar in nature but differ most fundamentally in that a salt results in the ionization of the drug through proton transfer, while a cocrystal does not.[17–20] Various other possibilities, such as zwitterionic structures or mixed salt/cocrystal systems, are also encountered, albeit less frequently. Formation of a salt or cocrystal is often driven by addition of a salt former (e.g., a strong acid) or cocrystal former to a solution followed by crystallization. Most salts and cocrystals are molecular crystals, although the possibility of forming an amorphous salt or cocrystal as a pure phase exists as well as the possibility of forming an amorphous salt or cocrystal using a complex-former tethered to a support (such as a salt-forming group tethered to a resin and commonly referred to as a *resinate*). Patent applications directed to cocrystal phases are fairly common and may offer advantages, in part from the diverse range of possible cocrystal former and general unpredictability of the results of cocrystal screening, at least with respect to the obtained properties of the discovered cocrystal phases (such as dissolution rate, physical stability, or chemical stability).[21–23]

SSNMR has many advantages in the analysis of cocrystals and other molecular complexes because of its ability to detect molecular association through both dipolar coupling and chemical shift effects (see Chapter 21). For the purposes of a patent application, it may be necessary to demonstrate that a novel molecular complex has formed. This can be accomplished using SCXRD structures to illustrate unique effects such as hydrogen bonding patterns or aromatic ring stacking effects.[22] In the absence of SCXRD data, SSNMR techniques are presently able to detect molecular association and intermolecular interactions in a cocrystal with results that are often superior to those obtained from vibrational spectroscopic methods.[24] For example, SSNMR studies can help provide support for patent claims directed to two known compounds in a 'hydrogen bonded' structure by identifying the specific functional groups engaged in hydrogen bonding.[22,24]

A patent application that claims a solid crystalline hydrate of foretinib by its ^{13}C and ^{19}F SSNMR chemical shifts, along with PXRD and Raman spectral features, provides a typical example of an approach to using SSNMR data.[25] This hydrate of foretinib can vary in the water content within its crystal structure relative to the drug substance as water exchanges with the surroundings. Claims are directed to the crystalline variable hydrate form of foretinib and are based on ^{13}C and ^{19}F SSNMR spectra as well as Raman spectra and PXRD patterns. In patents and patent applications of this type, a 'nesting' approach for claims is typically used, wherein the dependent claims become narrower by adding limitations such as additional ^{13}C SSNMR resonances or by combining SSNMR and PXRD peak positions in the same claim. Narrower claims have the disadvantage of being potentially more difficult to use to prove infringement, in that more testing may be needed or interferences (e.g., from excipients) may cause a problem. However, narrow claims have the advantage of being less susceptible to invalidation if prior art is later found that is encompassed by the claims. For example, a claim that recites a crystalline form of a particular drug with two ^{13}C SSNMR peaks at 120.3 and 112.6 ppm is relatively broad, and as most excipients do not yield ^{13}C resonances in the aromatic

region, it is readily enforced against finished drug samples potentially containing the form of interest. However, there is a risk that a previously known form of that same drug (before the patent application) also has resonances at those two positions, posing a risk that the patent claims will not be allowed or, if allowed, may be invalidated for lack of novelty. While the broad claim may be a useful independent claim, the additional use of narrower claims that list several more ^{13}C SSNMR signals, or several PXRD reflections in addition to the two ^{13}C SSNMR signals, improves the likelihood that the overall patent will not ensnare the prior art and can survive an invalidity challenge. This approach is similar to the typical situation for chemical structure claims described earlier, where a narrow species claim to a particular drug structure is dependent on a broad genus claim covering thousands of potential compounds with a particular core structure. SSNMR studies of the ^{13}C and ^{19}F nuclei often have sufficient sensitivity and specificity to be useful for detecting the form of drug within a formulated product such as an oral tablet or inhaled dosage form. Inclusion of claims directed to ^{13}C and ^{19}F spectra may be beneficial in detection of infringement of a crystalline form patent in a finished drug product, as access to the pure drug substance of a suspected infringer will generally be limited until well after litigation has commenced.

SSNMR can also be highly useful in cases where PXRD is limited by the lack of long-range order in a material. For example, in the case of a crystalline polymorphic form of atorvastatin calcium, claims directed to features in a broadened PXRD pattern (possibly the result of poor crystalline order and/or disorder within the repeating unit cell) were augmented by use of claims directed to features of a ^{13}C SSNMR spectrum.[26] In an amorphous material, no Bragg reflections are observed by PXRD and a broad diffuse scattering pattern is often obtained that lacks specificity. However, SSNMR is able to provide detailed information about systems such as amorphous drugs, amorphous drug-polymer dispersions, and dispersions of drugs in mesoporous materials (see Chapter 21).[27,28] Patent applications seeking to claim amorphous forms of a drug or an amorphous dispersion typically include detailed SSNMR and vibrational spectroscopic characterization, which can provide limitations for claims in the absence of strong PXRD support. Many larger molecules, such as oligonucleotides, oligopeptides, and oligosaccharides, as well as large proteins, are amorphous in the solid state or are phase mixtures of amorphous drug substance and amorphous or crystalline excipients and may be amenable to SSNMR analysis.

The use of SSNMR spectral data in patent applications, as with the use of solution-state NMR, typically benefits from a full disclosure of experimental parameters and measurement conditions as well as from careful calibration of instrumentation. Because most SSNMR instrumentation operates without a field-frequency lock, it is particularly important to make use of internal or external chemical shift reference materials before measurement of spectra of materials that might be later claimed in a patent application.[13] The disclosure of RF field strengths used for cross polarization, MAS rates, and other pulse sequence parameters and the avoidance of nonstandard experimental methods and processing conditions (unless fully disclosed) are also important for the reproducibility of results.

27.5 MAGNETIC RESONANCE IMAGING

The use of MRI has steadily increased in the development and assessment of pharmaceutical formulations (see Chapter 23), in studies of preclinical animal models, and in clinical applications. In studies of novel formulations, MRI data may be used to image drug release or other critical aspects of the technology, such as the ingress of dissolution medium. MRI studies of pharmaceutical dosage forms are often accompanied by other forms of imaging, such as optical microscopy, vibrational microscopy, and fluorescence or luminescence microscopy. Results from these studies may provide useful support for patent applications seeking to claim new formulations.

At present, MRI is more commonly utilized in drug development to support preclinical and clinical studies and particularly to assess pathology and safety in animal models and ultimately in human studies.[29,30] MRI is used, for example, in studies of drug effects in therapeutic areas such as oncology (especially with respect to solid tumors), cardiology, inflammation, and neurology. The MRI techniques used include a variety of image-weighting methods, use of contrast agents and paramagnetic probes, and use of NMR spectral acquisition across 3D voxels of an image (commonly referred to as magnetic resonance spectroscopy or MRS). MRI is also combined in multimodal approaches with other molecular or biomolecular imaging experiments. MRI and MRS results may be useful for demonstrating the potential efficacy and safety of drugs in a manner that helps support patent claims directed to methods for treating a particular disease. For example, MRI and MRS studies of an animal model for a solid tumor that observe shrinkage of the tumor upon treatment with a drug, or can distinguish cancerous tissue from other tissue, might be used to provide support for an eventual claim to treatment of solid tumors with that drug. Although results derived from MRI data are rarely encountered directly as a limitation in the claims of a patent application, the importance of the data in supporting a claim to efficacious treatment provides motivation to ensure reproducible results of high quality are used as described earlier for solution-state NMR and SSNMR results.

27.6 CONCLUSION

The pharmaceutical and biopharmaceutical patent application and prosecution process consists of a relatively complex series of events spanning many years. The process begins with an application that may rely on solution-state NMR, SSNMR, or MRI data to demonstrate an adequate description of the claimed subject matter, to enable others in the field to make and use the invention, and which also provides much of the support for future arguments that the invention is novel, useful, and not obvious. Selection of quality results for inclusion in patent applications and provision of sufficient information to allow others to reproduce experiments is thus a key aspect of this process. Solution-state NMR, SSNMR, or MRI results can also be critical in later infringement and invalidity arguments when a patent is enforced or challenged. These techniques are likely to continue to provide important support to patent applications and granted patents in the future.

27.7 DISCLAIMER

The views expressed in this chapter are those of the author and not of his employers or clients. The material provided in this chapter is for general informational and educational use only, should not be relied upon as legal advice, and should not be used as a substitute for consultation with a qualified legal practitioner in the relevant jurisdiction.

ACKNOWLEDGMENT

The author thanks his colleagues for helpful discussions relating to IP law in the field of life sciences.

RELATED ARTICLES IN EMAGRES

Stereochemistry and Long Range Coupling Constants

Determining the Stereochemistry of Molecules from Residual Dipolar Couplings (RDCs)

Magic Angle Spinning

Cross Polarization in Rotating Solids: Spin-1/2 Nuclei

REFERENCES

1. A. Kapczynski, C. Park, and B. Sampat, *PLoS One*, 2012, **7**, e49470.

2. R. E. Schechter and J. R. Thomas, *Intellectual Property: The Law of Copyrights, Patents, and Trademarks*, Thomson-West: St. Paul, Minnesota, 2003.

3. H. Grabowski, *Nat. Rev. Drug Discov.*, 2008, **7**, 479.

4. S. Braun, H. O. Kalinowski, and S. Berger, *150 and More Basic NMR Experiments*, 2nd edn, Wiley-VCH Verlag GmbH: New York, 1998.

5. A. G. Palmer, W. J. Fairbrother, J. Cavanagh, and N. J. Skelton, *Protein NMR Spectroscopy: Principles and Practice*, 2nd edn, Academic Press: New York, 2005.

6. B. D. Roth, U.S. Pat. 4,681,893 (Jul. 21, 1987).

7. V. Gilard, S. Trefi, S. Balayssac, M. A. Delsuc, T. Gostan, M. Malet-Martine, R. Martine, Y. Prigent, and F. Taulelle, in *NMR Spectroscopy in Pharmaceutical Analysis*, eds I. Wawer and B. Diehl, Elsevier, Oxford, 2008, 269.

8. J. M. Seco, E. Quiñoà, and R. Riguera, *Chem. Rev.*, 2004, **104**, 17.

9. D. S. Gregg, U.S. Pat. 8,173,441 (May 8, 2012).

10. S. W. Fesik, P. J. Hajduk, and E. T. Olejniczak, European Pat. EP 0870197 B2 (Nov. 6, 2006).

11. D. Tammo, International Pat. Appl. WO 2003/054532 A2 (Jul. 3, 2003).

12. C. Dalvit, C. Battistini, P. Caccia, P. Giodano, P. Pevarello, M. Sundström, and M. Tato, European Pat. 1224486 B1 (Dec. 12, 2005).

13. R. K. Harris, E. D. Becker, S. M. Cabral de Menezes, R. Goodfellow, and P. Granger, *Pure Appl. Chem.*, 2001, **73**, 1795.

14. J. A. Smith, T. R. Graumlich, R. P. Sabin, and J. W. Vigar, U.S. Pat. 6,265,008 (Jul. 24, 2001).

15. F. G. Vogt, in *New Applications of NMR in Drug Discovery and Development*, eds L. Garrido and N. Beckmann, The Royal Society of Chemistry, London, 2013, 43.

16. J. Bernstein, *Polymorphism in Molecular Crystals*, Oxford University Press, New York, 2002.

17. H. Stahl and C. G. Wermuth (eds), *Handbook of Pharmaceutical Salts: Properties, Selection, and Use*, Wiley-VCH Verlag GmbH, New York, 2002.

18. I. Miroshnyk, S. Mirza, and N. Sandler, *Expert Opin. Drug Deliv.*, 2009, **6**, 333.

19. H. Cannon, D. Igo, and T. Tran, U.S. Pat. 8,673,912 (March 18, 2014).

20. S. M. Reutzel-Edens, in *Pharmaceutical Salts and Co-crystals*, eds J. Wouters and L. Quéré, Royal Society of Chemistry, London, 2012, 212.

21. Ö. Almarsson, M. L. Peterson, and M. Zaworotko, *Pharm. Pat. Anal.*, 2012, **1**, 313.

22. M. Hoffman and J. A. Lindeman, in *Pharmaceutical Salts and Co-crystals*, eds J. Wouters and L. Quéré, Royal Society of Chemistry, London, 2012, 318.

23. A. V. Trask, *Mol. Pharmaceut.*, 2007, **4**, 301.

24. F. G. Vogt, J. S. Clawson, M. Strohmeier, A. J. Edwards, T. N. Pham, and S. A. Watson, *Cryst. Growth Des.*, 2009, **9**, 921.

25. H. Cannon, F. Kang, and F. G. Vogt, International Pat. WO 2011/112896 A1 (Sep. 15, 2011).

26. A. Ayalon, M. Levinger, S. Roytblat, V. Niddam, R. Lifshitz, and J. Aronhime, U.S. Pat. 7,411,075 (Aug. 12, 2008).

27. S. Janssens and G. Van den Mooter, *G. J. Pharm. Pharmacol.*, 2009, **61**, 1571.

28. K. K. Qian and R. H. Bogner, *J. Pharm. Sci.*, 2012, **101**, 444.

29. J. Ripoll, V. Ntziachristos, C. Cannet, A. L. Babin, R. Kneuer, H.-U. Gremlich, and N. Beckmann, *Drugs R&D*, 2008, **9**, 277.

30. D. M. Morris, J. P. B. O'Connor, and A. Jackson, in *New Applications of NMR in Drug Discovery and Development*, eds L. Garrido and N. Beckmann, The Royal Society of Chemistry, London, 2013, 490.

PART F
Drug Manufacture

Chapter 28

Analysis of Counterfeit Medicines and Adulterated Dietary Supplements by NMR

Myriam Malet-Martino and Robert Martino

Groupe de RMN Biomédicale, Laboratoire SPCMIB (UMR CNRS 5068), Université Paul Sabatier, 118 route de Narbonne, 31062 Toulouse cedex 9, France

28.1 INTRODUCTION

Recent events have been a reminder of the problem of counterfeit drugs. Indeed, in May 2014, Interpol launched Operation Pangea VII that led to the seizure of 9.4 million fake and illicit medicines.[1] Counterfeit medicines (CM) is the common term, although improper, that covers various expressions found in the literature such as spurious, substandard, falsified, falsely labeled, fraudulent, illicit, unregistered, unapproved, unregulated, counterfeit, and fake drugs.[2] All these denominations whose definitions are still debated[3] have been recently lumped together by the World Health Organization (WHO) member states in the new collective term SSFFC (substandard/spurious/falselylabeled/falsified/counterfeit) medical products[3,4] and later on SFFC (spurious/falsely labeled/falsified/counterfeit) medicines.[5] For more clarity, the term *CM* will be used in the present article. The most widely used definition of a CM is that proposed by the WHO, which states that 'A counterfeit medicine is one which is deliberately and fraudulently mislabeled with respect to identity and/or source. Counterfeiting can apply to both branded and generic products and counterfeit products may include products with the correct ingredients or with the wrong ingredients, without active ingredients, with insufficient active ingredients or with fake packaging.'[6] For several reasons (absence of clear internationally agreed definitions, variety of information sources, illegal character of the trade, etc.), there is no reliable data on the real scale of the CM problem. WHO estimated that the incidence of CM ranges from <1% in most industrialized countries with effective regulatory systems and market control to 10–30% in developing markets that have poor drug regulatory and enforcement systems, weak or no pharmacovigilance,

NMR in Pharmaceutical Sciences. Edited by Jeremy R. Everett, John C. Lindon, Ian D. Wilson, and Robin K. Harris
© 2015 John Wiley & Sons, Ltd. ISBN: 978-1-118-66025-6
Also published in eMagRes (online edition)
DOI: 10.1002/9780470034590.emrstm1423

and more than 50% from websites that conceal their physical address.[2,5] The same imprecision affects the value of this highly lucrative market. Available estimates are in the range US$75 to US$200 billion, which nevertheless highlights the significance of the problem.[2,7] Any kind of product can be and has been counterfeited[5]: expensive life-style medicines such as drugs for treating erectile dysfunction, fat reducing or sleeping remedies, antibiotics, anticancer drugs, medicines for hypertension or cholesterol lowering, and inexpensive versions of simple pain killers. In developing countries, the most disturbing issue is the common availability of CM for the treatment of life-threatening diseases such as malaria, tuberculosis, and AIDS. The impact on patients using CM ranges from treatment failure to drug resistance and even to death.[5] Therefore, obviously, CM poses a serious growing threat to public health particularly in low- and middle-income countries (LMICs).

Dietary supplements (DS) and herbal drugs are widely used for a variety of reasons such as improving or maintaining overall health, balancing the diet, perking up the appearance, boosting the performance, or delaying the onset of age-related diseases.[8] They are freely available to consumers in pharmacies, health food stores, department stores, groceries, on the Internet, and on the black market. The worldwide DS industry is a US$ multibillion enterprise with sales that surpassed US$ 95 billion in 2013.[9] If industrialized countries are little affected by the problem of CM, at least in the legitimate supply chain, they are faced with another problem currently less recognized by health authorities but which is another alarming emerging risk to public health. Indeed, there is a growing trend in the intentional adulteration of claimed natural DS with synthetic drugs.[10] This illegal practice is partly due to the lack of harmonized regulations between countries and to the difficulties of oversight, which enable some manufacturers to introduce hazardous compounds in a natural matrix. Moreover, DS are regarded by many consumers and physicians as being harmless because of their natural origin or of their presumed safe composition. It is thus tempting for unscrupulous manufacturers to fraudulently add synthetic actives in order to improve the effectiveness of the DS. Furthermore, the ability to purchase DS on the Internet easily, anonymously, and often cheaply has created a new channel for the distribution of adulterated products. Given the Internet's regulation-free environment and most often the unknown origin and composition of the DS bought, consumers are not aware of taking a

prescription drug that has contraindications. Another major issue is that some manufacturers, in an attempt to evade regulatory inspection and to reduce the risk of patent-infringement lawsuits, use not only approved active pharmaceutical ingredients (APIs) but also unapproved designer analogues in which minor modifications were brought to the parent structure. These derivatives may retain corresponding pharmacological actions. Nevertheless, they may possess slightly or entirely different properties, especially different toxicities, that are unknown as the analogues have not been subjected to clinical trials. Prescription drugs withdrawn from the market by health regulatory agencies because of safety concerns as well as drugs still in clinical trials are also found in DS. DS marketed for sexual enhancement, body-building, and weight loss are the most affected by illegal adulterations.[11,12]

Fighting CM and adulterated DS is a considerable public health challenge that requires legal measures (standardized definitions, international regulatory harmonization, centralized pharmacovigilance at the global scale, coordinated police, and customs operations, etc.) as well as technical solutions, especially the development of analytical methods.[3,13] Numerous analytical technologies, ranging from inexpensive field assays to sophisticated laboratory instruments and methods, have been described with much emphasis on the detection of CM.[14–17] The most commonly used method is high-performance liquid chromatography (HPLC), considered the gold standard, coupled with various detectors, from a classical ultraviolet (UV) detector to various mass spectrometry (MS) systems. A few fast screening and in-the-field techniques such as thin-layer chromatography (TLC) and colorimetry have also been developed. Due to the high selectivity of these methods, which only target the API, the complex nature of the drugs is neglected. Spectroscopic methods (Infrared (IR), Raman, and NMR) emphasize much more the integrative and holistic characteristics of the samples. IR and Raman techniques are very appealing because they do not require sample preparation, can be used by a technician with minimal training, and can be implemented as portable devices. However, they depend on the use of reference libraries of API and excipients (that must be regularly updated) to identify falsified and substandard products. In contrast, NMR has the great advantage of providing structural information on virtually all the components of the CM or the DS without resorting to libraries. Moreover, the method is intrinsically quantitative, which is important for ensuring that the API is present in the correct dosage, with no need for reference compounds.

The aim of this article is to present several examples of the use of NMR as a first-line analytical technique for uncovering CM and adulterated DS. We deliberately chose to ignore the numerous articles dedicated to the detection and identification of impurities in pharmaceutical products by NMR even though the method can be of interest to reveal substandard medicines. Starting with the heparin case, we present successively the studies devoted to the quality evaluation of essential antimalarial medicines, drugs and DS intended to treat erectile dysfunction, slimming DS, anti-inflammatory DS, and some cosmetic creams. We will end up with a paragraph describing the potential of isotopic NMR in the field of drug counterfeiting. Finally, we discuss the assets and drawbacks of NMR and we conclude with an example demonstrating the capabilities of a recently released bench-top cryogen-free low-field ^{1}H NMR spectrometer to detect tainted DS.

28.2 HEPARIN

Heparin, a polydisperse mixture of sulfated glycosaminoglycans (GAGs), has been in clinical use for over 75 years and is still the most widely employed anticoagulant and antithrombotic drug in medicine. Heparin is usually isolated by extraction from animal (bovine, ovine, and porcine) tissues, mainly porcine intestinal mucosa or whole porcine intestine (approximately one billion pigs a year are required to constitute the heparin supply). Besides heparin itself, crude heparin contains proteins, nucleic acids, and other related GAGs, including heparan sulfate (HS), dermatan sulfate also known as chondroitin sulfate B (CSB), chondroitin sulfate A (CSA), and hyaluronic acid. Subsequent clean-up proprietary processes that incorporate multiple precipitation and viral inactivation steps convert crude heparin into APIs. The differences in these processes as well as the diversity of the animal source material can lead to some variability in the amount of native impurities in the final product. Heparin is marketed in multiple forms, including natural or unfractionated heparin (UFH) with molecular weights ranging between 5 and 40 kDa (12–15 kDa for the vast majority of the preparations), and low molecular weight heparins (LMWH) (4–6 kDa) prepared from UFH by partial depolymerization with controlled chemical

or enzymatic methods.[18] Due to its highly complex, polydisperse and microheterogeneous structure, the pharmaceutical grade heparin has historically been defined based on units of anticoagulant activity specified by plasma clotting-time assays, rather than on the basis of molecular properties.[18,19] The controls have been strengthened recently because of the heparin contamination crisis. In late 2007 and early 2008, clusters of serious adverse events including acute hypotension, severe allergic-type reactions, and deaths (more than 200 around the world) were reported mainly for patients undergoing hemodialysis and receiving intravenous injections of heparin. The observed patient reactions were linked to suspect heparin lots containing a then unidentified contaminant that was subsequently determined to be oversulfated chondroitin sulfate A (OS-CSA) using ^{1}H and ^{13}C NMR as well as 2D [1]H–[13]C heteronuclear single quantum coherence (HSQC) NMR. Comparison of the NMR spectra of the isolated material with standard OS-CSA synthesized from CSA confirmed the identity of the contaminant.[18–20] OS-CSA is not a natural product arising from animal sources and then cannot be a native impurity of heparin. The presence of OS-CSA in heparin APIs was not due to an accidental contamination but was an act of purposeful adulteration.[18] Because the contaminated heparin samples passed the whole-blood screening tests in place at that time, illustrating their insufficiency, orthogonal, and complementary methods to those employed in the pharmacopeias have been rapidly added for assuring the quality of heparin APIs. Among all the chromatographic and spectroscopic methods developed for the detection and identification of heparin impurities and contaminants, the regulatory authorities in the United States and in Europe implemented ^{1}H NMR and capillary electrophoresis (CE) as mandatory tests in the first stage revision (2008) of the European and US Pharmacopoeias (PhEur and USP) 'heparin sodium' (UFH) monographs, whereby the CE method was replaced by strong anion-exchange high-performance liquid chromatography (SAX-HPLC) in the second stage revision of the USP monograph.[18,21,22] For more information, see Chapter 30.

Due to its high sensitivity to even minor structural variations, ^{1}H NMR is an information-rich technique for providing characteristic fingerprints of heparin samples. Indeed, ^{1}H NMR allows for differentiation between heparin, CSB, a native impurity often present at residual levels in heparin products due to incomplete purification, and OS-CSA, a contaminant that is

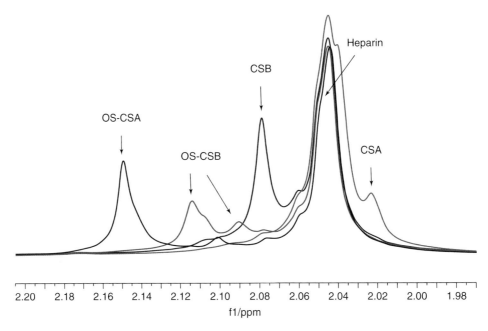

Figure 28.1. Overlay of the 500 MHz ^{1}H NMR spectra (2.20–1.98 ppm region) of a heparin sodium API spiked with 10.0 wt% CSA, OS-CSA, CSB, or OS-CSB. (Adapted with permission from Ref. 24. Copyright (2011) American Chemical Society)

synthesized by chemically oversulfating CSA, based on the chemical shifts of the *N*-acetyl methyl resonances. The ability of ^{1}H NMR to identify a range of possible native (CSA, CSB, and HS) or synthetic oversulfated (OS-CSA, oversulfated chondroitin sulfate B (OS-CSB), and oversulfated heparan sulfate (OS-HS)) GAGs in heparin was investigated. This is illustrated in Figure 28.1 showing superposed ^{1}H NMR spectra of heparin API samples spiked with CSA, OS-CSA, CSB, and OS-CSB. Their *N*-acetyl methyl resonances are well separated: 2.05 ppm for heparin, 2.08 ppm for CSB, 2.09 and 2.11 ppm for OS-CSB, 2.02 ppm for CSA, and 2.15 ppm for OS-CSA. Signals of HS, OS-HS, and OS-heparin appearing at 2.04 ppm are difficult to distinguish from that of heparin although HSQC analysis clearly identifies signals associated with OS-heparin from those of heparin.[19,23] Chemometric pattern recognition techniques have been successfully applied to discriminate heparin API from heparin containing various levels of CSB and OS-CSA and heparin spiked with other GAGs (see, for example, Ref. 24). Heparin formulations also contain residual processing solvents and reagents that can be readily identified by their typical ^{1}H chemical shifts: ethanol (triplet at 1.18 ppm), acetate (singlet (s) at 1.94 ppm), acetone (s at 2.23 pm), methanol (s at

3.36 ppm), and formic acid (s at 8.45 ppm), whereas their detection is not possible by means of HPLC or CE.[21] NMR has proven to be an invaluable method for the characterization of heparin impurities and contaminants and has been selected by the Food and Drug Administration (FDA) along with SAX-HPLC for controlling heparin batches. Samples with unusual resonances/peaks in these assays are highlighted for analysis by additional approaches such as 2D ^{1}H–^{13}C HSQC or LC–MS analysis of an enzymatic digest of the sample.[22]

28.3 ANTIMALARIAL DRUGS

More than one-third of antimalarials used in Southeast Asia and sub-Saharan Africa are counterfeit or substandard.[25] Artemisin-based combination therapies (ACTs) that combine an artemisin active derivative (artesunate, artemether, or dihydroartemisinin) with another antimalarial compound (amodiaquine, lumefantrine, piperaquine, etc.) have been the WHO-recommended option for first-line malaria treatment since 2001. Although WHO has called for an end to the production and marketing of artemisinin monotherapy since 2006, drugs such as artesunate and dihydroartemisinin are still available as monotherapy for purchase in Southeast Asia and sub-Saharan Africa.

A number of rapid and economic instruments such as the mobile GPHF Minilab® based on TLC test methods, the Truscan® handheld Raman, or the Counterfeit Detection Device version 3, a handheld electronic tool operating in the UV–visible and IR spectral regions have been developed to distinguish fake from genuine medicines and applied to artemisinin-based drugs.[14,26] However, only HPLC with MS detection and NMR allow characterizing the chemical composition of these medicines.[14]

^{1}H NMR was used to analyze 31 artemisinin-based medicines (15 artemether–lumefantrine, 10 artesunate, 5 of which were coformulated with amodiaquine, and 6 dihydroartemisinin, 2 of which were coformulated with piperaquine and 1 triformulated with piperaquine and trimethoprim) purchased near Accra in Ghana in 2009. Surprisingly, not only did all the formulations contain the active ingredients but also the drug quality was found to be uniformly high as only one formulation contained less than 80% of the listed amount and most of them were within 90% of the announced doses. These findings indicate that antimalarial drugs in Africa are not always of low quality.[27]

Fourteen artesunate tablets, five classified as genuine products and nine as counterfeit ones based on packaging inspection, were analyzed with both NMR and MS, each method processing one half of the same tablet. ^{1}H and 2D DOSY ^{1}H NMR chemical profiling showed that the expected API was detected in the five genuine formulations whereas five contained unexpected APIs (one the antimalarial drug artemisinin, two the analgesic and antipyretic agent acetaminophen, and two the nonsteroidal anti-inflammatory medicine dipyrone) and four no API at all (Figure 28.2). Moreover, common organic excipients such as dextrin and starch as tablet diluents, stearate-based lubricant, lactose, and sucrose were also detected. Direct analysis in real-time (DART) MS and desorption electrospray ionization (DESI) MS allowed the identification of all the APIs but not of the excipients cited above except lactose and sucrose that could be detected only by DESI MS and differentiated by their DESI MS[2] spectra as these disaccharides are structural isomers. These results emphasize the powerful of ^{1}H and 2D DOSY ^{1}H NMR to distinguish easily genuine and counterfeit artesunate medicines. Moreover, ^{1}H NMR spectra can also serve

as fingerprints of the manufacturers of genuine artesunate tablets if the excipient composition is different. For example, the artesunate formulation from Guilin Pharmaceutical (China) contained sucrose that was replaced by lactose in the products manufactured by Pharbaco and Mekophar (Vietnam), Yangon Pharmacy Industry (Myanmar), or Mepha (Switzerland).[28]

NQR spectroscopy is currently developed as a noninvasive, nondestructive, and quantitative method for authentication of medicines through their packaging. Nitrogen-14 (^{14}N) NQR was applied to the analysis of an antimalarial drug, Metakelfin®, suspected to be counterfeit. Genuine Metakelfin contains two APIs, sulfamethoxypyrazine and pyrimethamine at 500 and 25 mg per tablet, respectively. [14]N NQR spectra of the genuine and suspect drugs revealed the presence of sulfamethoxypyrazine signals in both types of drugs but with intensity markedly lower in the suspect tablets compared to genuine ones, leading to the conclusion that the suspect drug contained only 43% as much sulfamethoxypyrazine as the genuine. HPLC analysis established that the sulfamethoxypyrazine content in suspect Metakelfin was 42% of that in genuine tablets and that of pyrimethamine in accordance with the stated dose. [14]N NQR demonstrated that the suspected Metakelfin tablets were substandard concerning sulfamethoxypyrazine but provided no information on pyrimethamine that was not detected.[29]

28.4 PHOSPHODIESTERASE-5 INHIBITORS (PDE5i)

Phosphodiesterase-5 Inhibitors (PDE5i) are a class of medicines used primarily in the treatment of erectile dysfunction. There are four PDE5i approved in Europe and the United States (sildenafil (Viagra®), tadalafil (Cialis®), vardenafil (Levitra®), and avanafil (Spedra® in Europe and Stendra® in the United States)). Three other PDE5i medicines are also approved: udenafil (Zydena®) in Asia and Russia, mirodenafil (Mvix®) in South Korea, and lodenafil carbonate (Helleva®) in Brazil. These prescription-only medicines are widely available on the illegal market and are also often added to taint DS at least for the three first. Moreover, numerous undeclared and unapproved PDE5i designer analogues have been found in DS claimed to be 100% natural.[30–32]

A number of analytical methods and strategies have been developed for the detection, characterization, and quantification of PDE5i (see Ref. 31 for

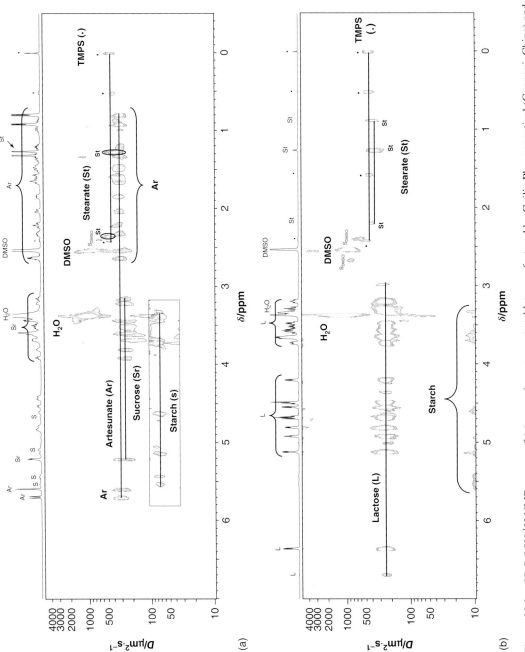

Figure 28.2. 2D DOSY ^{1}H NMR spectra of (a) a genuine artesunate tablet manufactured by Guilin Pharmaceutical (Guangxi, China) and (b) a counterfeit artesunate tablet, recorded at 500 MHz in DMSO-d_6 with trimethylsilylpropane sulfonic acid (TMPS) as an internal reference standard. (Adapted with permission from Ref. 14. Copyright (2010) Springer Science and Business Media and Ref. 28. Copyright (2009) American Chemical Society)

a recent comprehensive review). Chromatographic techniques, essentially HPLC or ultrahigh-pressure liquid chromatography (UHPLC) with various detectors (UV, MS, MS/MS, high-resolution mass spectrometry (IIR-MS), HR-MS/MS, combined with database library spectra search), are predominantly used. Vibrational spectroscopic techniques such as IR or Raman in conjunction with chemometrics are efficient tools to discriminate authentic Viagra® and Cialis® from counterfeit. They are fast screening methods requiring no or little sample preparation that can be performed by personnel with moderate training and expertise. However, their application implies the availability of reference standards for comparison of their spectral data to those of the suspected samples. Compared to these techniques, ¹H NMR is less rapid and sensitive but allows the detection of all the compounds with hydrogen atoms, their identification, and their quantification without the need for reference substances.

Eight commercial formulations of tadalafil (one being the brand formulation from Eli Lilly) and 17 of sildenafil (including two brand formulations from Pfizer (100 and 50 mg)) were analyzed with ¹H NMR and 2D DOSY ¹H NMR.[33,34] Active ingredients as well as organic excipients were detected, thus giving a precise and global chemical signature of a formulation and hence of its manufacturer. Six tadalafil and 13 sildenafil medicines were very good copies as they contained the active ingredient at the stated dose even though three sildenafil formulations were slightly overdosed (111–114% of the stated dose). They differed from the genuine drugs by their composition in excipients as, for example, hypromellose and polyethylene glycol in a copy of sildenafil, and hypromellose, lactose and triacetin in the genuine product (Figure 28.3). One purported tadalafil formulation contained no tadalafil but instead vardenafil and homosildenafil, a sildenafil analogue. These two compounds were also found in two sildenafil formulations together with a very low amount of sildenafil in one of them.

The power of NMR for the structural determination of isolated natural or synthetic compounds is well established. For instance, the structures of all the PDE5i analogues found as adulterants in DS or CM, more than 50 to date, were unambiguously elucidated by NMR and confirmed by MS (Ref. 32 and references quoted therein) until very recently when high-resolution Orbitrap MS under different fragmentation modes was used to identify two new sildenafil analogues.[35]

¹H NMR spectroscopy was used as a first-line screening method for detecting adulteration in ≈180 sexual enhancement DS. Besides the first characterization of three novel sildenafil analogues, it enabled the detection of (i) synthetic PDE5i (sildenafil and 12 analogues, tadalafil and 2 analogues, vardenafil), (ii) other drugs (yohimbine, flibanserin, DHEA, testosterone, and phentolamine) and substances extracted from plants (osthole and icariin) active against sexual dysfunction, and (iii) one drug (acetaminophen) and natural substances (honokiol, tetrahydropalmatine, salicin, curcumin, and piperine) with no known enhancing sexual performance activity. Among 150 DS tested, 61% were tainted with at least one synthetic PDE5i and 8% with other drugs or plant extracts used for improving sexual performance, whereas only 31% could be considered true herbal/natural products. Among the 92 samples adulterated with PDE5i, 64% contained only one drug while 36% more than one drug (26% two drugs as illustrated in Figure 28.4, 9% three PDE5i analogues, and 1% four). These results illustrate the manufacturers' trend to taint the formulations with a mixture of PDE5i rather than with just one. In an attempt to render more difficult the detection of adulterants in laboratories of regulatory health agencies or customs, the manufacturers have developed various strategies: (i) addition of PDE5i analogues not previously identified instead of well-known prescription-only medicines such as sildenafil, tadalafil, or vardenafil, (ii) modification of the adulterant in the same brand over time, and (iii) presence or absence of the adulterant(s) in the same batch of a given DS.[32]

28.5 SLIMMING PRODUCTS

In 2008, according to the WHO, more than 1.4 billion adults were overweight among which 500 million were obese.[36] Overweight and obesity are associated with premature mortality, chronic morbidity, diabetes, heart disease, osteoarthritis, and cancer. Given (i) the medical and social impact of being overweight or obese, (ii) the few drugs marketed for these indications, and (iii) the safety concerns of these medications, more and more patients turn to natural weight-loss DS. Unfortunately, a frequent illegal practice consists in adding undeclared drugs, such as anorexics, anxiolytics, antidepressants, diuretics, or stimulants, to improve their efficacy.

Several methods have been developed to screen pharmaceutical adulterants in weight-loss DS.[37,38]

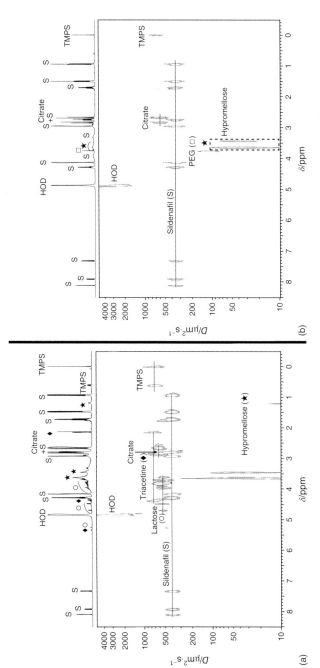

Figure 28.3.   2D DOSY ^{1}H NMR spectra of (a) genuine Viagra® and (b) a counterfeit formulation recorded at 500 MHz in D_2O with trimethylsilylpropane sulfonic acid (TMPS) as an internal reference standard. S: sildenafil; ◆: hypromellose; ★: sildenafil; □: polyethylene glycol (PEG). In part B, a deeper section of hypromellose signals is shown in a box. (Adapted with permission from Ref. 34. © John Wiley & Sons, Ltd., 2009)

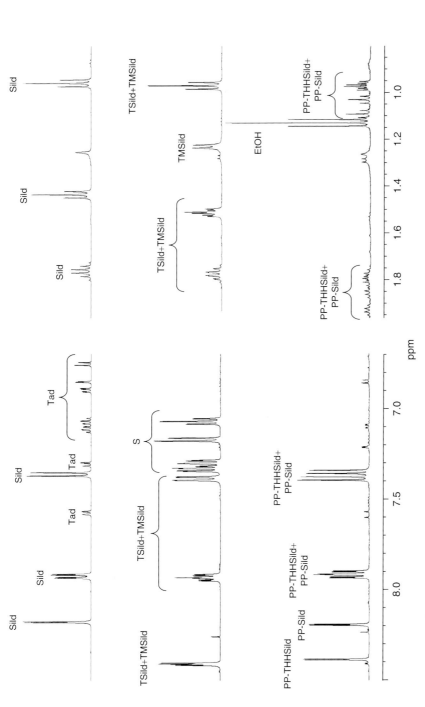

Figure 28.4. ^{1}H NMR spectra recorded at 500 MHz in CD$_3$CN : D$_2$O (80 : 20) of three '100% natural' sexual enhancement dietary supplements. Two characteristic regions are displayed: 8.5–6.7 and 1.95–0.8 ppm. Sild: sildenafil; Tad: tadalafil; TSild: thiosildenafil; TMSild: thiomethisosildenafil; S: salicin; PP-THHSild: propoxyphenyl-thiohydroxyhomosildenafil; PP-TSild: propoxyphenyl-sildenafil; EtOH: ethanol. (Adapted with permission from Ref. 32. Copyright (2014), with permission from Elsevier)

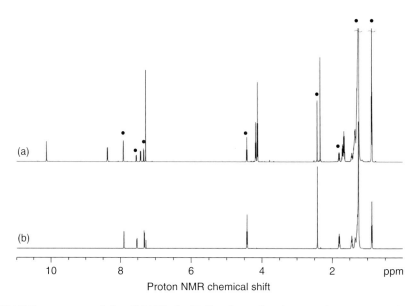

Proton NMR chemical shift

Figure 28.5. ^{1}H NMR spectra recorded at 600 MHz in CDCl$_3$ of two slimming over-the-counter products, Zenigal (a) and Orlislim (b). The Zenigal spectrum (a) shows the signals of cetilistat (●) and a cetilistat-related compound as well as other compounds. The Orlislim spectrum (b) shows the signals of cetilistat only. The residual CHCl$_3$ peak appears at 7.27 ppm. (Adapted with permission from Ref. 41. Copyright (2014), with permission from Elsevier)

Chromatographic methods, especially HPLC with UV, MS, or MS/MS detection, but also gas chromatography–MS and CE, are most widely used. Ion mobility spectrometry with bench-top or portable instruments serves as a high-throughput screening tool for the analysis of weight-loss products. Attenuated total reflectance-IR spectroscopy combined with chemometrics and X-ray powder diffractometry are also able to discriminate legal from adulterated products.

Twenty DS marketed as natural slimming products were analyzed by ^{1}H and 2D DOSY ^{1}H NMR.[39] Two were strictly herbal, four had a composition corresponding to declared ingredients, and fourteen were adulterated. Eight formulations contained sibutramine alone at doses between 4 and 31 mg per capsule, five sibutramine and phenolphthalein (5–20 and 4–66 mg per capsule or tablet, respectively) and the last synephrine (20 mg per capsule). The anorectic drug sibutramine, which was often found at levels higher than the maximum daily dosage of 15 mg, was withdrawn from the European and US markets in 2010 due to increased risk of heart attack and stroke. Phenolphthalein, a laxative compound, was banned as a medicine several years ago after concerns about carcinogenicity. Sibutramine was also found by ^{1}H NMR

in slimming coffee and tea products previously approved for the German market by several domestic private food-testing laboratories.[40]

The composition of the slimming over-the-counter products Zenigal and Orlislim, supposed to contain the antiobesity drug orlistat, was also investigated by ^{1}H NMR. Another antiobesity drug, cetilistat, was actually found in both products at doses of 22 and 121 mg per capsule, respectively. Zenigal also contained a ring-opened analogue of cetilistat (44 mg per capsule) and other unidentified compounds (Figure 28.5; Johansson).[41]

The weight-loss drug lorcaserin was identified very recently in a claimed 'pure natural fast slimming pill' at a dose of 6.6 mg per capsule by NMR. Its structure was confirmed, after purification by flash chromatography, with ^{1}H and ^{13}C NMR, MS, and MS/MS, and IR spectroscopy.[42]

28.6 ANTI-INFLAMMATORY HERBAL MEDICINES

The formulation Natural Life Harp-100 freely sold in Brazil both locally and on the Internet as an herbal medicine possessing anti-inflammatory and analgesic properties was subjected to ^{1}H NMR analysis.[43] The spectrum of a crude methanol extract provided a strong indication that synthetic substances might be present in the mixture. A 2D DOSY ^{1}H NMR experiment enabled the detection of at least four different major compounds in the mixture whose identification was not straightforward without chromatographic separation because of signal overlap. As only three peaks were clearly observed in a first HPLC-UV exploratory analysis, it was supposed that two or more components could have eluted at the same retention time. The compounds corresponding to the chromatographic peaks obtained from HPLC-UV coupled to an online solid-phase extraction (SPE) module were automatically trapped. The trapping of 20 injections was dried, redissolved in CD$_3$OD, and transferred into a 3 mm NMR tube to perform the 1D ^{1}H NMR, 2D ^{1}H–^{1}H COSY, and 2D ^{1}H–^{13}C HSQC and heteronuclear multiple bond correlation (HMBC) analyses. Citrate and a product derived from the hydrolysis of ranitidine (as confirmed from MS data) were identified in the first peak eluting, orphenadrine and piroxicam in the second, and dexamethasone in the final peak. The fractions obtained in the SPE procedure were also injected directly for MS analysis to confirm the chemical structures identified with NMR. The composition of the DS analyzed was not that of an herbal medicine for pain relief as claimed by the vendors, but corresponded to Rheumazin, an unregistered drug in Brazil, which contains a mixture of two anti-inflammatory drugs dexamethasone and piroxicam, one antalgic drug orphenadrine, and cyanocobalamin (not detected in this study) coformulated with the medicine for treating peptic ulcers, i.e., ranitidine or a product of ranitidine hydrolysis.

Another example concerns the analysis of a Vietnamese DS intended to treat rheumatic diseases and advertised as 100% natural.[44] The ^{1}H NMR spectrum of a crude extract was dominated by the signals of the analgesic drug acetaminophen. Four ingredients were detected in the HPLC analysis and, after isolation, identified by NMR as acetaminophen, sulfamethoxazole, trimethoprim (both antibiotics), and indomethacin (anti-inflammatory). Their total amounts represented ≈45% of the powder content of each sachet, which also contained phosphate and cinnamon bark. The formulation was obviously not natural. Moreover, purchasing these sachets was illegal, as indomethacin and the combination sulfamethoxazole/trimethoprim, called *co-trimoxazole*, are prescription-only drugs.

28.7 CORTICOSTEROIDS

Ten over-the-counter cosmetic products were screened by ^{1}H and ^{19}F NMR to detect their possible adulteration with corticosteroids.[45] As these APIs are normally present in cream and ointment medicines at low concentrations (0.025–0.1%), an extraction step was required prior to NMR analysis. Two formulations were found to contain corticosteroids. In one of the two positives, the presence of the API was declared while it was not in the other. The structure of the undeclared fluorocorticosteroid, triamcinolone acetonide-21-acetate, was deduced from (i) its molecular mass measured in a quadrupole time-of-flight liquid chromatography mass spectrometry (Q-TOF-LC–MS) experiment and (ii) the identification by spiking with reference compounds of its acidic hydrolysis products formed successively as triamcinolone acetonide and triamcinolone. The declared fluorocorticoid in the second positive sample was dexamethasone-21-acetate. Its identity was confirmed by spiking the cream extract with the authentic product. Its concentration measured with both ^{1}H and ^{19}F NMR (0.09%) was close to that mentioned on the package (0.075%).

28.8 ISOTOPIC NMR

It has been well established that isotope profiling is a powerful method for the authentication of the origin of products because it provides information about the chemical history of a given compound. Ratios of naturally occurring stable isotopes constitute chemical tracers that are generally available in measurable concentrations. They are determined by isotope ratio mass spectrometry (IRMS) or by NMR spectroscopy. Recently, several studies demonstrated the potential of isotopic NMR in the field of drug counterfeiting by focusing on the API. Deconinck et al.[46] analyzed various formulations of ibuprofen and aspirin collected from different worldwide manufacturers. They used the site-specific natural isotope fractionation (SNIF)-NMR method, a technique that quantitatively determines deuterium content (^{2}H/^{1}H ratio) at specific sites of a molecule. In addition, these authors performed ^{13}C, ^{18}O, and ^{2}H measurements by IRMS and then applied various chemometric tools to both IRMS and NMR data sets. It was possible to distinguish groups of samples based on their geographic origin and on the manufacturer. The discrimination was due to differences in synthetic pathways or reagents involved in the synthesis. Samples from the same company and origin were found close together, which indicates that the reproducibility of the measurements is satisfactory and that the inter-batch influence can be neglected for the same manufacturing process. The use of SNIF-NMR for gaining information on the synthetic sequence followed for the preparation of commercial drugs was also described by Acetti *et al.*[47] in the case of ibuprofen and naproxen and by Brenna *et al.*[48] for fluoxetine formulations (Prozac®).

Isotopic ^{13}C NMR at natural abundance is an emerging methodology in comparison to the well-established ^{2}H SNIF-NMR method. The variation in the ^{13}C content of natural products is within a range of about 5%, compared with 50% for ^{2}H. Therefore, small changes in the order of 0.1% have to be detected, i.e., the method must be able to detect a difference of 1‰ in the peak areas. Thus, the main difficulty for measuring site-specific ^{13}C/^{12}C ratios directly using quantitative ^{13}C NMR is meeting the requirement for such a high level of precision. Nevertheless, the approach has been successfully applied for isotopic profiling of aspirin and acetaminophen drugs collected from pharmacies in different countries.[49] The site-specific ^{13}C content ($\delta_i {}^{13}$C) determined by ^{13}C NMR was compared to the global ^{13}C content ($\delta_g {}^{13}$C) measured by IRMS. While the $\delta_g {}^{13}$C are comprised within an SD of $\approx 3‰$, the $\delta_i {}^{13}$C values showed a much greater variability for aspirin and acetaminophen both for inter-carbon-position and inter-sample comparisons. This variability enabled the characterization of each compound and thus of each batch of API. The isotopic fingerprints of different batches of aspirin from the same supplier were different. Conversely, the same batch of aspirin from different packaging presented the same profile.

Using the adiabatic INEPT sequence for ^{13}C isotopic measurements considerably reduces the recording time without any deterioration in short-(≈ 90 min) or long-time (over a 3-month period) stability of the NMR measurement. For the quantification of hydrogen-bearing carbons of ibuprofen detected with polarization transfer methods, the recording time was divided by a factor 7.9 compared to the one-pulse sequence. The ^{13}C intensities depend primarily on molar fractions but also, with the INEPT sequence, on the scalar coupling constants ^{1}J(^{13}C–^{1}H) and on the number of attached protons. Consequently, only relative values of the isotope ratios are obtained from the observed intensities, contrarily to ^{2}H NMR and

single pulse ^{13}C NMR, which provide 'true isotope ratios'. However, these relative values are sufficient for establishing an isotopic fingerprint for a given molecule and to compare accurately differences in the isotopic distribution between different samples. The results obtained on 13 commercial ibuprofen samples from 5 countries and 11 samples of naproxen purchased from pharmacies or from manufacturers in 4 countries showed that the individual ^{13}C profile determined is a mark of the raw material origin and the manufacturing process for each batch.[50,51]

As a conclusion, multinuclear NMR fingerprinting is an excellent tool for tracking APIs. Indeed, ^{2}H and ^{13}C NMR spectroscopy give direct access to site-specific isotope content at natural abundance and thus provide complementary and valuable information to detect counterfeiting practices in the pharmaceutical industry, as for instance the use of synthetic ingredients when a natural origin is labeled, the deliberate copying of processes or the infringement of current patents for generic medicines.[51]

28.9 CONCLUSION

What should be the ideal analytical technique for screening and characterizing CM and adulterated DS? The answer is not the same for both concerns because, as already stated, CM are more prevalent in developing countries while developed countries are more affected by the circulation of adulterated DS. So, the appropriate technologies can differ considerably in their characteristics: cost (inexpensive field assays or sophisticated laboratory instruments and methods), level of technical expertise of people in charge of the controls, type of data provided that can be simply qualitative (presence or absence of a specific API or of an adulterant) or more detailed (determination of the structure of the wrong API or adulterant), or quantitative.[17] Nonetheless, the ideal method should be rapid, sensitive, specific, accurate, low cost and would require no or minimum sample preparation and no reference standards. Does NMR meet these expectations?

Compared with conventional analytical methods such as chromatographic techniques, the intrinsic insensitivity of NMR spectroscopy remains its Achilles' heel. However, this limitation is not actually a real concern in the field of fraudulent medicines or DS detection as the concentration of the adulterant is usually substantial. Moreover, the technical evolution over the past 15 years (cryoprobes and microprobes, ultrahigh field instruments) has greatly lowered the limit of detection of the method and decreased the time spent for recording spectra. NMR spectroscopy has several advantages over other techniques. NMR spectroscopy is invasive but nondestructive (after NMR analysis, samples can be reused for further investigations), highly reproducible, robust, and easily quantitative (see Chapter 5). It requires minimal sample preparation by dissolution of solid samples (or dilution in case of liquids) in an adequate deuterated solvent. The main advantage of NMR spectroscopy relies on its holistic approach making it particularly suitable for the analysis of mixtures. Indeed, the method is nonselective, meaning that all the compounds present in the solution (provided they contain the nucleus under investigation) are detected simultaneously in a single run. NMR spectroscopy provides full structural elucidation through the use of a powerful toolkit including 1D and 2D NMR experiments (COSY, HSQC, HMBC, TOCSY, NOESY, etc.). In addition, NMR isotopic analysis affords accurate information on the traceability of the API (route of synthesis, origin of starting materials). The main interest of DOSY NMR lies in its ability to facilitate structural identification as it adds a second dimension enabling the virtual separation of components of complex mixtures according to their self-diffusion coefficients related to molecular weight at a rough estimate. Despite its great analytical power and the recent improvements in its automation that make it high throughput (automated sample

preparation, sample changer, automated data analysis, etc.), NMR is rarely used in quality control laboratories. Both high purchase and running costs and the high level of expertise needed by the operator to run the equipment and interpret the data are the primary reasons of this situation. However, the present price of an NMR spectrometer suitable for drug quality control (7.0/9.4 T corresponding to proton resonance frequencies of 300/400 MHz) does not greatly exceed that of sophisticated LC–MS equipment. Moreover, the magnets have long lifetimes ($\approx$15–20 years), and advances in computer technology can be implemented continuously in NMR spectrometers at a reasonable cost. Another limitation of NMR spectroscopy is that in its current configuration (high-field super-conducting magnets requiring a devoted and air-conditioned room, regular fillings with cryogenic fluids, etc), it is not feasible outside a research laboratory.

The recent release of bench-top cryogen-free low-field ^{1}H NMR spectrometers that are 'saving instruments' (saving purchase cost, saving maintenance and running costs, saving space, and saving user experience) and push-button devices might change things and these instruments may become part of quality control laboratories and governmental agencies (customs, antifraud organizations, health agencies, etc.) as an initial routine tool for screening falsified medicines and DS. Figure 28.6 illustrates the capability of a 60 MHz spectrometer in revealing the adulteration of a sexual enhancement DS.[52]

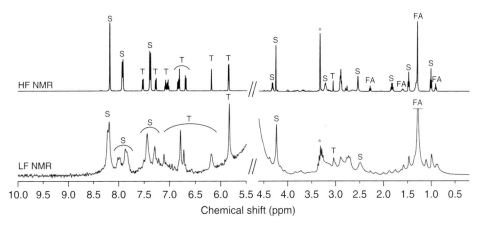

Figure 28.6. Comparison of the ^{1}H NMR spectra of a '100% natural' sexual enhancement dietary supplement recorded in CD$_3$OD with a high-field (HF) 500 MHz spectrometer (8 scans; recording time <1 min) and low-field (LF) 60 MHz spectrometer (256 scans; recording time: 22.5 min). S: sildenafil; T: tadalafil; FA: fatty acids; *: CD$_2$HOD. (Adapted with permission from Ref. 52. Copyright (2014) American Chemical Society)

In the frame of CM and adulterated DS detection, NMR is obviously a powerful method that would deserve a more generalized use. However, as all the analytical techniques have their own advantages and limitations, one must be mindful of the need to use a combination of two or more orthogonal techniques for an in-depth characterization of samples analyzed.

ACKNOWLEDGEMENTS

The authors wish to acknowledge the French National Agency for the Safety of Medicines and Health Products (Agence Nationale de Sécurité du Médicament et des produits de santé: ANSM) for financial support (projet AAP-2012-082, convention ANSM/UPS n°2012S071).

RELATED ARTICLES IN EMAGRES

Analysis of High-Resolution Solution State Spectra

Diffusion Measurements by Magnetic Field Gradient Methods

Polysaccharides and Complex Oligosaccharides

Quantitative Measurements

Glycosaminoglycan Structural Characterization

REFERENCES

1. Interpol, Thousands of illicit online pharmacies shut down in the largest-ever global operation targeting fake medicines, 2014. http://www.interpol.int/News-and-media/News/2014/N2014-089.

2. T. K. Mackey and B. A. Liang, *BMC Med.*, 2013, **11**, 233.

3. A. Attaran, D. Barry, S. Basheer, R. Bate, D. Benton, J. Chauvin, L. Garrett, I. Kickbusch, J. C. Kohler, K. Midha, P. N. Newton, S. Nishtar, P. Orhii, and M. McKee, *BMJ*, 2012, **345**, e7381.

4. World Health Organization, WHO's role in the prevention and control of medical products of compromised quality, safety and efficacy such as substandard/spurious/falsely labelled/falsified/counterfeit medical products, 2011. http://apps.who.int/gb/ssffc/pdf_files/A_SSFFC_WG2_3-en.pdf.

5. World Health Organization, Medicines: spurious/falsely-labelled/falsified/ counterfeit (SFFC) medicines, Fact sheet n°275, 2012. http://www.who.int/mediacentre/factsheets/fs275/fr/.

6. World Health Organization, Guidelines for the development of measures to combat counterfeit drugs, 1999. http://www.who.int/medicines/publications/counterfeitguidelines/en/.

7. World Health Organization, *Bull. World Health Org.*, 2010, **88**, 247.

8. R. L. Bailey, J. J. Gahche, P. E. Miller, P. R. Thomas, and J. T. Dwyer, *JAMA Intern. Med.*, 2013, **173**, 355.

9. newhope, 2014 Global Supplement and Nutrition Industry Report. http://newhope360.com/2014-global-supplement-and-nutrition-industry-report.

10. A. Petroczi, G. Taylor, and D. P. Naughton, *Food Chem. Toxicol.*, 2011, **49**, 393.

11. M. Nuccitelli, Bodybuilding supplements, male enhancement and weight loss diet pills tainted, 2011. http://www.amazines.com/view_author.cfm?authorid=327961.

12. Z. Harel, S. Harel, R. Wald, M. Mamdani, and C. M. Bell, *JAMA Intern. Med.*, 2013, **173**, 926.

13. K. Dégardin, Y. Roggo, and P. Margot, *J. Pharm. Biomed. Anal.*, 2014, **87**, 167.

14. R. Martino, M. Malet-Martino, V. Gilard, and S. Balayssac, *Anal. Bioanal. Chem.*, 2010, **398**, 77.

15. Counterfeit Medicines: Volume II: Detection, Identification and Analysis, eds P. G. Wang and A. I. Wertheimer, ILM Publications: Hertfordshire, UK, 2012.

16. Committee on Understanding the Global Public Health Implications of Substandard, Falsified, and Counterfeit Medical Products, Countering the Problem of Falsified and Substandard Drugs, eds L. O. Gostin and G. J. Buckley, The National Academies: Washington, DC, 2013, 255.

17. S. Kovacs, S. E. Hawes, S. N. Maley, E. Mosites, L. Wong, and A. Stergachis, *PLoS ONE*, 2014, **9**, e90601.

18. S. Beni, J. F. K. Limtiaco, and C. K. Larive, *Anal. Bioanal. Chem.*, 2011, **399**, 527.

19. M. Guerrini, Z. Zhang, Z. Shriver, A. Naggi, S. Masuko, R. Langer, B. Casu, R. J. Linhardt, G. Torri, and R. Sasisekharan, *Proc. Natl. Acad. Sci. U.S.A.*, 2009, **106**, 16956.

20. M. Guerrini, D. Beccati, Z. Shriver, A. Naggi, K. Viswanathan, A. Bisio, I. Capila, J. C. Lansing, S. Guglieri, B. Fraser, A. Al-Hakim, N. S. Gunay,

Z. Zhang, L. Robinson, L. Buhse, M. Nasr, J. Woodcock, R. Langer, G. Venkataraman, R. J. Linhardt, B. Casu, G. Torri, and R. Sasisekharan, *Nat. Biotechnol.*, 2008, **26**, 669.

21. T. Beyer, M. Matz, D. Brinz, O. Rädler, B. Wolf, J. Norwig, K. Baumann, S. Alban, and U. Holzgrabe, *Eur. J. Pharm. Sci.*, 2010, **40**, 297.

22. D. A. Keire, H. Ye, M. L. Trehy, W. Ye, R. E. Kolinski, B. J. Westenberger, L. F. Buhse, M. Nasr, and A. Al-Hakim, *Anal. Bioanal. Chem.*, 2011, **399**, 581.

23. D. A. Keire, D. J. Mans, H. Ye, R. E. Kolinski, and L. F. Buhse, *J. Pharm. Biomed. Anal.*, 2010, **52**, 656.

24. Q. Zang, D. A. Keire, R. D. Wood, L. F. Buhse, C. M. Moore, M. Nasr, A. Al-Hakim, M. L. Trehy, and W. J. Welsh, *Anal. Chem.*, 2011, **83**, 1030.

25. G. M. L. Nayyar, J. G. Breman, P. N. Newton, and J. Herrington, *Lancet Infect. Dis.*, 2012, **12**, 488.

26. N. Ranieri, P. Tabernero, M. D. Green, L. Verbois, J. Herrington, E. Sampson, R. D. Satzger, C. Phonlavong, K. Thao, P. N. Newton, and M. R. Witkowski, *Am. J. Trop. Med. Hyg.*, 2014, DOI: 10.4269/ajtmh.13-0644.

27. E. Y. Klein, I. A. Lewis, C. Jung, M. Llinas, and S. A. Levin, *Malar. J.*, 2012, **11**, 110.

28. L. Nyadong, G. A. Harris, S. Balayssac, A. S. Galhena, M. Malet-Martino, R. Martino, R. M. Parry, M. D. Wang, F. M. Fernández, and V. Gilard, *Anal. Chem.*, 2009, **81**, 4803.

29. J. Barras, D. Murnane, K. Althoefer, S. Assi, M. D. Rowe, I. J. F. Poplett, G. Kyriakidou, and J. A. S. Smith, *Anal. Chem.*, 2013, **85**, 2746.

30. B. J. Venhuis and D.de Kaste, *J. Pharm. Biomed. Anal.*, 2012, **69**, 196.

31. D. N. Patel, L. Li, C. L. Kee, X. Ge, M. Y. Low, and H. L. Koh, *J. Pharm. Biomed. Anal.*, 2014, **87**, 176.

32. V. Gilard, S. Balayssac, A. Tinaugus, N. Martins, R. Martino, M. Malet-Martino 2015 *J. Pharm. Biomed. Anal.*, **102**, 476.

33. S. Trefi, C. Routaboul, S. Hamieh, V. Gilard, M. Malet-Martino, and R. Martino, *J. Pharm. Biomed. Anal.*, 2008, **47**, 103.

34. S. Trefi, V. Gilard, S. Balayssac, M. Malet-Martino, and R. Martino, *Magn. Reson. Chem.*, 2009, **47**, 5163.

35. C. L. Kee, H. L. Koh, B. C. Bloodworth, Y. Zeng, K. H. Kiang, M. Y. Low, and X. Ge, *J. Pharm. Biomed. Anal.*, 2014, **98**, 153 and references quoted in.

36. World Health Organization, Obesity and Overweight, Fact Sheet n°311, 2014. http://www.who.int/mediacentre/factsheets/fs311/en/.

37. L. M. de Carvalho, M. Martini, A. P. Moreira, A. P.de Lima, D. Correia, T. Falcão, S. C. Garcia, A. V.de Bairros, P. C.do Nascimento, and D. Bohrer, *Forensic Sci. Int.*, 2011, **204**, 6.

38. F. Song, D. Monroe, A. El-Demerdash, and C. Palmer, *J. Pharm. Biomed. Anal.*, 2014, **88**, 136.

39. J. Vaysse, S. Balayssac, V. Gilard, D. Desoubdzanne, M. Malet-Martino, and R. Martino, *Food Addit. Contam.*, 2010, **27**, 903.

40. Y. B. Monakhova, T. Kuballa, S. Löbell-Behrends, S. Maixner, M. Kohl-Himmelseher, W. Ruge, and D. W. Lachenmeier, *Drug Test. Anal.*, 2013, **5**, 400.

41. M. Johansson, D. Fransson, T. Rundlöf, N. H. Huynh, and T. Arvidsson, *J. Pharm. Biomed. Anal.*, 2014, **100**, 215.

42. R. Hachem, M. Malet-Martino, and V. Gilard, *J. Pharm. Biomed. Anal.*, 2014, **98**, 94.

43. L. M. A. Silva, E. G. A. Filho, S. S. Thomasi, B. F. Silva, A. G. Ferreira, and T. Venâncio, *Magn. Reson. Chem.*, 2013, **51**, 541.

44. J. Wiest, C. Schollmayer, G. Gresser, and U. Holzgrabe, *J. Pharm. Biomed. Anal.*, 2014, **97**, 44.

45. I. McEwen, A. Elmsjö, A. Lehnström, B. Hakkarainen, and M. Johansson, *J. Pharm. Biomed. Anal.*, 2012, **70**, 245.

46. E. Deconinck, A. M.van Nederkassel, I. Stanimirova, M. Daszykowski, F. Bensaid, M. Lees, G. J. Martin, J. R. Desmurs, J. Smeyers-Verbeke, and Y. V. Heyden, *J. Pharm. Biomed. Anal.*, 2008, **48**, 27.

47. D. Acetti, E. Brenna, G. Fronza, and C. Fuganti, *Talanta*, 2008, **76**, 651.

48. E. Brenna, G. Fronza, and C. Fuganti, *Anal. Chim. Acta*, 2007, **601**, 234.

49. V. Silvestre, V. Maroga Mboula, C. Jouitteau, S. Akoka, R. J. Robins, and G. S. Remaud, *J. Pharm. Biomed. Anal.*, 2009, **50**, 336.

50. U. Bussy, C. Thibaudeau, F. Thomas, J. R. Desmurs, E. Jamin, G. S. Remaud, V. Silvestre, and S. Akoka, *Talanta*, 2011, **85**, 1809.

51. G. S. Remaud, U. Bussy, M. Lees, F. Thomas, J. R. Desmurs, E. Jamin, V. Silvestre, and S. Akoka, *Eur. J. Pharm. Sci.*, 2013, **48**, 464.

52. G. Pagès, A. Gerdova, D. Williamson, V. Gilard, R. Martino, and M. Malet-Martino, *Anal. Chem.*, 2014, **86**, 11897.

Chapter 29

Pharmaceutical Industry: Regulatory Control and Impact on NMR Spectroscopy

Andrea Ruggiero and Sarah K. Branch

Medicines and Healthcare Products Regulatory Agency (MHRA), 151 Buckingham Palace Road, Victoria, London SW1W 9SZ, UK

29.1 INTRODUCTION

NMR spectroscopy has proved to be a powerful tool in a number of fields, including the research and development of medicines. The technical advances in instrumentation and the wide range of techniques have been exploited by the pharmaceutical industry during development of drug substances and new drug products as described in the other articles of this work. It is inevitable therefore that a number of NMR methods have made their way through into the regulatory dossiers that accompany applications for regulatory approval. This chapter sets out the principles of regulatory approval and the requirements for any analytical methods used to control drug substances and drug products. It highlights areas where NMR spectroscopy is particularly advantageous while acknowledging the challenges associated with using NMR methods. It gives examples of pitfalls encountered with presentation of NMR data during the assessment of dossiers at the MHRA. Regulatory control through pharmacopoeial monographs is addressed elsewhere (see Chapter 30). A variety of synonyms are used to describe a substance used as a drug in a medicine, e.g., drug substance, active substance, or active ingredient. Likewise, a number of terms are used for a drug in a preparation for administration to a patient, e.g., drug product, medicinal product, dosage form, or formulation. In this chapter, we use the terms adopted by the International Conference for Harmonization of Technical Requirements for the Registration of Pharmaceuticals for Human Use (ICH) i.e., *drug substance* and *drug product*.[1]

The views expressed in this chapter are those of the authors and do not necessarily represent the views or the opinions of the Medicines and Healthcare products Regulatory Agency, other regulatory authorities, or any of their advisory committees.

NMR in Pharmaceutical Sciences. Edited by Jeremy R. Everett, John C. Lindon, Ian D. Wilson, and Robin K. Harris
© 2015 John Wiley & Sons, Ltd. ISBN: 978-1-118-66025-6
Also published in eMagRes (online edition)
DOI: 10.1002/9780470034590.emrstm1419

29.2   THE REGULATORY DOSSIER

While jurisdictions world-wide vary in their regulatory procedures for approval of applications to market drug products, all are based on the principle of establishing the pharmaceutical quality of the product, its safety – both toxicological and clinical – and its efficacy, i.e., demonstration of a positive benefit-risk in clinical use. The need to avoid duplication of tests and clinical trials in different jurisdictions has led to a number of initiatives to harmonize requirements for regulatory approval, the most important of which is the ICH that was initiated in 1990. In 2000, ICH introduced the Common Technical Document (CTD), followed in 2002 by its electronic counterpart, the eCTD.[2] The CTD provides a common structure for the main body of scientific and technical data supporting applications to market drug products in the three sponsoring regions (USA, EU, and Japan). As the CTD is a format recognized by jurisdictions other than the three sponsoring regions, it has led to a significant reduction in the administrative aspects of submitting regulatory dossiers globally. The CTD is supported by a range of scientific and technical guidelines developed by ICH. They are designed to set out common requirements for gathering the data required to support a marketing authorization application and for the tests and criteria to be applied for regulatory approval. Additional region-specific guidance may also be available. Use of NMR spectroscopic methods tends to be associated with establishing the pharmaceutical quality of the product in question.

29.3   INFORMATION REQUIRED TO ESTABLISH THE QUALITY OF DRUG SUBSTANCE AND DRUG PRODUCT

The information required about the quality of a drug substance and corresponding drug product is set out in Module 3 of the Common Technical Document. Table 29.1 outlines the expected content.

29.3.1   ICH Quality Guidelines

Table 29.2 gives a summary of ICH guidelines that are relevant to the pharmaceutical quality section of the regulatory dossier.[3]

Applications of NMR spectroscopy in establishing the quality of drug substances and drug products are discussed in Section 29.4.

29.3.2   Pharmacopoeial Drug Substances and Drug Products

Where there is a relevant monograph in the pharmacopoeia of a particular jurisdiction, a drug substance or drug product must comply with that monograph. It should also be noted that general monographs for dosage forms may also apply to the drug product. Where a pharmacopoeial monograph for a drug substance, pharmaceutical excipient, or a drug product employs NMR spectroscopy in a test method, the substance must comply with this test. On the other hand, a non-pharmacopoeial test may be used if proof is supplied that the substance or product meets the requirements of the relevant pharmacopoeia, if tested. If a pharmacopoeial monograph is applicable to a substance or drug product, then there is no need for the applicant to provide full details of the analytical tests, or their validation, as reference to the pharmacopoeia in question is deemed sufficient. Applications of NMR

Table 29.1.   Content of Module 3 of the CTD[a]

3.1	Table of Contents
3.2	Body of Data
3.2.S	Drug Substance
3.2.S.1	General Information
3.2.S.2	Manufacture
3.2.S.3	Characterization
3.2.S.4	Control of Drug Substance
3.2.S.5	Reference Standards or Materials
3.2.S.6	Container Closure System
3.2.S.7	Stability
3.2.P	Drug Product
3.2.P.1	Description and Composition of Drug Product
3.2.P.2	Pharmaceutical Development
3.2.P.3	Manufacture
3.2.P.4	Control of Excipients
3.2.P.5	Control of Drug Product
3.2.P.6	Reference Standards or Materials
3.2.P.7	Container Closure System
3.2.P.8	Stability
3.2.A	Appendices (specified)
3.2.R	Regional Information
3.3	Literature References

[a] Further subsections are not listed here.

Table 29.2. Selected ICH guidelines relevant to Module 3 of the CTD

Title	Relevant CTD section
Q1A(R2)[a] *Stability Testing of New Drug Substances and Products*	3.2.S and 3.2.P
Q5C *Stability Testing of Biotechnological/Biological Products*	3.2.P
Q2(R1) *Validation of Analytical Procedures: Definitions and Terminology*	3.2.S and 3.2.P
Q2B, *Validation of Analytical Procedures: Methodology*	3.2.S and 3.2.P
Q3A *Impurities in New Drug Substances*	3.2.S
Q3B *Impurities in New Drug Products*	3.2.P
Q6A *Specifications: Test Procedures and Acceptance Criteria for New Drug Substances and New drug Products: Chemical Substances*	3.2.S and 3.2.P
Q6B *Test Procedures and Acceptance Criteria for New Drug Substances and New drug Products: Biotechnological/Biological products*	3.2.S and 3.2.P

[a] R denotes a revision.

spectroscopy in pharmacopoeias are examined in more detail in Chapter 30.

It should be noted that where a drug substance or excipient used in the formulation of a drug product has been prepared by a method that may lead to impurities not controlled by the monograph, these impurities and their maximum limits must be declared and a suitable test procedure described in the regulatory dossier. Changing a route of synthesis might give rise to different solvent impurities, catalysts, or related substances. Residual solvents and metals should comply with limits in the relevant guidelines.[4,5] The limits proposed for the related substances must be qualified, that is, must consider the amounts present in batches used in toxicology studies (see below). NMR techniques are, of course, particularly useful for identifying new organic impurities if they can be isolated in sufficient quantities.

A competent authority (i.e., the regulatory body responsible for approving an application to place a drug product on the market) may also request more appropriate specifications if it considers that the monograph is out-of-date and requires revision or is insufficient to assure adequate quality of the substance or drug product. Thus, additional controls are needed if different synthetic methods are used for the drug substance, further tests may be required for purity, particle size, polymorphic form, microbial contamination, and sterility as necessary to ensure its correct performance in the drug product. Limits that are tighter than the pharmacopoeial specification may be imposed, e.g., if appropriate for the substance in question considering the dosage form in which it will be used or for a drug product release specification to allow for degradation on storage. In addition, the

competent authority will require stability data for drug substances and drug products so that suitable storage conditions can be established. Stability data are also used to specify retest period for a drug substance and the shelf-life or expiry date for the drug product. The retest period for a drug substance is the period of time for which it is expected to remain within specification and after which it must be retested for compliance and used immediately for manufacture of drug product.

In the European Union, there is a mechanism whereby a manufacturer can request a Certificate of Suitability (CEP) for their substance from the European Directorate for the Quality of Medicines and HealthCare (EDQM).[6,7] Such a certificate confirms that a European Pharmacopoeia (Ph. Eur.) monograph is suitable for the control of that pharmaceutical substance produced by a particular manufacturer. A certificate provides assurance that the tests in the monograph are adequate to control the drug substance manufactured by a particular synthetic route. Additional tests (e.g., gas chromatography for residual solvents) and relevant limits that are not listed in the Ph. Eur. monograph are evaluated by the EDQM and annexed to the CEP.

29.3.3 Drug Master Files

Alternatively in the European Union, an applicant can submit the drug substance dossier to competent authorities following the Active Substance Master File (ASMF) procedure.[8] In particular, the ASMF procedure can be used for the following drug substances:

1. New drug substances;
2. Existing drug substances not included in the Ph. Eur. or a pharmacopoeia of an EU Member State;
3. Pharmacopoeial drug substances that are included in the Ph. Eur. or in the pharmacopoeia of an EU Member State.

An existing drug substance is one that has already been approved in a marketing authorization. The ASMF procedure cannot be used for biological drug substances.

It is possible that the ASMF holder may have an ASMF and a CEP issued by EDQM for a single drug substance. Generally, it is not acceptable that the applicant or marketing authorization holder refers to an ASMF and to a CEP for the same drug substance in a particular marketing authorization. In cases where the CEP does not contain key information (e.g., stability), the competent authorities may decide that additional information should be provided in the dossier. In such cases it may be acceptable to refer both to an ASMF and a CEP.

The main objective of the ASMF procedure is to allow valuable confidential intellectual property or 'know-how' of the manufacturer of the active substance to be protected, while at the same time allowing the applicant or marketing authorization holder to take full responsibility for the drug product and the quality and quality control of the drug substance. The competent authorities thus have access to the complete information that is necessary for an evaluation of the suitability of the use of the drug substance in the drug product.

The scientific information in the ASMF should be physically divided into two separate parts, namely the Applicant's Part (AP) and the Restricted Part (RP). The AP contains the information that the ASMF holder regards as nonconfidential and has shared with the marketing authorization holder, whereas the RP contains the information that the ASMF holder regards as confidential, usually relating to details of the synthesis of the drug substance. It is emphasized that the AP is still a confidential document that cannot be submitted by anyone to third parties without the written consent of the ASMF holder. In all the cases the AP should contain sufficient information to enable the applicant or marketing authorization holder to take full responsibility for ensuring that the drug substance specification is suitable for the control of its quality and its use in the manufacture of a specified drug product.

The RP may contain the remaining information, such as detailed information on the individual steps of the manufacturing method (reaction conditions, temperature, validation and evaluation data of critical steps) and the quality control during the manufacture of the active substance. The competent authorities may not accept that particular information has not been disclosed to the applicant or marketing authorization holder. In such cases, they may ask for an amendment to the AP.

The ASMF may be provided in a format that is consistent with the CTD format. With regards to NMR spectra, these are normally provided in the AP, within the characterization sections for the drug substance and relevant impurities (i.e., CTD 3.2.S.3.1 *'Elucidation of Structure and Other Characteristics'* and CTD 3.2.S.3.2 *'Impurities'*).

29.4 NMR SPECTROSCOPY IN THE REGULATORY DOSSIER

Non-pharmacopoeial drug substances (or pharmaceutical excipients in certain circumstances) that are not listed in a pharmacopoeia should be described in a similar format to a monograph. The two ICH guidelines on *Specifications: Test Procedures and Acceptance Criteria for New Drug Substances and New Drug Products: Chemical Substances* and *Biotechnological/Biological substances* are of particular relevance here.[9,10] Testing against a specification is normally done post-manufacture but before the batch release. However, NMR spectroscopy is also used as part of less conventional drug substance routine testing, for instance, real-time release testing (RTRT).[11] RTRT can be applied to discrete unit operations and to chemical reactions or separations (e.g., of diastereoisomers) during drug substance synthesis. The use of NMR is beneficial for identification purposes and establishing purity in such cases.

In either post-manufacture or real-time testing, ^{1}H, ^{13}C, or multinuclear NMR spectroscopy may be used as appropriate. The wider spectral width and the ability to distinguish the drug substance from compounds not containing the observed heteroatom, thus increasing specificity, are the possible advantages of multinuclear techniques. The ability of NMR to provide greater structural specificity than the other spectroscopic techniques is well-recognized, and it has therefore been used in identity tests for more complex molecules such as peptides and proteins as well as heparins. NMR has also been used to confirm that the drug substance is present in the correct polymorphic form (see the following section).

29.4.1 Drug Substance

The data required in a regulatory dossier for the drug substance includes general information (such as nomenclature, general properties, and structure), details of its manufacture and any in-process controls, its characterization, information on impurities, quality control specifications, and results of batch analyses. Information should also be given on reference standards or materials used for quality control and analytical validation. Finally, stability data is required for storage in specified containers and closure systems.

29.4.1.1 Manufacture

Control of the drug substance manufacturing process requires the application of specifications to starting materials, intermediates, solvents and reagents as well as the final drug substance. Any of these may include NMR tests, usually to provide identification of the substance in question, and these may also be used in real-time (see the earlier section).

29.4.1.2 Characterization

NMR spectroscopy comes into its own in characterization of the drug substance that is expected to include evidence of structure, discussion of potential isomerism, polymorphism, and physicochemical characterization.

The structural evidence provided for new drug substances should be related to the actual material to be used in the marketed product. This is particularly the case when complex molecules are involved where the exact final structure is more dependent on the specific method of manufacture. If the data provided relates to substance produced by a different route of synthesis, then it will be necessary to reconfirm the structural identity of the resulting material. A similar principle applies to existing drug substances. It is a matter of routine that NMR spectroscopy, together with other spectroscopic techniques and elemental analysis, is used to confirm the structure of the drug substance. Proof of structure can be supported by the route of synthesis, and NMR spectroscopy has

a role in confirming that the expected reactions have taken place through establishing the identity of intermediate products.

A wide range of ^{1}H, ^{13}C, and multinuclear FT-NMR techniques have been presented in regulatory dossier applications for marketing authorizations to support the structures proposed for drug substances, both chemical and biological. Polarization transfer experiments such as DEPT are used to indicate carbon multiplicity, and the use of two-dimensional techniques to facilitate signal assignment is seen more-or-less routinely, particularly in the case of more complex structures. These include ^{1}H–^{1}H correlation (COSY and related phase sensitive experiments), ^{1}H–^{13}C proton–carbon heteronuclear correlation, including inverse spectroscopy, to identify short and long-range couplings, and Nuclear Overhauser Enhancements experiments (NOESY and its rotating frame equivalent, ROESY) to identify intramolecular interactions.

NMR spectroscopy can be of value in establishing the stereochemistry of the molecule, examples being the use of difference NOE spectroscopy to identify cis–trans isomerism or to distinguish between endo and exo isomers of bicyclic molecules. Single-crystal X-ray diffraction is the best method of establishing the absolute configuration of molecules containing chiral centers but NMR methods such as those employing chiral shift reagents may also be of value in establishing the presence or absence of optical isomers.

NMR has been of substantial value in establishing the identity and conformation of macromolecules and this type of data is now being seen regularly in regulatory dossiers. Conformational data obtained through the analysis of coupling constants, use of NOE and relaxation measurements, among other methods, is of particular interest where the correct conformation is essential for activity.

The bioavailability of a drug substance (its absorption and uptake by the body from the formulated drug product) is dependent on its physicochemical parameters. Physicochemical data will be needed (even if the drug substance is listed in a pharmacopoeia) if they are critical for bioavailability. The information provided may include data on the crystalline form and solubility of the drug substance, its particle size (after pulverization if necessary), state of solvation, partition coefficient, pH, and pK_a, even if these tests are not included in the final specification. Spectroscopic and thermal characteristics of material recrystallized in a variety of solvents and conditions may indicate the existence of different polymorphic forms of a drug substance that could have different solubility characteristics. Control of polymorphs is particularly important for drug substances of low solubility where dissolution of the drug may have a significant effect on its bioavailability. Solid-state NMR, particularly the ^{13}C-MAS technique, has been used to characterize the polymorphic forms of drug substances and, in some cases, has been the routine test method chosen for the specification where its specificity and sensitivity have proved superior to alternative methods such as IR spectroscopy or X-ray powder diffraction (see Chapter 5).

29.4.1.3 Specifications

Specifications are needed to control drug substances and drug products once they are being manufactured for release onto the market. Two ICH guidelines comprehensively cover the test procedures and acceptance criteria necessary in specifications for new chemical and biological substances and their associated drug products.[8,9] NMR spectroscopy is most frequently used as an identity test in drug substance specifications.

A discussion of potential impurities should be provided in the regulatory dossier and details provided for any that have been synthesized or detected. Impurities may arise from a number of sources: raw materials, solvents, and reagents used in the manufacture of the drug substance; intermediates or by-products of the synthesis; and from degradation of the substance (see the following section). For chemical substances, structural characterization is required for organic impurities at identification thresholds set out in the ICH guidelines on impurities in new drug substances.[12] For example, the threshold for a drug administered at a maximum daily dose of up to 2 g is 0.10% or 1.0 mg per day intake of the impurity, whichever is lower. However, lower limits may be applicable if the impurities are particularly toxic. Identification of organic impurities normally includes NMR methods in conjunction with other spectroscopic techniques and confirmation by independent synthesis.

The impurity and related substance limits proposed in a drug substance specification should be *qualified*, that is they should consider not only the *quality* of the drug substance, i.e., the actual levels of impurities and related substances found in batch analysis but also their *safety*. Their safety is established through the testing of preclinical or clinical batches containing the specified levels of impurities or related substances if they

are above the dose-dependent qualification thresholds prescribed in the ICH guidelines, e.g., 0.15% or 1.0 mg per day, whichever is lower, for drugs with a maximum daily dose of up to 2 g.[11] Below these thresholds, qualification is not required.

Similar principles apply to the qualification of impurities in biological and biotechnological products except that quantitative thresholds are not given in ICH guidance because of the more complex nature of these substances and their manufacture.

Batch analysis data is required to show the results that have been obtained from routine quality control of the drug substance, including data from any NMR tests, and to demonstrate compliance with the proposed specification. Earlier batches may have been tested against slightly different specifications, e.g., with wider assay or higher impurity limits, and an explanation should be given in these circumstances. Data for batches used in toxicity studies and clinical trials should be reported, including the test results for impurities and related substances to facilitate assessment of the qualification process.

29.4.1.4 Stability

Data must be submitted on the stability of the drug substance and fall into two categories. The first is 'stress' studies in which the drug substance is subjected to extreme heat, light, acidic, basic, and oxidative conditions with the intention of forcing degradation. These data are combined with the isolation and identification of degradation products to elucidate degradation pathways in the drug substance. NMR spectroscopy plays an important role in the identification of degradation products. Degraded samples may also be used to test the specificity of the analytical methods applied to the drug substance.

Secondly, stability studies are needed to establish the retest period for the drug substance. The ICH guideline on stability testing for new drug substances and products stipulates standard conditions for storage applicable to the three sponsoring regions (EU, US, and Japan): 25 °C/60%RH or 30 °C/65%RH for long-term studies (at least 12 months) and 40 °C/75%RH for accelerated stability studies (at least 6 months).[13] The guideline give recommendations on the number of batches, frequency of testing, and evaluation of results. Guidance on stability testing of biotechnological/biological products is given separately.[14] The tests used should be the same as those in the drug substance specification, including any NMR spectroscopy methods. However, additional stability-indicating methods may be included in the protocol. Solid-state NMR might be an example of the latter for identifying the occurrence of polymorphs in drug substance, even though such a test might not be necessary in the final drug substance specification. The dossier submitted with a marketing authorization application must include full information on the batches tested, their packaging, the test methods (both description and validation), and results, together with proposals for the retest period and details of any on-going studies.

29.4.2 Excipients

Some of the information outlined earlier for new drug substances may also be needed to characterize novel excipients, i.e., those which have never been included in a drug product before. Often, such excipients have been used previously in the cosmetic or food industry and therefore will already have substantial data packages or established safety data that can be used to support an application for a marketing authorization for a drug product. NMR spectroscopy has been used particularly to control the composition of polymeric excipients used in the dosage form, the proportion of monomeric components, the proportion of functional groups, or their positional distribution. Pharmacopoeial monographs provide useful examples in this respect (see Chapter 30).

29.4.3 Drug Product

Batches of drug product intended for release to the market must comply with an approved drug product specification or relevant pharmacopoeial monographs. The ICH guidelines on specifications for new drug substances and new drug products for either chemical substances or biotechnological/biological products set out the typical requirements.[8,9] They include details of the tests that might be expected for the control of different types of dosage forms. In some cases, by agreement with the competent authority, certain tests do not need to be carried out routinely; however, the frequency of such skip or periodic testing must be stated in the specification.

NMR spectroscopy, including multinuclear, has mainly been used as an identity test in the specifications for authorized products to ensure the presence of the drug substance. However, where there is a pharmacopoeial monograph for both the drug substance and the drug product, it would not normally be necessary to use an NMR test in the identification of the latter if another simpler test is also included. This is due to the more complex sample preparation required for dosage forms given their multicomponent nature and difficulties associated with masking of signals. If the drug substance contains a paramagnetic agent, NMR relaxation measurements may be used in the control of the quality of finished products (see example in Chapter 30).

As for drug substances, stressed stability studies are used to establish likely degradation products in the drug product, and NMR spectroscopy is used in the characterization of any compounds that can be isolated. The causes of degradation may be either the usual chemical breakdown mechanisms or specific interactions with excipients. These may occur during manufacture of the drug product, e.g., owing to heating, in which case a limit may be required in the specification for release, or on storage of the product, thus needing control in the shelf-life specification. Those impurities that are included in the drug substance specification and only occur in the synthetic process do not need further control in the drug product specification. For new drug products, degradation products should be identified and qualified as indicated above for drug substances, using the thresholds described in the ICH guidelines on impurities in new drug products.[15] The reporting, identification, and qualification thresholds appropriate for different maximum daily doses of drug substances are presented in this guidance.

ICH provides guidance on stability testing protocols for drug products following similar principles to the drug substance.[12,13] Allowance is made for reduced testing of new dosage forms containing an existing drug substance already present in other authorized drug products.[16]

29.4.4 Analytical Validation

The ICH guideline on analytical validation should be applied to any NMR method used in the control of drug substances or drug products.[17] It deals first with definitions and terminology and then with validation methodology (originally two separate guidelines). It should be borne in mind that revalidation of methods may be necessary if changes are made to the synthesis of the drug substance, manufacture of the drug product, or to the analytical method itself, e.g., to reassess specificity.

Identity tests need to be specific for the substance being tested, and the power of NMR spectroscopy is well established in this regard. Validation of identification methods must ensure lack of interference from related substances or other impurities. The specialized equipment needed for NMR tests means that it is not more frequently used as an identity test in drug substance specifications but its specificity is of particular value for complex molecules not amenable to more routine methods of identification. NMR spectroscopy, including hyphenated methods, has been used to advantage in the validation of test methods for other techniques, for example, to ensure peak purity in chromatography.

The use of NMR spectroscopy for quantitative determinations is addressed in Chapter 5. It is particularly important to ensure reproducibility of the magnetic field or compensation by the use of appropriate standards. Such standards must be fully characterized both qualitatively and, where necessary, quantitatively. Although traditionally limited by relative lack of sensitivity, current NMR technology has allowed NMR methods to become feasible options for quantitative methods, particularly where difficulties are encountered with alternative techniques. In certain cases, NMR spectroscopy may offer an advantage in simultaneous measurement and identification of compounds, such as related substances, present in the sample.

29.5 PRESENTATION OF NMR DATA

Figures 29.1 and 29.2 present illustrative examples of partial ^{1}H and ^{13}C NMR spectra provided in regulatory dossiers using an anonymized drug substance,

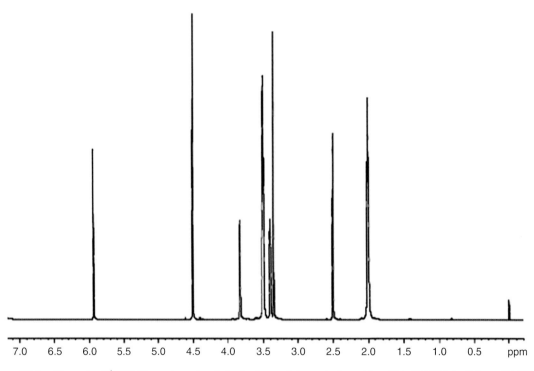

Figure 29.1. Example of ^{1}H NMR spectrum (partial) of a drug substance (as bromide salt) in DMSO-d_6. Solvent DMSO-d_6 and water resonances appear at 2.50 and 3.35 ppm, respectively. Propan-1-ol impurity resonances observed at 0.83, 1.41, and 4.37 ppm

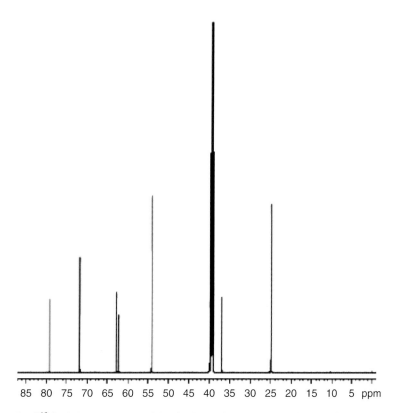

Figure 29.2. Example of ^{13}C NMR spectrum (partial) of a drug substance (as bromide salt) in DMSO-d$_6$. Solvent DMSO-d$_6$ resonance appears at 39.4 ppm

Table 29.3. Interpretation of example ^{1}H NMR spectrum

Chemical shift $(\delta)^a$	Multiplicity	No. of H atoms	Assignment
2.01	t	6	1, 2, 3
3.40	m	2	4
3.50	t	6	5, 6, 7
3.83	m	2	8, 9
4.51	s	2	10, 11
5.94	s	1	12

^{a} ppm referenced to tetramethylsilane at 0 ppm.
s = singlet, d = doublet, t = triplet, and m = multiplet.

Table 29.4. Interpretation of example ^{13}C NMR spectrum

Chemical shift $(\delta)^a$	Assignment
24.9	1, 2, 3
37.0	4
54.2	5, 6, 7
62.4	8
62.8	9
71.9	10
79.3	11

^{a} ppm referenced to tetramethylsilane at 0 ppm.

which is a bromide salt. The corresponding interpretations as usually presented (chemical shifts, multiplicities, and assignments) are provided in Tables 29.3 and 29.4. A polar, aprotic solvent, dimethylsulfoxide-d$_6$ (DMSO-d$_6$), has been used in this case to achieve full dissolution of the bromide salt in the NMR tube and

ultimately to provide sound characterization data to the competent authorities.

29.6 COMMON PITFALLS

Some frequent NMR deficiencies encountered by competent authorities during the assessment of marketing authorization applications are provided below.

- *Illegible and unassigned spectra*: Care should be taken to ensure that reproductions of spectra are completely legible (a common problem in dossiers), and full assignments should be given where possible.
- *Vicinal coupling constants*: Incorrect assignment of *J* values for vicinal protons present on cis/trans geometrical isomers (i.e., the highest value of the *J* assigned to the cis isomer instead of the trans isomer).
- *Deuterated solvents not corresponding*: The deuterated solvent used to run the proton and carbon NMR is CDCl$_3$. However, the table of chemical shifts' values state d$_6$-DMSO.
- *Proton integrations*: Inaccurate description of the overall number of hydrogens present in the drug substance. Mismatch between spectra and table with protons' integration.
- *Tautomers*: Drug substances containing primary amides functionalities may give rise to tautomers (i.e., keto or enol forms). In this scenario, the applicant is asked to run dynamic NMR experiments (proton and carbon) with a view to confirming the exact structure of the drug substance.
- *Carbon multiplicity*: A DEPT-135 spectrum is useful to complement a standard ^{13}C NMR spectrum as it allows distinction between CH, CH$_3$ (positive signals), and CH$_2$ carbons (negative signals). However, this information is often missing in the initially submitted dossier and is therefore sometimes required from applicants.
- *Ipso carbons on aromatic rings*: ^{13}C NMR spectra of drug substances containing aromatic rings do not clearly show peaks belonging to Ipso carbons. The applicant is requested to run the NMR spectrum either with a more concentrated sample of drug substance or to run the NMR for a longer period (e.g., overnight).
- *Lack of multinuclear NMR*: For drug substances containing fluorine, phosphorous, carbon, and hydrogen atoms, only NMR spectra for the latter two are provided. In this context, the applicant is asked to provide ^{19}F and ^{31}P NMR spectra also.
- *Lack of 2D NMR spectra*: Spectra not provided for more complex drug substances (e.g., containing bicyclic structure with endo end exo protons). Only one-dimensional ^{1}H and ^{13}C carbon NMR spectra are provided, which are not considered sufficient to characterize the molecule satisfactorily.

29.7 CONCLUDING REMARKS

This chapter has reviewed the requirements for the registration of pharmaceuticals and the role of NMR spectroscopy in the regulatory dossier supporting applications to market medicines. The power of NMR spectroscopy in structure elucidation is well recognized but the specialized equipment required has prohibited its widespread use in routine drug substance or drug products specifications. However, advances in NMR technology and techniques mean that NMR spectroscopy has become more accessible and the method of choice in certain circumstances, particularly for the control of more complex molecules, including natural products and biologicals.

More sophisticated NMR techniques making optimal use of latest advances are more likely to be used during the drug discovery and development process, without necessarily finding their way into the final regulatory dossier that accompanies an application to market a medicine. The full range of structure elucidation tools are applied to candidate drugs and their related substances especially as the number of biological drug substances under development increases, e.g., for sequencing and conformational analysis of peptides and proteins. Solid-state NMR is used for the detection of polymorphic forms essential to understanding dissolution behavior and bioavailability of drug substances when administered. NMR imaging techniques in the clinical setting have become part of the armory of investigations to understand disease progression and the physiological or pharmacological influence of therapeutic agents. Imaging methods have also been used to investigate the physical properties of dosage forms to facilitate development of manufacturing processes or understand their behavior on administration. Regulatory dossiers will increasingly include examples of drug substances and drug products that have relied on NMR techniques for part of their research and development or for their control during routine manufacture for the marketplace.

REFERENCES

1. International Conference on Harmonization of Technical Requirements for Registration of Pharmaceuticals for Human Use (ICH) http://www.ich.org/.

2. ICH M4(R3): The Common Technical Document, September 2002 http://www.ich.org/products/ctd.html.

3. ICH Quality Guidelines http://www.ich.org/products/guidelines/quality/article/quality-guidelines.html.

4. ICH Q3C(R5) Impurities: Guideline for Residual Solvents, February 2011.

5. ICH Q3D Draft Consensus Guidelines for Elemental Impurities, October 2013.

6. Certification of suitability to the monographs of the European Pharmacopoeia (revised version), Council of Europe Public Health Committee Resolution AP-CSP (07) 1, February 2007.

7. Council of Europe, European Directorate for the Quality of Medicines and Healthcare, Certification of Suitability, http://www.edqm.eu/en/certification-new-applications-29.html.

8. Guideline on Active Substance Master File Procedure, CHMP/QWP/227/02 Rev 3/Corr*, May 2013 http://www.ema.europa.eu/docs/en_GB/document_library/Scientific_guideline/2012/07/WC500129994.pdf.

9. ICH Q6A Specifications: Test Procedures and Acceptance Criteria for New Drug Substances and New Drug Products: Chemical Substances, October 1999.

10. ICH Q6B Specifications: Test Procedures and Acceptance Criteria for Biotechnological/Biological Products, March 1999.

11. Guideline on Real Time Release Testing (formerly Guideline on Parametric Release), EMA/CHMP/QWP/811210/2009-Rev1, March 2012 http://www.ema.europa.eu/docs/en_GB/document_library/Scientific_guideline/2012/04/WC500125401.pdf.

12. ICH Q3A Impurities in New Drug Substances, October 2006.

13. ICH Q1A(R2) Stability Testing of New Drug Substances and Products, February 2003.

14. ICH Q5C Stability Testing of Biotechnological/Biological Products, November 1995.

15. ICH Q3B Impurities in New Drug Products, June 2006.

16. ICH Q1C Stability Testing of New Dosage Forms, November 1996.

17. ICH Q2(R1) Validation of Analytical Procedures: Text and Methodology, November 2005.

Chapter 30

NMR Spectroscopy in the European and US Pharmacopeias

Helen Corns and Sarah K. Branch

Medicines and Healthcare Products Regulatory Agency (MHRA), 151 Buckingham Palace Road, Victoria, London SW1W 9SZ, UK

30.1 INTRODUCTION

The remit of a pharmacopoeia is to provide a publicly available set of quality standards for medicinal

The views expressed in this article are those of the authors and do not necessarily represent the views or the opinions of the Medicines and Healthcare products Regulatory Agency, other regulatory authorities, or any of their advisory committees.

NMR in Pharmaceutical Sciences. Edited by Jeremy R. Everett, John C. Lindon, Ian D. Wilson, and Robin K. Harris
© 2015 John Wiley & Sons, Ltd. ISBN: 978-1-118-66025-6
Also published in eMagRes (online edition)
DOI: 10.1002/9780470034590.emrstm1403

products. These quality standards take the form of monographs for individual active substances or for their formulated preparations. Compliance is mandatory where a pharmacopoeia is referred to in drug regulations and a product or article must comply (if tested) throughout its period of use or lifetime. The active substances, excipients, pharmaceutical preparations, and other articles described in the monographs are intended for human and veterinary use (unless explicitly restricted to one of these uses).

The Council of Europe's Convention on the Elaboration of a European Pharmacopoeia (1964) and the amendment protocol of 1994 together with the European Union Directives on human and veterinary medicine (2001/82/EC, 2001/83/EC and 2003/63/EC, as amended) give the legal basis under which European Pharmacopoeia (Ph. Eur.) monographs are brought into effect. The net result of this means that the Ph. Eur. houses legally enforceable quality control standards for pharmaceutical articles in its 37 member states (including all those in the European Union), and all manufacturers of substances for pharmaceutical use must comply with these standards if they wish to market their products in those states that are signatory to the Convention.[1]

The United States Pharmacopeia Convention is responsible for publishing USP–NF, which is a combination of two compendia, the United States Pharmacopeia (USP) and the National Formulary (NF). The official standards in the USP-NF are recognized within US legislation under the Federal

Food, Drug, and Cosmetic Act.[2] The USP contains monographs for drug substances, dosage forms, and compounded preparations, while excipient monographs are in the NF.

Outside the legal framework under which they were developed, both the Ph. Eur. and USP are widely recognized internationally and are adopted in the regulatory framework and legislation of a number of countries worldwide.

The Ph. Eur. and USP contain a number of general texts that support the specific monographs, as the latter do not stand alone. The general notices, methods, and monographs provide the framework of pharmacopoeias. It is worth noting that even if there is not a specific monograph for a particular product, there will still be mandatory general tests with which an article must comply. Thus, for example, an oral suspension will need to comply with the general method for Liquid Preparations for Oral Use Ph. Eur., if marketed in the EU.

30.2   PHARMACOPOEIAL MONOGRAPHS

The monographs for pharmaceutical preparations (drug product) monographs in the pharmacopoeias will feature a number of specific tests that are in addition to those given in general monographs for certain dosage forms or for individual substances for pharmaceutical use. The tests in individual monographs cover aspects that are more specific to that product, for example, identification or related substances.

Table 30.1 lists the typical contents of a monograph for a drug substance or drug product. The list is not exhaustive, nor will each test apply to every monograph.

The tests highlighted in bold in Table 30.1 are those in which the use of NMR has been directed in either the Ph. Eur. or the USP or both. It is perhaps worth touching on the overall role of NMR spectroscopy in pharmacopoeial monographs before looking at some specific examples of these tests.

The tests given in pharmacopoeias are designed for use by a wide range of stakeholders, from pharmaceutical manufacturers to independent analysts assessing whether a product is of the appropriate quality. There is an acceptance that not everyone using the tests within a monograph has access to complex equipment and so, where possible, pharmacopoeias tend to include methods using readily available technologies. At the same time, pharmacopoeias do not want to constrict the use of emerging or more complex techniques and most of

Table 30.1.   Content of a typical monograph

Title

Definition – for example, a systematic chemical name or a pharmacopoeial quality. The name of the drug substance, meeting pharmacopoeial requirements, that a formulation should contain

Content – assay limits

Characters – which could include appearance, solubility, stability factors, hygroscopicity, solid-state properties, behavior in solution, or half-life for a radiopharmaceutical

Identification

Tests – the following are often listed under 'Tests' and are predominantly aimed at identifying the presence of impurities

    Appearance of solution

    pH

    Optical rotation

    Related substances

    Foreign anions/cations

    Heavy metals

    Loss on drying

    Sulfated ash

    Residue on evaporation

    Residual solvents

Assay

Labeling – for example, content

Impurities – with structures given in drug substance monographs

Functionality-related characteristics

them allow the use of alternative methods, provided that they produce a comparable result and are fully validated. However, in the case of dispute, the pharmacopoeial method will alone be authoritative. This is of particular relevance to NMR spectroscopy, which requires sophisticated instrumentation not necessarily available in all quality control laboratories. For this reason, NMR spectroscopy is relatively infrequently specified in tests in the pharmacopoeias. However, in support of those occurrences and the option of using alternative analytical techniques to those specified in a pharmacopoeia, both the Ph. Eur. and USP contain general chapters and supporting documentation detailing the use of NMR spectroscopy.[3,4]

30.3   NMR SPECTROSCOPY IN THE EUROPEAN PHARMACOPOEIA

The relevant chapter of the Ph. Eur. is method 2.2.33 – Nuclear Magnetic Resonance Spectrometry.[5]

This method sets out general requirements for applying NMR spectroscopy to tests used for pharmacopoeial purposes. Through the introduction and general principles sections, the chapter introduces the various scenarios under which NMR spectroscopy may be employed by including the various nuclei that can produce an NMR spectrum and the use of the technique for both qualitative and quantitative analyses. The chapter continues with the expectations of the apparatus, i.e., features that the equipment should include, like a magnet to deliver a constant magnetic field, or may include, such as a system for pulsed field gradient NMR, which sets a baseline capability for the instrumentation used. This is followed by a section on Fourier transform NMR spectroscopy, as most modern NMR instruments operate in this way, which covers the parameters that should be controlled and that can be adjusted, as well as the optimization of parameters (both in acquisition and processing) when carrying out quantitative experiments. Some detail is then given on the NMR of samples in solution, which includes choice of solvents, referencing, and lock. After setting out how experimental features should be considered and controlled, the chapter then centers on the practicalities of a qualitative and a quantitative test. In qualitative tests, the sample should ideally be compared to a reference material. The quantitative test section details two different procedures for quantitation – the use of an internal standard and normalization. There follows a discussion of sample handling and the measurement procedure. The chapter finishes with a short section on solid-state NMR, which offers the potential for inclusion of this technique in pharmacopoeial methods in the future (see Chapter 5 for further discussion of quantitative NMR methods). The pharmacopoeial chapter sets out a framework for a high-quality NMR spectroscopy method to be developed and validated for a test to control an aspect of a drug substance or drug product without being overly directive, thus allowing a freedom for the technique to evolve.

In addition to the general chapter, the Ph. Eur. also contains a more specific chapter on Peptide Identification by Nuclear Magnetic Resonance Spectrometry (Ph. Eur. method 2.2.64). The scope of this general method is limited to the use of one-dimensional experiments for the qualitative identification of small peptide products up to approximately 15 amino acids using ^{1}H NMR spectroscopy. It can be applied to ^{13}C experiments with appropriate modification. The Ph.

Eur. states that this procedure should be used in conjunction with general method Nuclear Magnetic Resonance Spectrometry 2.2.23 and sets out more specific instructions around certain features of the experiment than are found in the general method. Particular attention is given to acquisition conditions, chemical shift referencing, sample preparation, and verification of the identification.

NMR spectroscopy is further mentioned in the general method for Crystallinity (Ph. Eur. method 5.16), specifically under methods for monitoring and determining crystallinity, where it is listed as one of the techniques that can be used (in combination with other techniques also listed) to elucidate polymorphism and related relative molecular conformations. It does, however, include a note of caution that it may not be clear from the results whether the sample is a mixture of crystals in different forms or whether the sample crystals are exhibiting disorder.

Further information relating to the use of NMR spectroscopy in Ph. Eur. tests is available in the technical guides, which are freely available on the European Directorate for Quality of Medicines and HealthCare (EDQM) website.[6] The EDQM laboratory supports the development of Ph. Eur. reference substances. The Ph. Eur. Technical guide for the elaboration of monographs gives information to stakeholders on the foundation of a monograph. The general information for identification tests sets out that a pharmacopoeial identification test is a means of confirming identity rather than determining an unknown, and as such weight is given to the simplicity and speed with which such tests can be carried out. With that in mind, it is perhaps apparent why NMR is not specified in a larger number of monographs.

Information specific to Ph. Eur. method Peptide Identification by Nuclear Magnetic Resonance Spectrometry 2.2.64 can be found in both the Ph. Eur. Technical guide for the elaboration of monographs and the Ph. Eur. Guide for the elaboration of monographs on synthetic peptides and recombinant DNA proteins. In the former, there is guidance on method validation covering spectral consistency, specificity, and other variability that should be addressed. In the latter, under the identification guidance for synthetic peptides, it is noted that synthetic peptides are more amenable for NMR-based identification tests as they tend to be smaller than recombinant DNA products and can have structures that do not naturally occur in proteins or peptides. An example is described in Section 30.6.7. Particularly for the second point, the guide suggests that NMR spectroscopy may be appropriate as an additional means of identification.

It has been recognized through the publication of the general chapter on the use of NMR for peptide identification that NMR spectroscopy is a preferred technique for this purpose. There is another area where NMR proffers an advantage over other analytical techniques, which is in the monographs for fatty oils and derivatives. NMR is given specific mention in the Ph. Eur. technical guide for the elaboration of monographs on fatty oils and derivatives under tests for the determination of a propylene oxide:ethylene oxide ratio (see the example described in Section 30.6.5), should the inclusion of that test be meaningful for a polymer monograph. It is also given as an example assay technique as used under Farmed Salmon Oil (Ph. Eur. monograph 1910) for monographs on animal oils, fats, and waxes.

30.4 NMR SPECTROSCOPY IN THE UNITED STATES PHARMACOPEIA

The USP takes a slightly different approach with its two general NMR texts; General Chapter <761> Nuclear Magnetic Resonance Spectroscopy[7] and general information chapter <1761> Applications of Nuclear Magnetic Resonance Spectroscopy.[8] It splits out the process of qualification and validation or verification from NMR applications. General chapter <761> uses examples focused on ^{1}H NMR and to a lesser extent ^{13}C NMR spectroscopy, which is to be expected given the relative prevalence of these two techniques. The text covers requirements for the qualification of the instrument, covering installation, operational, and performance qualification (IQ, OQ, and PQ), including sample test conditions for common PQ tests such as resolution and lineshape, and signal-to-noise measurements. The chapter then details qualitative and quantitative applications followed by validation and verification procedures. General information chapter <1761>, however, encompasses a much broader range of NMR techniques. Starting off with the theory and principles of NMR spectroscopy, the chapter progresses to information on spectrometers and features of experiments, such as relaxation and resolution, before discussing structural elucidation, quantitative applications, and both solid-state and low-field NMR spectroscopy. Under

each of these, detailed information is given on the types of NMR techniques (such as COSY and total correlation spectroscopy (TOCSY) experiments) that can be used in these applications and why they might be applied.

Similarly to the Ph. Eur., NMR spectroscopy is predominantly specified for the control of biological products in the USP. A further general information chapter <1238> 'Vaccines for Human Use – Bacterial Vaccines', under the protein purification section, includes NMR spectroscopy as a means of determining depolymerized polysaccharides using certain functional groups, for example, *O*-acetyl groups. It is also given as an example technique for identifying and determining the *O*-acetylation level in carbohydrate-based intermediates and drug substances, along with a mention that for certain vaccines, particularly those using purified polysaccharide immunogens, identity, and immunogen quality could potentially be demonstrated by high-field NMR spectroscopy.

One routine use of NMR spectroscopy, which is perhaps not so well-known to users of the pharmacopoeias, is in the establishment of the chemical reference substances (CRS) that support the monographs in the pharmacopoeias. Unlike the tests in the pharmacopoeias, NMR techniques, usually ^{1}H, sometimes ^{13}C and on occasion 2D experiments, are carried out as part of the characterization of the reference materials used as primary standards for pharmacopoeial methods.

30.5 HISTORICAL PERSPECTIVE

NMR spectroscopy first appeared in specific monographs in the mid 1970s when the British Pharmacopoeia (BP) included an NMR procedure for the identification and quantification of the components of gentamicin in the monograph for Gentamicin Sulfate. At the same time, it published an appendix on NMR Spectroscopy as a general method. This was followed by the use of NMR to quantitatively determine moisture content in monographs for Cloprostenol and Fluprostenol in the BP (Vet) 1977 and by NMR identification tests in BP monographs for some corticosteroid sodium phosphates and aminoglycosidic antibiotics in the BP 1980. These monographs have since been superseded by Ph. Eur. monographs. The Ph.

Eur. Appendix on NMR Spectroscopy was first published in the second edition in 1980. It was included in specific Ph. Eur. monographs in the early 1980s, including the monograph for Gentamicin Sulfate, although the NMR procedure was later replaced by a liquid chromatography (LC) test, which quantitatively determined the relative proportions of the four main components.

30.6 EXAMPLES OF NMR SPECTROSCOPY IN MONOGRAPHS FOR DRUG SUBSTANCES AND DRUG PRODUCTS

NMR spectroscopic methods are found in the following tests in pharmacopoeial monographs:

- Identification using ^{1}H and ^{13}C NMR spectroscopy
- Assay for content or composition, frequently for oligo- or polymeric materials, using ^{1}H and ^{13}C NMR spectroscopy
- Related substances, including isomers, and other impurities using ^{1}H NMR spectroscopy
- Specific tests, usually to determine composition of oligo- or polymeric substances
- Relaxivity for paramagnetic lanthanide diagnostic agents

Some individual examples of these tests are described in the following section. They have been selected to illustrate the use of both ^{1}H and ^{13}C NMR techniques over a range of tests and the different methods employed for qualitative and quantitative analysis, e.g., comparison to reference spectra or use of internal standards, absolute, or relative methods of quantitation.

30.6.1 Identification and Assay by ^{13}C NMR Spectroscopy: Heparin Sodium (USP and Ph. Eur.)

Heparin sodium is the sodium salt of a sulfated glycosaminoglycan that is present in mammalian tissue. It is an anticoagulant and enhances the action of antithrombin III. The USP and Ph. Eur. both employ ^{1}H NMR for the identification of heparin sodium.

The substance to be tested is dissolved in a solution of 20 µg ml^{-1} of deuterated trimethylsilylpropionate (TMSP) in deuterium oxide. In the event that the signal at 5.21 ppm is smaller than 80% of the signal at 5.42 ppm, 12 µg ml^{-1} of sodium edetate can be included in the solution that should remove signal broadening effects due to paramagnetic relaxation. Such relaxation can be seen in some samples and is caused by small amounts of manganese metal ions from a bleaching procedure in the heparin manufacturing process. The EDTA chelates the manganese ions, which prevents the occurrence of these paramagnetic effects. The identification of the substance is confirmed by means of a qualitative comparison to a reference substance, prepared to an equivalent concentration in the same solvent (Figure 30.1). The USP method specifies that the following signals should be present (Table 30.2).

The NMR methods are also capable of determining the presence of oversulfated chondroitin sulfate (OSCS, (**1**)). This contaminant was thought to be the cause of severe allergic reactions, in some cases fatal, in a crisis that emerged in 2008. The adverse reactions were associated with drug products manufactured with drug substance sourced from China where it was extracted from pig intestines. An NMR method of determining OSCS and another contaminant, dermatan sulfate, was developed in response to the request for a pharmacopoeial method.[9]

There is a difference in NMR operating frequency in the two pharmacopoeial methods: not less than 300 MHz in the Ph. Eur. and not less than 500 MHz in the USP. This difference arises because the USP test is used to demonstrate absence of OSCS in Heparin Sodium, in conjunction with an HPLC (high-performance liquid chromatography) procedure. Under 'Absence of oversulfated chondroitin sulfate' in

Table 30.2. ^{1}H NMR chemical shifts of heparin sodium used for identity confirmation

	Resonances (ppm)
Signal 1: H1 of *N*-acetylated glucosamine/*N*-sulfated glucosamine 6S	5.42 ± 0.03
Signal 2: H1 of iduronic acid 2S	5.21 ± 0.03
Signal 3: H2 of glucosamine sulfate	3.28 (doublet) ± 0.03
Signal 4: Methyl of *N*-acetylated glucosamine	2.05 ± 0.03

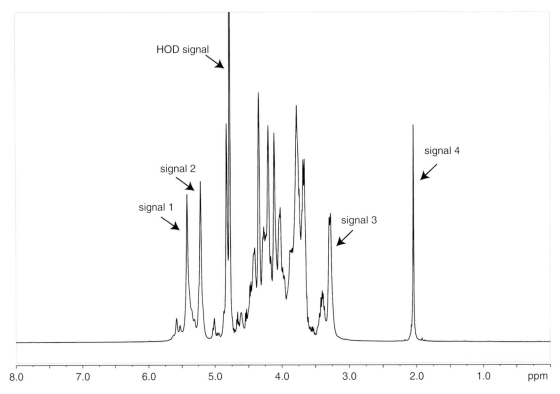

Figure 30.1.   ¹H NMR spectrum of heparin sodium. (Reproduced with permission. © The United States Pharmacopeial Convention, 2014)

the Impurities section of the USP monograph, there is a cross-reference to the NMR identification result and a requirement that no signals associated with OSCS are detected between 2.12 and 3.00 ppm, confirmed by comparison to an OSCS reference standard. It may not be possible to detect the lower concentration of OSCS in the system suitability solution used in the USP NMR method (0.3% w/w compared to Ph. Eur. 1.0% w/w) with a 300 MHz NMR spectrometer. The Ph. Eur. does not explicitly state that the NMR test controls OSCS, but it does include a requirement that 'no unidentified signals larger than 4% compared to the height of the heparin signal at 5.42 ppm are present in the ranges 0.10–2.00 ppm, 2.10–3.10 ppm, and 5.70–8.00 ppm' in the NMR identification result. The HPLC procedure under Related Substances in the Ph. Eur. monograph states that the presence of OSCS should not be detected in Heparin Sodium.

(1)

30.6.2   Identification and Assay by ¹³C NMR Spectroscopy: Farmed Salmon Oil and Farmed Cod Liver Oil (Ph. Eur.)

Fish liver oil is a source of triglycerides of omega-3 fatty acids. Omega-3 fatty acids need to be obtained from the diet because they cannot be synthesized by the body; they are contributors to many metabolic processes and are used medicinally as a lipid-regulating drug. Omega-3 fatty acids are long-chain polyunsaturated fatty acids containing 18–22 carbon atoms where a double bond is located at the third C–C position from the terminal methyl group.

Cervonic acid

Timnodonic acid

2

(3)

Moroctic acid

(4)

Both the fish species itself and whether it is from farmed or wild stock will affect the $\beta(2)$-acyl positional distribution in cervonic acid (C22:6 n-3; docosahexaenoic acid/DHA, (**2**)), timnodonic acid (C20:5 n-3; eicosapentaenoic acid/EPA, (**3**)), and moroctic acid (C18:4 n-3, (**4**)). ^{13}C NMR is a useful tool for the assay of positional distribution ($\beta(2)$-acyl) groups in fatty acids and therefore suitable for differentiating between fish oils. The Ph. Eur. contains monographs for both Farmed Salmon Oil and Farmed Cod Liver Oil, which use a similar ^{13}C NMR procedure for the determination of positional distribution of ($\beta(2)$-acyl) of fatty acids and identification of the drug substance. In the Ph Eur. monograph for Farmed Salmon Oil, the positional distribution test appears under Assay but in the Farmed Cod Liver Oil monograph it is listed as a specific test.

The fish liver oil sample is dissolved in $CDCl_3$ to give a solution of approximate concentration $0.4\,g\,l^{-1}$. The test is carried out in triplicate. A spectrometer operating at a minimum of 300 MHz is used to obtain a spectrum from -5 to 195 ppm using the parameters given in the monograph. The $CDCl_3$ signal is used for shift referencing, the central signal of the 1:1:1 triplet being set to 77.16 ppm. A system suitability test is conducted: the signal-to-noise ratio for

Table 30.3. ($\beta(2)$-acyl) positional distribution in farmed salmon oil and farmed cod liver oil

	($\beta(2)$-acyl) positional distribution (%)	
	Farmed Salmon Oil	Farmed Cod Liver Oil
Cervonic acid	60–70	71–81
Timnodonic acid	25–35	32–40
Moroctic acid	40–55	28–38

the smallest relevant peak (α C18:4 in the range δ 172.95–172.99 ppm) should be a minimum of 5 and the peak width at half-height for the central $CDCl_3$ signal at 77.16 ppm should be a maximum of 0.02.

The 171.5–173.5 ppm region of the resulting spectrum is compared to a reference spectrum that is given in each monograph (see the example for farmed salmon oil in Figure 30.2). The NMR identification test is in part completed if the sample spectrum contains peaks between 172 and 173 ppm with chemical shifts similar to those in the reference spectrum. The other part of the NMR identification requirement is fulfilled if the sample complies with the test for Positional distribution ($\beta(2)$-acyl) of fatty acids (see below). Table 30.3 gives the limits for Positional distribution ($\beta(2)$-acyl) of fatty acids for Farmed Salmon Liver and Farmed Cod Liver Oils.

The data for the calculation of the positional distribution ($\beta(2)$-acyl) of omega-3 fatty acids in fish lipid triglycerides is also extracted from the carbonyl region (171–174 ppm) of the spectrum. The calculation of ($\beta(2)$-acyl) positional distribution is carried out using equation 30.1:

$$\frac{\beta}{\alpha + \beta} \times 100 \qquad (30.1)$$

where α is the peak area of the corresponding α-carbonyl peak and β is the peak area of the β-carbonyl peak from C22:6 n-3, C20:5 n-3, or C18:4 n-3, respectively, using the tabulated shift ranges for the signals given in Table 30.4. The limits that should be met are those given in Table 30.3.

30.6.3 Assay for Drug Substance Content by ^{1}H NMR Spectroscopy: Amyl Nitrite and Amyl Nitrite Inhalant USP

Amyl nitrite is a mixture of the 2-methylbutyl and 3-methylbutyl esters of nitrous acid. It is used as an inhalant for the relief of pain caused by angina owing

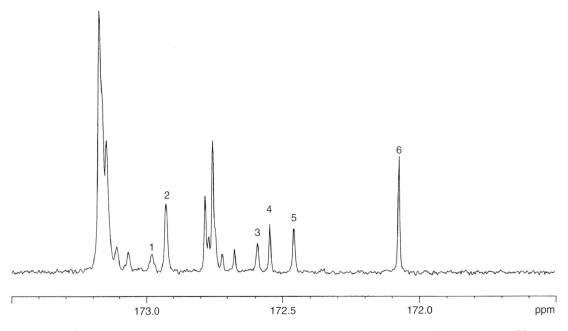

Figure 30.2. ^{13}C NMR spectrum for farmed salmon oil. (Reproduced with permission from Ref. 3. © European Directorate for the Quality of Medicines & HealthCare, Council of Europe, 2014)

Table 30.4. ^{13}C NMR chemical shifts for fish oil omega-3 fatty acids

Signal	Shift range (ppm)
β-Cervonic acid	172.05–172.09
α-Cervonic acid	172.43–172.47
β-Timnodonic acid	172.52–172.56
α-Timnodonic acid	172.90–172.94
β-Moroctic acid	172.56–172.60
α-Moroctic acid	172.95–172.99

to its smooth muscle relaxant properties that dilate cardiac blood vessels thus increasing blood supply to the heart.

The USP monographs for both Amyl Nitrite drug substance and Amyl Nitrite Inhalant contain an Assay that uses ^{1}H NMR spectroscopy to determine the content of drug substance. For the drug substance, 4–5 mEq of benzyl benzoate reference substance, accurately weighed, are dissolved in 2–3 ml of carbon tetrachloride in a semimicro sampling tube, which is then sealed with a sampling valve and septum. The sealed assembly is weighed. About 500 µl of the sample is introduced through the valve and the assembly reweighed. About 500 µl is used to obtain a ^{1}H NMR spectrum without spinning, or with the spinning adjusted so that the spinning side bands of neither the substance nor the internal standard interfere with the regions to be integrated. The absolute method of quantitation is used as described in the general chapter on Nuclear Magnetic Resonance Spectroscopy <761>. The average area of the multiplet with a band centered at about 4.8 ppm (A_U), representing the α-CH$_2$ protons of amyl nitrite, is compared to the average area (A_S) of the internal standard singlet appearing at about 5.3 ppm due to the methylene protons of benzyl benzoate. The content of $C_5H_{11}NO_2$ in the amyl nitrite sample is calculated using 58.57 and 106.12 as the equivalent weights of amyl nitrite (EW_U) and benzyl benzoate (EW_S), respectively. It should be 85.0–103.0%. The procedure for Amyl Nitrite inhalant is similar but there are additional instructions for the preparation of samples from ampoules containing the inhalant. The content of $C_5H_{11}NO_2$ in the Inhalant should be 80.0–105.0%.

The Amyl Nitrite monograph contains an NMR identification test, which is cross-referenced to in the Amyl Nitrite Inhalant monograph. The Identification test states that the NMR spectrum recorded in the

Assay exhibits, among other peaks, a doublet with a band centered at about 1 ppm and a multiplet with a band centered at about 4.8 ppm representing the α-CH_3 and α-CH_2 of amyl nitrite, respectively.

30.6.4 Assay for Control of Functional Group Substitution by ^{1}H NMR Spectroscopy: Hydroxypropyl Starch Ph. Eur. and USP

Hydroxypropyl starch is widely used, along with other starches, as an excipient in drug products particularly as a filler and disintegrant for solid oral dosage forms such as tablets. It is a partially substituted 2-hydroxypropylether of maize, potato, cassava, rice, or pea starch. It may also be partially hydrolyzed using acids or enzymes to produce a 'thinned starch' with reduced viscosity.

Starch is a challenging material to analyze and the Ph. Eur. and USP assays employ similar ^{1}H FT-NMR methods. The Ph. Eur. assay is described in more detail here. A spectrometer operating at a minimum of 300 MHz is specified using parameters set out in the monograph. The procedure uses an internal standard solution comprising 50.0 mg of 3-trimethylsilyl-1-propanesulfonic acid sodium salt reference substance in about 5 g of deuterium oxide (weighed to the nearest 0.1 mg). The mass fraction (W_1) of internal standard is calculated in milligrams per gram. For the test solution, the hydroxypropyl starch sample is first washed by dispersing 20 g in 200.0 ml carbon dioxide-free water, agitating, then filtering and repeating this operation twice. The sample must then be dried under specified conditions. The moisture content ($W\%$) of this sample is determined separately on a 5 g portion. 12.0 mg of the dried sample is weighed (m) and mixed with 0.1 ml of deuterium chloride solution (7.6%) and 0.75 ml of deuterium oxide in a 5 mm NMR tube, which is then placed in a boiling water bath until a clear solution is obtained. The tube is cooled, dried, and weighed, and then reweighed after adding 0.05 ml of the internal standard solution to obtain the mass of standard solution added (m_1 g) and percentage content of internal standard (P).

The peak area (A_2) of the doublet arising from the methyl groups in the hydroxypropyl function at 1.2 ppm is compared to the area (A_1) of the methyl group singlet of the internal standard at 0 ppm without

^{13}C satellites. The content of hydroxypropyl groups as a percentage of the dried substance is calculated using equation 30.2:

$$\frac{3A_2}{A_1} \times \frac{P}{100} \times \frac{W_1 \times m_1}{218} \times 59 \times \frac{100}{m} \times \frac{100}{100 - W} \tag{30.2}$$

where 218 and 59 are the molar masses of the internal standard and hydroxypropyl groups, respectively, expressed in grams per mole. The content of hydroxypropyl groups should be 0.5–7.0%.

30.6.5 Assay for Monomeric Composition by ^{1}H NMR Spectroscopy: Poloxamer USP and Poloxamers Ph. Eur.

Poloxamer

(5)

Poloxamers (**5**) are synthetic block copolymers of ethylene oxide and propylene oxide, available in several types including both liquid and solid forms. They are functional excipients in pharmaceutical dosage forms used as surfactants, emulsifiers, solubilizers, dispersing agents, and to enhance in vivo absorption. Both the USP-NF, under Assay, and Ph. Eur. use an NMR method to control the relative proportion of the two monomers in the different types of poloxamers. The limit is expressed as the percentage of oxyethylene by weight in both cases, although in the Ph. Eur., it is a specific test named as the oxypropylene:oxyethylene ratio. The limit is dependent on the type according to Table 30.5 as given in Ph. Eur. monograph. The USP-NF limits for percent oxyethylene are nearly identical.

The USP-NF uses the relative method of quantitation (referred to as the *normalization procedure* in the Ph. Eur.). The poloxamer sample is dissolved in either deuterated water or deuterochloroform, according to solubility of the particular type being examined, with sodium 2,2-dimethyl-2-silapentane-5-sulfonate (DMSS) or tetramethylsilane (TMS), respectively, as the reference (1%) to give a solution of 0.1–0.2 g

Table 30.5.   Poloxamer characteristics (Ph. Eur.)

Poloxamer type	Ethylene oxide units (a)	Propylene oxide units (b)	Content of oxyethylene (%)	Average relative molecular mass
124	10–15	18–23	44.8–48.6	2090–2360
188	75–85	25–30	79.9–83.7	7680–9510
237	60–68	35–40	70.5–74.3	6840–8830
338	137–146	42–47	81.4–84.9	12 700–17 400
407	95–105	54–60	71.5–74.9	9840–14 600

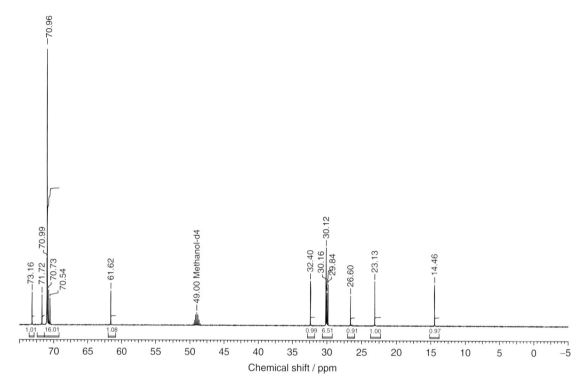

Figure 30.3. ^{13}C NMR spectrum of lauromacrogol 400. (Reproduced with permission from Ref. 3. © European Directorate for the Quality of Medicines & HealthCare, Council of Europe, 2014)

in 1 ml. If the nonaqueous NMR solvent is used, one drop of deuterated water is added to the sample solution. The region 0–5 ppm is scanned. The average area of the oxypropylene methyl group doublet at about 1.08 ppm compared to the reference standard (A_1) is measured against the average area of the composite band in the range 3.2–3.8 ppm (A_2), which is due to CH$_2$O groups of both the oxyethylene and oxypropylene units and the CHO groups of the oxypropylene units. Equations 30.3 and 30.4 are used to calculate the percentage of oxyethylene by weight:

$$\alpha = (A_2/A_1) - 1 \qquad (30.3)$$

$$\text{Result} = 3300 \times \alpha/(33 \times \alpha + 58) \qquad (30.4)$$

The Ph. Eur. method specifies a 100 g l^{-1} solution of the substance to be examined in deuterated chloroform only and records the same signals as the USP-NF method. The use of TMS as the reference standard is not mentioned specifically in the Poloxamers Ph. Eur. monograph, but the general method for NMR spectroscopy indicates that

TMS is the conventional chemical shift reference. The same equations as in the USP-NF are used to calculate the percentage of oxyethylene by weight.

30.6.6 Test for Composition by ^{13}C NMR Spectroscopy: Lauromacrogol 400 Ph. Eur.

Lauromacrogol 400 is a mixture of lauryl alcohol (dodecanol) monoethers of mixed macrogols (polyethylene glycols). The number in the name corresponds to the average molecular mass of the macrogol portion. It may include some free macrogols and contains various amounts of free lauryl alcohol. The ratio of moles of ethylene oxide reacted with moles of lauryl alcohol is 9. While macrogols are frequently used as excipients in drug products, the Ph. Eur. monograph for Lauromacrogol 400 applies to its use as an active substance. Medicines containing Lauromacrogol 400 are used for the symptomatic relief of rectal conditions such as hemorrhoids.

A specific test in the Ph. Eur. controls the average chain length of the fatty alcohol and average number of moles of ethylene oxide in the substance using a ^{13}C NMR spectroscopy method. A spectrometer operating at a minimum field of 300 MHz is specified with parameters set out in the monograph. The substance may need to be warmed before sampling if solid at room temperature. The test solution is prepared by dissolving 0.4 ml of the substance in 0.3 ml of a mixture of 1 : 2 volumes of $CD_3OD:CDCl_3$ containing 0.1 mol l^{-1} chromium(III) chloride as a relaxation aid. The CD_3OD multiplet at 49.0 ppm is used as the chemical shift reference. The spectrum of the test sample is compared with the spectrum published in the monograph (Figure 30.3).

The shift values should lie near those given in Table 30.6. A system suitability test should also be conducted on the smallest relevant peak (CH$_2$ at 73.1 ppm) for signal-to-noise ratio (minimum 150) and on the central CDCl$_3$ signal at 78.6 ppm for peak width at half-height (maximum 0.05 ppm).

For the test to calculate the average chain length of the fatty alcohol and the average number of moles of ethylene oxide, the signal at 23.2 ppm is defined as 1.000 and the integrals of the other signals listed in Table 30.6 are normalized. The average chain length

Table 30.6. ^{13}C NMR chemical shifts for lauromacrogol 400

CH$_3$	14.4	0.989
CH$_2$ (alkyl chain)	23.2	1.000
CH$_2$ (alkyl chain)	25.5	1.001
CH$_2$'s (alkyl chain)	30	7.410
CH$_2$ (alkyl chain)	32.5	0.963
CH$_2$ (−CH$_2$−OH) (end CH$_2$-group of macrogol)	61.6	1.001
CH$_2$'s (macrogol)	70.7	16.25
CH$_2$ (R−CH−O-macrogol) (CH$_2$ in alpha position)	72.6	0.998
CH$_2$ (macrogol)	73.1	0.929

of the fatty alcohol is calculated using equation 30.5:

$$\sum_{14-33} I_{n,i} + I_{n,72.6} \qquad (30.5)$$

where $\sum_{14-33} I_{n,i}$ is the sum of the normalized integrals of the signals from 14 to 33 ppm (due to the lauryl alkyl chain) and $I_{n,72.6}$ is the normalized integral of the signal at 72.6 ppm (due to the macrogol alpha-CH$_2$).

The average number of moles of ethylene oxide is calculated using equation 30.6:

$$0.5 \times (I_{n,62} + I_{n,71} + I_{n,73}) \qquad (30.6)$$

where $I_{n,62}$, $I_{n,71}$, and $I_{n,73}$ are the normalized integrals of the signals at 62, 71, and 73 ppm, respectively. The sum of these integrals corresponds to the average number of methylene groups in the macrogol part of the substance. The average chain length of the fatty alcohol should be 10.0–14.0 and the average number of moles of ethylene oxide should be 7.0–11.0.

30.6.7 Specific Test for Amino Acid Content by ^{13}C NMR Spectroscopy: Goserelin Acetate USP

Goserelin

(6)

Goserelin (**6**) is a synthetic nonapeptide analog of the hypothalamic decapeptide gonadotropin-releasing hormone (gonadorelin) used in the treatment of

Table 30.7. ^{13}C chemical shifts of goserelin acetate

Amino acids	Resonances (ppm)
Azo-glycine	162.2
Histidine	118.4
Tyrosine	116.7
tert-Butyl serine	62.5
Serine	62.2
Tryptophan	55.7
Arginine	41.8
Pyroglutamic acid	26.3
Proline	26.0
Leucine	23.5

prostate cancer. It is manufactured by chemical synthesis and is available as an acetate salt.

The USP monograph uses a ^{13}C NMR method in its test for Amino Acid Content. The spectrum of the sample solution is compared to a solution of goserelin reference substance at the same concentration (within 5%) but adjustments can be made depending on the quality of the ^{13}C spectra obtained. The same conditions must be used for the two solutions and the quality must be sufficient to allow quantification of integrals of the specified signals (see below). Repeating and averaging of integrals and spectra is permitted.

The sample and standard solutions have a concentration of about 10% w/v in deuterium oxide and are adjusted to pH 4 with deuterated acetic acid-d$_4$. The ^{13}C proton-decoupled NMR spectrum of both the sample and standard solutions should be qualitatively similar and all the resonances of the standard solution should be present in the sample solution, within 0.1 ppm for the goserelin moiety and 0.5 ppm for the acetate. Other resonances in the spectrum of the sample solution should be identified. The relative amino acid ratio is calculated by integrating resonances assigned to each amino acid as in Table 30.7.

The ratio of each of the amino acids from the integrals of the standard and sample solutions is calculated using equation 30.7:

$$\text{Result} = r_\text{U}/r_\text{S} \qquad (30.7)$$

where r_U is the integral of the resonance of a designated amino acid from the sample solution and r_S is the integral of the resonance of a designated amino acid from the standard solution. The acceptance criteria specified in the monograph are that the resulting ratios fall within the following limits: tyrosine, *tert*-butyl

serine, serine, tryptophan, arginine, pyroglutamic acid, proline, and leucine 0.9–1.1; azo-glycine 0.8–1.2.

The Ph. Eur. contains a monograph for Goserelin (not the acetate), which does not contain an equivalent ^{13}C NMR test. It specifies amino acid analysis for control of amino acid content (as one of the three identification tests, two of which must be carried out) using ion-exchange chromatography with post-column ninhydrin derivatization following hydrolysis of the sample. The limits specified in the Ph. Eur. monograph are glutamic acid, histidine, tyrosine, leucine, arginine, and proline 0.9–1.1; serine 1.6–2.2. Not more than traces of other amino acids should be present, with the exception of tryptophan. The differences in the limits compared with those of the USP monograph arise from the different treatment of the samples in the two analytical methods used, some of the derivatized amino acids being cleaved by hydrolysis in the amino acid analysis method of the Ph. Eur. The USP ^{13}C NMR method has the advantage of measuring directly the individual amino acids in the nonapeptide chain of goserelin.

However, the Ph. Eur. monograph does contain a ^{1}H NMR identification test using a 13 mg ml^{-1} solution of the sample in 0.2 M deuterated sodium phosphate buffer solution at pH 5.0 containing 20 µg ml^{-1} of deuterated sodium trimethylsilylpropionate. This is compared to an equivalent solution of goserelin for NMR identification CRS. A minimum operating field strength of 300 MHz is specified and a temperature of 25 °C to obtain the spectra of each solution. The two spectra should be qualitatively similar.

30.6.8 Isomeric-related Substances by ^{1}H NMR Spectroscopy: Orphenadrine Citrate USP

Orphenadrine is an anticholinergic drug used to treat Parkinson's disease (**7**). Both the Ph. Eur. and USP monographs for orphenadrine citrate limit the content of the *meta*- and *para*-methylbenzyl positional isomers, Impurities A and B (**8, 9**).

The Ph. Eur. uses a gas chromatography (GC) technique after extraction of orphenadrine base and specifies a limit of 0.3% for each isomer. The same method is used for Orphenadrine Hydrochloride Ph. Eur. However, the current USP (USP 32-NF 32 Supplement 1) uses ^{1}H NMR spectroscopy and the relative method of quantitation to limit the combined *meta-* and *para-*isomers to 3.0%. It should be noted, however, that the USP has proposed to replace the NMR procedure in January 2015 with a GC procedure similar to the one in the Ph. Eur. to allow quantification of each isomer.

Orphenadrine and its impurities A and B

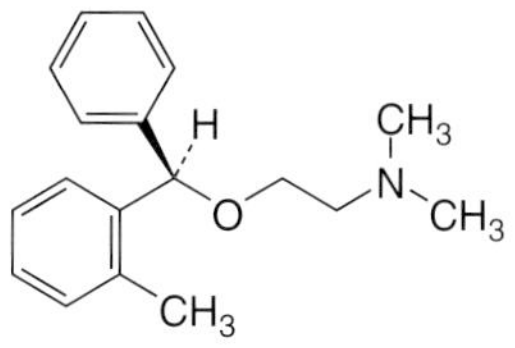

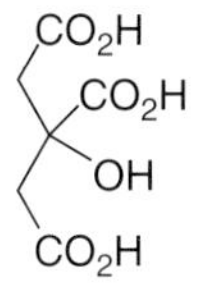

(7)

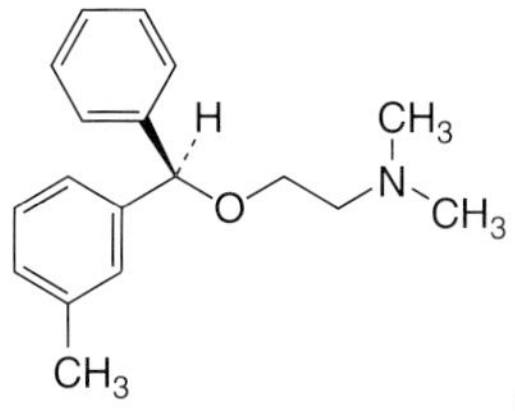

(8)

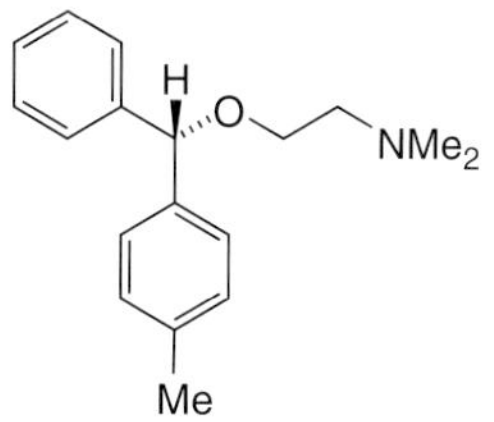

(9)

In the USP NMR method, the sample is prepared using ether to extract orphenadrine base ether from an aqueous solution to which sodium hydroxide has been added. Using quantitative procedures, the ether is dried with sodium sulfate and then evaporated. About 400 mg of the resulting orphenadrine base is then dissolved in 0.5 ml carbon tetrachloride to give the sample solution with the addition of one drop of TMS as the reference standard.

Using the Relative Method of Quantitation, which appears in Nuclear Magnetic Resonance Spectroscopy <761>, the sum of the average areas of the combined methine peaks associated with the *meta-* and *para-*methylbenzyl isomers (A_1), which appears at about 5.23 ppm, is compared with the area of the methine peak associated with the *ortho-*methylbenzyl isomer (A_2), appearing at about 5.47 ppm, using a normalization factor of 1 for the integrals of each peak.

30.6.9 Impurities by ^{1}H NMR Spectroscopy: Medronic Acid for Radiopharmaceutical Preparations Ph. Eur.

Medronic acid and its impurities A and B

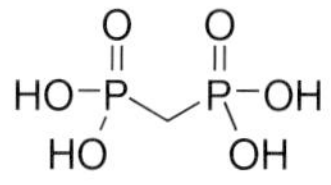

(10)

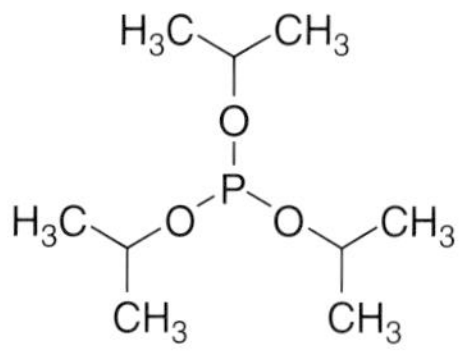

(11)

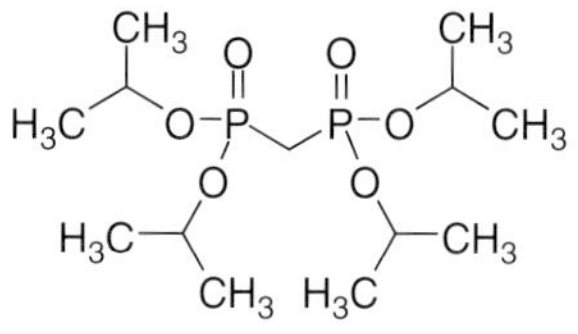

(12)

Medronic acid (**10**) is a bisphosphonate compound used in bone scanning as a diagnostic agent following complexation with radioactive sodium pertechnetate (^{99m}Tc) to form technetium (^{99m}Tc) medronate. The complex is administered as an injection. Medronic Acid for Radiopharmaceutical preparations is supplied as a freeze–dried powder with suitable excipients in 10 mg vials. ^{1}H NMR spectroscopy is used both as an identification test (together with infrared spectroscopy) and for the specific test for Impurities A and B (**11, 12**). For the identity test, a 100 g l^{-1} solution in deuterium oxide is compared with a 100 g l^{-1} solution of medronic acid CRS.[10]

Impurity A is tris(1-methylethoxy)phosphate and Impurity B is tetrakis(1-methylethyl)methylenediphosphonate. Each of these two impurities is limited to not more than 1%. The ^{1}H NMR spectrometry test employs the method of standard additions. A test solution is prepared by adding 10 ml deuterated chloroform to 1.0 g of test substance, stirring for 1 h, and then filtering and concentrating the filtrate to about 0.5 ml. Reference solutions are prepared using

Impurity A and Impurity B reference substances 10 µl each with 1.0 ml deuterated chloroform. After recording the NMR spectrum of the test solution (using a spectrometer operating at a minimum of 250 MHz), 10 µl aliquots of the two reference impurity solutions are added and the spectrum rerecorded. TMS may be used as a chemical shift internal reference if necessary. The positions of the signals are given in the Ph. Eur. monograph: Impurity A about 4.4 and 1.3 ppm; Impurity B about 4.7, 2.4, and 1.3 ppm. The reference impurity solutions are also used for a system suitability test: the positions of the signals owing to impurities A and B should not differ significantly from those in the spectrum obtained with the test solution after addition of the reference impurity solutions. The limits are calculated by integrating the multiplet at 4.4 ppm owing to impurity A and the multiplet at 2.4 ppm owing to impurity B in the spectrum obtained with the test solution and comparing the integrals with those obtained from the spectrum of the test solution with reference impurities added. The peak areas for each impurity in the spectrum for the test solution should be not more than 0.5 times the area of the corresponding peak for the reference solution equivalent to a limit of 1%. A reference spectrum is not included in the monograph itself but has been made available on the EDQM website.[11]

30.6.10 Relaxivity: Gadoversetamide Injection USP

Gadoversetamide

(13)

Gadoversetamide (**13**) is a gadolinium complex used as a diagnostic agent in MRI, in particular for contrast enhancement and visualizing lesions and abnormal structures in the central nervous system (CNS) and liver. It is administered as an intravenous injection. Its

mode of action depends on the relaxivity of the paramagnetic gadolinium species and the USP monograph for the drug product includes a test for this property. Relaxivity is the magnitude of a substance's capacity to enhance the relaxation rate of a nucleus, expressed in units of s^{-1} mM^{-1}. Relaxivity of a substance is determined experimentally by measuring the spin-lattice relaxation time (T_1) of a test substance and plotting $1/T_1$ against the concentration in units of mM. The slope of the curve is the numerical relaxivity.[8]

In the method described in the USP monograph for Gadoversetamide injection, standard solutions are prepared from accurately weighed quantities of manganese (II) chloride tetrahydrate dissolved and diluted with water to obtain solutions having known concentrations of 0.9, 2.7, and 4.5 mM. Test solutions are prepared by diluting 5.0 ml of the injection in water to give 0.504, 1.008, 2.016, and 3.024 mM, respectively.

A low-field NMR spectrometer with suitable sensitivity is used after checking the system suitability with the standard solutions in 10 mm tubes, warmed to 40°C for not less than 10 min. The average T_1 at 20 MHz for replicate measurements must be within 5% of 156 ms for the 0.9 mM, 52 ms for the 2.7 mM, and 32 ms for the 4.5 mM standard solutions, respectively.

T_1 of each test solution is measured under the same conditions. $1/T_1$ is plotted against the molarities of the test solutions, and a regression analysis performed. The slope of the plotted line, the relaxivity, should be between 4.0 and 5.0 s^{-1} mM^{-1}.

30.7 CONCLUSION

Pharmacopoeias provide quality standards and methods for a wide range of users. Pharmacopoeial methods have to be generally accessible and this has not always been the case with NMR techniques that, historically, have required sophisticated and relatively expensive equipment. For this reason, NMR spectroscopy has tended to be included in monographs only when other methods are not feasible or when an NMR technique has something unique to offer. As illustrated in the previous examples, the feature of NMR spectroscopy that is exploited in monographs is its well-recognized ability for structural discrimination, particularly for polymeric compounds of natural or biological origin. The inclusion of other test methods in a monograph does not preclude the use of NMR methods for control of drug substances or drug products where these are considered more appropriate by users in their own particular circumstances, e.g., during product research and development. Likewise, other analytical techniques may be used in place of an NMR pharmacopoeial method. However, the substance or product should comply with the relevant monograph, if tested.

ACKNOWLEDGMENT

The authors would like to thank Matilda Vallender, Editor-in-Chief, British Pharmacopoeia, for reviewing the manuscript Dr Ravi Ravichandran and Dr Anita Y. Szajek of the US Pharmacopeia for helpful comments and the spectrum for Figure 30.1 and the Department of Publications and Multimedia at EDQM for providing the spectra for Figures 30.2 and 30.3.

REFERENCES

1. Background and Mission of the European Pharmacopoeia. http://www.edqm.eu/site/european-pharmacopoeia-background-50.html.

2. USP in U.S. Law, US Pharmacopeial Convention. http://www.usp.org/about-usp/legal-recognition/usp-us-law.

3. European Pharmacopoeia, 8th edn (8.2), Council of Europe, Strasbourg, 2014.

4. United States Pharmacopeia – National Formulary, USP 37- NF 32.

5. Appendix II C. Nuclear Magnetic Resonance Spectrometry, Ph. Eur. method 2.2.33. http://www.pharmacopoeia.co.uk/bp2014updated/ixbin/bp.cgi?a=display&r=aQtax38780l&id=1014&tab=search.

6. Technical Guides, European Directorate for the Quality of Medicines and HealthCare. http://www.edqm.eu/site/technical-guides-589.html.

7. General Chapter <761> Nuclear Magnetic Resonance Spectroscopy, USP 37- NF 32.

8. General Chapter <1761> Applications of Nuclear Magnetic Resonance Spectroscopy, USP 37- NF 32.

9. I. McEwen, B. Mulloy, E. Hellwig, L. Kozerski, T. Beyer, U. Holzgrabe, A. Rodomonte, R. Wanko, and J.-M. Spieser, *Pharmeuropa Bio*, 2008, **1**, 31.

10. Appendix I E. Reference Materials, British Pharmacopoeia, 2014. http://www.pharmacopoeia.co.uk/bp2014/ixbin/bp.cgi?id=1013&a=display#WIXLINKcrs07361.

11. European Directorate for Quality of Medicines and Healthcare, Knowledge Database. http://www.edqm.eu/site/Knowledge-Database-707.html.

Chapter 31

NMR in Pharmaceutical Manufacturing

Edwin Kellenbach[1] and Paulo Dani[2]

[1]*Aspen Pharmacare, PO Box 20, 5340BH Oss, Oss, The Netherlands*
[2]*Former employee from Merck Sharp & Dohme (MSD), Oss, The Netherlands*

31.1 INTRODUCTION

The development of a pharmaceutical product goes through basically three phases: a discovery phase, in which a range of chemical compounds are screened against receptors in an attempt to identify potential leads; a development phase, in which the lead candidates previously found are more thoroughly tested and where the first attempts to formulate a product are carried out in preparation for the last phase, which is the manufacture of the pharmaceutical product.

NMR is present at all moments of this trajectory, although not in the same way and frequency. For instance, the classical NMR analyses in solution are intensively present during discovery especially for the structure elucidation of the candidate molecules and of those from the medicinal chemistry route. Quantitation by NMR (or simply qNMR application) is also often present at this stage, albeit high precisions and accuracies are not required.

In the development phase, structure elucidation and quantitation tasks remain occupying NMR time, but now in a different manner. The molecular structures to be elucidated are mostly from impurities, degradation products, and metabolites from the lead molecules. This means that, on the one hand, a reduced number of NMR analyses are necessary compared to the discovery phase. On the other hand, however, aspects related to *quality* are far more stringent at this stage often requiring qualified spectrometers kept under GLP (Good Laboratory Practice) or GMP (Good Manufacturing Practice) regimes and the use of validated methods. Furthermore, far better accuracies and precisions are expected for quantitation work, meaning that the analyst has to pay a much closer attention to the guidelines dictating how to perform proper qNMR analyses (see Chapter 5). In this phase, issues related to drug substance crystallinity (e.g., polymorphism, solid-state transformations, and formation of solvates) are more intensively addressed, meaning that solid-state NMR applications, basically absent during discovery, are often utilized.

In manufacturing, NMR applications can be divided into two main groups: quality control (QC) and troubleshooting. NMR is basically applied in the control of the identity, composition, and (physical) purity of

NMR in Pharmaceutical Sciences. Edited by Jeremy R. Everett, John C. Lindon, Ian D. Wilson, and Robin K. Harris
© 2015 John Wiley & Sons, Ltd. ISBN: 978-1-118-66025-6
Also published in eMagRes (online edition)
DOI: 10.1002/9780470034590.emrstm1397

Written by Edwin Kellenbach in his capacity as an employee of Aspen Pharmacare.

the substances used in the preparation of the pharmaceutical product and of the pharmaceutical product itself, including here the detection of counterfeits. The reliable characterization and purity control of (bio)molecules – not least the creation of analytical methods that can signal cases of counterfeiting or the presence of unexpected endogenous or exogenous substances – are crucial tasks in assuring the quality of a medicine, and are great challenges to analytical chemists, particularly when dealing with large biomolecules of a polymeric nature. NMR has been playing an important role in this area, owing to its holistic capability as a 'general' detector, facile collection of quantitative data, limited sample pretreatment, its capacity to interrogate and correlate different nuclei in multidimensional techniques, and its sensitivity toward spatial structural alterations. A sign of the recognition of NMR importance is its increasing presence in pharmacopoeias.[1]

NMR is an excellent technique for troubleshooting both products and processes. A simple proton NMR spectrum, for instance, is easily recorded and quickly offers an abundance of information on the presence of process-related impurities, e.g., solvents, seal oil, and other molecules used in the process. In addition, the presence of metal ions and changes in pH may be detected indirectly by their effect on the spectrum. In this manner, a number of possible root causes of product and production problems can quickly be asserted or excluded. Moreover, the speed of analysis and the lack of sample preparation allow monitoring of the pharmaceutical production process. No (or limited) method development is required, speeding up the process of root cause analysis.

Not many articles have addressed these roles of NMR. This is probably because of the low academic profile of pharmaceutical manufacturing and the confidential nature of the investigations. This chapter aims, on the one hand, to encourage NMR spectroscopists to extend their techniques and skills to manufacturing and also to enhance the application of NMR as a root cause analysis tool in pharmaceutical manufacturing. Even if not all manufacturing sites have their own NMR facilities, a number of contract research organizations (CROs) offer GLP compliant and cGMP NMR services for small molecules, macromolecules, and product impurities nowadays, facilitating the access to NMR. Although NMR spectrometers are expensive investments, an NMR analysis itself is not expensive because of the lack of sample preparation, method development, and the low amount and/or price

of consumables such as deuterated solvents, NMR tubes, liquid nitrogen, and helium. It is important to note that the acquisition of a simple ^{1}H NMR spectrum only takes a few minutes, and a large number of spectra can be acquired in a short time, thanks to user-friendly, intuitive software and robust hardware allowing 24/7 unsupervised acquisition of spectra from many samples. In addition, sample amount is rarely a limitation in manufacturing.

There follow examples of the application of NMR in manufacturing. Applications related to time domain NMR and process analytical technology (PAT) are outside the scope of this chapter.

31.2 QUALITY CONTROL

31.2.1 ^{1}H NMR as an Alternative for the Amino Acid Analysis of Peptides

Currently, a number of monographs for synthetic peptides included in the European Pharmacopoeia (Ph.Eur.) require amino acid analysis (AAA) as an identity test.[2] Briefly, AAA entails the acid hydrolysis of a peptide to yield the related amino acids followed by chromatographic separation, ninhydrin derivatization, and ultraviolet (UV) detection. The relative amount (ratio) and absolute content of amino acids in a peptide can be determined, making AAA an effective technique to be used both as an identity test and to quantify peptides. Some drawbacks of this method are the degradation of amino acids by the harsh hydrolysis conditions (typically low pH overnight at 100 °C) and the laborious and time-consuming sample preparation.

On the other hand, NMR analysis in solution is widely recognized, accepted, and established as a powerful tool for the structural elucidation and characterization of peptides and proteins.[3,4] ^{1}H and ^{13}C NMR spectra, for instance, readily distinguish all 20 naturally occurring amino acids and unnatural amino acids present in synthetic peptides. An example of the selectivity of NMR is given in Figure 31.1, in which a simple comparison of ^{1}H NMR spectra allows the differentiation of the peptide goserelin acetate from its serine diastereoisomer named *goserelin-related compound A* that differs from the former by the epimerization of a single serine alpha carbon (marked with an arrow in Figure 31.1).[5] Note the signal dispersion and the differences in the spectra of the two compounds (major differences are in the boxed areas).

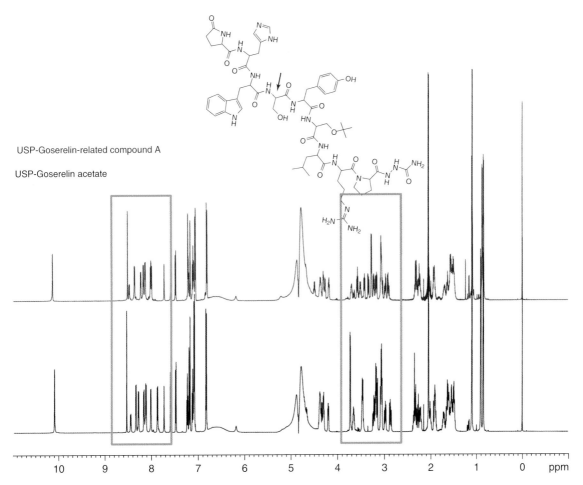

USP-Goserelin-related compound A

USP-Goserelin acetate

Figure 31.1. ¹H NMR (600 MHz, 95% H₂O) spectrum of goserelin acetate and from a serine diastereoisomer (*goserelin-related compound A*). Regions with clear spectral differences are within boxes. (Reproduced from Ref. 5. © Canadian Crown, 2010)

A ¹H NMR spectrum of a peptide depends primarily on the relative amounts and the identities of the amino acids present in the sequence, i.e., the same information obtained from an AAA. Thus, from a scientific point of view, NMR is a suitable alternative for AAA. In this way, ¹H NMR at 400 MHz was recently introduced as an identity test for eight peptides in the Ph. Eur. The identification consists of a straightforward and simple comparison with the spectrum of a peptide standard, and no calculations are performed. Even peptides differing very slightly can be readily distinguished. Moreover, as no hydrolysis is performed before analysis, a distinction can be made between Glu and Gln, and Asp and Asn, something that AAA

is not able to do. In addition, amino acids prone to hydrolysis and/or oxidation such as Trp and Cys in AAA can be detected without any problems by NMR. Table 31.1 summarizes the benefits and drawbacks of NMR and AAA.

Figure 31.2 demonstrates the strength of the ¹H NMR method by showing the spectra of eight different peptides, all recorded in the same way at 400 MHz,[6] a moderate field strength widely available within pharmaceutical industry. All peptides can be readily distinguished, even in the case of the two salt forms of gonadorelin (acetate and diacetate), using the areas of the acetate counter ion (Figure 31.3), and in the case of goserelin and buserelin, two peptides showing only

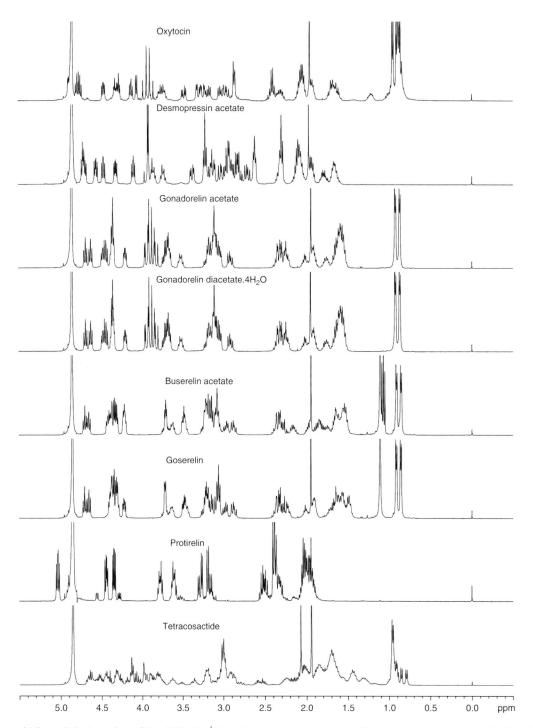

Figure 31.2. Aliphatic region of the 400 MHz ^{1}H NMR spectra of the eight different peptides listed in Table 31.2. The spectra were recorded in D_2O in a 200 mM phosphate buffer at pH 5. (Reproduced with permission from Ref. 6. © European Directorate for the Quality of Medicines & HealthCare (EDQM), Council of Europe, 2008)

Table 31.1. A comparison between NMR and AAA

Method	Advantages	Drawbacks
AAA	Widely available; classical, accepted technique; also applicable for larger peptides/proteins	Destruction of Asn and Gln; Cys, Met and Trp; hydrolysis conditions differ from one peptide to the other; time-consuming and laborious sample preparation
NMR	General, robust technique; no hydrolysis; data acquisition fast and interpretation straightforward; allows identification/ quantitation of proton-bearing organic counter ions	Less widely available[a] High once-off investment

[a] NMR analyses can, however, be readily outsourced to laboratories working under GMP for prices in the order of hundreds of Euros, i.e., far less than the price for outsourcing of a classical AAA.
Reproduced with permission from Ref. 6. © European Directorate for the Quality of Medicines & HealthCare (EDQM), Council of Europe, 2008.

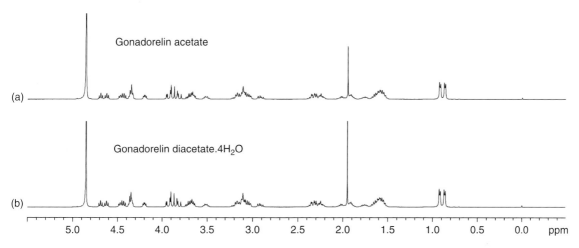

Figure 31.3. ^{1}H NMR spectra of Gonadorelin acetate (a) and Gonadorelin diacetate (b). The intensity of the acetate peak (singlet at about 1.9 ppm) allows distinction between the two peptides. (Reproduced with permission from Ref. 6. © European Directorate for the Quality of Medicines & HealthCare (EDQM), Council of Europe, 2008)

very small differences in their sequence (Table 31.2). The method is robust and reproducible and brought great simplification to the final release testing of the manufactured peptides.

It is expected that in the future other NMR methods will be used (e.g., two-dimensional ones) allowing the replacement of AAA by NMR in the identification of larger peptides. In this context, it has been shown that human, porcine, and bovine insulins (51 amino acids, differing in 1 to 4 amino acids) can be readily distinguished by two-dimensional NMR.[7]

31.2.2 Heparin

Perhaps one of the most clear-cut examples showing the strength of NMR in manufacturing as a broad nonspecific detection technique able to observe expected and unexpected contaminants is related to the monitoring of the quality of heparin sodium. Heparin sodium is a complex, polydisperse glycosaminoglycan (GAG) polysaccharide that has been widely used as a safe anticoagulant for decades (Figure 31.4). In March 2008, a number of adverse effects (including death of patients) associated with heparin sodium

Table 31.2. Composition of the peptides from Figure 31.2

Peptide	Number of AA	Sequence
Oxytocin	9	H-Cys-Tyr-Ile-Gln-Asn-Cys-Pro-Leu-Gly-NH$_2$
Desmopressin acetate	9	Mpa-Tyr-Phe-Gln-Asn-Cys-Pro-D-Arg-Gly-NH$_2$
Gonadorelin acetate	10	Glp-His-Trp-Ser-Tyr-Gly-Leu-Arg-Pro-Gly-NH$_2$
Gonadorelin diacetate	10	Glp-His-Trp-Ser-Tyr-Gly-Leu-Arg-Pro-Gly-NH$_2$
Buserelin acetate	10	Glp-His-Trp-Ser-Tyr-D-Ser(tBu)-Leu-Arg-Pro-Gly-NHEt
Goserelin	10	Glp-His-Trp-Ser-Tyr-D-Ser(tBu)-Leu-Arg-Pro-Azagly-NH$_2$
Protirelin	3	Glp-His-Pro-NH$_2$
Tetracosactide	24	H-Ser-Tyr-Ser-Met-Glu-His-Phe-Arg-Trp-Gly-Lys-Pro-Val-Gly-Lys-Lys-Arg-Arg-Pro-Val-Lys-Val-Tyr-Pro-OH

Reproduced with permission from Ref. 6. © European Directorate for the Quality of Medicines & HealthCare (EDQM), Council of Europe, 2008.

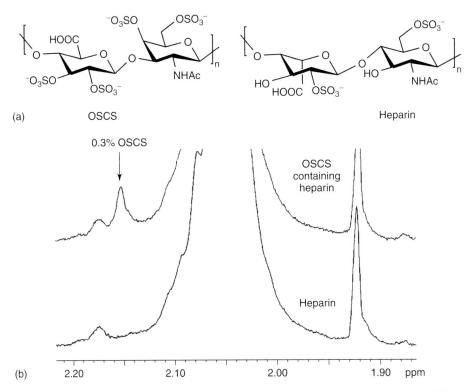

Figure 31.4. (a) The chemical structure of the major repeating disaccharide units from heparin and OSCS and (b) the ^{1}H NMR (600 MHz, D$_2$O) spectrum of the acetyl region from heparin and from the USP system suitability test sample containing 0.3% OSCS. OSCS resonates at 2.15 ppm. The signal at 1.92 ppm is from acetate. The signal at 2.175 ppm is the C-13 satellite of the heparin large *N*-acetyl signal at 2.04 ppm (the other satellite is under the acetate signal)

prompted the Food & Drug Administration (FDA) to publish a proton NMR quality test method for the screening of a contaminant behind these effects.[8] The contaminated heparin batches displayed a signal in the *N*-acetyl region at about 2.15 ppm, which was absent in the spectrum of uncontaminated batches (Figure 31.4). The nature of this contaminant was later identified – using, among others, multidimensional

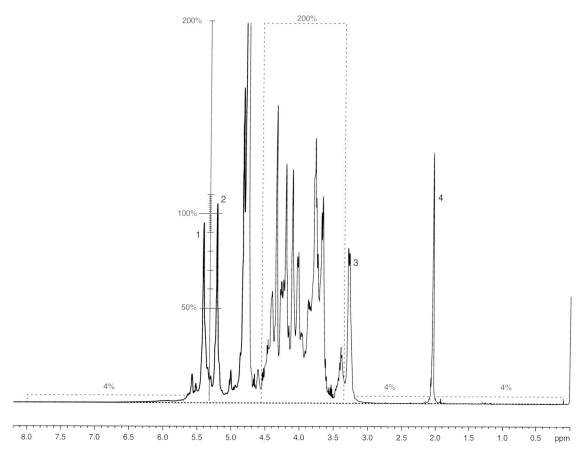

Figure 31.5. ^{1}H NMR spectrum (600 MHz, D_2O) of heparin sodium according to the USP monograph. Among other aspects, the spectrum is divided into areas or *boxes* with intensity thresholds of 4% and 200% (taking as base the average intensities of the anomeric signals 1 and 2), which no unidentified signals should exceed

heteronuclear magnetic resonance[9,10] – as oversulfated chondroitin sulfate (OSCS), a chemically modified and related GAG, able to deceive the limited release tests at the time (until 2008).

Since then, the updated United States Pharmacopoeia, the European Pharmacopoeia, and the Japanese Pharmacopoeia heparin sodium monographs have grown in scope, requiring manufacturers to perform ^{1}H NMR analysis for identity and purity testing.[11–13] Accordingly, a single ^{1}H NMR spectrum tests not only for the presence of the OSCS *N*-acetyl signal but also takes other parts of the spectrum into account aiming to establish the heparin identity and also to control the existence of other impurities. The sample preparation consists simply of

dissolution in D_2O containing as reference standard TSP (3-(trimethylsilyl)propionic acid-d^4 sodium salt). An example from the USP Pharmacopoeia is given in Figure 31.5.

31.2.3 Quantification of Chondroitin Sulfate and Dermatan Sulfate in Danaparoid Sodium by ^{1}H NMR spectroscopy and PLS Regression

Danaparoid sodium, the active pharmaceutical ingredient (API) in Orgaran, is a biopolymeric heparinoid used as anticoagulant and antithrombotic agent for patients allergic to heparin. It consists of a mixture

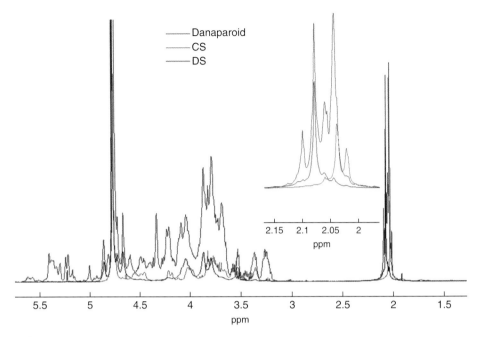

Figure 31.6. ¹H NMR (600 MHz, D₂O) spectra of danaparoid sodium, CS, and DS

of three GAGs in a specific ratio: heparan sulfate, 8–16% dermatan sulfate (DS) and <8.5% chondroitin sulfate (CS). The amounts of CS and DS are traditionally quantified by means of a laborious and time-consuming enzymatic method. As an alternative, a nondestructive multivariate regression method (partial least squares, PLS) was developed for quantifying CS and DS in danaparoid sodium based on ¹H NMR analysis.[14] Figure 31.6 shows the ¹H NMR spectra of CS, DS, and danaparoid sodium in D₂O.

The combination of ¹H NMR and PLS in this particular case did require extensive method development efforts; however, on the other hand, it has delivered key advantages over the traditional test. It resulted in a faster, more accurate and robust quantification process for both CS and DS in danaparoid sodium, and in a methodology that does not depend on enzymes and (calibrated) CS and DS standards. In this way, the analysis time (including PLS prediction) was reduced to 35 min/sample instead of 60 h for a maximum of 16 samples. Accuracies of ±0.7% (w/w) and ±1.1% (w/w) for CS and DS, respectively, are obtained by this method – better than the ±1.0% (w/w) and ±1.3% (w/w) ones observed in the enzymatic method.

31.2.4 Quality Control During Vaccine Production

As the earlier sections made clear, NMR is very suitable and it finds an important role in the characterization and quality control of complex and heterogeneous materials of biological origin. Therefore, it comes as no surprise that NMR has been applied in the analysis of vaccines, especially in the light of the technological developments in NMR from the last decade. There are a number of publications addressing the advantages and shortcomings of NMR in this area and future trends, and the interested reader is redirected to them and their references.[15,16]

31.2.5 Vorapaxar Sulfate Disproportionation

Vorapaxar sulfate (Figure 31.7) disproportionation in a drug product offers an example of the application of NMR in tackling the quality control of an API from a small molecule. Vorapaxar sulfate is indicated in the reduction of thrombotic cardiovascular events in some groups of patients. Upon storage of the drug product, it has been observed that there is partial

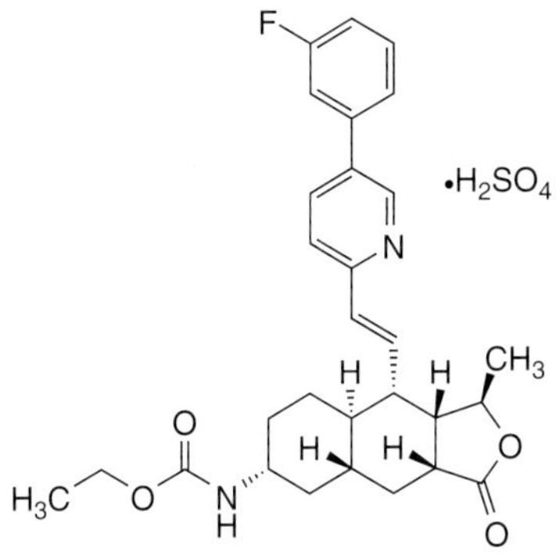

Figure 31.7. Vorapaxar sulfate molecule[17]

disproportionation of the API into the related free base.[17] The detection and quantitation of this process in the drug product has been successfully attained by a validated method based on ^{19}F solid-state NMR owing to the fact that the ^{19}F resonance from the Vorapaxar aromatic fluorine is sensitive to the disproportionation process.

31.3 TROUBLESHOOTING

31.3.1 Clarity Test Problem

A number of batches of a biological API presented clarity issues when dissolved in water: they did not yield a clear solution. Analysis of clarity is a routine pharmacopoeia test on many APIs and it probes for insoluble and/or extraneous matter. A ^{1}H NMR analysis, carried out as part of the efforts for finding the root cause of the problem, revealed the presence of unexpected aliphatic signals, later identified as coming from C14–C18 linear alkanes inadvertently introduced earlier in the process. As shown in Figure 31.8, the intensity of these signals strongly correlated with the clarity issue (expressed in nephelometric turbidity units – NTUs).

The hydrophobic alkanes probably formed tiny droplets in the aqueous solution causing light scattering, which in turn resulted in the clarity problem. The important point here is that the analysis of these batches by a simple ^{1}H NMR experiment promptly led to the identification of residual alkane material, making it possible to correlate alkane amounts with the severity of the clarity problem. This quickly elected the alkanes as the likely cause of the problem allowing

the manufacturing technical staff to concentrate their efforts on this specific issue and so in taking corrective and preventive actions.

31.3.2 Filtration Problem

Another example of the added value of performing NMR analysis rather early in the process of troubleshooting concerns investigations related to filtration problems arising during the manufacturing of an intermediate of an API. By comparing the ^{1}H NMR spectra from the intermediate having no issues with those having filtration problems, it was possible to identify a variation in the amounts of citrate present as a likely cause (Figure 31.9).

Indeed, an increase in the amount of citrate eliminated the filtration problems, possibly by a process of complexation/solubilization of divalent metal cations, which otherwise would precipitate causing filtration issues. Again, the straightforward comparison of one-dimensional proton spectra of batches with and without issues (requiring no method development) allowed the identification of the root cause of the problem at first glance.

It is important to emphasize that in both cases, NMR offered an important lead to the solution of the problems at hand, helping to identify what were then unforeseen root causes, and consequently narrowing the scope of the investigations into a manageable workload. Furthermore, NMR was excellently suited for a timely, accurate, and reproducible analysis of large numbers of samples, thanks to robust hardware (e.g., sample changers working on a 24/7 basis), sensitive probes, and that no time and efforts were spent on activities related to method development and optimization. Once the nature of the problem becomes recognized, more specific and sensitive methods with higher accuracy or capable of handling larger number of samples may subsequently be applied if desired (e.g., mass spectrometry for alkane detection or titration for citrate quantitation).

In conclusion, NMR spectroscopy seems currently underutilized in pharmaceutical manufacturing being confined to a few QC and troubleshooting applications despite its broad capabilities.

However, aspects such as the lack of laborious sample pretreatment and/or derivatization, straightforward and accurate quantitation, possibility to interrogate and correlate different nuclei, and the competitive price of

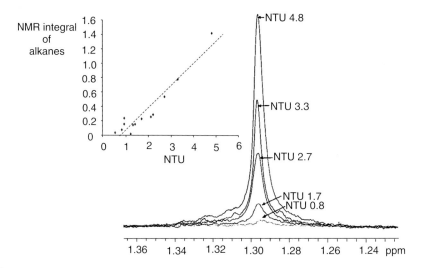

Figure 31.8. ^{1}H NMR spectrum (600 MHz, D_2O) correlating the intensity of the C14–C18 alkane signals at about 1.29 ppm with the clarity problem (expressed in nephelometric turbidity units – NTUs) in several batches of the API. The greater the clarity issue, the larger the NTU is, and the more intense the signals. As is shown in the inset, the clarity problem and the presence of the alkanes have a fairly linear correlation

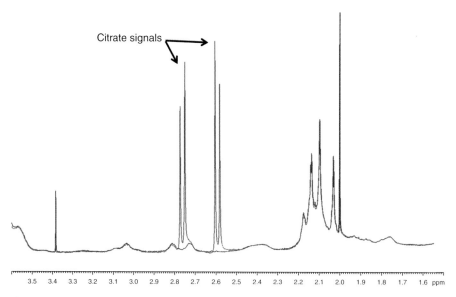

Figure 31.9. ^{1}H NMR spectra (in D_2O) of two batches of a production intermediate of a biological API. The blue trace is from the material having a filtration problem. The red trace, from a problem-free material, clearly shows the presence of citrate

an individual NMR analysis make NMR a fast, robust, and economic QC technique, which is rapidly gaining recognition, especially for biologicals. Moreover, the inherent holistic quality of the NMR analyses allied to a general lack of method development make NMR spectroscopy suitable for the identification of unexpected root causes, increasing the efficiency while reducing the time needed in the troubleshooting of processes and products in the pharmaceutical manufacturing.

ACKNOWLEDGMENT

The authors would like to thank Käthy Sanders, Maurice Seegers, Bülent Üstün, Maarten Lunenburg, and Marc Baremans for their contributions to the work described here and Carel Funke for carefully reading the manuscript.

REFERENCES

1. H. Corns and S. K. Branch, *Encycl. Magn. Reson.*, 2015, in press.

2. *European Pharmacopoeia 8.0*, European Department for the Quality of Medicines, Strasbourg, France. General chapter 2.2.64 Peptide Identification by Nuclear Magnetic Resonance Spectroscopy.

3. K. Wüthrich, *NMR in Biological Research: Peptides and Proteins*, North Holland/American Elsevier: Amsterdam/New York, 1976, 15.

4. K. Wüthrich, *NMR of Proteins and Nucleic Acids*, 1st edn, John Wiley & Sons, Inc.: New York, 1986, 13.

5. Y. Aubin, C. Jones, and D. I. Freedberg, *Biopharm. Int.*, 2010, **23**, 28.

6. E. Kellenbach, K. Sanders, G. Zomer, and P. L. A. Overbeeke, *Pharmeur. Sci. Notes*, 2008, **2008**, 1.

7. C. W. Funke, J. R. Mellema, P. Salemink, and G. N. Wagenaars, *J. Pharm. Pharmacol.*, 1988, **40**, 78.

8. USP U.S. Pharmacopeial Convention. Key Issue: Heparin [Online]. Available: http://www.usp.org/ usp-nf/key-issues/heparin. [Accessed 16 Dec 2014].

9. M. Guerrini, D. Beccati, Z. Shriver, A. Naggi, K. Viswanathan, A. Bisio, I. Capila, J. C. Lansing, S. Guglieri, B. Fraser, A. Al-Hakim, N. S. Gunay, Z. Zhang, L. Robinson, L. Buhse, M. Nasr, J. Woodcock, R. Langer, G. Venkataraman, R. J. Linhardt, B. Casu, G. Torri, and R. Sasisekharan, *Nat. Biotechnol.*, 2008, **26**, 669.

10. T. K. Kishimoto, K. Viswanathan, T. Ganguly, S. Elankumaran, S. Smith, K. Pelzer, J. C. Lansing, N. Sriranganathan, G. Zhao, Z. Galcheva-Gargova, A. Al-Hakim, G. S. Bailey, B. Fraser, S. Roy, T. Rogers-Cotrone, L. Buhse, M. Whary, J. Fox, M. Nasr, G. J. Dal Pan, Z. Shriver, R. S. Langer, G. Venkataraman, K. F. Austen, J. Woodcock, and R. Sasisekharan, *N. Engl. J. Med.*, 2008, **358**, 2457.

11. Information on Heparin, U. S. Food and Drug Administration, 2012. [Online]. Available: http://www. fda.gov/Drugs/DrugSafety/PostmarketDrugSafety InformationforPatientsandProviders/ucm112597.htm. [Accessed 30 November 2014].

12. European Pharmacopoeia heparin sodium monograph PA/PH/Exp. 6/T(0) 42 PUB monograph number 333. EDQM, Strasbourg, 2010. [Online]. Available: http://www.edqm.eu/medias/fichiers/NEW_Heparin_ sodium_0820100333.pdf. [Accessed 29 Nov 2014].

13. Japanese Pharmacopoeia 16th Edition, JP Online, Official Monographs D to K, p. 916, [Online]. Available: http://www.pmda.go.jp/english/pharmacopoeia/pdf/ sixteenth_edition/JP16%20Monograph%20D%20to% 20K.pdf. [Accessed 14 November 2014].

14. B. Üstün, K. B. Sanders, P. Dani, and E. R. Kellenbach, *Anal. Bioanal. Chem.*, 2011, **399**, 629.

15. D. S. Wishart, *Trends Anal. Chem.*, 2013, **48**, 96.

16. C. Abeygunawardana, T. C. Williams, J. S. Sumner, and J. P. Hennessey Jr, *Anal. Biochem.*, 2000, **279**, 226.

17. ZONTIVITY™ (vorapaxar) Tablets 2.08 mg, for oral use. Full Prescribing Information, Merck, Sharp & Dohme Corp., 2013. [Online]. Available: http://www. merck.com/product/usa/pi_circulars/z/zontivity/ zontivity_pi.pdf. [Accessed 08 September 2014].

Index

NPC	National Phenome Centre
NRA	Nuclear Reaction Analysis
NSAID	NonSteroidal Anti-Inflammatory Drug
NTUs	Nephelometric Turbidity Units
NUS	NonUniform Sampling
OA	OsteoArthritis
OMERACT	Outcome Measures in Rheumatoid Arthritis Clinical Trials
OPLS-DA	Orthogonal Partial Least Squares Discriminant Analysis
OS	OverSulfated
P/HA	Protons to Heavy Atoms
PASS	Phase-Adjusted Spinning Sidebands
PAT	Process Analytical Technology
PBDS	Plate-Based Diversity Screening
PC	Principal Component
pCASL	pseudo-Continuous Arterial Spin Labeling
PCA	Principal Component Analysis
PCR	Polymerase Chain Reaction
PCT	Patent Cooperation Treaty
PDE5i	PhosphoDiEsterase-5 inhibitors
PDF	Pair Distribution Function
PEI	Percentage Efficiency Index
PET	Positron Emission Tomography
PFG	Pulse Field Gradient
PGSE	Pulsed-field Gradient Spin Echo
Ph.Eur.	European Pharmacopoeia
PI	Principal Investigator
PII	Personal Identifiable Information
PK	PharmacoKinetics
PLS	Partial Least Squares
PLSDA	Projections to Latent Structures - Discriminant Analysis
PPI	Protein-Protein Interaction
PROJECT	Periodic Refocusing Of J-Evolution by Coherence Transfer
PROJECTED	Periodic Refocusing Of J-Evolution by Coherence Transfer Extended to Diffusion-ordered spectroscopy
PSA	Polar Surface Area
PSA	Prostate Specific Antigen
PTK	Protein Tyrosine Kinase
PTP	Protein Tyrosine Phosphatase
PTSD	PostTraumatic Stress Disorder
PULCON	PUlse Length-based CONcentration determination
PXRD	Powder X-Ray Diffraction
QC	Quality Control
QCPMG	Quadrupolar Carr-Purcell-Meiboom-Gill
qNMR	quantitative Nuclear Magnetic Resonance
RA	Rheumatoid Arthritis
RAAS	Renin-Angiotensin-Aldosterone System
RAMRIS	Rheumatoid Arthritis Magnetic Resonance Image scoring System
RCDs	Respiratory Chain Deficiencies

R&D	Research and Development
RD	Relaxation Delay
RDC	Residual Dipolar Coupling
RMMC	Recursive Maximum Margin Criterion
ROC	Receiver Operating Characteristic
RP	Restricted Part
RSPA	Recursive Segment-wise Peak Alignment
RTK	Receptor Tyrosine Kinases
RTRT	Real-Time Release Testing
SAR	Structure-Activity Relationships
SAX-HPLC	Strong Anion-eXchange High-Performance Liquid Chromatography
SCXRD	Single Crystal X-Ray Diffraction
SDV	Standard DeViation
SE	Spin Echo
SEI	Surface-binding Efficiency Index
SET	Solubility-Enhancement Tag
SFC	Supercritical Fluid Chromatography
SGF	Simulated Gastric Fluid
SHM	Statistical Health Monitoring
SHY	Statistical HeterospectroscopY
SIF	Simulated Intestinal Fluid
SIMCA	Soft Independent Modeling of Class Analogy
SMART	SMAll Recovery Times
SMI	Small Molecule Inhibitor
S/N	Signal-to-Noise
SNIF	Site-specific Natural Isotope Fractionation
SNP	Single-Nucleotide Polymorphism
SNR	Signal-to-Noise Ratio
SOFAST	band-Selective Optimized Flip-Angle Short-Transient
SOP	Standard Operating Procedure
SOS	Son-Of-Sevenless
SPECT	Single-Photon Emission Computed Tomography
SPE	Solid-Phase Extraction
SPINAL	Small Phase INcremental ALternation
SPI	Single Point Imaging
SPR	Surface Plasmon Resonance
SR	Study Reference
SSFFC	Substandard/Spurious/Falselylabeled/Falsified/Counterfeit
SSRI	Selective Serotonin Reuptake Inhibitor
STD	Saturation Transfer Difference
STIR	Short Tau Inversion Recovery
STOCSY	Statistical TOtal Correlation SpectroscopY
STORM	SubseT Optimization by Reference Matching
SVM	Support Vector Machines
T2D	Type-2 Diabetes
TBA	Total Brain Atrophy
TLC	Thin-Layer Chromatography
TNF-α	Tumor Necrosis Factor-α
TOCSY	TOtal Correlation SpectroscopY
ToF	Time of Flight
TOSS	TOtal Sideband Suppression
TPPM	Two-Pulse Phase Modulation